YEAST PHYSIOLOGY AND BIOTECHNOLOGY

To Pia, Roy and Danica

yours aye, Graeme.

YEAST PHYSIOLOGY AND BIOTECHNOLOGY

By

GRAEME M. WALKER
Reader in Biotechnology
School of Molecular and Life Sciences
University of Abertay Dundee
Dundee
Scotland

JOHN WILEY & SONS

Chichester · New York · Weinheim · Brisbane · Singapore · Toronto

Baffins Lane, Chichester,
West Sussex PO19 1UD, England

National 01243 779777
International (+44) 1243 779777
e-mail (for orders and customer service enquiries): cs-books@wiley.co.uk
Visit our Home Page on http://www.wiley.co.uk
or http://www.wiley.com

Reprinted February 1999, January 2000

Other Wiley Editorial Offices

John Wiley & Sons, Inc., 605 Third Avenue,
New York, NY 10158-0012, USA

WILEY-VCH Verlag GmbH, Pappelallee 3,
D-69469 Weinheim, Germany

Jacaranda Wiley Ltd, 33 Park Road, Milton,
Queensland 4064, Australia

John Wiley & Sons (Asia) Pte Ltd, 2 Clementi Loop #02-01,
Jin Xing Distripark, Singapore 129809

John Wiley & Sons (Canada) Ltd, 22 Worcester Road,
Rexdale, Ontario M9W 1L1, Canada

Library of Congress Cataloging-in-Publication Data
Walker, Graeme M.
Yeast physiology and biotechnology / by Graeme M. Walker.
p. cm.
Includes bibliographical references and index.
ISBN 0-471-96447-6 (hbk. : alk. paper). — ISBN 0-471-96446-8
(pbk. : alk. paper)
1. Yeast fungi—Biotechnology. 2. Yeast fungi—Physiology.
I. Title.
TP248.27.Y43W35 1998
571.2'9563—dc21 97-40557
CIP

British Library Cataloguing in Publication Data

A catalogue record for this book is available from the British Library

ISBN 0-471-964476 (cased)
ISBN 0-471-964468 (paper)

Typeset in 10/12pt Times from the author's disks by Acorn Bookwork, Salisbury, Wilts.

CONTENTS

Preface xi

1 Introduction to Yeasts **1**
1.1 The World of Yeasts 1
1.1.1 A Glance at Yeast Taxonomy 1
1.1.2 Yeast Biodiversity 3
1.1.3 Yeast Habitats 4
1.2 Importance of Yeasts to Mankind 5
1.2.1 Yeasts in Human Culture 5
1.2.2 Industrial Exploitation of Yeasts 5
1.2.3 Yeasts and Fundamental Biological Research 5
1.2.4 Yeasts in Human Health and Disease 6
1.3 Yeast Physiology and Biotechnology 9
1.4 References 9

2 Yeast Cytology **11**
2.1 Introduction 11
2.2 General Cellular Characteristics of Yeasts 12
2.3 Cytological Methods for Yeasts 13
2.3.1 Light and Fluorescence Microscopy 13
2.3.2 Image Analysis 15
2.3.3 Flow Cytometric Analysis 15
2.3.4 Electron Microscopy 16
2.3.5 Yeast Cell Fractionation and Organelle Isolation 16
2.4 Yeast Cell Architecture and Function 17
2.4.1 The Cell Envelope 19
2.4.1.1 The Plasma Membrane 19
2.4.1.2 The Periplasm 21
2.4.1.3 The Cell Wall 22
2.4.1.4 The Yeast Spore Wall 28
2.4.1.5 Fimbriae 28
2.4.1.6 Capsules 28

2.4.2 The Cytoplasm and Cytoskeleton 29
2.4.3 The Nucleus and Extrachromosomal Elements 31
2.4.3.1 The Nucleus 32
2.4.3.2 Extrachromosomal Elements 34
2.4.4 The Secretory System and Vacuoles 35
2.4.5 Mitochondria 38
2.5 Summary 41
2.6 References 42

3 Yeast Nutrition **51**
3.1 Introduction 51
3.2 Yeast Nutritional Requirements 52
3.2.1 Yeast Cell Chemistry and Essential Elements 52
3.2.2 Sources of Utilizable Nutrients for Yeasts 52
3.2.2.1 Carbon 52
3.2.2.2 Hydrogen 53
3.2.2.3 Oxygen 55
3.2.2.4 Nitrogen 55
3.2.2.5 Sulphur 55
3.2.2.6 Phosphorus 56
3.2.2.7 Mineral Elements 56
3.2.2.8 Growth Factors 57
3.2.3 Yeast Cultivation Media 58
3.3 Nutrient Acquisition by Yeasts 61
3.3.1 General Nutritional Strategies Adopted by Yeasts 61
3.3.1.1 Physiological Responses to Nutrient Availability 61
3.3.1.2 Translocation of Nutrients into the Yeast Cell 63
3.3.1.3 Intracellular Fates of Nutrients in Yeasts 65
3.3.2 Specific Nutritional Strategies Adopted by Yeasts 66
3.3.2.1 Transport of Water 66
3.3.2.2 Transport of Sugars 66
3.3.2.3 Transport of Alcohols 74
3.3.2.4 Transport of Hydrocarbons 75
3.3.2.5 Transport of Organic Acids 75
3.3.2.6 Transport of Sterols 75
3.3.2.7 Transport of Fatty Acids 76
3.3.2.8 Transport of Nitrogenous Compounds 77
3.3.2.9 Transport of Anions 81
3.3.2.10 Transport of Cations 83
3.3.2.11 Transport of Divalent Metal Cations 86
3.3.2.12 Transport of Heavy Metals 90
3.4 Summary 91
3.5 References 92

4 Yeast Growth **101**
4.1 Introduction 101
4.2 Cellular Growth of Yeasts 102

4.2.1 Vegetative Reproduction in Yeasts 102
4.2.1.1 Budding 102
4.2.1.2 Fission 106
4.2.1.3 Filamentation 110
4.2.2 Control of Yeast Growth and Cell Division 112
4.2.2.1 The Cell Cycle in Yeasts 112
4.2.2.2 Molecular Aspects of Cell Cycle Control in Yeasts 120
4.2.3 Sexual Reproduction in Yeasts 124
4.3 Population Growth of Yeasts 127
4.3.1 Colonial Yeast Growth 127
4.3.2 Population Yeast Growth 131
4.3.2.1 Batch Growth of Yeasts 131
4.3.2.2 Continuous Growth of Yeasts 135
4.3.2.3 Synchronous Yeast Growth 137
4.3.3 Cultural Yeast Growth 139
4.3.3.1 Yeast Pure Culturing and Maintenance Strategies 140
4.3.3.2 Assessments of Yeast Viability and Vitality 141
4.3.3.3 Cultivation Strategies in Yeast Biotechnology 143
4.4 The Physicochemical Environment and Yeast Growth 144
4.4.1 Physical Requirements for Yeast Growth 144
4.4.1.1 Temperature 146
4.4.1.2 Water 147
4.4.1.3 Media pH and pO_2 148
4.4.2 Effects of Physical Stresses on Yeast Growth 149
4.4.2.1 Temperature Stress 149
4.4.2.2 Water Stress 154
4.4.2.3 Other Physical Stresses 160
4.4.3 Effects of Chemical Stresses on Yeast Growth 162
4.5 Biotic Factors Influencing Yeast Growth 169
4.5.1 Yeast–Plant Interactions 169
4.5.2 Yeast–Animal Interactions 171
4.5.3 Yeast–Microbe Interactions 173
4.6 Yeast Cell Death 175
4.6.1 Physical Parameters and Yeast Cell Death 176
4.6.2 Chemical Factors and Yeast Cell Death 177
4.6.3 Biological Factors Influencing Yeast Cell Death 179
4.7 Summary 181
4.8 References 183

5 Yeast Metabolism **203**
5.1 Introduction 203
5.2 Carbon and Energy Metabolism 205
5.2.1 Sugar Catabolism and its Regulation 205
5.2.1.1 Sugar Catabolism 205
5.2.1.2 Regulation of Sugar Catabolism 211
5.2.1.3 Respiration versus Fermentation 213
5.2.1.4 Oscillatory Metabolism in Yeasts 220

5.2.2 Gluconeogenesis and Carbohydrate Biosynthesis 221
5.2.2.1 Structural Polysaccharide Synthesis 223
5.2.2.2 Storage Carbohydrate Synthesis 225
5.2.3 Metabolism of Non-Hexose Carbon Sources 227
5.2.3.1 Biopolymer Metabolism 228
5.2.3.2 Pentose Sugar Metabolism 228
5.2.3.3 Lower Aliphatic Alcohol Metabolism 231
5.2.3.4 Sugar Alcohol Metabolism 233
5.2.3.5 Hydrocarbon Metabolism 234
5.2.3.6 Fatty Acid and Lipid Metabolism 235
5.2.3.7 Sterol Biosynthesis 238
5.2.3.8 Organic Acid Metabolism 239
5.3 Nitrogen Metabolism 241
5.3.1 Nitrogen Assimilation by Yeasts 241
5.3.2 Amino Acid Metabolism 244
5.3.3 Protein Metabolism 244
5.4 Phosphorus and Sulphur Metabolism 247
5.4.1 Phosphorus Metabolism 247
5.4.2 Sulphur Metabolism 250
5.5 Specialized Metabolism 254
5.6 Summary 254
5.7 References 255

6 Yeast Technology **265**
6.1 Introduction 265
6.2 Applied Molecular Genetics of Yeasts 268
6.2.1 Genetics of 'Industrial' *Saccharomyces* Yeasts 268
6.2.2 Recombinant DNA Technology in Yeasts 268
6.2.2.1 Molecular Genetic Aspects 270
6.2.2.2 Cellular Aspects 276
6.2.2.3 Technical Aspects 280
6.2.2.4 Commercial Aspects 281
6.3 Developments in Yeast Technologies 283
6.3.1 Alcoholic Beverages 283
6.3.1.1 Developments in Brewing Yeasts and Fermentation 283
6.3.1.2 The Brewing Process 283
6.3.1.3 Brewing Yeasts 284
6.3.1.4 Yeast Fermentation Technology Developments 288
6.3.1.5 Developments in Yeasts Producing other Alcoholic Beverages 290
6.3.2 Industrial Alcohols 294
6.3.2.1 Bioethanol 294
6.3.2.2 Other Alcohols from Yeast 299
6.3.3 Yeast Biomass-Derived Products 300
6.3.3.1 Baker's Yeast 300
6.3.3.2 Other Applications of Whole-Cell Yeast Biomass 303
6.3.3.3 Extracted Yeast Cell Products 303

6.3.4 Industrial Enzymes and Chemicals 305
6.3.5 Therapeutic Proteins 306
6.4 Yeasts in Biomedical Research 308
6.5 Summary 310
6.6 References 311

Index **321**

PREFACE

Yeasts are truly fascinating organisms. The diverse and dynamic activities of yeasts impinge on many areas of science, technology and medicine. Some species play beneficial roles in the production of foods, beverages and pharmaceuticals, while others play detrimental roles as spoilage organisms and agents of human disease. Increasingly significant roles are being found for yeasts as model eukaryotic cells in furthering our fundamental knowledge in the biological and biomedical sciences.

In recent years, knowledge of yeast cell physiology has considerably lagged behind that of yeast genetics and molecular biology. For example, important aspects of the regulation of yeast growth and metabolism, which are crucial for successful exploitation of yeasts in industry, are still poorly understood from a fundamental viewpoint. My principal aim in writing this book, therefore, was to link salient areas of yeast cell physiology – cytology, nutrition, growth and metabolism – with yeast biotechnology. Rather than devoting a separate chapter to yeast genetics, which is expertly covered in other texts and reviews, I have integrated relevant aspects of classical and molecular genetics into other chapters (e.g. section 6.2). The book is aimed at those who study and exploit yeasts: bioscience students, yeast researchers and yeast technologists.

The literature on yeasts is vast. Rather than providing a comprehensive coverage of yeast science and technology, I have attempted to review broadly those aspects of cell physiology which I have deemed pertinent to the practical uses of yeasts in industry. The literature survey was completed in April, 1997 and I apologise in advance to colleagues in the yeast community for overlooking certain areas of yeast research.

I wish to thank Roger Jones and John Duffus for firstly introducing me to, respectively, yeast technology and yeast physiology. I also wish to acknowledge numerous undergraduate and postgraduate students in Dunedin, Dublin and Dundee for many stimulating discussions over the years on yeast matters. I am indebted to the following individuals who cast their expert eyes over certain sections of this book: Alessandro Martini, Ann Vaughan-Martini and Gianluigi Cardinali in Perugia; Colin Slaughter and Graham Stewart in Edinburgh; and Derek Jamieson in Dundee. I am also grateful to Malcolm Sratford, Cecilia Laluce, Julia Douglas, Alexander Rapoport, Marc-Andre Lachance, Marten Veenhuis, Chris Kaiser, Masako Osumi and Ricardo de Souza Pereira for providing me with yeast micrographs and to numerous scientists and publishers for granting permission to reproduce published figures. Finally, I wish to most sincerely thank Maggie Glenday, Johann Anderson and Pia Walker for their expert processing of the manuscript and to my family for their support in enabling me to complete this project.

Graeme Walker
Dundee
May 1997

1

INTRODUCTION TO YEASTS

1.1 THE WORLD OF YEASTS
1.1.1 A Glance at Yeast Taxonomy
1.1.2 Yeast Biodiversity
1.1.3 Yeast Habitats
1.2 IMPORTANCE OF YEASTS TO MANKIND
1.2.1 Yeasts in Human Culture
1.2.2 Industrial Exploitation of Yeasts
1.2.3 Yeasts and Fundamental Biological Research
1.2.4 Yeasts in Human Health and Disease
1.3 YEAST PHYSIOLOGY AND BIOTECHNOLOGY
1.4 REFERENCES

1.1 THE WORLD OF YEASTS

1.1.1 A Glance at Yeast Taxonomy

The word 'yeast' is not easily defined, but basically yeasts are recognized as being unicellular fungi. More definitively: 'Yeasts are ascomycetous or basidiomycetous fungi that reproduce vegetatively by budding or fission, and that form sexual states which are not enclosed in a fruiting body' (Boekhout and Kurtzman, 1996).

The subdivisions (as depicted in Table 1.1) are based on aspects of yeasts' sexuality (Ascomycotina or Basidiomycotina) or lack of it (Deuteromycotina) and the lower taxonomic categories are based on various morphological, physiological and genetic characteristics (Table 1.2). Barnett (1992) has reviewed the taxonomy of the genus *Saccharomyces*. Although yeast taxonomists aim to classify yeasts to *species* level, identification of individual *sub-species*, or *strains*, is more the focus of yeast technologists (see Lachance, 1987).

Yeast identification and characterization is very important in yeast biotechnology. For example, the ability to distinguish between *wild* yeasts and *cultured* yeasts in industrial processes is essential. This is exemplified in brewing fermentations where the presence of wild yeasts may impart undesirable off-flavours to the product and also in baker's yeast propagations where wild yeasts like *Candida utilis* may out-grow pure baking strains of *Saccharomyces cerevisiae* due to the more efficient sugar transport capabilities of the former (see Chapter 3).

Yeast classification is based on the following hierarchical system:

Taxonomic category	Example
Subdivisions	*Ascomycotina*
Families	*Saccharomycetaceae*
Subfamilies	*Saccharomycetoideae*
Genera	*Saccharomyces*
Species	*cerevisiae*

Table 1.1. An overview of yeast genera[1].

Teleomorphic[2] Ascomycetous Genera (Ascomycotina)	Anamorphic[2] Ascomycetous Genera (Deuteromycotina)	Teleomorphic Heterobasidiomycetous Genera (Basidiomycotina)	Anamorphic Heterobasidiomycetous Genera (Basidiomycotina)
Ambrosiozyma (2 species);	*Aciculoconidium* (1)	*Bulleromyces* (1)	*Bensingtonia* (10);
Arthroascus (4);	*Arxula* (2);	*Chinosphaera* (1);	*Bullera* (14);
Arxiozyma (1);	*Brettanomyces* (3);	*Cystofilobasidium* (4);	*Cryptococcus* (40);
Ascoidea (6);	*Candida* (152);	*Erythrobasidium* (1);	*Fellomyces* (4);
Ashbya (1);	*Kloeckera* (1);	*Filobasidiella* (1);	*Itersonilia* (1);
Botryoascus (1);	*Myxozyma* (9);	*Filobasidium* (5);	*Kockovaella* (2);
Cephaoloascus (2);	*Oosporidium* (1);	*Leucosporidium* (3);	*Kurtzmanomyces* (2);
Citeromyces (1);	*Saitoella* (1);	*Kondoa* (1);	*Malassezia* (7);
Clavispora (2);	*Schizoblastosporion* (2);	*Mrakia* (4);	*Phaffia* (1);
Coccidiascus (1);	*Sympodiomyces* (1);	*Rhodosporidium* (9);	*Rhodotorula* (37);
Cyniclomyces (1);	*Trigonopsis* (1)	*Sporidiobolus* (3);	*Sporobolomyces* (27);
Debaryomyces (10);		*Sterigmatosporidium* (1);	*Sterigmatomyces* (2);
Dekkera (2);		*Tilletiaria* (1);	*Sympodiomycopsis* (1);
Eremothecium (2);		*Tremella* and	*Tilletiopsis* (6);
Galactomyces (2)		*Sirobasidium* (12);	*Trichosporon* (20);
Guilliermondella (1);		*Udeniomyces* (3);	*Tsuchiyaea* (1)
Hanseniaspora (6);		*Xanthophyllomyces* (1)	
Hansenula[3] (1);			
Hormoascus (3);			
Hyptopichia (1);			
Issatchenkia (4);			
Kluyveromyces (17);			
Lipomyces (5);			
Lodderomyces (1);			
Metschnikowia (10);			
Nadsonia (3);			
Nematospora (1);			
Pachysolen (1);			
Pichia (87);			
Saccharomyces[4] (16);			
Saccharomycodes (2);			
Saccharomycopsis (6);			
Saturnispora (4);			
Schizosaccharomyces (3);			
Schwanniomyces (1);			
Sporopachydermia (3);			
Stephanoascus (2);			
Torulaspora (3);			
Wickerhamia (1);			
Wickerhamiella (1);			
Williopsis (5);			
Yarrowia (1);			
Zygoascus (1);			
Zygosaccharomyces (9);			
Zygozyma (4)			

[1] Information from Boekhout and Kurtzman (1996) and A. Vaughan-Martini and A. Martini (University of Perugia, personal communication, 1997). The latest (4th) edition of *The Yeasts. A Taxonomic Study* (edited by C.P. Kurtzman and J.W. Fell and published by Elsevier, 1997) was in press at the time of writing.

[2] Teleomorphic and anamorphic refers to, respectively, meiosporic and mitosporic expression of yeast species.

[3] The sole *Hansenula* species is *H. misumaiensis*. Other *Hansenula* species, including *H. polymorpha* which is referred to elsewhere in this book, are not listed above due to the fact that Kurtzman (1984) transferred *Hansenula* species with hat-shaped spores to *Pichia* and those with saturn-shaped spores to *Williopsis*. (Thus, *H. polymorpha* is now *Pichia angusta*.)

[4] *Saccharomyces sensu stricto* have been separated into four species: *S. bayanus*, *S. cerevisiae*, *S. paradoxus* and *S. pastorianus* (see Vaughan-Martini and Martini, 1993).

Table 1.2. Criteria used in yeast species classification and strain identification.

Morphological characteristics	Physiological characteristics	Immunological characteristics	Molecular characteristics
Giant colony morphology	Fermentation of sole C sources	Serology: agglutination	rRNA and rDNA phylogeny
Cell morphology in liquid media	Assimilation of sole C sources	Immuno-electrophoresis	RFLP of mitDNA
Mode of vegetative and/or sexual reproduction	Assimilation of sole N sources	Immunofluorescence microscopy	DNA base composition (mol % G + C)
Spore characteristics	Pigment production		Karyotype analysis
Presence/absence of hyphae or pseudohyphae	Acid production		DNA hybridization
Pellicle formation at liquid surfaces	Osmophilia		Random Amplification of Polymorphic DNA (RAPD)
Flocculation in liquid media			

Identification of yeast genera can often be achieved by morphological tests supplemented with a few physiological tests. With regard to the latter, sole carbon and nitrogen source assimilation by yeasts may be determined by *auxanography* which nowadays can be conveniently carried out using commercially available kits: for example, Analytical Profile Index (API) strips (BioMérieux, France) or the automated/computerized BCCM/Allev 2.00 system (Louvain-la-Neuve, Belgium). Sugar assimilation and fermentation tests are commonly accomplished using glucose, galactose, maltose, sucrose, lactose, raffinose, trehalose and xylose. With regard to fermentation of these sugars, Scheffers (1987) has argued that the anaerobic liberation of CO_2 into Durham tubes is not very accurate for detecting slowly fermenting yeast species. Ethanol production assays are deemed to be more appropriate determinants of sugar fermentation by yeasts.

1.1.2 Yeast Biodiversity

To date, around 700 species of yeast have been described, but this represents only a fraction of yeast biodiversity on this planet. For ascomycetes in general, the numbers of undescribed genera and species have been calculated at 62 000 and 669 000, respectively (Hawksworth and Mouchacca, 1994). At the current rate of new fungal species descriptions each year, it would take mycologists several hundred years to document all new species thought to exist in numerous habitats (Hawksworth and Mouchacca, 1994). Even with conventional selective media isolation approaches, Lachance (1990) has noted that it is highly likely that a considerable portion of the yeast community is being ignored.

It is important for yeast biologists not only to appreciate the immense untapped yeast biodiversity but also to develop ways of characterizing and preserving remaining species, especially those of biotechnological potential. Several molecular biological approaches are now assisting in the detection of yeasts in the environment, and together with input from yeast cell physiological studies, this will provide means to preserve and exploit yeast biodiversity (Roberts and Wildman, 1995).

In addition to our ignorance of yeast species' genetic biodiversity, there is also considerable ignorance within known species relating to cell physiological biodiversity. For example, a considerable percentage (around 50%; see Oliver, 1996) of the 6000 genes identified by the *Saccharomyces cerevisiae* genome project are of unknown function. The 'hidden potential' of a yeast which has been exploited for thousands of years is clearly enormous. International effort is now underway to analyse functionally the yeast genome in order not only to reveal the full biotechnological potential of yeasts, but also to achieve an understanding of how simple eukaryotic cells work (Bassett *et al.*, 1996; Goffeau *et al.*, 1996).

1.1.3 Yeast Habitats

Although not as ubiquitous as bacteria, yeasts are widespread in the natural environment (Table 1.3). Yeast cells lack chlorophyll (are non-photosynthetic) and are strictly chemoorganotrophic, meaning that they require fixed, organic forms of carbon for growth. Sources of carbon for yeast metabolism (which are discussed further in Chapters 3 and 5) are quite diverse and include: simple sugars, polyols, organic and fatty acids, aliphatic alcohols, hydrocarbons and various heterocyclic and polymeric compounds. Different yeasts are able to utilize different carbon sources and nutritional selectivity determines yeast species diversity in particular niches. In other words, yeasts exhibit great specialization for habitat (Phaff *et al.*, 1978). In addition, being non-motile, yeasts rely on aerosols, animal vectors and human activity for their natural dispersal.

Table 1.3. Natural yeast habitats.

Habitat	Comments
Plants	Plants are common niches for yeasts, especially the interface between soluble nutrients (sugars) and the septic world (e.g. surface of grapes). The spread of yeasts on the phyllosphere is aided by insects (e.g. *Drosophila* spp.). Some yeasts are plant pathogens (e.g. *Ashbya, Nematospora* spp. and the yeast-like *Ophiostoma ulmii*, the causative agent of Dutch elm disease).
Animals	Several non-pathogenic yeasts are associated with the intestinal tract (e.g. *Candida pintolopesii, Candida slooffii, Cyniclomyces guttulatus*) and skin (e.g. *Pityrosporum* spp.) of warm-blooded animals. Several yeasts (see Table 1.6) are pathogenic towards animals. Numerous yeasts are commensally associated with insects (e.g. fruit flies, bark beetles) which act as important vectors in the natural distribution of yeasts (see Phaff, Miller and Mrak, 1978).
Soil	For many yeasts, soil may only be a reservoir for their long-term survival, rather than a habitat for free growth. *Lipomyces* and *Schwanniomyces* spp. appear to be examples of genuine soil genera (isolated exclusively from soil). Poor (e.g. sandy) soils may harbour very few yeasts, but rich (e.g. agricultural) soils may have as many as 40 000 viable yeasts per gram.
Water	Yeasts are widely distributed in both fresh water and seawater (*Candida, Cryptococcus, Rhodotorula* and *Debaryomyces* spp.). Some marine yeasts exist at temperatures of −3 to +13°C, salinities of around 35% and at depths of 4000 m, but others are more narrowly restricted. Seawater normally contains 10–100 yeasts per litre, but numbers can increase dramatically in estuarine regions.
Atmosphere	From the vegetative layer above soil surfaces, yeasts (e.g. *Cryptococcus, Rhodotorula, Sporobolomyces* and *Debaryomyces* spp.) are dispersed by air currents. A few viable yeast cells may be expected per m^3 of air.
Extreme habitats	Some halotolerant yeasts (e.g. *Debaryomyces hansenii*) can grow in nearly saturated brine solutions. Osmophilic yeasts (*Zygosaccharomyces rouxii* growing in 40–70% sugar) were found (during Soviet Antarctic expeditions) in 3250-year-old glacier horizons (see Beker and Rapoport, 1987). Vishniae (1996) has reviewed yeast biodiversity in Antarctic soils.

Yeasts can be isolated from terrestrial, aquatic and aerial environments. Preferred habitats are plant tissues but a few species are found in commensal or parasitic relationships with animals. Some yeasts may be regarded as extremophiles, particularly certain osmophilic yeasts which are able to thrive in solute-rich environments. Several of these yeast species are encountered as food spoilage organisms.

In addition to natural habitats, some yeasts have found niches in man-made environments. For example, *S. cerevisiae* is almost the only yeast species found colonizing surfaces in wineries (Martini, 1993), whereas in hospitals, several opportunistically pathogenic yeasts may be found. With regard to the latter, *Candida* spp. (especially *C. albicans*) may account for over 80% of all nosocomial (hospital-derived) fungal infections (Schaberg *et al.*, 1991) and these yeasts are frequently isolated from inanimate surfaces in hospital environments which have been implicated as important reservoirs of yeast infections (reviewed by Mahayni *et al.*, 1995).

Further aspects of yeast ecology in the natural world have been discussed by Phaff and Starmer (1980) and Spencer and Spencer (1997), and methods for isolating and characterizing yeasts from environmental samples have been described by Beech and Davenport (1971), Phaff *et al.* (1978) and Spencer and Spencer (1996).

1.2 IMPORTANCE OF YEASTS TO MANKIND

1.2.1 Yeasts in Human Culture

Yeasts are of major economic, social and health significance in human culture. They have often been described as mankind's oldest 'domesticated' organisms, having been used to produce alcoholic beverages and leaven bread dough for millennia. In fact, the brewing of beer probably represented the world's first biotechnology. In modern times, yeasts have found numerous other roles besides traditional food fermentations. In particular, genetically manipulated yeasts can now be exploited to produce many different biopharmaceutical agents for preventing and treating human disease. Several of the traditional, modern and emerging yeast technologies are discussed further in Chapter 6, and Table 1.4 provides an historical overview of the human study and usage of yeasts.

1.2.2 Industrial Exploitation of Yeasts

In the future, yeast exploitation is likely to make significant impacts in relation to renewable energy supply, environmental biotechnology including biological control and in health-care issues, particularly the study of human genetic disorders and cancer (see Figure 1.1). With regard to energy, yeast metabolism is capable of providing fuel ethanol from renewable carbohydrate fermentation feedstocks. Horecker (1978) even went as far as to suggest that yeast fermentation (for the production of bioethanol as a replenishable energy source) represented 'a great hope for the survival of civilization on this planet'. In the field of medicine, great advances have been made in the production of human therapeutic proteins with genetically engineered yeasts. Incidentally, the use of yeasts in therapeutic medicine is not solely linked to modern recombinant DNA technology. For example, since earliest times, yeasts were used in the 'biological control' of human bacterial infections (Guilliermond, 1920). Other exciting developments are linked with the functional analysis of yeast genomes and how such analysis may provide insight into human gene structure and function.

1.2.3 Yeasts and Fundamental Biological Research

Yeasts, in particular *S. cerevisiae* and *Schizosaccharomyces pombe* have been extensively studied in several areas of fundamental biological science and these two species serve as valuable model eukaryotic cells in such studies. Table 1.5

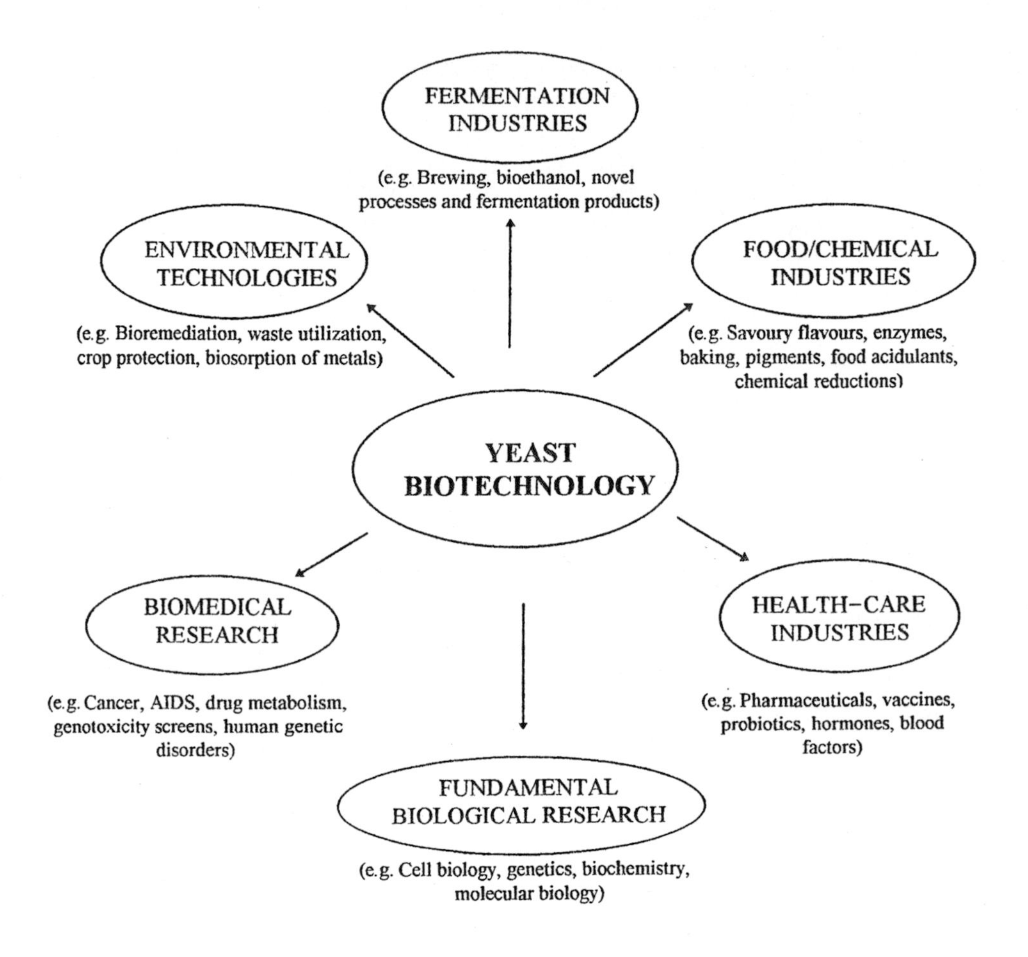

Figure 1.1. Diversity of outlets involving yeast biotechnology.

summarizes some of the areas where studies with yeasts have contributed significantly to the furtherance of biological knowledge. Although many biochemical and genetic features of *S. cerevisiae* are shared with higher eukaryotic cells, it should be stressed that this yeast is not an *ideal* model eukaryote in terms of certain cytological and growth characteristics. For example, budding is a unique mode of reproduction and the asymmetry of cell division (see Chapter 4) is not a feature of higher eukaryotic cells. Other non-animal cell features are the presence of a thick cell wall (see Chapter 2) and the continuity of the nuclear membrane during mitosis. *S. cerevisiae* also displays some peculiarities with respect to its mode of sugar transport (see Chapter 3). The fission yeast, *Sch. pombe*, displays a symmetrical mode of cell division and this, together with several other characteristics, have led some scientists to argue that *Sch. pombe* is a more representative eukaryote than its distant cousin, *S. cerevisiae* (e.g. Glick, 1996).

1.2.4 Yeasts in Human Health and Disease

While the vast majority of industrial and medical activities of yeasts are beneficial to human life,

Table 1.4. Some milestones in the study and exploitation of yeasts.

Chronology	Milestones
2000–6000 BC	Brewing (Sumeria, Babylonia); grape cultivation (Georgia); dough leavening (Egypt)
1680	Description of the microscopic appearance of yeasts (van Leeuwenhoek).
1830s	Alcoholic fermentation associated with yeast budding (Cagnaird-Latour in 1835). The name, *Saccharomyces cerevisiae*, created for a yeast observed in malt (Meyen in 1837). Sugar acting as food source for yeast growth (Schwann and Kützing in 1839).
Late 1800s	'All true fermentations are correlative with physiological phenomena' and fermentations associated with yeast metabolism (Pasteur in 1857). Etudes sur la bière (Pasteur in 1876). The term 'enzyme' (Greek; 'in yeast') introduced by Kuhne in 1877. Isolation of single yeast cells and employment of pure yeast strains for brewing (Hansen, 1880–1883). The production of alcohol and CO_2 from sugar by cell-free yeast extracts (E. and H. Buchner) – birth of biochemistry as a new scientific discipline.
1915	Production of glycerol by 'steered' yeast fermentations (Neuberg).
1920	Knowledge of yeast physiology, sexuality and phylogeny reviewed (see Guilliermond, 1920).
1930s–1940s	Genetic studies commenced on brewing yeast (Winge, Lausten, Lindegren). Demonstration of sexual reproduction and mating type system.
1930–1960	Yeast taxonomy studies by Kluyver and the Delft School.
1970s	Rare mating, cytoduction and protoplast fusion in brewing yeasts. First transformation of yeast in 1978 (Hinnen, Hicks and Fink, USA; Beggs, UK).
1980s–1990s	First commercial pharmaceutical (hepatitis B vaccine) from recombinant yeast. UK Government approval of use of genetically engineered baker's yeast (1990) and brewer's yeast (1994). Completion of S. *cerevisiae* genome project (1996).

several detrimental aspects of yeasts exist. This is particularly apparent when considering the excessive consumption of yeast-derived ethanol and the occasionally pathogenic relationship between yeasts and humans. Food spoilage may also be considered a detrimental influence if one considers biodeterioration of food nutritive quality by yeasts (especially in high-sugar foods and dairy produce). It is important to note, however, that yeasts are not causative microbial agents of food *poisoning*. Some humans appear to exhibit immune hypersensitivities to dietary yeasts. For example, sufferers of Crohn's disease display elevated antibody responses to baking and brewing strains of *S. cerevisiae* (McKenzie *et al.*, 1990; Barclay *et al.*, 1992; Colombel *et al.*, 1995). In such cases, hypersensitivity to yeast antigens may be involved in the pathogenesis of an important intestinal disorder.

Many yeasts are also opportunistically pathogenic towards humans (Table 1.6). Predisposing factors to yeast infections include: natural factors (e.g. endocrine dysfunctions, malignant and microbial diseases); dietary factors (e.g. sugar-rich foods, iron and vitamin deficiencies); mechanical factors (e.g. burns, wearing of dentures); iatrogenic and nosocomial factors (e.g. antibiotic and immunosuppressive therapy, surgical procedures such as heart-valve replacements and indwelling catheters) and viral factors (e.g. HIV infection). Mycoses caused by *Candida albicans* (i.e. candidosis) are the most common opportunistic yeast infections. Immunocompromised individuals appear particularly susceptible

Table 1.5. Some contributions of yeasts to biological science.

Discipline	Contributions
Biochemistry	Elucidation of glycolysis. Nature and function of enzymes involved in respiration, fatty acid metabolism, proteolysis etc. Signal transduction mechanisms (e.g. the roles of cAMP, inositol phosphates and protein kinases).
Cytology	Mechanisms of mitosis and meiosis. Biogenesis of organelles (e.g. mitochondria, vacuoles, peroxisomes). Cytoskeletal structure and function. Elucidation of protein secretory pathways.
Genetics and molecular biology	Mating type switching phenomena. Nucleic acid and genome structure. Mechanisms of recombination. Control of gene expression. Gene mapping and sequencing. Control of the cell cycle. Functions of oncogenes.

Table 1.6. Frequently encountered and emerging yeast pathogens.

Yeast species	Disease	Common manifestations
Blastomyces dermatidis	*Blastomycosis*	Granulomatous inflammations
Candida albicans and other *Candida* spp. (e.g. *C. glabrata, C. krusei, C. parapsilosis, C. tropicalis, C. lusitaniae)*	*Candidosis*	Various superficial ('Thrush') and systemic mycotic lesions and candidaemia (see Odds, 1988).
Cryptococcus neoformans	*Cryptococcosis*	Meningitis, pneumonitis
Histoplasma capsulatum	*Histoplasmosis*	Pulmonary infections
Sporotrichum schenckii	*Sporotrichosis*	Pigmented ulcerations, abscesses
Emerging pathogens: species of *Malassezia* (e.g. *M. furfur*), *Hansenula* (e.g. *H. anomala*), *Rhodotorula* (e.g. *R. rubra*), *Trichosporon* (e.g. *T. beigelii*)	Various	See Hazen (1995)
Pneumocystis carinii[1]	*Pneumocystosis*	Severe pneumonia in immunocompromised individuals (e.g. AIDS patients)

[1] This organism is included in the list because it shows DNA homology with Ustomycetous red yeasts (Wakefield *et al.*, 1992).

to candidosis and in AIDS patients, candidal oesophagitis may be employed as a marker for underlying cellular immunodeficiency. Yeasts previously only considered to be of industrial importance or present as innocuous inhabitants of the environment have now been shown to be capable of attacking human hosts. Hazen (1995) has discussed some of these *emerging* yeast pathogens.

Even *S. cerevisiae* has been implicated in human pathogenesis in certain compromised individuals (McCusker *et al.*, 1994).

1.3 YEAST PHYSIOLOGY AND BIOTECHNOLOGY

Yeast *physiology* may be defined as the understanding of growth and metabolism of yeast cells. In general terms, this relates to how yeasts interact with their biotic and abiotic environment; in specific terms, yeast physiology relates to how yeast cells feed, metabolize, grow, reproduce (sexually and asexually), survive and ultimately die.

Yeast *biotechnology* may be defined as the exploitation of yeast physiology (the late Professor Tony Rose, personal communication, 1984).

Yeast physiology and yeast biotechnology are therefore inextricably linked and it is only through a thorough understanding of yeast growth and metabolic processes that the full industrial potential of yeast cells will materialize. Unfortunately, there is a considerable gap between the body of scientific knowledge on yeasts and practical knowledge of yeast biotechnology. This statement is particularly evident when one considers the current wealth of information on genetics and molecular biology of 'scientific' yeasts, compared with the relative paucity of information on the physiology of 'industrial' yeasts. It should be emphasized in this respect that in yeast recombinant DNA technology, successful commercial exploitation of transformed cells can only be realized following complete analysis of the physiological factors which regulate growth, cell division, metabolism and genetic stability of yeast populations in bioreactors.

The following chapters aim to cover salient aspects of yeast physiology that pertain to the exploitation of yeasts in traditional and modern biotechnologies. The aspects in question include: **yeast cytology** (ultrastructural form and function of yeast cells and organelles); **yeast nutrition** (translocation and assimilation of essential yeast nutrients); **yeast growth** (modes of reproduction, growth and coordination of growth with cell division and factors influencing survival of yeast cells in culture); **yeast metabolism** (major catabolic and anabolic routes for nutrients in yeast cells and the regulation of principal metabolic pathways) and, finally, how these aspects of yeast cell physiology impact on **yeast technology** (the industrialization and commercialization of *Saccharomyces* and non-*Saccharomyces* yeasts, including recombinant strains). The overall 'flow' of this book is outlined in Figure 1.2.

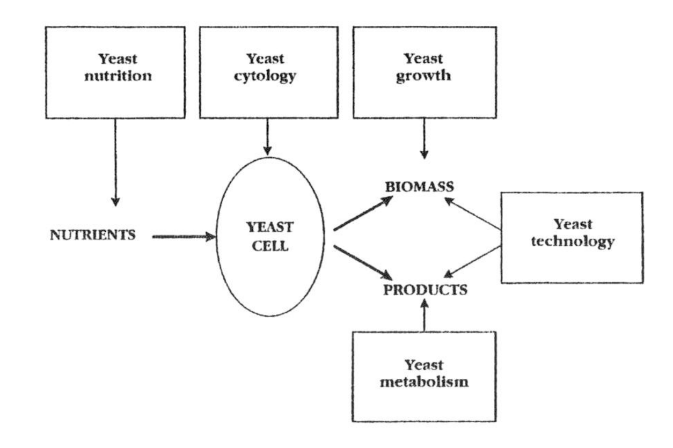

Figure 1.2. Integration of chapters in this book.

1.4 REFERENCES

Barclay, G.R., McKenzie, H., Pennington, J., Parratt, D. and Pennington, C.R. (1992) The effect of dietary yeast on the activity of stable chronic Crohn's disease. *Scandinavian Journal of Gastroenterology*, **27**, 196–200.

Barnett, J.A. (1992) The taxonomy of the genus *Saccharomyces* Meyen *ex* Reess: a short review for non-taxonomists. *Yeast*, **8**, 1–23.

Bassett, D.E., Basarai, M.A., Connelly, C., Hyland, K.M., Kitagawa, K., Mayer, M.L., Morrow, D.M., Page, A.M., Resto, V.A., Skibbens, R.V. and Hieter, P. (1996) Exploiting the complete yeast genome sequence. *Current Opinion in Genetics and Development*, **6**, 763–766.

Beech, M.J. and Davenport, R.R. (1971) Isolation, purification and maintenance of yeasts. In *Methods in Microbiology*, Vol. 4 (ed. C. Booth), pp. 153–182. Academic Press, London, New York.

Beker, M.J. and Rapoport, A.I. (1987) Conservation of yeasts by dehydration. *Advances in Biochemical Engineering/Biotechnology*, **35**, 128–171.

Boekhout, T. and Kurtzman, C.P. (1996) Principles and methods used in yeast classification, and an overview of currently accepted yeast genera. In *Nonconventional Yeasts in Biotechnology. A Handbook* (ed. K. Wolf), pp. 1–99. Springer-Verlag, Berlin, Heidelberg.

Colombel, J.F., Sendid, B., Jacquinot, P.M., Cortot, D. and Poulain, D. (1995) Evidence for a specific antibody response to *Saccharomyces cerevisiae* oligomannosidic epitopes in Crohn's disease. *Gastroenterology*, **108**, A800.

Glick, B.S. (1996) Cell biology: alternatives to baker's yeast. *Current Biology*, **6**, 1570–1572.

Goffeau, A. *et al.* (15 others) (1996) Life with 6000 genes. *Science*, **274**, 546–567.

Guilliermond, A. (1920) *The Yeasts* (Translated by F.W. Tanner). J. Wiley, New York.

Hawksworth, D.L. and Mouchacca, J. (1994) Ascomycete systematics in the nineties. In *Ascomycete Systematics: Problems and Perspectives in the Nineties* (ed. D.L. Hawksworth), pp. 3–11. Plenum Press, New York.

Hazen, K.C. (1995) New and emerging yeast pathogens. *Clinical Microbiology*, **8**, 462–478.

Horecker, B.L.(1978) Yeast enzymology: retrospectives and perspectives. In *Biochemistry and Genetics of Yeasts* (eds M. Bacila, B.L. Horecker and A.O.M. Stoppani), pp. 1–15. Academic Press Inc., New York.

Kurtzman, C.P. (1984) Synonomy of the yeast genera *Hansenula* and *Pichia* demonstrated through comparisons of deoxyribonucleic acid relatedness. *Antonie van Leeuwenhoek*, **50**, 209–217.

Lachance, M.-A. (1987) Approaches to yeast identification. In *Yeast Biotechnology* (eds D.R. Berry, G.G. Stewart and I. Russell), pp. 33–51. Allen and Unwin, London.

Lachance, M.-A. (1990) Yeast selection in nature. In *Yeast Strain Selection* (ed. C.J. Panchal), pp. 21–41. Marcel Dekker Inc., New York.

Mahayni, R., Vazquez, J.A. and Zervos, M.J. (1995) Nosocomial candidiasis: epidemiology and drug resistance. *Infectious Agents and Disease*, **4**, 248–253.

Martini, A. (1993) Origin and domestication of the wine yeast *Saccharomyces cerevisiae*. *Journal of Wine Research*, **4**, 165–176.

McCusker, J.H., Clemons, K.V. Stevens, D.A. and Davis, R.W. (1994) Genetic characterization of pathogenic *Saccharomyces cerevisiae* isolates. *Genetics*, **136**, 1261–1269.

McKenzie, H., Main, J., Pennington, C.R. and Parratt, D. (1990) Antibody to selected strains of *Saccharomyces cerevisiae* (baker's and brewer's yeast) and *Candida albicans* in Crohn's disease. *Gut*, **31**, 536–538.

Odds, F.C. (1988) *Candida and Candidosis*. Baillière Tindall, London.

Oliver, S.G. (1996) From DNA sequence to biological function. *Nature*, **379**, 597–600.

Phaff, H.J. and Starmer, W.T. (1980) Specificity of natural habitats for yeasts and yeast-like organisms. In *Biology and Activities of Yeasts* (eds F.A. Skinner, S.M. Passmore and R.R. Davenport), pp. 79–101. Academic Press, London.

Phaff, H.J., Miller, M.W. and Mrak, E.M. (1978) *The Life of Yeasts*. 2nd edn. Harvard University Press, Cambridge (Mass.) and London.

Roberts, I.N. and Wildman, H.G. (1995) The diverse potential of yeast. *Bio/technology*, **13**, 1246.

Schaberg, D.R., Culver, D.H. and Gaynes, R.P. (1991) Major trends in the microbial etiology of nosocomial infection. *American Journal of Medicine*, **91**(3B), S727–S755.

Scheffers, W.A. (1987) Alcoholic fermentation. *Studies in Mycology*, **30**, 321–332.

Spencer, J.F.T. and Spencer, D.M. (1996) Isolation and identification of yeasts from natural habitats. In *Methods in Molecular Biology. Vol. 53: Yeast Protocols* (ed. I.H. Evans), pp. 1–4. Humana Press, Totowa, NJ, USA.

Spencer, J.F.T. and Spencer, D.M. (1997) Ecology: where yeasts live. In *Yeasts in Natural and Artificial Habitats*. (eds J.F.T. Spencer and D.M. Spencer), pp. 33–58. Springer–Verlag, Berlin, Heidelberg.

Vaughan-Martini, A. and Martini, A. (1993) A taxonomic key for the genus *Saccharomyces*. *Systematic and Applied Microbiology*, **16**, 113–119.

Vishniac, H.S. (1996) Biodiversity of yeasts and filamentous microfungi in terrestrial Antarctic ecosystems. *Biodiversity and Conservation*, **5**, 1365–1378.

Wakefield, A.E., Peters, S.E., Banerji, S., Bridge, P.D. and Hopkin, J.A. (1992) *Pneumocystis carinii* shows DNA homology with the Ustomycetous red yeast fungi. *Molecular Microbiology*, **6**, 1903–1911.

2

YEAST CYTOLOGY

2.1 INTRODUCTION
2.2 GENERAL CELLULAR CHARACTERISTICS OF YEASTS
2.3 CYTOLOGICAL METHODS FOR YEASTS
  2.3.1 Light and Fluorescence Microscopy
  2.3.2 Image Analysis
  2.3.3 Flow Cytometric Analysis
  2.3.4 Electron Microscopy
  2.3.5 Yeast Cell Fractionation and Organelle Isolation
2.4 YEAST CELL ARCHITECTURE AND FUNCTION
  2.4.1 The Cell Envelope
    2.4.1.1 The Plasma Membrane
    2.4.1.2 The Periplasm
    2.4.1.3 The Cell Wall
    2.4.1.4 The Yeast Spore Wall
    2.4.1.5 Fimbriae
    2.4.1.6 Capsules
  2.4.2 The Cytoplasm and Cytoskeleton
  2.4.3 The Nucleus and Extrachromosomal Elements
    2.4.3.1 The Nucleus
    2.4.3.2 Extrachromosomal Elements
  2.4.4 The Secretory System and Vacuoles
  2.4.5 Mitochondria
2.5 SUMMARY
2.6 REFERENCES

2.1 INTRODUCTION

Yeast cytology refers to the cellular anatomy of yeasts, or 'a morphological inventory of their structural components' (Robinow and Johnson, 1991). Yeast cells have fascinated cytologists for many years. This is mainly due to the fact that they are easily grown unicellular eukaryotes which portray the ultrastructural features of higher eukaryotic cells. Yeasts are also amenable to a variety of biochemical, genetic and molecular biological investigations of their cytology. Yeast cell biology has been comprehensively reviewed in Volume 4 of *The Yeasts* (Rose and Harrison, 1991) and by Kocková-Kratochvilová (1990). Rather than provide an exhaustive coverage of yeast cell anatomy, the present chapter will focus on the relationship between subcellular organization and physiological function.

General cellular characteristics of yeasts, including macromolecular constituents and morphological diversity, will be considered first. Cytological methods, including visualization and isolation of cell organelles, will then be reviewed before coverage of yeast cell architecture. Ultrastructural features of yeasts will be described and their functions discussed in relation to growth and metabolic behaviour. Several yeast organelles and macromolecular structures play key roles in yeast biotechnology and particular attention will be paid to the role of the yeast cell wall in cell–cell interactions and the role of organelles in carbon metabolism and protein secretion. Such activities are directly pertinent to the production of industrial commodities, including recombinant proteins, by yeast fermentation.

2.2 GENERAL CELLULAR CHARACTERISTICS OF YEASTS

Yeast biomass primarily comprises macromolecules which are assembled into the structural components of the cell. The macromolecules in question mainly comprise proteins, polysaccharides, lipids and nucleic acids (Table 2.1). The structural chemistry of *S. cerevisiae* has been described in detail by Kreutzfeldt and Witt (1991). The relative concentrations of various macromolecular constituents will vary from species to species and will also be greatly influenced by conditions of growth.

Yeast cells exhibit great diversity with respect to cell size, shape and colour. Even individual cells from a pure strain of a single species can display morphological and colorimetric heterogeneity. Although a certain degree of morphological variation can be expected within a yeast culture, profound effects in individual cell morphology are induced by alterations in physical and chemical conditions.

Yeast cell size can vary widely. Phaff *et al.* (1978) have stated that some yeasts may be only 2–3 μm in length, whereas other species may attain lengths of 20–50 μm. Cell width appears less variable, between 1–10 μm. *S. cerevisiae* cells are generally ellipsoidal in shape with a large diameter ranging from 5–10 μm, and a small diameter 1–3 to 1–7 μm. The mean cell volumes for a haploid and diploid cell are 29 and 55 μm^3, respectively (Tschopp *et al.*, 1987). Brewing strains of *S. cerevisiae* are generally bigger than laboratory strains; for example, an ale yeast (NCYC 1006) possesses a mean diameter of 13.4 μm (Hough *et al.*, 1982). Note that mean cell size of *S. cerevisiae* also increases with cell age (see Chapter 4).

Table 2.1. Macromolecular constituents of yeast cells.

Class of macromolecule	Comments
Proteins	Structural proteins are mainly actin and tubulin of the cytoskeleton, histones and membrane proteins. Ribosomal proteins are found in the 60S and 40S ribonucleoprotein subunits. Hormonal proteins are mating pheromones. Functional proteins include hundreds of enzymes.
Glycoproteins	For example, structural mannoproteins in cell walls and functional glycoprotein enzymes (e.g. invertase).
Polysaccharides	Structural polysaccharides are mainly cell wall glucan, mannan and chitin and capsular heteropolysaccharides. Storage saccharides are glycogen and trehalose.
Polyphosphates	Mainly as storage polyphosphate in the vacuole.
Lipids	Structural phospholipids are free sterols in membranes; storage lipid particles are sterol esters and triglycerides; functional lipids include phosphoglyceride derivatives (in signal transduction) and free fatty acids (in growth and metabolic processes).
Nucleic acids	Deoxyribonucleic acid: approx. 80% of total is nuclear genomic DNA, 10–20% is mitochondrial genomic DNA and 1–5% is extrachromosomal (e.g. 2μm circles in *S. cerevisiae*, killer plasmids in *Kluyveromyces* spp.). Ribonucleic acid: approx. 80% of total is in ribonucleoprotein (rRNA); approx. 5% is mRNA in cytoplasm, rough ER and mitochondria. Some dsRNA is found in killer plasmids (*S. cerevisiae*) and some small nuclear RNA (snRNA) is found in the nucleus.

Some yeasts have distinguishable cell **types**, the best known of which are the **a**, α, and **a**/α cells of *S. cerevisiae* (see Chapter 4). The **a** and α haploid cells are able to undergo mating, a process which culminates in nuclear fusion and creation of the **a**/α diploid.

With regard to cell shape, yeasts can be: ellipsoidal/ovoid (e.g. *Saccharomyces* spp.), cylindrical with hemi-spherical ends (e.g. *Schizosaccharomyces*), apiculate/lemon-shaped (e.g. *Hanseniaspora* and *Saccharomycodes* spp.), ogival (elongated cell is rounded at one end and pointed at other, e.g. *Dekkera* and *Brettanomyces* spp.), flask-shaped (cells dividing by bud-fission, e.g. *Pityrosporum* spp.), triangular (e.g. *Trigonopsis* spp.), curved (e.g. *Cryptococcus cereanus*), filamentous (with pseudohyphae and septate hyphae, e.g. *Candida albicans*, *Yarrowia lipolytica*), stalked (e.g. *Sterigmatomyces* spp.), spherical (e.g. *Debaryomyces* spp.) or elongated (many yeasts, depending on growth conditions). Cellular differentiation of vegetative yeast cell forms into filamentous forms (e.g. germ-tubes, pseudohyphae and true hyphae) occurs in several yeast species. Polymorphic behaviour is often triggered by changes in the nutritional environment and even occurs in the normally ellipsoidal *S. cerevisiae* (see Chapter 4).

Several yeasts are pigmented and the following colours may be visualized in surface-grown colonies:

Colour	Examples
Cream	Many yeasts, including *S. cerevisiae*
White	*Geotrichum* spp., albino mutants of *Phaecoccomyces* spp.
Black	*Phaecoccomyces* spp. *Aureobasidium pullulans*
Pink	*Phaffia* spp., *Oosporidium* spp.
Red	*Rhodotorula* spp., adenine mutants of *S. cerevisiae*
Orange	*Rhodosporidium* spp.
Yellow	*Cryptococcus laurentii*, *Bullera* spp.

Some pigmented yeasts, such as *Phaffia rhodozyma*, have uses in biotechnology. The astaxanthin (3,3′-dihydroxy-β,β-carotene-4,4′-dione) pigments of this yeast have applications as fish food colorants for farmed salmonids, which have no means of synthesizing these red compounds (Johnston and Gil-Hwan, 1991). High pigment-yielding strains of *P. rhodozyma* can be propagated on inexpensive feedstocks like molasses to produce economic alternatives to chemically synthetic astaxanthin.

2.3 CYTOLOGICAL METHODS FOR YEASTS

2.3.1 Light and Fluorescence Microscopy

Very limited structural information can be gleaned on unstained yeast cells with the light microscope. At 1000-fold magnification, it may be possible to visualize the yeast vacuole and several cytoplasmic 'inclusion bodies'. The use of phase-contrast microscopy, together with several staining techniques, enables a number of cellular structures of yeasts to be visualized (Table 2.2). Several fluorochromic dyes can also be used with a fluorescence microscope to highlight features both within yeast cells and on the cell surface (Table 2.2). Pringle *et al.* (1989) have reviewed fluorescence microscopy methods for yeast cells.

A relatively recent development (Stearns, 1995) involves the use of the green fluorescent protein (GFP) from the jellyfish (*Aequorea victoria*) as a reporter molecule for intracellular localization and *in vivo* gene expression studies in *S. cerevisiae* (e.g. Niedenthal *et al.*, 1996; Waddle *et al.*, 1996); *Candida albicans* (Cormack *et al.*, 1997) and *Sch. pombe* (Sawin and Nurse, 1996). Genes of interest may be fused with the *GFP* gene and the subcellular destiny of the expressed fusion proteins followed by fluorescence microscopy (blue light at 395 nm). Such techniques employing GFP are likely to assist in the functional analysis of yeast genomes.

When fluorescent dyes such as fluorescein isothiocyanate and Rhodamine B are conjugated with monospecific antibodies raised against yeast structural proteins, then the range of cellular features visualized is greatly increased. The

Table 2.2. Some cytochemical and cytofluorescent dyes for yeast microscopy.

Dye	Structures visualized	Comments/References
Methylene Blue	Whole cells	Non-viable cells stain blue (see Chapter 4)
FUN-1™	Vacuoles/cell walls	Non-viable cell vacuoles fluoresce red-orange; cell walls blue (Haugland, 1996)
ABT[1]	Cell walls	Fischer (1977)
9-Aminoacridine	Cell walls	Sensitive indicator of surface electrostatic potential (Jones *et al.*, 1995)
F-C ConA[2]	Cell walls	Binds specifically to mannan (Tkacz and Lampen, 1972)
Alcian blue	Cell walls	Binds to mannoprotein of yeast cell surface (Friis and Ottolenghi, 1970)
Calcofluor white	Bud scars	Chitin in scar tissue fluoresces
Aniline blue	Cell walls/septa	β1,3 glucan is stained, but not chitin which appears as black holes in bud scars (Brown *et al.*, 1994; Kippert and Lloyd, 1995)
India ink	Capsules	Capsules appear as a clear light zone between the cell wall and the dark ink (Golubev, 1991)
HCl–Giemsa	Nuclei	Yeast chromosomes may also stain (Robinow, 1981)
Lomofungin	Nuclei	Chromosomes stained red (Kopecka, 1977)
DAPI[3]	Nuclei	Renders DNA fluorescent (Williamson and Fennell, 1975)
Toluidine blue	Nucleoli	Colour intensity diminishes following ribonuclease treatment (Kocková-Kratochvilová, 1990)
CDCFDC[4]	Vacuoles	Pringle *et al.* (1991)
Neutral red	Vacuoles	Vacuoles stain red-purple (Streiblová, 1988)
CMAC[5]	Vacuoles	Vacuolar lumen stained blue (Haugland, 1996)
Lucifer yellow	Endocytic vesicles	Pringle *et al.* (1989)
DASPMI[6]	Mitochondria	Visser *et al.* (1995)
Rhodamine 123	Mitochondria	Alfa *et al.* (1993); Skowronek *et al.* (1990)
DAPI	Mitochondria	Mitochondria fluoresce pink-white (Williamson and Fennell, 1975)
Janus green	Mitochondria	Mitochondria stained green-blue (Streiblová, 1988)
$DiOC_6$[7]	ER and nuclear envelopes	Koning *et al.* (1993)
CFDA[8]	Cytoplasmic pH	Cimprich *et al.* (1995)
Iodine	Glycogen deposits	Glycogen stained red brown; proteins yellow (Streiblová, 1988)
Sudan black	Lipid granules	Lipid stained blue-grey; cytoplasm pale pink (Streiblová, 1988)
DAB[9]	Microbodies	Stain reacts with catalase (Carson and Cooney, 1990)
R-C P[10]	F-Actin	Alfa *et al.* (1993)
GFP[11]	Subcellular localization of specific proteins	e.g. Microtubules (Kahana *et al.*, 1995); spindle pole bodies (Stearns, 1995)

[1] ABT, aldehyde bisulphite -toluidine blue; [2] F-C ConA, fluorescein-conjugated ConcanavalinA; [3] DAPI, 4,6-diamidino-2-phenylindole; [4] CDCFDC, 5,[6] - carboxy-2,7-dichlorodihydrofluorescein diacetate; [5] CMAC, 7-amino-4-choromethyl coumarin; [6] DASMPI, dimethyl-amino styryl-methylpyridinium iodine; [7] $DiOC_6$, 3,3-dihexyloxacarbocyanide iodide; [8] CFDA, carboxyfluoresceine diacetate; [9] DAB, 3,3-diaminobenzidine; [10] R-C P, Rhodamine-conjugated phalloidin; [11] GFP, Green Fluorescent Protein (Jelly fish).

cellular sites of immunochemical reactions can be differentiated by brilliant fluorescence using a dark-ground microscope. For example, immunofluorescence has enabled tubulin and actin to be localized during the *S. cerevisiae* cell division cycle (Kilmartin and Adams, 1984). Confocal scanning-laser immunofluorescence microscopy can also be used to detect intracellular distribution of proteins within yeast cells. For example, Wu *et al.* (1996) have employed such techniques to study the spatial organization of mitosis-regulating protein kinases and phosphatases in fission yeast. Confocal microscopy has also proved very useful in vizualizing organellar morphology in yeasts and Visser *et al.* (1995) have used this method to produce high-resolution three-dimensional images of fluorescently labelled yeast mitochondria (see Fig 2.15). Although, in general, immunofluorescence microscopy can provide simple mapping information (e.g. localization of proteins in the cytoplasm, nucleus or vacuole), higher resolution ultrastructural information on yeast cells can only be obtained with immunoelectron microscopy techniques (Clark, 1994).

Note that endogenous, metabolic fluorescence in yeasts (mainly due to NADH and NADPH) can be continuously determined using commercially available fluorosensors. Such devices enable process control of yeast bioreactions by feedback regulation of biomass levels via oxygen and nutrient supply. This is particularly advantageous in automated control of yeast fed-batch propagations (Beyeler *et al.*, 1981; Beyeler and Meyer, 1984). Chemiluminescence methods can also be employed to analyse redox states and vitality of yeast cultures (see Nishimoto and Yamashoji, 1994).

2.3.2 Image Analysis

Information on gross cell morphology (e.g. mean cell volumes) of yeasts may be obtained using electronic particle counters and size analysers (e.g. Coulter Multisizer). Digital image processing can also be employed to provide computer-automated characterization of yeast cell morphology. This has been accomplished during growth and fermentation with several yeast species (Huls *et al.*, 1992; Pons *et al.*, 1993; Hashida *et al.*, 1995; O'Shea and Walsh, 1996). For example, O'Shea and Walsh (1996) have quantitatively monitored the geometry of cells and filaments of *Kluyveromyces marxianus* during fermentation of cheese whey using image analysis. In *S. cerevisiae*, Zalewski and Buckholz (1996) have used digital image process technology for on-line estimates of morphological heterogeneity. Information on vacuolar formation in *S. cerevisiae* during growth can also be obtained by image processing. Such analyses may be especially useful for process observation and control of industrial yeast fermentations. This is because structural heterogeneity of individual cells may reflect the physiological condition (or 'metabolic fitness') of the culture and this, in turn, may impact on fermentation productivity.

2.3.3 Flow Cytometric Analysis

Flow cytometry has several applications in yeast cytological and physiological studies. Davey and Kell (1996) have reviewed these applications for assessment of yeast cell physiological states and have additionally discussed the value of cell sorting technology for isolating high-yielding strains for biotechnological uses. Fluorescence-activated cell sorting (FACS) has been employed by Gift *et al.* (1996) for isolating slowly growing variants of *S. cerevisiae*. Such procedures are deemed very useful in biotechnology for enriching cultures with hyperproducing cells.

In studies of the yeast cell cycle and organelle biogenesis, flow cytometers have proved particularly useful. For example, Sazer and Sherwood (1990) have analysed DAPI-stained *Sch. pombe* cells to monitor changes in the development of mitochondria. Discrimination between respiratory-competent and respiratory-deficient cells of *S. cerevisiae* stained with Rhodamine 123 has also been achieved by flow cytometry (Skowronek *et al.*, 1990). Cell cycle progression in *S. cerevisiae* has been tracked by Porro *et al.* (1995) using a flow cytometer. In this method, developed by Porro and Srienc (1995), yeast cell walls were labelled with Concanavalin A conjugated to fluorescein isothiocyanate (FITC) and cell protein with tetramethylrhodamine isothiocyanate (TRITC). This double-label cytometric tag enables quantitative information on the growth properties of individual yeast cells as they progress through their cell cycle. Figure 2.1 shows the type of information which can be

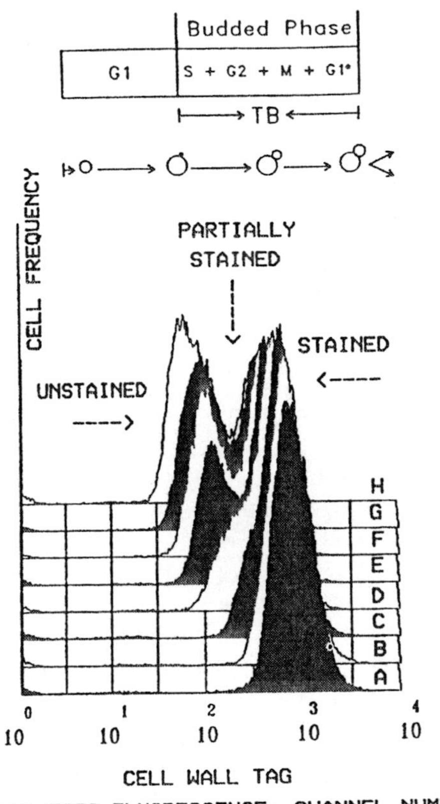

Figure 2.1. Flow cytometric analysis of fluorescently labelled *S. cerevisiae* cells. The figure shows the dynamics of a cell wall-fluorescent tag (ConAFITC) for *S. cerevisiae* cells. The evolution of partially stained and unstained cells over time is clearly visible. The specific growth rate of the overall population and the length of budded phase were 0.125 h-1 and 1.98 h1, respectively. T = 0 h, (A); T = 1 h, (B); T = 2 h, (C); T = 3 h (D); T = 4 h, (E); T = 5 h, (F); T = 6 h, (G); T = 7 h, (H). The inset on the top shows the cell cycle phases of a growing yeast cell. Reproduced with permission from Porro *et al.* (1995) © J. Wiley and Sons Ltd.

obtained from flow cytometric analysis of fluorescently labelled *S. cerevisiae* cells.

2.3.4 Electron Microscopy

Detailed information on organelle ultrastructure and macromolecular architecture can only be obtained with the aid of an electron microscope. Scanning electron microscopy (SEM) is useful in studies of cell topology, while transmission electron microscopy (TEM) of ultrathin yeast sections is essential for the visualization of intracellular fine structure. SEM and TEM methods for yeasts have been described by Streiblová (1988).

High-contrast, nanometre resolution of yeast cell surfaces by non-destructive means may be achieved with the use of an *atomic force microscope*. De Souza Pereira *et al.* (1996) have used this technique on uncoated, unfixed preparations of *S. cerevisiae* and have demonstrated the possibility of imaging cell surface polysaccharide molecules at the distance of atoms. Figure 2.2 reveals interesting topographical differences of the cell walls of several baker's strains of *S. cerevisiae* as imaged by atomic force microscopy.

2.3.5 Yeast Cell Fractionation and Organelle Isolation

Yeast cell wall disruption approaches have been reviewed by Kurtzman (1987) and Middleberg (1993) has discussed various physical, chemical, enzymatic and mechanical methods for process-scale disruption of *S. cerevisiae* cells. Evaluation of such procedures is particularly important when considering the purification of unsecreted heterologous proteins from yeasts. The preparation of cell wall-less yeast protoplasts has been reviewed by Freeman and Peberdy (1983) and Zimmerman and Sipiczki (1996) have discussed methods employed for protoplast fusion in yeasts, including non-*Saccharomyces* species. The isolation of yeast cell organelles for biochemical analyses of subcellular functions has been reviewed by Lloyd and Cartledge (1991) and Zinser and Daum (1995). Table 2.3 provides examples of published methods for yeast organelle isolation and purification.

The purity of different yeast cell fractions can be determined by electron microscopy, by monospecific antibodies or biochemically using marker enzymes (Table 2.4).

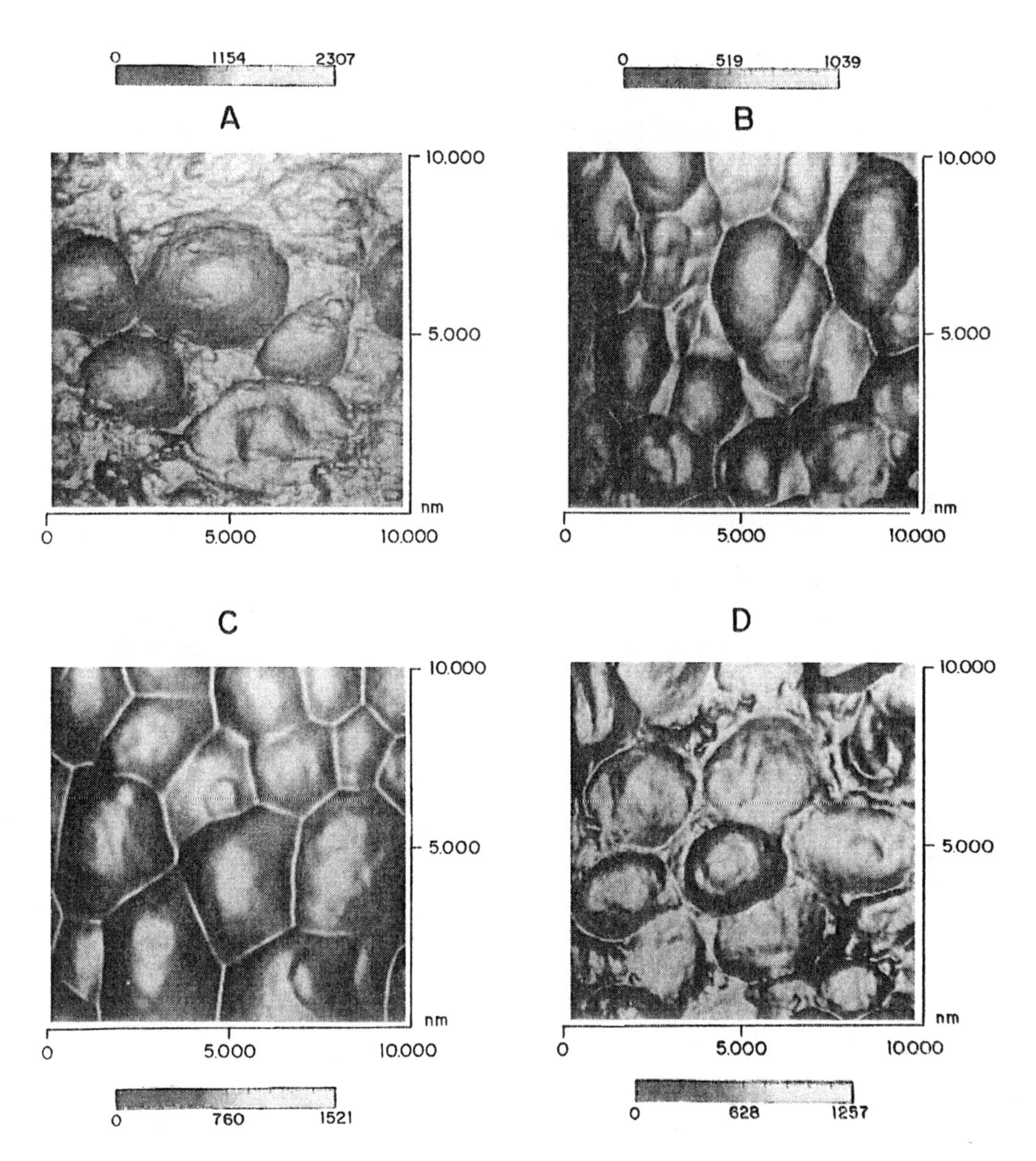

Figure 2.2. Atomic force microscopy of yeast cell surfaces. The micrographs show different morphological aspects between commercial baking strains of *S. cerevisiae*. A, Fermipan yeast (Holland); B, Itaiquara yeast (Brazil); C, Nishin Seifun Co. Yeast (Japan); D, Fleischman yeast (Brazil). The micrographs were kindly provided by Dr Ricardo de Souza Pereira, Universidade Estadual Paulista, Sao Paulo, Brazil.

2.4 YEAST CELL ARCHITECTURE AND FUNCTION

Subcellular compartmentalization in yeast typifies that of eukaryotic cells with the following structures present: nucleus, mitochondria, Golgi apparatus, secretory vesicles, endoplasmic reticulum, vacuoles and microbodies. Note that several organelles derive from an extended intramembranous system and are not completely indepen-

Table 2.3. Examples of methods for yeast organelle isolation.

Organelle	References
Cell walls	Catley (1988); Fleet (1991)
Plasma membranes	Rose and Veazey (1988); Panaretou and Piper (1996)
Plasma membrane vesicles	Menendez *et al.* (1995)
Mitochondria	Rickwood *et al.* (1988); Herrmann *et al.* (1994); Glick and Pon (1995)
Nuclei	Mann and Mecke (1982); Lohr (1988)
Nuclear envelopes	Stambio-de-Castilla *et al.* (1995)
Spindle pole bodies	Rout and Kilmartin (1994)
Rough ER membranes	Sanderson and Meyer (1994)
ER-derived vesicles	Lupashin *et al.* (1996)
Vacuoles	Wiemken *et al.* (1979)
Microsomes	Käppeli *et al.* (1982)
Peroxisomes	Distel *et al.* (1996)
Golgi membranes	Lupashin *et al.* (1996)

Table 2.4. Marker enzymes for isolated yeast organelles.

Organelle/compartment		Marker enzyme
Cell wall:	periplasm	Invertase
	secretory	Acid phosphatase
Plasma membrane		Vanadate-sensitive Mg^{2+}-ATPase
Cytoplasm		Glucose 6-phosphate dehydrogenase
Nucleus:	nucleoplasm	DNA-dependent RNA polymerase
	nuclear envelope	TEM or 38 kDa protein antiserum
Endoplasmic reticulum:	light microsomal fraction	NADPH: cytochrome c oxidoreductase
Vacuole:	membrane	α-Mannosidase
	sap	Proteases A and B; carboxypeptidase Y
Golgi apparatus		β-Glucan synthetase; mannosyltransferase
Mitochondrion:	matrix	Aconitase; fumarase
	intermembrane space	Cytochrome c peroxidase
	inner membrane	Cytochrome c oxidase
	outer membrane	Kynurenine hydroxylase
Peroxisome		Catalase; isocitrate lyase; flavin oxidases

Adapted from Kreutzfeldt and Witt (1991) and Griffin (1994), who provide references for assaying the different marker enzymes.

dent from each other. The cytoplasm contains ribosomes and occasionally plasmids and the structural organization of the intracellular milieu is maintained by a cytoskeleton. The cellular contents are encased by an envelope comprising plasma membrane, periplasm and cell wall. A further capsular and fibrillar layer may also be present in some yeasts.

Figure 2.3(a) provides a diagrammatic view of the ultrastructural features of an idealized yeast cell. Note that some yeasts (e.g. *Candida albicans*, see Figure 2.3(b)) may not possess some of the

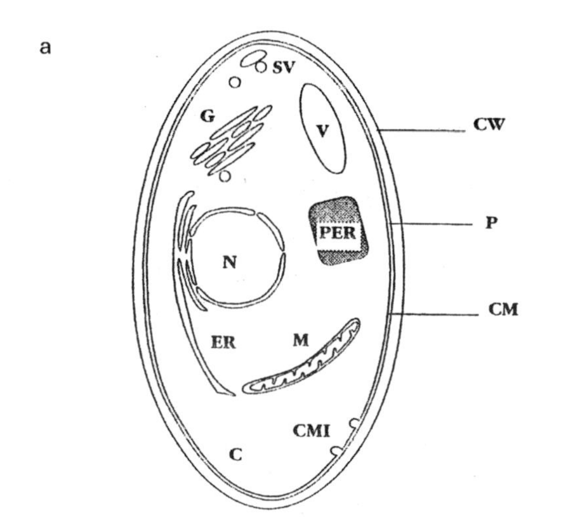

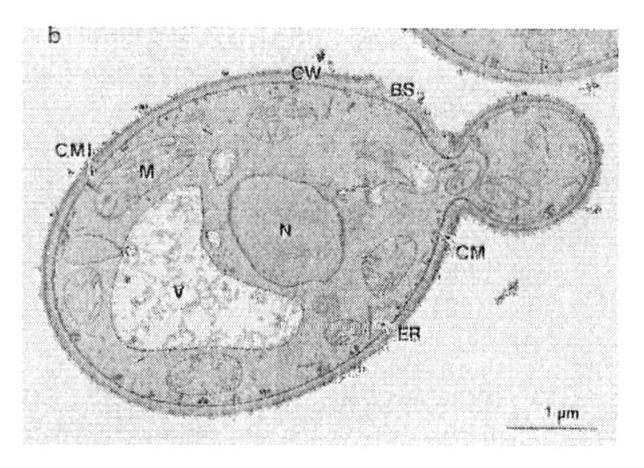

Figure 2.3. Yeast cell ultrastructural features. (a) Idealized yeast cell. (b) *Candida albicans* cell. CW, cell wall; P, periplasm; CM, plasma membrane; CMI, invagination; BS, birth scar; C, cytoplasm; N, nucleus; M, mitochondrion; ER, endoplasmic reticulum; G, Golgi apparatus, S, secretory vesicles; V, vacuole; PER, peroxisome. The *Candida albicans* transmission electron micrograph (×21 000) was kindly provided by Professor Masako Osumi, Japan Women's University, Tokyo.

structures indicated and also that some structures may only appear under certain specialized conditions of growth (e.g. peroxisomes).

Because yeast cells share many of the structural (and functional) features of higher eukaryotes, yeasts have been considered *models* for eukaryotic cell science. Of course, the presence of a cell wall is *the* principal feature which distinguishes a yeast cell from an animal cell.

2.4.1 The Cell Envelope

The yeast cell envelope is considered as the structure which surrounds and encases the yeast cytoplasm. That is, from the inside looking out: the plasma membrane, the periplasmic space, the cell wall and, in certain yeasts, the capsule and other extracellular features. In *S. cerevisiae*, the cell envelope occupies about 15% of the total cell volume and plays a major role, as it does in all yeasts, in controlling the osmotic and permeability properties of the cell. Structural and functional aspects of these cell envelope components will now be described in turn, with reference being made to their practical significance.

2.4.1.1 The Plasma Membrane

Yeasts would not exist as unicellular entities without plasma membranes. This is because these structures represent the primary barriers for passage of hydrophilic molecules and they therefore prevent yeast cytoplasmic contents mixing freely with the aqueous environment. The following briefly describes yeast plasma membrane ultrastructure and discusses membrane functional aspects relating to yeast physiology and biotechnology. More comprehensive coverage of the chemical composition, molecular anatomy and functional characteristics of yeast plasma membranes, especially those of *S. cerevisiae*, is to be found in Henschke and Rose (1991), van der Rest *et al.* (1995) and Höfer (1997).

The *S. cerevisiae* plasma membrane is about 7.5 nm thick with occasional invaginations protruding into the cytoplasm. Like other biological membranes, it can be described as a lipid bilayer interspersed with globular proteins which form a fluid mosaic. The lipid components comprise mainly phospholipids (principally phosphatidylcholine and phosphatidylethanolamine with minor proportions of phosphatidylinositol, phosphatidylserine and phosphatidyl-glycerol) and

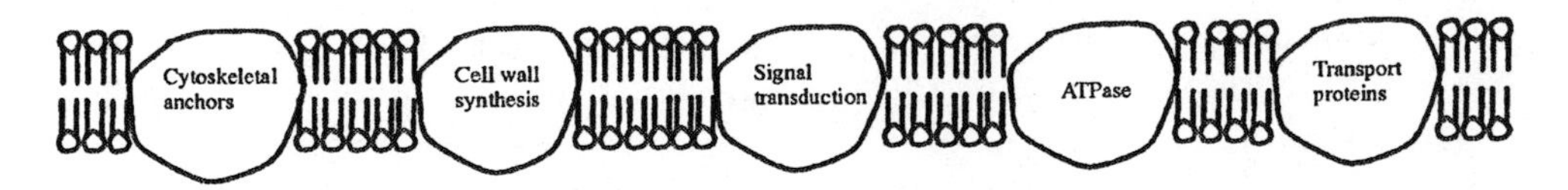

Figure 2.4. Classes of yeast membrane proteins. Adapted from van der Rest *et al.* (1995).

sterols (principally ergosterol and zymosterol with minor proportions of fecosterol and lanosterol). It is likely that the phospholipids confer fluidity and the sterols rigidity to the membranes. The protein components include those involved in: solute transport (ATPase, permeases, channels); cell wall biosynthesis (glucan and chitin synthases); transmembrane signal transduction (adenylate cyclase, G-proteins) and cytoskeletal anchoring. Figure 2.4 schematically depicts the main classes of yeast plasma membrane proteins. In addition, *S. cerevisiae* membranes contain ATP binding cassette (ABC) transporter proteins which are implicated in multidrug transport (Kolaczkowski *et al.*, 1996).

Yeast species differ with regard to their plasma membrane structural make-up and even different strains of a single species can exhibit variations in the membrane lipid composition. For example, strains of brewing yeast (*S. cerevisiae*) possess a much higher phosphatidylcholine content (approx. 10-fold) compared with baking strains (Vendramin-Pintar *et al.*, 1995). The yeast plasma membrane should also not be regarded as a fixed, static feature of a pure yeast strain. This is because it changes both structurally and functionally depending on the conditions of growth. For example, lipid composition, particularly the unsaturated fatty acid constitution, can alter quite dramatically with changing growth rates, temperature and oxygen availability. Such changes in lipid composition of the membrane change, in turn, its functional properties – especially with regard to amino acid and sugar transport (see Henschke and Rose, 1991).

The primary functions of yeast plasma membranes are to dictate what enters and what leaves the cytoplasm. These selective permeability properties are mediated by specialized membrane proteins. Their role in yeast nutrition, that is, in the uptake of sugars, nitrogenous sources, ions, etc. is discussed in Chapter 3. Of prime importance in active transport of solutes into yeast cells is the activity of the plasma membrane proton-pumping ATPase. This enzyme hydrolyses ATP and generates an electrochemical proton gradient, which together with the membrane potential, provides the driving force for uptake of essential solutes.

Another very important physiological function of the yeast plasma membrane is in signal transduction of external stimuli to mediate (via second messengers) a number of internal biochemical reactions. Henschke and Rose (1991) have discussed the plasma membrane phosphoinositide second messenger system which operates in *S. cerevisiae*. Further aspects of signal transduction in control of yeast growth and metabolic processes are discussed in Chapters 4 and 5, respectively, and its role specifically in *S. cerevisiae* has been extensively reviewed (Uno, 1992; Thevelein, 1994; van Dam, 1996).

Other transport-related functions of the yeast plasma membrane relate to exocytosis and endocytosis. In the former, secretory vesicles derived from the endoplasmic reticulum and Golgi apparatus fuse with the plasma membrane to deliver proteins (e.g. mannoproteins) through the cell envelope (see section 2.44.). The vesicles in question are also believed to contain constituents required for the biosynthesis and assembly of the cell envelope (Tschopp *et al.*, 1987). With regard to endocytosis, this is a system of internalizing and localizing certain molecules with the aid of specialized membranous structures known as **endosomes** (Figure 2.5).

In the first stages of endocytosis, plasma membrane invaginations pinch off to form vesicles which deliver their extracellular cargo to the endosomes. Endocytosis in *S. cerevisiae* is known

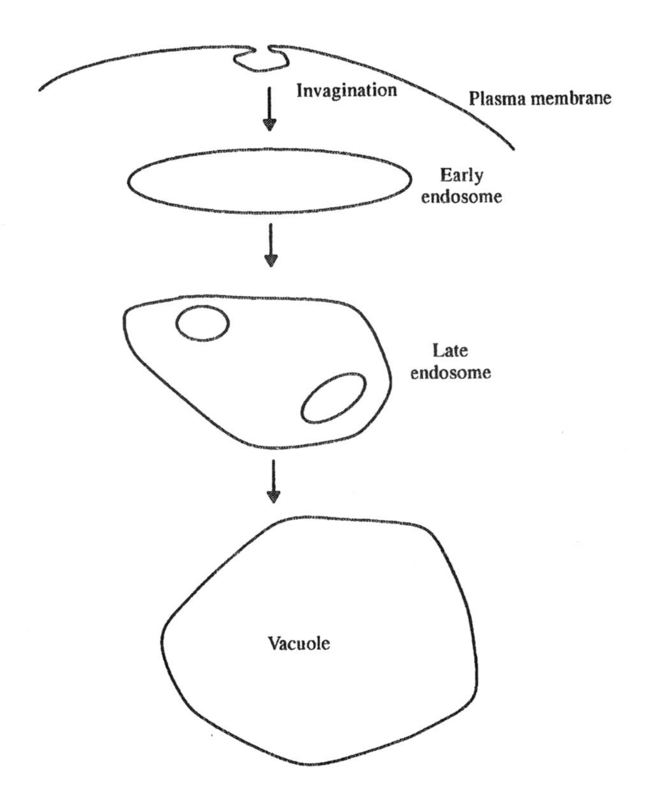

Figure 2.5. Schematic representation of endocytosis in yeast cells. Adapted from Riezman (1993) who has discussed the molecular events of the endocytic pathway in *S. cerevisiae*

to play an important role in the internalization of mating pheromones and their subsequent transport from the plasma membrane to the vacuole. Endocytic pathways may also be involved in other aspects of yeast physiology, including stress responses and sporulation (Riezman, 1993). In addition, Govindan and Novick (1995) have discussed the possible involvement of endocytosis in regulating budding patterns in *S. cerevisiae*. Horazdovsky *et al.* (1995) have reviewed molecular biological aspects of endocytosis and intracellular protein sorting in yeast cells.

The plasma membrane is especially significant when considering the physiology of industrial yeasts. For example, the ability of brewing yeast strains to produce and tolerate ethanol is intimately linked to the nature of unsaturated fatty acids and sterols in their plasma membranes. Thus, cell membranes enriched in linoleyl ($C_{18:2}$) fatty acid residues and with ergosterol (as opposed to other sterols) exhibit a greatly enhanced tolerance to ethanol (Rose, 1993). Oxygen availability in fermentations plays a key role in dictating the fatty acid/sterol make-up of brewing yeast membranes. This is because oxygen is absolutely required for the synthesis of unsaturated fatty acids and sterols. In practical terms, this means that brewer's yeast requires initial oxygenation before fermentation to ensure correct plasma membrane biosynthesis. Failure to provide sufficient molecular oxygen will render cell membranes more susceptible to ethanol toxicity and this restricts yeast growth and fermentative capability.

2.4.1.2 The Periplasm

Continuing in the inside-to-outside direction, the next 'structure' encountered as a component feature of the yeast cell envelope is the periplasm. This is a thin (35–45Å), cell wall-associated region external to the plasma membrane and internal to the wall. It was described by Arnold (1991) as the **periplasmic space**, and by Robinow and Johnson (1991) as the **never-never land** of the yeast cell.

The periplasm mainly comprises secreted proteins (e.g. mannoproteins) which are unable to permeate the cell wall. These include the glycoprotein enzymes invertase and acid phosphatase which catalyse the hydrolysis of substrates that do not cross the plasma membrane. Other enzymes which may be located in the periplasm of certain yeasts are melibiase and trehalase (Arnold, 1991).

The biotechnological significance of the yeast periplasm is that invertase is an enzyme commercially prepared from baker's yeast following cell wall autolysis or hydrolysis. Yeast invertase has applications in the confectionery industry where it is used to hydrolyse crystalline sucrose to fructose and glucose in the manufacture of soft-centred chocolates and candies.

2.4.1.3 The Cell Wall

The wall of yeast cells represents quite a thick (generally, 100–200 nm) structure comprising between 15–25% of the total dry mass of the cell and is a prominent distinguishing feature of all yeasts. The composition, structure and function of wall components in *C. albicans* and *S. cerevisiae* have been reviewed, respectively, by Shepherd (1987) and Fleet (1991).

The main structural constituents of yeast cell walls are polysaccharides, which account for 80–90% of the wall. These are principally glucans and mannans, with a minor proportion of chitin. Yeasts derive their strength from the glucan components of the cell wall which are partly arranged in a microfibrillar network. Both β-1,6- and β-1,3-linked glucans are present, distinguished by their solubility properties in acid and alkali. Mannans are present as an α-1,6-linked inner core with α-1,2 and α-1,3 side chains. Chitin, a polymer of *N*-acetylglucosamine, is present in small quantities in *S. cerevisiae* (approx. 2–4%), mainly in bud scars. Some filamentous yeasts (e.g. *C. albicans*) possess higher contents of chitin whereas other yeasts appear to lack chitin completely. Other components of yeast cell walls include variable proportions of protein, lipids and inorganic phosphate. The precise proportion of the chemical constituents of yeast cell walls will vary depending on yeast strain, cell age and growth conditions.

The precise molecular arrangements of the *S. cerevisiae* wall are unresolved and several structural models exist (e.g. Zlotnick *et al.*, 1984; Stratford, 1994; Vukovic and Mrsa, 1995). Nevertheless, the general consensus is that the cell wall is a layered structure, the outermost section of which comprises cross-linked mannoproteins (Figure 2.6). These are linked to each other by hydrophobic interaction or by dis-

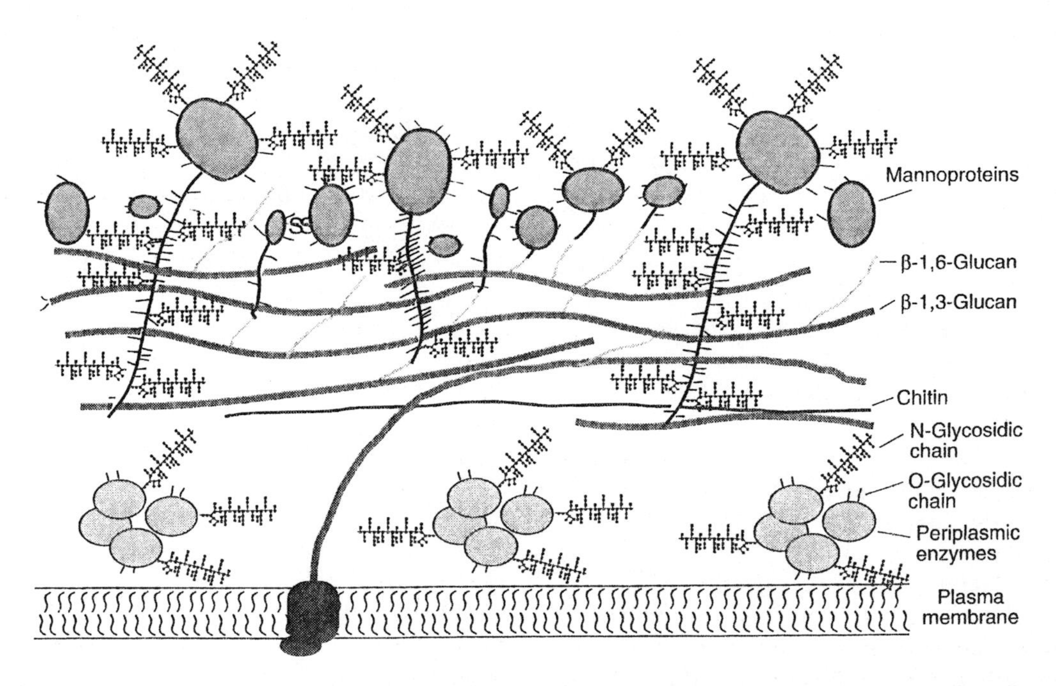

Figure 2.6. Composition and structure of the cell wall of *Saccharomyces cerevisiae.* The cell wall, which is located outside the plasma membrane, consists of two layers. The inner layer provides cell wall strength, and is made of β-1,3- and β-1,6-glucan that is complexed with chitin. The outer layer consists of mannoproteins, and determines most of the surface properties of the cell. The majority of mannoproteins are covalently linked to the inner glucan layer. Periplasmic enzymes are trapped between the plasma membrane and the inner skeletal layer. Reproduced with permission from Schreuder *et al.* (1996a) and Elsevier Science Ltd.

ulphide bonds. They are also linked to the inner fibrillar glucan network by covalent bonds. The mannoproteins may be important in determining the porosity of the yeast cell wall which is known to exclude molecules greater than about 600 Da. Some β-glucan of the inner layer is cross-linked to chitin (Brown *et al.*, 1993). Chitin itself is a major constituent of bud scars in budding yeasts and different chitin synthase isozymes play key roles in septum formation and morphogenesis (Silverman, 1989). Chitin is also located in smaller amounts throughout the cell wall of *S. cerevisiae* where it serves various functions as: a killer toxin receptor (Takita and Castilho-Valavicius, 1993) and in the maintenance of osmotic and morphological integrity (see Stratford, 1994). Secreted proteins may interact with cell wall polysaccharides (glucan) and Stratford (1994) has discussed the structural role of a new class of cell wall mannoproteins which anchor in the plasma membrane, span the wall and protrude into the medium. In *S. cerevisiae*, these proteins are involved in flocculation and sexual aggregation.

Stratford (1994) has made the following interesting analogy between yeast cell wall structure and reinforced concrete: 'Steel reinforcing rods are represented by enmeshed alkali-insoluble (1,3)-β-glucan fibrils, comprising some 35% of the wall. The reinforcing is surrounded by concrete, pebbles in a sand/cement matrix; secreted mannoproteins represent pebbles, some 25–50% of the wall, encased and bonded to the reinforcing fibrils by a matrix of amorphous β-glucan and chitin'.

Bud scars are chitin-rich, convex, ringed protrusions which remain on the mother cell surface of budding yeasts following birth of daughter cells (Figure 2.7).

The localization of bud sites on the cell wall of *S. cerevisiae* is discussed in Chapter 4. **Birth scars** are concave indentations which remain on the daughter cell surface following budding. These structures do not contain as much chitin as bud scars and are less well visualized using fluorescent dyes.

Fission yeasts similarly leave scar tissue on their cell surfaces following division. In *Sch. pombe*, these fission scars or **scar plugs** are formed during cleavage of mother cells. During the next round of growth these scar plugs grow out and a circular scar is retained between the original cell and the new growth (see Chapter 4 for a fuller description of fission in *Sch. pombe*).

With regard to the physiological function of the yeast cell wall, it should be stressed from the outset that the wall is not merely an inert exoske-

a

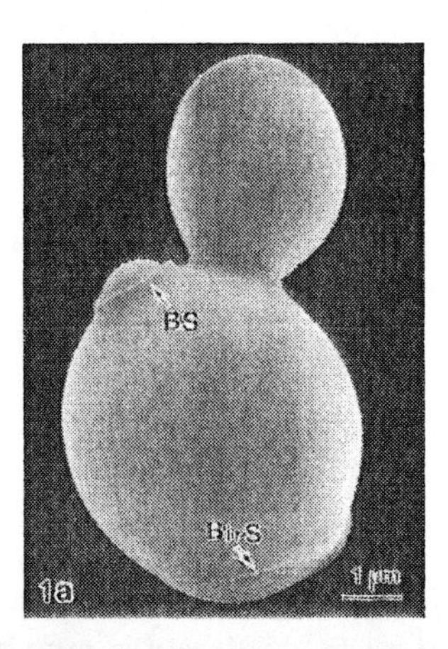

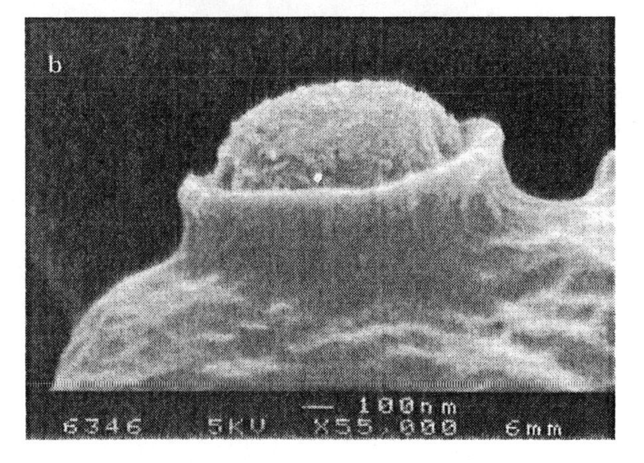

Figure 2.7. Scanning electron micrographs showing yeast bud and birth scars. (a) *Saccharomyces cerevisiae*, showing scar tissue on the surface of *S. cerevisiae* cells (×10 000). BS, bud scar; BirS, birth scar. Micrograph kindly provided by Professor Masako Osumi, Japan Women's University, Tokyo. (b) *Zygosaccharomyces bailii*, showing a bud scar on the surface of *Z. bailii* NCYC1766 cells (×55 000). Micrograph taken by Mark Kirkland and kindly provided by Dr Malcolm Stratford, Unilever Research, Bedford, UK.

leton whose only role is to protect the protoplast. On the contrary, the yeast cell wall should be recognized as a living organelle whose functions change during yeast growth and metabolism. In *S. cerevisiae*, the availability of several cell wall mutants have enabled physiological function studies to be undertaken. For example, mutants have been isolated with defects in: killer toxin reception, protein glycosylation, mannose biosynthesis, chitin deposition and glucan fibril assembly. Such mutants have revealed that the wall serves vital roles in yeast cell physiology (Table 2.5).

In short, the yeast cell wall is a multifunctional organelle involved in cell protection, shape maintenance, cellular interactions, reception, attachment and specialized enzymatic activities (Fleet, 1991).

Removal of the yeast cell wall with lytic enzymes in the presence of osmotic stabilizers produces **spheroplasts**. The lytic enzymes in question are usually commercially available preparations such as 'Helicase' from snail digestive juice or 'Zymolyase', 'Novozyme' and 'Lyticase' from microbial sources. Studies of cell wall regeneration in yeast spheroplasts has provided valuable information on the cytology and biochemistry of cell wall assembly. Figure 2.8 shows cell wall regeneration in spheroplasts of the fission yeast, *Schizosaccharomyces pombe*.

Inter- and intrageneric fusion of yeast spheroplasts is also an extremely valuable procedure for both basic genetic studies and for producing hybrids with biotechnological potential (see Mann and Jeffery, 1986; and Chapter 6).

Several structural and functional aspects of the yeast cell wall have practical implications in yeast biotechnology and in medicine (see Schreuder *et al.*, 1996a,b; also Table 2.6). However, in addition to being a barrier for the yeast cell itself, the wall is also a barrier for the yeast technologist. For example, in yeast recombinant DNA technology, both the influx of foreign DNA and the efflux of expressed heterologous proteins are restricted by the presence of a thick cell wall. There are several biochemical and genetic approaches which can be employed to overcome this barrier. These include the use of lithium acetate to transform intact cells and the use of secretory signal sequences to direct protein export. Several cell wall mutants of yeast may also be of value in yeast biotechnology:

Fragile mutants are yeasts with defects in cell wall mannoprotein biosynthesis which are osmo-

Table 2.5. Physiological functions of the yeast cell wall.

Function	Comments
Physical protection	As well as protecting the protoplast, the cell wall maintains the shape of yeast cells.
Osmotic stability	Removal of cell walls results in protoplast lysis in the absence of osmotic stabilizers.
Permeability barriers	Solutes larger than about 600 Da fail to permeate through the wall. The wall also plays a role in controlling the entry of water into yeast.
Enzyme support	Wall-softening enzymes (e.g. glucanases) and hydrolases (e.g. invertase) are immobilized in the matrix of the cell wall.
Cation binding	Several cations are known to be effectively sequestered by the cell wall, including heavy metals.
Cell-cell recognition	Recognition sites for mating pheromones and killer toxins are located in the cell wall.
Cell-cell adhesion	Yeast-yeast sexual agglutination, flocculation and agglomeration are wall-related phenomena. Yeast–human cell adhesion is relevant in pathogenicity of some species (e.g. *Candida albicans*).

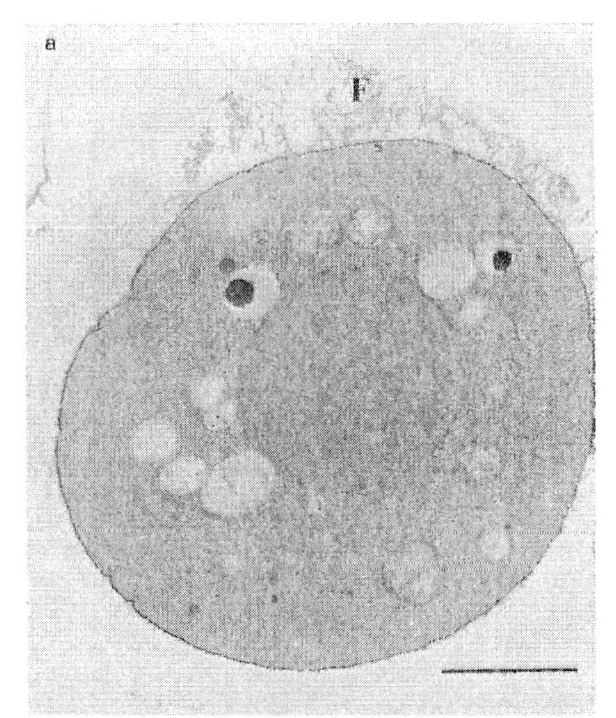

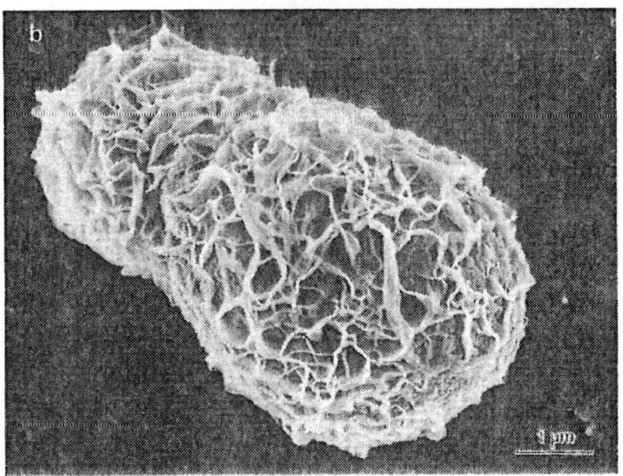

Figure 2.8. Cell wall regeneration in spheroplasts of *Schizosaccharomyces pombe.* (a) Transmission EM micrograph of a 3-h-reverting protoplast of *Sch. pombe* (× 20 000; bar = 1 μm). Regeneration of new cell wall β-glucan from the cell surface with polarity is demonstrated (F). (b) Scanning EM micrograph of a 5-h-reverting protoplast of *Sch. pombe* (× 18 000; bar = 1 μm). The protoplast is covered with a network of β-glucan microfibrils. The micrographs were kindly provided by Professor Masako Osumi Japan Women's University, Tokyo. Further information on the formation of the glucan network in fission yeast can be found in Osumi *et al.* (1995).

tically fragile. Such mutants have been described in several species, including *S. cerevisiae* (Venkov *et al.*, 1974; Blagoeva *et al.*, 1991) and *Sch. pombe* (Belda and Zarate, 1996) and some have been considered useful in both the uptake of DNA (Venkov and Ivanov, 1982) and in the recovery of recombinant proteins (Bröker, 1994).

Autolytic mutants of *S. cerevisiae*, which are thermosensitive cells with an osmotically unstable cell wall (Cabib and Durán, 1975), may be of value in biotechnology to facilitate release of intracellular proteins into the extracellular medium (Alvarez *et al.*, 1994) following temperature upshift.

Permeability mutants of *S. cerevisiae*, which possess enhanced cellular permeability to macromolecules due to the presence of truncated mannoprotein side chains in the cell wall, have been shown to have practical potential in testing for chemical mutagens in environmental samples (Staleva *et al.*, 1996).

The cell wall confers flocculant, buoyant or adhesive properties on yeast which are of practical relevance to the fermentation industries and also in medicine.

Flocculation is a phenomenon of particular importance in brewing and physiological, biochemical and genetic aspects of brewing yeast flocculence have been widely discussed in the literature (e.g. Speers *et al.*, 1992; Stratford, 1992; Straver *et al.*, 1993; Rose, 1993; Teunissen and Steensma, 1995; Dengis and Rouxhet, 1997). Yeast flocculation is the phenomenon of asexual cellular aggregation when cells adhere, reversibly, to one another to form macroscopic flocs which sediment out of suspension. Historically, brewing yeast strains were distinguished as highly flocculant **bottom yeasts** (for lager fermentations) or weakly flocculant or buoyant **top yeasts** (for ale fermentations). However, brewing strains of *S. cerevisiae* are quite heterogeneous with respect to flocculation and individual strains may be classed into several flocculence types based on their sedimentary and flotation characteristics (Hough *et al.*, 1982). These distinctions are primarily reflections of differences in cell wall composition between yeast strains. Yeast flocculation can be

Table 2.6. Biotechnological and medical significance of the yeast cell wall.

Practical significance	Comments/References
Biosorption of metals	Yeast cell walls have uses in bioremediation of heavy metals (e.g. Cd, Cu, Ag, Zn) in industrial wastewaters (Volesky *et al.*, 1993; Blackwell *et al.*, 1995; Simmons *et al.*, 1995).
Antigenicity properties	May be exploited in yeast taxonomy and serological diagnosis of pathogenic yeasts.
Yeast 'glycan'	Non-nutritive food stabilizer (Hay, 1993).
Heterologous protein binding	Hepatitis B virus surface antigens have been expressed in the outer mannoprotein of *S. cerevisiae* which may have potential as a live oral vaccine (Schreuder *et al.*, 1996b).
Cell adhesion	Flocculation and agglomeration of *S. cerevisiae* is important during fermentation (see text), while adhesion of *C. albicans* is medically important (Calderone and Braun, 1991).
Protein secretion barrier	Fragile and other mutants overcome this (see text).
Killer toxin reception	Killer yeasts have several applications in biotechnology (see Chapter 4, Table 4.33).
Immunomodulation, anti-tumour agent	Wall components such as (1,3)-β-glucans may act as biological response modifiers and anti-tumour agents (Bohn and BeMiller, 1995).

readily quantified (Soares and Mota, 1997) and in brewing fermentations, it is now possible to measure on-line the intensity of yeast flocculation for process control purposes (Podgornik *et al.*, 1997).

In general, the flocculation of brewing yeast is associated with the onset of stationary phase, but the timing and degree of flocculation is not precisely regulated. This is an important consideration in brewing practice because if flocculation occurs too early, fermentation will cease prematurely, leaving residual sugar in the wort. This can cause microbiological stability problems and can adversely affect the flavour characteristics of the final product. Conversely, if it takes an unduly long time for cells to flocculate, this can cause downstream processing problems in beer clarification. For these and other reasons, the biochemical mechanisms and genetic basis of brewing yeast flocculation have been intensively studied over the years. Flocculation also occurs in other yeasts and has been investigated in *Kluyveromyces*, *Zygosaccharomyces*, *Pichia*, *Candida* and *Schizosaccharomyces* species (see Stratford, 1994 for references).

Although far from being completely understood, it is now widely recognized that yeast flocculation conforms to the *lectin-like hypothesis* originally proposed by Miki *et al.* (1982). Thus, interactions occur between calcium-activated lectins (sugar-binding proteins) and α-mannan receptors on neighbouring yeast cell walls (Figure 2.9).

Cell wall lectins in yeast may be the *FLO* gene-encoded **flocculins** which are surface glycoproteins capable of directly binding mannoproteins of adjacent cells (Teunissen and Steensma, 1995). Cell surface hydrophobicity of yeast cell walls has also been implicated in playing a distinct or complementary role with lectins to mediate cell–cell flocculation (Straver *et al.*, 1993). Brewing yeast hydrophobicity may be readily assayed using a novel magnetic separation procedure (Straver and Kijne, 1996). Smart and her co-workers (Smart *et al.*, 1995) have studied cell surface hydrophobicity of brewing yeast strains in relation to physiological stress and their influences on flocculation. Research has suggested that cell surface *charge*, rather than hydrophobicity, is the primary determinant of cell floc-

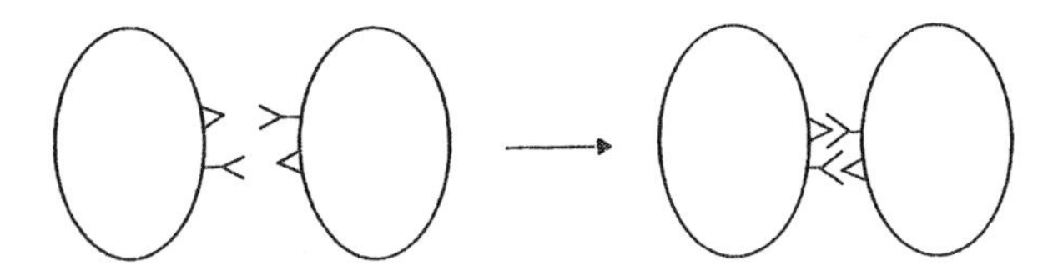

Figure 2.9. Lectin hypothesis for flocculation of yeast cells. Surface proteins with lectin properties specifically bind to mannose residues in the wall of neighbouring cells. Calcium ions are required to maintain the lectins in active conformations. Yeast flocs will be progressively built up by simultaneous inter-binding of many cells.

culence in brewing yeasts (Wilcocks and Smart, 1995).

Yeast flocculation is genetically determined and different flocculation phenotypes have been identified based on their sensitivities to sugar inhibition and proteolytic enzymes (Stratford and Assinder, 1991). These phenotypes, named Flo and NewFlo, also possess different sensitivities to yeast growth conditions, most notably culture pH (Stratford, 1996), temperature (Soares *et al.*, 1994) and glucose availability (Soares and Mota, 1996). Lo and Dranginis (1996) have identified a novel flocculin-encoding gene, *FLO11*, in *S. cerevisiae* which exhibited homology to the *STA* genes which encode secreted glucoamylase. The product of *STA1* was implicated in Ca-dependent flocculation activity.

The converse of flocculation in yeasts is **flotation** which is associated with cell wall hydrophobicity. The flotation characteristic relevant in brewing is the ability of yeasts to trap CO_2 bubbles in a fermenting liquid and form a yeast 'head' at the top of fermentation vessels. This behaviour is typical of top-fermenting yeasts employed in traditional ale breweries. Note that unlike flocculant cells, *S. cerevisiae* strains prone to flotation do not form cell–cell aggregates (see Chapter 4 and Figure 4.25).

Cell wall hydrophobicity is important not only in influencing flocculation and flotation properties of brewer's yeast, but also in adhesion of yeasts. Yeast cells can adhere to:

- inert surfaces (e.g. walls of bioreactors, surfaces of medical implants);
- soft tissues (e.g. human epithelial cells); and
- water-immiscible carbon sources (e.g. hydrocarbon globules).

There is particular fundamental and practical interest in the hydrophobic nature of the surface of *Candida albicans* cells, due to the medical significance of this yeast as the causative agent of human and animal mycoses (collectively referred to as candidosis). The ability of *C. albicans* to adhere to and colonize host cells and surgical devices is an important determinant in the pathogenicity of this yeast (Calderone and Braun, 1991). The hydrophobic adherence properties of *C. albicans* have been studied using spectroscopic techniques by Hobden *et al.* (1995).

True flocculation in yeast cells is reversible and can be distinguished from chain formation or aggregation by the ability of the chelating agent EDTA to disperse flocs, but not chains or aggregates. The phenomenon of **agglomeration** is different from flocculation and may be regarded as an extensive, non-reversible cell aggregation process and is particularly relevant in baking strains of *S. cerevisiae*. Some baker's yeast strains have a tendency to agglomerate into macroscopic aggregates which do not resuspend when cells are mixed with water. The appearance of granular yeast material is often referred to by bakers as 'grit' or 'sand'. Grittiness is detrimental to baker's yeast quality since it results in inadequate mixing into bread dough leading to limited leavening ability. The formation of yeast grit is strain-dependent but is also influenced by the ionic strength of the suspending solvent (Guinard and Lewis, 1993). Therefore, although agglomeration (or grit formation) is distinct from flocculation, both phenomena are influenced by a variety of genetic and physiological factors and both are cell wall-mediated processes. The mechanism of yeast agglomeration is unknown.

Guinard and Lewis (1993) found no significant difference between gritty and non-gritty yeasts in terms of cell wall mannoproteins, surface hydrophobicity or flocculence. Phosphorus and lipid concentrations did appear to be higher in non-gritty cells and Guinard and Lewis (1993) have proposed a model for yeast agglomeration based on Ca^{2+}-activated protein–mannan binding between adjoining cells which is facilitated by reduced phosphorus and lipid concentrations on gritty yeast cell walls.

2.4.1.4 The Yeast Spore Wall

The development of spores by yeast represents a process of morphological, physiological and biochemical differentiation of sexually reproductive cells. Sporulation involves meiosis and ascopore development in cells and in *S. cerevisiae* can be initiated by depriving diploid cells of nitrogen and providing acetate as a respiratory carbon source. The cytology of meiosis, ascospore formation and spore germination in fission and budding yeasts has been discussed by Robinow and Johnson (1991) and by Williamson (1991).

Meiotic division in *S. cerevisiae* produces four-lobed nuclei from which four spores are formed. The spindle pole body (SPB) in each nuclear lobe polarizes the meiotic spindles and forms thickened plaques in the outer nuclear membrane which serve as the beginnings of spore wall formation. Similar modifications of the outer membrane plaque of meiotic SPBs occurs in *Sch. pombe* (Hirata and Shimoda, 1994). These outer plaques in lobed yeast nuclei initiate the development of spore walls which eventually grow round to envelope the lobes of the mitotic nucleus and to delimit spores.

After meiosis is completed, yeast spore walls develop within the ascus from a structure known as the **forespore membrane**. Robinow and Johnson (1991) have discussed the biogenesis and elaboration of multilayered spore walls in *S. cerevisiae*, *Sch. pombe* and other spore-forming yeasts. Generally speaking, however, knowledge of spore wall synthesis is lacking in comparison with that of vegetative cell walls of yeasts. Griffin (1994) has discussed physiological, biochemical and genetic aspects of sporulation in *S. cerevisiae*. With respect to external signals involved in sporulation, Suizu *et al.* (1995) found an elevation in intracellular Ca^{2+} ions during spore formation and have discussed the possible regulatory role of Ca^{2+}-dependent signalling for both meiosis and sporulation in *S. cerevisiae*.

The genetics of *S. cerevisiae* meiosis and sporulation have been discussed, respectively, by Malone (1990) and Dawes (1983). Several genes of both the mating type control pathway and the nutritional control pathway are involved in regulating meiosis and sporulation in *S. cerevisiae* (e.g. Kawaguchi *et al.*, 1992).

2.4.1.5 Fimbriae

Fimbriae are long (variable from 0.1–10 μm), thin (5.7 nm diameter), proteinaceous protrusions emanating from the surface of several basidiomycetous and ascomycetous yeast species (Gardiner *et al.*, 1982). They appear to be mainly involved in cell–cell interactions before sexual conjugation. However, fimbrial protein is also present during vegetative growth in some yeasts which plays a different role from conjugal fimbriae. For example, short fimbriae in *S. cerevisiae* are implicated in flocculation. Smart *et al.* (1995) have illustrated fimbriae-like 'hairy' protrusions from the surface of flocculant cells of brewing yeast. By comparison, non-flocculant cells appeared smooth.

2.4.1.6 Capsules

Golubev (1991) has reviewed the structure and function of yeast capsules. These slimy extramural layers are regarded as yeast *organelles* due to their distinctive structural and functional characteristics. Capsules are prevalent in basidiomycetous yeasts such as *Cryptococcus*, *Rhodotorula* and *Sporobolomyces* species where they may serve in protecting cells from physical and biological stresses encountered in their natural habitats. Golubev (1991) has discussed the suggestions that yeast capsules act as buffers to prevent water loss from cells and to enhance acquisition of trace levels of nutrients in oligotrophic environments. These views are supported by the fact that capsule-forming yeasts predominate in poor habitats such as polar soils.

Most biochemical and physiological information on yeast capsules has come from studies with the pathogenic yeast, *Cryptococcus neoformans*. Capsules in this yeast may be important virulence determinants. Other yeasts, for example, *C. laurentii* and *Hansenula capsulata*, produce extracellular polysaccharide materials exhibiting plastic rheological characteristics which have potential biotechnological applications (Slodki and Cadmus, 1978; Sutherland and Elwood, 1979). Growth conditions will greatly influence the amount and type of capsule produced by yeasts. Nevertheless, Phaff *et al.* (1978) have noted that the principal extracellular/capsular compounds produced by yeasts are: phosphomannans (e.g. in *Hansenula*, *Pichia* and *Pachysolen* spp.); β-linked mannans (e.g. *Rhodotorula* spp.); heteropolysaccharides (e.g. *Cryptococcus*, *Lipomyces*, *Candida* and *Trichosporon* spp.) and sphingolipid-type compounds (e.g. *Hansenula* spp.).

2.4.2 The Cytoplasm and Cytoskeleton

The yeast cytoplasm is an aqueous acidic colloidal fluid containing low and intermediate molecular weight compounds, dissolved proteins, glycogen and other soluble macromolecules. Also suspended in the cytosol are membrane-delimited microbodies and macromolecular aggregations such as ribosomes, proteasomes and lipid particles. The cytoskeletal network providing structural organization to the yeast cytoplasm comprises microtubules and microfilaments. The cytosolic (non-organellar) enzymes of yeasts include glycolytic enzymes, the fatty acid synthetase complex and the enzymes of protein biosynthesis. In *S. cerevisiae*, the mean cytosolic pH of exponential-phase cells suspended in water was calculated at 5.25 (Cimprich *et al.*, 1995).

Freely-suspended yeast **ribosomes**, as opposed to endoplasmic reticulum-associated or mitochondrial ribosomes, consist of large 60S and small 40S ribonucleoprotein subunits (as in other eukaryotic cells) and exist as single ribosomes and as multiple mRNA-linked aggregates called polysomes (see Lee, 1991). The latter are the sites of protein biosynthesis. Unlike membrane-delimited organelles, cytoplasmic ribosomes are synthesized *de novo* (Warner, 1989; Planta *et al.*, 1995).

Lipid particles (or 'sphaerosomes') function as storage vesicles which may serve as lipid reservoirs for yeast membrane biosynthesis (Clausen *et al.*, 1974). They contain mainly sterol esters, but not triglycerides, and small amounts of phospholipid, protein and unsaturated free fatty acids. The content of the latter are known to increase in baker's yeast at the end of the growth phase in batch culture.

Microbodies in yeast cells include **peroxisomes** and **glyoxysomes** which are single membrane-delimited organelles distinct from endoplasmic reticulum-derived vesicles. Characteristic features of microbodies are: the presence of catalase, the occasional presence of crystalloids in their matrix and their osmotic fragility (Carson and Cooney, 1990; Veenhuis and Harder, 1991). The appearance of these organelles is largely determined by yeast growth conditions, in particular, the sources of available carbon and nitrogen.

Peroxisomes perform a variety of metabolic functions in eukaryotic cells. In yeasts, peroxisomes are ubiquitous organelles which contain, in addition to catalase, several oxidases which are involved in the oxidative utilization of specific carbon and nitrogen sources. For example, *Hansenula polymorpha* and *Pichia pastoris* can utilize methanol using peroxisomal enzymes such as alcohol oxidase; *Yarrowia lipolytica* and *Candida tropicalis* can utilize *n*-alkanes using several oxidases and *Candida utilis* can utilize D-alanine and uric acid using, respectively,

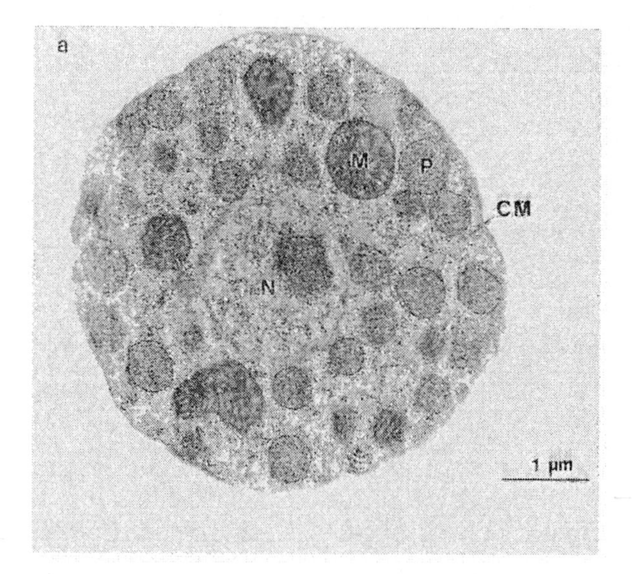

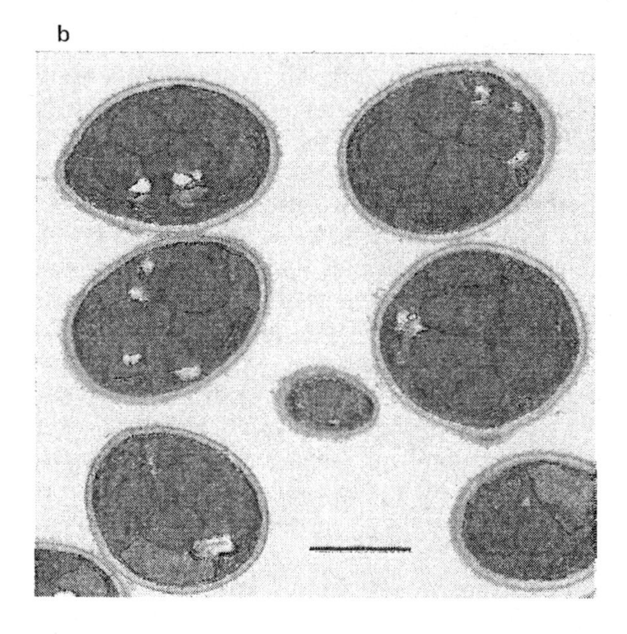

Figure 2.10. Yeast peroxisomes. (a) *Candida tropicalis*. The micrograph shows the ultrastructure of a *C. tropicalis* protoplast of *n*-alkane-grown cells. N, nucleus; P, peroxisomes; M, mitochondria; CM, cell membrane. Bar = 1 µm. Micrograph kindly provided by Professor Masako Osumi, Japan Women's University, Tokyo. (b) *Hansenula polymorpha*. The micrograph shows the ultrastructure of *H. polymorpha* cells grown in the presence of methanol. Bar = 1 µm. Kindly provided by Dr Marten Veenhuis, University of Groningen, The Netherlands.

peroxisomal D-alanine oxidase and uricase. When methylotrophic yeasts such as *H. polymorpha* are transferred from a glucose-containing medium to one containing methanol as sole carbon and energy source, there is a rapid increase in the size and crystalline ultrastructure of peroxisomes (Figure 2.10).

The organelles develop from the small peroxisomes originally present in glucose-grown cells mainly as a result of rapid synthesis of enzymes such as alcohol oxidase and catalase. A more detailed discussion of peroxisome biogenesis has been provided by Waterham and Cregg (1997). Several gene (*PEX*) products which are required for peroxisome biogenesis have now been identified in several yeasts. Molecular biological work on yeast *pex* mutants is relevant to our understanding of Zellweger syndrome, a congenital cerebro-hepato-renal disorder in humans which is characterized by an absence or deficiency of peroxisomes in cells from Zellweger patients (Waterham and Cregg, 1997).

In methanol-limited chemostats, *H. polymorpha* peroxisomes exhibit a completely crystalline substructure and the organelles can reach a cell volume fraction as high as 80%. In addition to offering an ideal system in which to study the biochemistry and genetics of organelle biogenesis and function (Titorenko *et al.*, 1993; Sudbery, 1994), peroxisome-producing yeasts also have biotechnological applications – for example, in the use of alcohol oxidase-promoted gene expression and intracellular packaging of heterologous proteins. The gene for this peroxisomal enzyme is one of the most powerful promoters known and has stimulated great interest in the use of methylotrophic yeasts such as *H. polymorpha* and *P. pastoris* in recombinant DNA technology (see Chapter 6).

Glyoxysomes in yeast contain, in addition to catalase, enzymes of the glyoxylate cycle (see Chapter 5) and amine metabolism. The organelles become especially prominent in ethanol-grown cells of *H. polymorpha* and *C. utilis*. Veenhuis and Harder (1991) have discussed the possibility that glyoxysomes and peroxisomes develop from each other in response to specific growth conditions and may not in fact be distinct organelles. The term **glyoxy-peroxisome** has been suggested to describe these yeast microbodies.

Proteasomes are large multisubunit protease enzyme complexes found in the cytoplasm and nucleoplasm of yeast cells with no apparent association with intracellular structures. In *S. cerevisiae*, 20S and 26S proteasomes exist whose function in regulating protein levels is essential for cell viability. The 26S particle is assembled from the 20S units which are cylindrical particles composed of four stacked rings. The 26S proteasome acts as an ATP-dependent protease in the ubiquitin pathway which is responsible for the rapid degradation of short-lived, abnormal proteins that are detrimental to the cell. Specific yeast physiological functions in which proteasomes are involved include cell cycle control, signal transduction, mating and adaptive stress responses (reviewed by Hilt and Wolf, 1995). These multifunctional roles of proteasomes have led Hilt and Wolf (1996) to consider these complexes as 'functionally sophisticated counterparts of the ribosome'.

The **cytoskeleton** of yeast cells comprises **microtubules** and **microfilaments**. These are dynamic structures which perform mechanical work in the cell through assembly and disassembly of individual protein subunits. Thus, α and β tubulin monomers polymerize as heterodimers into 25 nm-thick microtubules, while G-actin globular monomers polymerize into 7 nm-thick double-stranded filaments of F-actin. Cytoskeletal structures are assembled in an energy-dependent manner and the process is regulated by an extensive system of associated structural proteins and enzymes (Dustin, 1978; Ayscough and Drubin, 1996). Ayscough and Drubin (1996) have discussed the value of *S. cerevisiae* in molecular biological studies of the structure and function of the eukaryotic actin cytoskeleton.

Yeast microtubules and microfilaments are involved in several aspects of yeast physiology including mitosis and meiosis, organelle motility, and septation (see Heath, 1995). For example, the actin cytoskeleton in *S. cerevisiae* is intimately involved in bud-site selection (Yang *et al.*, 1997), and in *Sch. pombe*, microtubules play a key role in polarized growth and fission (see Chapter 4). The role of microtubules in governing the motility and positioning of mitochondria in *Sch. pombe* has been studied by Yaffe *et al.* (1996). Cytoplasmic microtubules appear directly apposed to the SPBs (Spindle Pole Bodies) on the nuclear envelope. Their role in maintaining cell shape of yeasts is evident when, for example, *Sch. pombe* cells are treated with inhibitors of microtubule biogenesis (e.g. thiabendazole) they develop abnormal filamentous cell morphologies (Walker, 1982).

2.4.3 The Nucleus and Extrachromosomal Elements

Comprehensive information on structural and functional aspects of yeast nuclei and extrachromosomal elements is available from recent published literature. For example, detailed coverage has been provided on: the nucleus by Williamson (1991); karyogamy (nuclear fusion) by Rose (1996); nuclear pore complexes by Bucci and Wente (1997); genome structure by Kaback (1995); chromosome replication by Theis and Newlon (1996); chromatin by Pérez-Ortin *et al.* (1989); histones by Ushinsky *et al.* (1997); the nucleolus by Léger-Silvestre *et al.* (1997), centromeres and kinetochores by Hyman and Sorger (1995) and Lechner and Ortiz (1996); telomeres by Louis (1995) and Zakian (1996); spindles by McDonald *et al.* (1996); spindle pole bodies by Kahana *et al.* (1995) and Sobel (1997); 2 μm plasmids by Wickner (1995); linear DNA plasmids by Fukuhara (1995); RNA viruses by Wickner (1996); prions by Tuite and Lindquist (1996) and Wickner *et al.* (1996) and retrotransposons by Kingsman and Kingsman (1988a). The following represents a very brief account of the structure and function of the yeast nucleus and genetic material.

2.4.3.1 The Nucleus

The **nucleus** in yeasts is a round–lobate organelle of around 1.5 μm diameter which is located in the centre of the cell or excentrically. The nucleoplasm is separated from the cytoplasm by a double membrane containing pores 50–100 nm in diameter. Figure 2.11 shows a diagrammatic representation of the yeast nucleus. The precise size and number of nuclear membrane pores in yeast cells varies with growth conditions and with phases of the cell cycle. Bucci and Wente (1997) have studied the *in vivo* assembly dynamics of nuclear pore complexes in *S. cerevisiae*. In *S. cerevisiae*, nuclear membranes are occasionally contiguous with and have similar chemical composition to the endoplasmic reticulum (Zinser *et al.*, 1991). Unlike most eukaryotic cells, the yeast nuclear membrane does not break down during mitosis. Within the nucleus there is a dense crescent-shaped region corresponding to the **nucleolus** which disappears during mitosis and reappears in interphase.

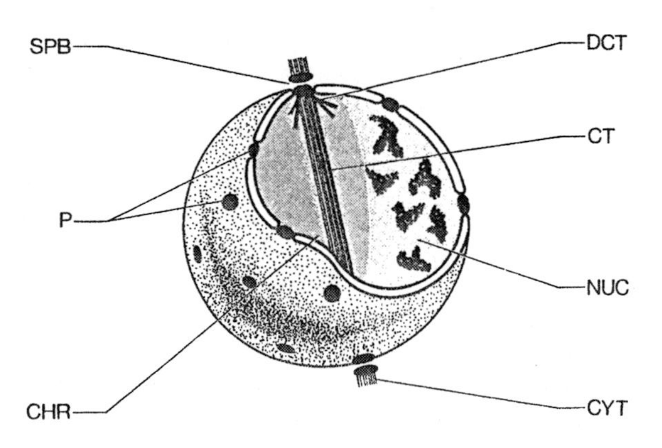

Figure 2.11. Diagram of the yeast nucleus. The figure is a diagrammatic cut-away view of a (budding) yeast nucleus in the late S or early G_2 phase of the cell cycle. The spindle is complete, but still 'short', and the nucleus has not yet started to move into the bud. For clarity only a few microtubules are shown. SPB, spindle pole body; DCT, discontinuous microtubules; CT, continuous microtubules; CYT, cytoplasmic microtubules; NUC, nucleolus; CHR, chromatin; P, pores. Reproduced with permission from Williamson (1991) and Academic Press.

The nucleolar organizer in the nucleolus contains ribosomal DNA repeats and is the site of rRNA transcription and processing. It is also involved in pre-mRNA processing and in the assembly of ribosomal subunits.

The nucleoplasm contains DNA, RNA, basic proteins (protamines and histones) and non-histone proteins. Some extrachromosomal elements (e.g. 2 μm circles) may also be present. The condensed basic nucleoprotein material, consisting of double-helical DNA–histone complexes, is **chromatin** which is organized in structures called chromosomes. During S phase (DNA synthetic phase) of the cell cycle, each chromosome is replicated and during M phase (mitosis) the duplicated chromosomes are segregated to daughter cells. Asexual mitotic nuclear division in yeasts occurs within the confines of intact nuclear membranes, unlike mammalian cells in which the nuclear membrane breaks down during mitosis. Detailed cytological descriptions of the stages of mitosis (prophase, metaphase, anaphase and telophase) and of meiosis in various yeast species are provided by Kocková-Kratochvilová (1990), Williamson (1991) and Robinow and Johnson (1991). The regulation of meiosis in budding and fission yeasts has been discussed by Malone (1990) and Yamamoto (1996), respectively.

Several general statements can be made about yeast chromosomes:

- Genome sizes are relatively constant and generally range from 10–15 Mb and encode between 5000 and 10 000 genes.
- Haploid chromosome numbers are variable. For example, from 2 (*Guilliermondella selenospora*) to 16 (*S. cerevisiae*). Most yeasts have fewer, larger chromosomes compared with *S. cerevisiae*.
- Individual yeast chromosomes vary in size by more than an order of magnitude from approx. 0.2 to 6 Mb.

Yeast chromosomes can be analysed by pulse-field electrophoretic techniques. Johnston (1994) has described electrophoretic karyotyping methods for *Saccharomyces* and Zimmerman and Fournier

Table 2.7. Analysis of some yeast chromosomes

Yeast	No. of haploid chromosomes	Haploid nuclear size (Mb)	Comments/References
Saccharomyces cerevisiae	16	12.07 (Range: Chromosomes I = 0.23; IV = 1.53)	Genome now completely sequenced (Goffeau *et al.*, 1996; Dujon, 1996) comprises non-ribosomal DNA encoding over 6000 proteins.
Schizosaccharomyces pombe	3	I = 5.7 II = 4.6 III = 3.5	Chromosomes large enough to condense and be visible during mitosis (Robinow and Hyams, 1989). Genome sequencing project ongoing (1997)
Guilliermondella selenospora	2	I = 2.0 II = 2.8	Dr G. Cardinali, University of Perugia, Italy (personal communication)
Candida albicans	8 pairs	1.2–3.0	Species normally diploid (Poulter, 1995)
Kluyveromyces spp.	6–13	12 ± 3.5 (*K. waltii* = 7) (*K. marxianus* = 8)	Genus shows large variation in their electrophoretic karyotypes (Kaback, 1995)
Hansenula spp.	4	0.7–1.0	Information from Griffin (1994) and Kaback (1995)
Cryptococcus neoformans	10–12	15–17	
Rhodotorula mucilaginosa	9	0.2–1.0	
Histoplasma capsulatum	7	2.0–5.7	
Rhodosporidium toruloides	10	0.4–4.0	

(1996) for non-*Saccharomyces* yeasts. Table 2.7 provides a summary of chromosomal number and size analysis in several yeasts. Note that considerable chromosome length polymorphism can occur among yeast isolates and karyotyping has proved a valuable method of differentiating between industrial strains of yeast which have their own characteristic karyotype patterns (e.g. Johnston, 1988; Naumov *et al.*, 1992; Querol *et al.*, 1992; Codon and Benitez, 1995).

The structural organization of chromosomes plays a crucial role in the regulation of gene function. Work with *S. cerevisiae*, in particular, has provided a great deal of information regarding functional elements of eukaryotic chromosomes (Kaback, 1995). For example, the ability to transform this yeast with chiameric plasmid DNA has led to the discovery of:

- **Origins of replication:** autonomously replicating sequences (ARS) enable plasmid DNA to replicate autonomously in the nucleus.
- **Centromeric DNA** sequences: centromeres are the sites of kinetochore formation and chromosome attachment to mitotic and meiotic spindles (see Hyman and Sorger, 1995; Lechner and Ortiz, 1996).
- **Telomeric DNA** sequences: telomeres are specialized sequences that enable complete replication of chromosome ends and prevent their degradation in the cell (see Louis, 1995). Zakian (1996) has discussed extrapolating information about yeast telomeric DNA to the ageing of human cells. Data from *S. cerevisiae* suggests that the ability to maintain telomere length may be critical for the tumorigenic state of mammalian cells.

2.4.3.2 Extrachromosomal Elements

Several non-chromosomal (non-Mendelian) genetic elements are known to exist in yeast cell mitochondria, cytoplasm and nucleus (Table 2.8).

Non-chromosomal traits in yeasts can be distinguished from their chromosomal counterparts by several genetic and biochemical criteria (Wickner, 1995). These are, for example, 4:0 meiotic segregation patterns, efficient transfer by cytoduction (mixed cytoplasm, but non-fused nuclei) and by 'curing' (elimination of plasmids) by chemical mutagens or temperature shock.

2 μm DNA is a stably maintained, circular nuclear plasmid in *S. cerevisiae* which replicates exactly once in each S phase of the cell cycle. No easily recognizable phenotypic change has been associated with loss of 2 μm circles from wild-type *S. cerevisiae* cells. 2 μm plasmids are present at a high copy number and are very useful in biotechnology in the construction of cloning vectors in recombinant DNA technology (Futcher, 1988). Plasmids which are structurally related to *S. cerevisiae* 2 μm DNA have been found in *Zygosaccharomyces* and *Kluyveromyces* spp. (Bianchi *et al.*, 1987; Wickner, 1995).

The double-stranded RNA and linear DNA plasmids are found in 'killer' strains of yeast. The physiology and biotechnology of killer yeasts and plasmid-encoded toxins is discussed in Chapter 4.

Certain yeast extrachromosomal elements such as [Psi] and [URE3] behave as **prions** (Derkatch *et al.*, 1996; Tuite and Lindquist, 1996; Wickner *et al.*, 1996). Prions are infectious protein particles which appear to be the causative agents of bovine spongiform encephalopathy (BSE), scrapie in sheep, and kuru, fatal familial insomnia and Creutzfeldt–Jacob disease in humans. Yeast prions are now being studied at the molecular genetic level to provide basic information on mammalian prions and for designing novel approaches for the treatment of prion infections.

Other genetic elements which violate Mendelian rules are the yeast retrotransposons, or Ty elements, which move from one genomic location to another via an RNA intermediate using reverse transcriptase enzymes (Kingsman and Kingsman, 1988a). *S. cerevisiae* cells contain

Table 2.8. Some extrachromosomal genetic elements in yeasts.

Element	Comments/References
Mitochondrial DNA	Involved in mitochondrial biogenesis and respiratory metabolism. Genetics of yeast mitochondria has been reviewed by Tzagoloff and Myers (1986).
2 μm DNA	Around 60 copies of these circular histone-covered plasmids are found in the nucleoplasm of *S. cerevisiae* cells. Although growth is marginally slower in 2 μm-containing strains, no immediately recognizable function has been identified for these elements. Nevertheless, they are very useful as cloning vectors (reviewed by Futcher, 1988; Parent and Bostian, 1995).
Killer plasmids: dsRNA	These encode killer toxins of *S. cerevisiae* and some other killer yeasts (Wickner, 1996). Other dsRNA viruses related to totiviruses have been described in *S. cerevisiae* (Park *et al.*, 1996).
linear DNA	Typically found in killer strains of *Kluyveromyces* spp. and some *Pichia* spp. (see Chapter 4, Table 4.32 for references).
20s and 23s RNA	These naked, circular, single-stranded RNA molecules are independent replicons with no clear phenotype in *S. cerevisiae* (Wickner, 1995).
[Psi], [Eta] and [URE3]	Ψ Increases ochre (UAA) nonsense codon suppression. Eta is a similar replicon to Ψ, but with a different phenotype (Wickner, 1995). [URE3] enables cells of *S. cerevisiae* to take up ureidosuccinate in the presence of high concentrations of ammonia (Wickner, 1995). Evidence now exists that these elements are protein **prions**, rather than nucleic acid replicons (e.g. Tuite and Lindquist, 1996).

several Ty copies that are found at variable locations on the chromosome. The Ty-encoded proteins have an ability to self-assemble into retrovirus-like particles (VLPs). However, like other yeast 'viruses' (e.g. dsRNA killer plasmids and ssRNA replicons), but unlike most animal cell viruses, they do not infect other cells by an extracellular route. Infectious transmission is solely by cell–cell fusion. By fusing and expressing additional protein-coding sequences to the retrotransposon, it is possible to produce hybrid Ty-virus like particles (Ty-VLPs) (Adams *et al.*, 1987) which have several applications in recombinant DNA technology including the production of monoclonal antibodies and viral antigenic determinants for synthetic vaccine production (Kingsman and Kingsman, 1988b; Kingsman *et al.*, 1994). Adams *et al.* (1991) have described the construction of hybrid Ty-VLPs for use in heterologous gene expression. Kingsman *et al.* (1994) have discussed the production of several different hybrid Ty-VLPs which carry protein components of several viruses including: HIV, influenza virus, feline leukaemia virus and bovine papillomavirus.

2.4.4 The Secretory System and Vacuoles

Various membrane-delimited compartments in the yeast cytoplasm play key roles in the trafficking of proteins both into and out of the cell. Transport of proteins between membrane compartments is a process common to all eukaryotic cells and yeasts have served as important models for our fundamental understanding of such processes.

The export of proteins by secretion from yeast cells involves intra- and inter-membranous trafficking (via vesicles) in which endoplasmic reticulum (ER), Golgi apparatus and the plasma membrane all participate. Proteins destined for the vacuole, rather than the extracellular medium are also transported by secretory organelles (see below). Cytoskeletal elements are also involved and the actin cytoskeleton is responsible for establishing the directionality of the secretory process. The import of proteins into yeast cells by endocytosis is similarly mediated by membrane-bound vesicles which deliver their cargo to the vacuole for proteolytic processing. Endocytosis in yeasts was discussed in section 2.4.1, and has been reviewed by Riezman (1993). Figure 2.12 represents a schematic overview of the main cytological events of exocytosis (secretion) and endocytosis in yeasts.

The secretory pathway has been extensively studied in *S. cerevisiae* by Schekman and his colleagues (e.g. Pryer *et al.*, 1992; Salama and Schekman, 1995; Lupashin *et al.*, 1996). The following represents a brief outline of the major stages in protein secretion by budding yeasts:

- Proteins destined for secretion are firstly synthesized on ER-associated polysomes.
- These proteins are then discharged into the lumen of the ER (see Lyman and Schekman, 1996).
- In the ER, proteolytic cleavage of a signal peptide and chaperone-assisted protein folding takes place along with glycosylation.
- Proteins are then directed from the ER by vesicles which fuse to the *cis*-Golgi apparatus which is arranged in parallel arrays or stacks (Rambourg *et al.*, 1995). Note that 'retrograde' transport of vesicles from the Golgi back to

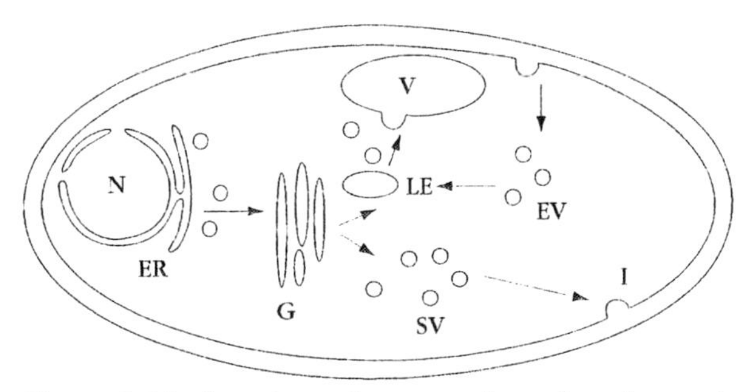

Figure 2.12. Overview of exocytosis and endocytosis in yeast cells. N, nucleus; ER, endoplasmic reticulum; G, Golgi apparatus; LE, late endosome (prevacuolar compartment); EV, endocytic vesicle; SV, secretory vesicle; I, invagination on plasma membrane; V, vacuole.

the ER may also take place (Lewis and Pelham, 1996).

- In the Golgi, further modification of carbohydrate side chains (outer-chain mannosylation) on the proteins takes place.
- Vesicles, derived from budding of the late-Golgi, then carry the proteins to the actively growing bud region of the cell in an actin filament-directed process. Different vesicles are involved in transporting plasma membrane components and secreted enzymes (Harsay and Bretscher, 1995).
- Fusion of secretory vesicles with the plasma membrane then delivers the proteins to the periplasm. Note that specific receptor proteins referred to as 'SNAREs' (see Rothman, 1994) in both vesicles (v-SNAREs) and the target membrane (t-SNAREs) facilitate interaction or 'docking' in the final stage of the secretory pathway when vesicles fuse with their target membranes (see Jiang *et al.*, 1995; Novick *et al.*, 1995; Pelham *et al.*, 1995).

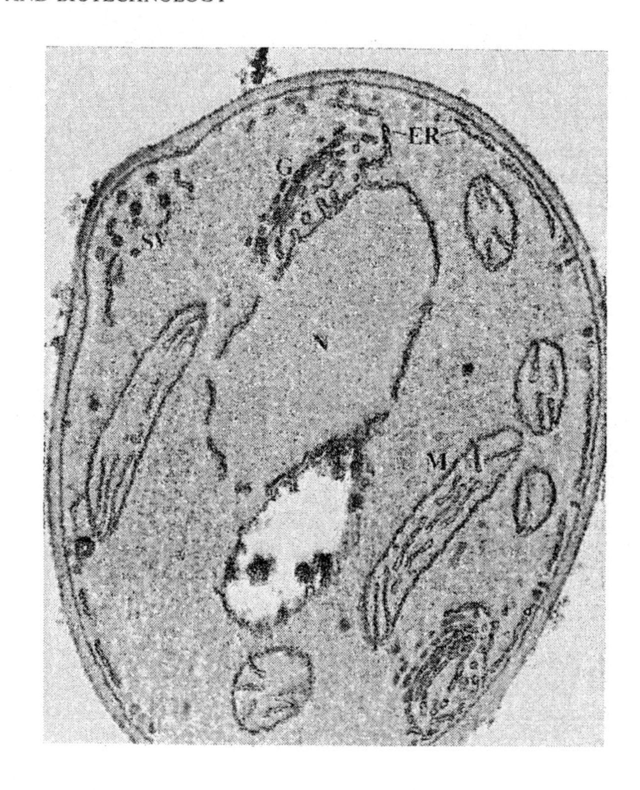

Figure 2.13. Secretion in *Pichia pastoris*. The micrograph is of a thin section of *P. pastoris* fixed and stained with $KMnO_4$. G, Golgi apparatus; N, nucleus; ER, endoplasmic reticulum; SV, secretory vesicles; M, mitochondria. Kindly provided by Dr Chris Kaiser, Massachusetts Institute of Technology, Cambridge, USA.

Figure 2.13 depicts components of the secretory system in the budding yeast, *Pichia pastoris*.

The molecular biology of such a pathway has been particularly well documented in *S. cerevisiae* for the secretion of the periplasmic enzymes invertase and acid phosphatase following the biochemical characterization of several *Sec* (secretory) mutants defective in various stages of the process (Rothblatt and Schekman, 1989).

Not all secreted yeast proteins are trafficked in the above manner and alternative pathways exist for some proteins. For example, Cleves *et al.* (1996) have identified a new protein export pathway in *S. cerevisiae* which is distinct from the classical ER–Golgi–vesicle route. By expressing a non-classically exported mammalian protein (galectin-1) in *S. cerevisiae*, Cleves *et al.* (1996) were able to identify genes which encode protein export *portals* in the plasma membrane. Such portals enable more direct secretion of proteins in a manner which by-passes the classical secretory pathway.

Understanding protein secretion mechanisms and their regulation is directly pertinent in those yeast biotechnologies concerned with expression of heterologous genes. This is because secretion by yeasts simplifies downstream purification procedures (by circumventing the need to disrupt cells) and also because it avoids potential toxicity of intracellularly accumulated protein. Additional advantages, particularly when considering the production of mature human polypeptides for pharmaceutical applications, is that passage of proteins through the yeast pathway enables proteolytic maturation and glycosylation of the heterologous protein to occur. Knowledge of the moleculai biology of secretion signals is of practical importance. Yeasts generally secrete very few proteins and at low abundance (Table 2.9).

Nevertheless, heterologous protein secretion by

Table 2.9. Some naturally secreted proteins from yeasts.

Protein	Comments
Mating pheromones	**a**-Factor and α-factor by haploid mating types of *S. cerevisiae* (see Chapter 4)
Killer toxins	Several yeasts secrete plasmid or chromosomally encoded killer toxins (see Table 4.32)
Invertase and acid phosphatase	These enzymes may not be 'secreted,' but are held extracellularly in the periplasmic space
Exoglucanase	This cell wall-modifying enzyme is involved in wall expansion during growth.
Hydrolases	Some yeasts possess an ability to secrete proteases, pectinases, lipases, amylases etc. (see Tables 5.9 and 6.33)

S. cerevisiae can be driven by signal (or 'leader') peptide sequences which derive from naturally secreted proteins such as mating factors, killer toxins or invertase. These sequences are specifically cleaved by ER cisternal proteases (e.g. KEX 2) as the protein is matured on its way out of the cell. The specificity of ER proteases allows *authentic* N-termini to be obtained on secreted proteins, whereas this is often hard to obtain with cytoplasmic expression. Exploitation of the molecular machinery in yeast cells for protein processing, glycosylation and secretion has enabled the production of purified human polypeptides by several transformed yeast strains, including *S. cerevisiae*, *Pichia pastoris*, *Hansenula polymorpha*, *Kluyveromyces lactis*, *Schizosaccharomyces pombe* and *Yarrowia lipolytica* (see Chapter 6). Several examples of successful secretion of therapeutic polypeptides by yeasts in which the chemical and biological properties of the proteins are comparable with their authentic, native counterparts have been described (e.g. Eckart and Bussineau, 1996; Rallabhandi and Yu, 1996).

The **vacuole** is a key organelle involved in intracellular protein trafficking in yeasts. Yeast vacuoles may not always be clearly distinct and independent organelles (like mitochondria), but form an integral component of the intramembranous system which includes the ER, Golgi and vesicles. The term **vacuolar compartment** has been proposed to reflect more accurately the structure of yeast vacuoles (Schwencke, 1991). Vacuoles of yeast cells are also dynamic structures which may exist in cells as a single large compartment or as several smaller compartments, or **provacuoles.** For example, vacuolar volume changes in a dynamic fashion with cell growth and in budding yeasts, small fragmented vesicular vacuoles which accompany bud initiation fuse to form larger vacuoles when budding proceeds. Stationary phase cells thus characteristically only contain one or two large vacuoles. Like some other cytoplasmic organelles, vacuoles arise from growth, multiplication and separation of pre-existing structures rather than by *de novo* synthesis (Raymond *et al.*, 1992). Haas and Wickner (1996) have discussed the regulation of vacuole inheritance during the life cycle of *S. cerevisiae*.

Vacuoles are bounded by a single membrane called the **tonoplast**. This membrane has a different phospholipid, unsaturated fatty acid and sterol content compared with the plasma membrane and is more elastic than it. The tonoplast therefore remains intact when yeast protoplasts lyse under hyposmotic conditions. However, tonoplasts will disintegrate under certain conditions which results in autolysis of the cell. For example, prolonged nutrient starvation, high temperature, acidic pH and high K^+ concentrations will result in release of vacuolar enzymes and digestion of yeast macromolecular contents.

Such autolytic processes are actively encouraged in the manufacture of yeast extracts for use in the food industry.

The vacuole is a lysosome-like acidic compartment which is primarily responsible for non-specific intracellular proteolysis in yeasts. These processes are catalysed by the activities of intravacuolar endopeptidases, aminopeptidases and carboxypeptidases. Delivery of these enzymes to the vacuole is mediated by a portion of the secretory pathway (see Figure 2.12; Conibear and Stevens, 1995). Vacuolar proteases have multi-functional roles in yeast cell physiology. The synthesis, processing and function of yeast vacuolar enzymes (e.g. in exocytosis and endocytosis) has been reviewed by Klionsky *et al.* (1990), Schwencke (1991), Horazdovsky *et al.* (1995) and Van den Hazel *et al.* (1996). Other proteolytic compartments in yeast cells include the ER, Golgi apparatus and mitochondria. Specific and more rapid protein degradation in yeasts, for example of ubiquitinated proteins, takes place in the cytoplasm in **proteasomes** which are responsible for digestion of proteins which may be detrimental to the cell (see Chapter 5).

Besides their role in degradative (lysosomal) processes, vacuoles are involved in several other physiological functions in yeasts. For example, vacuoles are the primary storage compartments for basic amino acids, polyphosphate and certain metal cations (Table 2.10). They are also involved in osmoregulation and the homeostatic regulation of cytosolic ion concentrations and pH. Although the latter is undoubtedly influenced by the activities of the tonoplast proton-translocating ATPase, the plasma membrane ATPase is thought to be more important in determining cytosolic pH (Serrano, 1991).

2.4.5 Mitochondria

For many years, yeasts have been employed as model eukaryotic organisms in fundamental studies of mitochondrial structure, function and biogenesis (e.g. Tzagoloff and Myers, 1986; Wilkie, 1993). Yeast cells contain mitochondria which structurally resemble those organelles found in higher eukaryotes. That is, they comprise:

Table 2.10. Component functions of yeast vacuoles.

Component	Functions
Vacuolar sap proteases: carboxypeptidase Y; proteinases A and B; aminopeptidase I	Non-specific proteolysis (e.g. during starvation, sporulation) and specific proteolysis (e.g. catabolite inactivation)
Vacuolar sap trehalase	Trehalose turnover
Tonoplast enzymes: α-mannosidase; dipeptidyl aminopeptidase B; alkaline phosphatase; ATPase	Various enzymatic functions, including cellular pH control (in the case of ATPase)
Vacuolar sap metabolites/ions:	
basic amino acids	Large arginine pool present
metal ions	Control of physiologically useful ions (K^+, Mg^{2+}, Ca^{2+}) and toxic ions (Sr, Cd, Co, Pb, etc.)
polyphosphates	Polyphosphate deposits ('volutin') represent a macromolecular storage of inorganic phosphate

Adapted from Schwencke (1991), who reviews additional information on yeast vacuolar structure and function.

- an outer membrane – containing enzymes involved in lipid metabolism;
- a matrix – containing enzymes of fatty acid oxidation and the citric acid cycle, together with protein synthetic machinery (including mitochondrial ribosomes) and mitochondrial DNA; and
- an inner membrane – containing cytochromes of the respiratory chain, NADH and succinate dehydrogenases and H^+-ATPase.

Detailed ultrastructure and function of yeast mitochondria has been reviewed by Guérin (1991) and Chapter 5 discusses aspects of mitochondrial respiratory metabolism. The biogenesis of yeast mitochondria, which involves genetic cooperativity between nuclear and mitochondrial genomes, has been widely studied in *S. cerevisiae* (e.g. Hurt and Van Loon, 1986; Tzagoloff and Myers, 1986; Pon and Schatz, 1991). The following discussion will focus on structure–function relationships of mitochondria in relation to yeast growth conditions, including those encountered by yeasts exploited in industrial fermentations.

Yeast mitochondria are dynamic organelles whose size, shape and number change according to the particular yeast species or strain, cell cycle phase and growth conditions. Several three-dimensional modelling studies (e.g. Davison and Garland, 1977; Visser *et al.*, 1995) have indicated that generally yeast cells contain one or very few large, occasionally branched mitochondria. These structures are perhaps more accurately described as **mitochondrial reticula**, to distinguish them from the more vesicular-shaped organelles frequently portrayed in biochemistry text books. However, mitochondrial morphogenesis, including definitions of inner membrane cristae, very much depends on external physiological factors which influence yeast growth; for example, oxygen partial pressure, glucose concentration, presence of unfermentable substrates, availability of sterols and fatty acids and levels of certain metal ions such as magnesium (see Walker *et al.*, 1982; Figure 2.14).

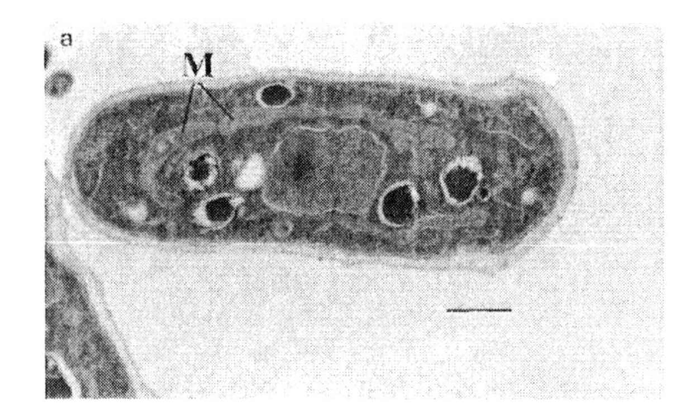

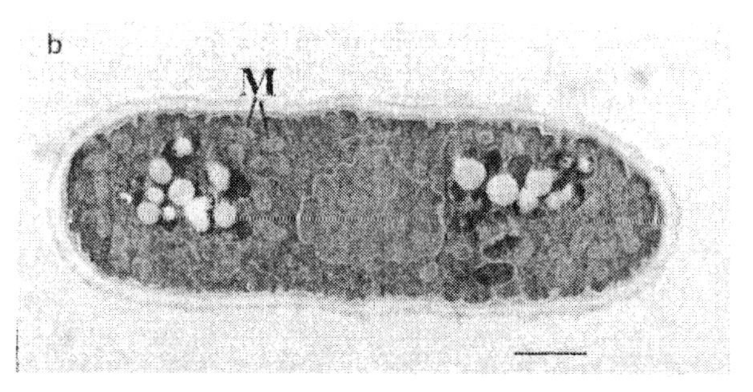

Figure 2.14. Mitochondrial polymorphism in the fission yeast *Schizosaccharomyces pombe*. The electron micrographs (× 15 000 magnification) are of longitudinal sections of *Sch. pombe* cells grown in the presence of mM (a) or μM (a) concentrations of magnesium ions. Mitochondria (M) change from being long, and branched reticular organelles to small, round vesicular organelles solely on the basis of altering magnesium ion availability in the growth medium (G.M. Walker, unpublished results). Bar = 1 μm. Micrographs were taken by Dr Aksel Birch-Anderson of the State Serum Institute, Copenhagen, Denmark. Further discussion on the role of magnesium in dictating mitochondrial morphology and metabolism in fission yeast is given by Walker *et al.* (1982).

Visser *et al.* (1995) have conducted a detailed investigation of mitochondrial morphology of *S. cerevisiae* in relation to growth conditions. By using a fluorescent probe (DASMPI, see Table 2.2) specifically to stain mitochondria and a confocal scanning laser microscope, Visser *et al.* (1995) were able to build up three-dimensional pictures of mitochondria in *S. cerevisiae* under conditions which circumvent the need for serial thin sectioning (as in transmission electron

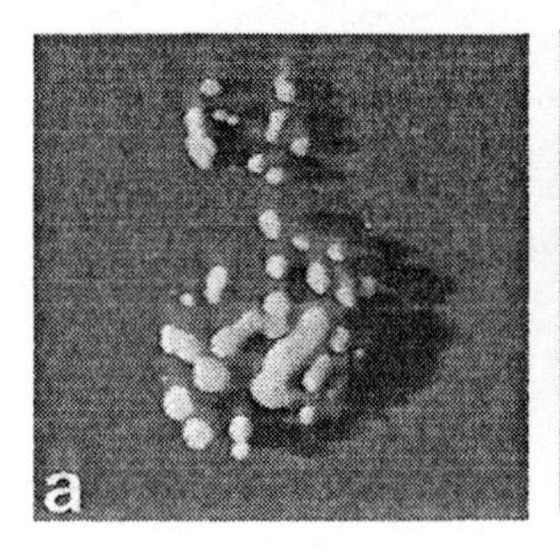

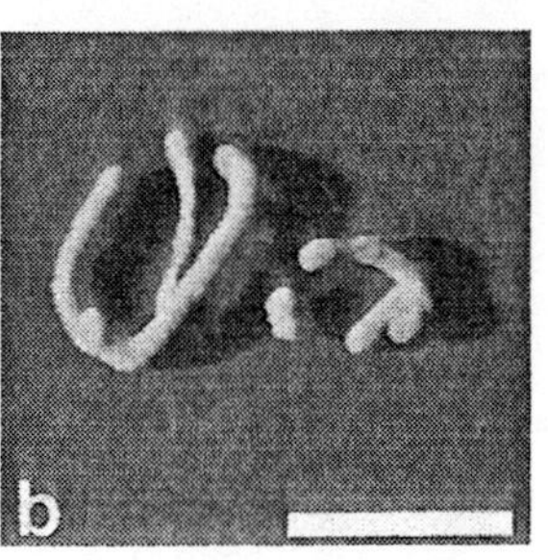

Figure 2.15. Reprojected confocal scanning laser microscopy images of the mitochondria of budding *S. cerevisiae* cells. The cells were grown in repeated batch cultures with ethanol as a non-fermentable carbon source (a) and with a relatively high concentration of glucose (b). Bar = 5 µm. Reproduced from Visser *et al.* (1995), with kind permission from Kluwer Academic Publishers.

microscopy). Figure 2.15 shows images of mitochondria in both respiratory and respiro-fermentative cells of *S. cerevisiae* as visualized by confocal scanning laser fluorescence microscopy.

Table 2.11 indicates the different mitochondrial morphologies observed by this technique when *S. cerevisiae* cells are grown under conditions of variable oxygen supply and carbon source availability.

Under aerobic conditions, yeast mitochondria are primarily involved in ATP synthesis during respiration. However, oxidative reactions of the citric acid cycle and the transfer of electrons along the respiratory chain (on inner membrane cristae of mitochondria) will depend on the particular yeast species and the expression of a phenomenon known as the Crabtree effect (see Chapter 5). Under anaerobic conditions, mitochondria are redundant in the respiratory sense due to the absence of oxygen as terminal electron acceptor. Nevertheless, mitochondria do perform other functions in anaerobic yeast cell physiology and these are summarized, in the case of brewing yeasts, in Table 2.12. O'Connor-Cox *et al.* (1996) have reviewed the relevance of mitochondrial metabolism in brewing yeasts. Although mito-

Table 2.12. Non-respiration functions of brewing yeast mitochondria.

- Synthesis and desaturation of fatty acids and membrane lipids.
- Mitochondrial cytochromes involved in ergosterol biosynthesis.
- General physiological adaptation to stresses caused by ethanol, toxic oxygen radicals and high sugar.
- Modification of cell surface characteristics involved in flocculation and cell partitioning.
- Enzymes for synthesis of amino acids, some dicarboxylic acids, pyrimidine/purine bases, porphyrin and pteridins.
- Mobilization of glycogen.
- Production of flavour and aroma compounds.

Information from: O'Connor-Cox *et al.* (1996).

Table 2.11. Mitochondrial morphology of *S. cerevisiae* cells under different cultivation conditions.

Growth substrate	Concentration	Oxygen	Respiratory enzymes	Mitochondrial morphology
glucose	excess	present	repressed	few, large
ethanol	excess	present	derepressed	many, small
glucose	excess	absent	repressed	few, large
glucose	limiting	absent	repressed	few, large
glucose	limiting	present	derepressed	many, small

Mitochondria were observed using confocal scanning laser fluorescence microscopy of DASPMI-stained *S. cerevisiae* cells. Substrate excess refers to cells grown in batch cultures on 4 $g\,l^{-1}$ carbon source. Substrate-limited conditions refer to chemostat cultures (D = 0.10 h^{-1}) fed with a medium in which the carbon source was the growth-limiting factor. Absence of oxygen refers to a dissolved oxygen concentration below the detection limit of commercially available oxygen probes. Adapted with permission from Visser *et al.* (1995).

chondria are assumed to be energetically non-functional in the anaerobic environment which prevails during brewing fermentations, the essentiality of minimal mitochondrial development should be recognized for provision of critical metabolic intermediates and cellular components. O'Connor-Cox *et al.* (1996) have concluded that mitochondrial functionality plays a critical role in brewing yeast physiology and fermentation performance.

The isolation and characterization of mitochondrial mutants of *S. cerevisiae* has provided fundamental biochemical, genetic and molecular biological information on the biogenesis and function of mitochondria (e.g. Trivedi *et al.*, 1985). Moreover, respiratory-deficient mutants of yeasts have been studied with regard to their possible exploitation in fermentation biotechnology. For example, the use of 'petite' mutants of *S. cerevisiae* in alcohol fermentations has been suggested since alcohol yields can be quite high due to their cells' insensitivity to oxygen and their minimal diversion of carbon to cell growth (Brown *et al.*, 1984). However, respiratory-deficient petites grow more slowly than wild-type 'grande' strains of *S. cerevisiae*. In fact, petite mutations are undesirable in brewing yeast and their frequency (typically 1 to several percent) is monitored in some brewery laboratories. Their undesirability stems from adverse affects on flavour (e.g. diacetyl) development, altered flocculation properties and inabilities to utilize certain sugars (e.g. maltotriose). Hammond (1993) has discussed the importance of the mitochondrial genome in determining the brewing characteristics of yeasts.

2.5 SUMMARY

Being eukaryotic and unicellular, yeasts represent model organisms for studying cellular structural organization and metabolic function which is pertinent to our fundamental understanding of the cytology of higher eukaryotes. Particular benefits of yeasts in cytological studies are evident when considering the ease of yeast cell growth, the availability of structurally defective mutants and the variety of microscopic and cytometric techniques available for visualizing and evaluating cells and organelles. Numerous colorimetric and fluorescent dyes can be used to see yeast organelles at relatively low magnification. More detailed information on cellular localization of certain proteins can be obtained using fluorescently labelled antibodies and by the use of the Green Fluorescent Protein reporter system. Even three-dimensional ultrastructure can now be probed using confocal scanning laser microscopic procedures. Scanning and transmission electron microscopy is still very useful in studies of yeast cell surface topology and intracellular ultrastucture, respectively. Quantitative cytological information can be obtained using techniques such as image analysis or flow cytometry.

Although yeast cells share the ultrastructural features of other eukaryotes, the individual organelles should not be viewed as being structures which act independently. There is structural and functional interdependence among the various membranous networks in yeast cells. For example, movement and positioning of organelles depends on the cytoskeleton and trafficking of proteins in and out of cells relies on vesicular communication between the ER, Golgi, vacuole and plasma membrane.

Cell surface characteristics of yeasts reflect aspects of their physiological condition. Yeast cell envelope structure–function relationships have therefore been discussed in this chapter with particular emphasis on flocculation behaviour of yeast cells. Many factors are now known to influence flocculation in industrial strains of *S. cerevisiae*. These include genetic determinants, cell wall composition, hydrophobicity, topography and charge. External factors are also important, including pH, temperature, oxygen, nutrient and metal ion availability.

Cellular diversity of the yeasts has been a recurring theme in this chapter. Gross morphological and ultrastructural organization is not only very diverse between yeast species, but even within particular species dynamism is exhibited in the shape of cells and organelles. Such dynamic changes in yeast cell morphology are highly dependent on the prevailing conditions of growth. This is particularly well exemplified when considering the size, shape and ultrastructure of yeast mitochondria.

Knowledge of the form and function of yeast cells is relevant in several areas of yeast biotechnology. Understanding structure–function relationships of yeast organelles and macromolecular structural organization is crucially important for the industrial exploitation of yeasts. For example, mechanistic information on the protein secretory pathway in *S. cerevisiae* has proved extremely beneficial in the production of purified human polypeptides for therapeutic use. In addition, knowledge of yeast cell wall characteristics has been of value for controlling the timing and extent of yeast cell–cell flocculation during brewing fermentations.

Further research into the relationship between yeast physiology and yeast cytology will prove valuable not only in fundamental eukaryotic cell investigations, but also when considering how to successfully exploit yeasts in biotechnology.

2.6 REFERENCES

Adams, S.E., Dawson, K.M., Gull, K., Kingsman, S.M. and Kingsman, A.J. (1987) The expression of hybrid Ty virus-like particles in yeast. *Nature*, **329**, 68–70.

Adams, S.E., Mark, S., Richardson, H., Kingsman, S.M. and Kingsman, A.J. (1991) Expression vectors for the construction of hybrid Ty-VLPs. In *Methods in Molecular Biology, Vol. 8. Practical Molecular Virology: Viral Vectors for Gene Expression* (ed. M. Collins), pp. 265–276. The Human Press Inc., Clifton, N.J., USA.

Alfa, C., Fantes, P., Hyams, J., McLeod, M. and Warbrick, E. (1993) *Experiments with Fission Yeast. A Laboratory Course Manual.* Cold Spring Harbor Laboratory Press, Cold Spring Harbor, New York.

Alvarez, P., Sampedro, M., Molina, M. and Nombela, C. (1994) A new system for the release of heterologous proteins from yeast based on mutant strains deficient in cell integrity. *Journal of Biotechnology*, **38**, 81–88.

Arnold, W.N. (1991) Periplasmic space. In *The Yeasts, 2nd edn, Vol. 4: Yeast Organelles* (eds A.H. Rose and J.S. Harrison), pp. 279–295. Academic Press, London.

Ayscough, K.R. and Drubin, D.G. (1996) Actin: general principles from studies in yeast. *Annual Reviews in Cell and Developmental Biology*, **12**, 129–160.

Belda, F. and Zarate, V. (1996) Isolation and characterization of *Schizosaccharomyces pombe* fragile mutants. *Yeast*, **12**, 555–564.

Beyeler, W. and Meyer, C. (1984) Control strategies for continuous bioprocesses based on biological activities. *Biotechnology and Bioengineering*, **26**, 916–925.

Beyeler, W., Einsele, A. and Fiechter, A. (1981) On-line measurement of culture fluorescence: method and application. *European Journal of Applied Microbiology and Biotechnology*, **13**, 10–14.

Bianchi, M.M., Falcone, C., Jie, C.X., Weslowski-Louvel, M., Frontali, L. and Fukuhara, H. (1987) Transformation of the yeast *Kluyveromyces lactis* by new vectors derived from the 1.6 μm circular plasmid pKD1. *Current Genetics*, **12**, 185–192.

Blackwell, K.J., Singleton, I. and Tobin, J.M. (1995) Metal cation uptake by yeast: a review. *Applied Microbiology and Biotechnology*, **43**, 579–584.

Blagoeva, J., Stoer, G. and Venkov, P. (1991) Glucan structure in a fragile mutant of *Saccharomyces cerevisiae*. *Yeast*, **7**, 455–461.

Bohn,J.A. and BeMiller, J.N. (1995) (1→)-β-D-Glucans as biological response modifiers: a review of structure–functional activity relationships. *Carbohydrate Polymers*, **28**, 3–14.

Bröker, M. (1994) Isolation of recombinant proteins from *Saccharomyces cerevisiae* by use of osmotically fragile mutant strains. *BioTechniques*, **16**, 604–610.

Brown, J.L. Kossaczka, Z., Jiang, B. and Bussey, H. (1993) A mutational analysis of killer toxin resistance in *Saccharomyces cerevisiae* identifies new genes involved in (1-6)-β-glucan synthesis. *Genetics*, **133**, 837–849.

Brown, J.L. Roemer, T., Lussier, M., Sdicu, A.-M. and Bussey, H. (1994) The K1 killer toxin: molecular and genetic applications to secretion and cell surface assembly. In *Molecular Genetics of Yeast. A Practical Approach* (ed. J.R. Johnston), pp. 217–232. IRL Press, Oxford.

Brown, S.W., Sugden, D.A. and Oliver, S.G. (1984) Ethanol production and tolerance in grande and petite yeasts. *Journal of Chemical Technology and Biotechnology*, **34B**, 116–120.

Bucci, M. and Wente, S.R. (1997) *In vivo* dynamics of nuclear pore complexes in yeast. *Journal of Cell Biology*, **136**, 1185–1199.

Cabib, E. and Durán, A. (1975) Simple and sensitive procedure for screening yeast mutants that lyse at nonpermissive temperatures. *Journal of Bacteriology*, **124**, 1604–1606.

Calderone, R.A. and Braun, P. (1991) Adherence and receptor relationships of *Candida albicans*. *Microbiological Reviews*, **55**, 1–20.

Carson, D.B. and Cooney, J.J. (1990) Microbodies in fungi: a review. *Journal of Industrial Microbiology*, **6**, 1–18.

Catley, B.J. (1988) Isolation and analysis of cell walls. In *Yeast. A Practical Approach* (eds I. Campbell and J.H. Duffus), pp. 163–183. IRL Press, Oxford and Washington.

Cimprich, P., Slavik, J. and Kotyk, A. (1995) Distribution of individual cytoplasmic pH values in a population of the yeast *Saccharomyces cerevisiae*. *FEMS Microbiology Letters*, **130**, 245–252.

Clark, M.W. (1994) Immuno-electron microscopy. In *Molecular Genetics of Yeast. A Practical Approach* (ed. J.R. Johnston), pp. 233–246. IRL Press, Oxford.

Clausen, M.K., Christiansen, K., Jensen, P.K. and Behnke, O. (1974) Isolation of lipid particles from baker's yeast. *FEBS Letters*, **43**, 176–179.

Cleves, A.E., Cooper, D.N.W., Barondes, S.H. and Kelly, R.B. (1996) A new pathway for protein export in *Saccharomyces cerevisiae*. *Journal of Cell Biology*, **133**, 1017–1026.

Codon, A.C. and Benitez, T. (1995) Variability of the physiological features of the nuclear and mitochondrial genomes of bakers' yeasts. *Systematic and Applied Microbiology*, **18**, 343–352.

Conibear, E. and Stevens, T.H. (1995) Vacuolar biogenesis in yeast sorting out the sorting proteins. *Cell*, **83**, 513–516.

Cormack, B.P., Bertram, G., Egerton, M., Gow, N.A.R., Falkow, S. and Brown, A.J.P. (1997) Yeast-enhanced green fluorescent protein (yEGFP): a reporter of gene expression in *Candida albicans*. *Microbiology (UK)*, **143**, 303–313.

Davison, M.T. and Garland, P.B. (1977) Structure of mitochondria and vacuoles of *Candida utilis* and *Schizosaccharomyces pombe* studied by electron microscopy of serial thin sections and model building. *Journal of General Microbiology*, **98**, 147–153.

Davey, H.M. and Kell, D.B. (1996) Flow cytometry and cell sorting of heterogeneous microbial populations: the importance of single-cell analyses. *Microbiological Reviews*, **60**, 641–696.

Dawes, I.W. (1983) Genetic control of gene expression during meiosis and sporulation in *Saccharomyces cerevisiae*. In *Yeast Genetics, Fundamental and Applied Aspects* (eds J.F.T. Spencer, D.M. Spencer and A.R.W. Smith), pp. 29–64. Springer-Verlag, New York.

Dengis, P.B. and Rouxhet, P.G. (1997) Flocculation mechanisms of top and bottom fermenting brewing yeast. *Journal of the Institute of Brewing*, **103**, 257–261.

De Souza Pereira, R., Parizotto, N.A. and Baranauskas, V. (1996) Observation of baker's yeast strains used in biotransformation by atomic force microscopy. *Applied Biochemistry and Biotechnology*, **59**, 135–143.

Derkatch, I.L., Chernoff, Y.O., Kushnirov, V.V., Inge-Vechtomov, S.G. and Liebman, S.W. (1996) Genesis and variability of [PSI] prion factors in *Saccharomyces cerevisiae*. *Genetics*, **144**, 1375–1386.

Distel, B., van der Leij, I. and Kos, W. (1996) Peroxisome isolation. In *Methods in Molecular Biology. Vol. 53: Yeast Protocols* (ed. I.H. Evans), pp. 133–138. Humana Press, Totowa, NJ, USA.

Dujon, B. (1996) The yeast genome project: what did we learn? *Trends in Genetics*, **12**, 263–270.

Dustin, P. (1978) *Microtubules*. Springer-Verlag, Berlin.

Eckart, M.R. and Bussineau, C.M. (1996) Quality and authenticity of heterologous proteins synthesized in yeast. *Current Opinion in Biotechnology*, **7**, 525–530.

Fischer, J. (1977) Optical polarization reveals different ultrastructural molecular arrangement of polysaccharides in the yeast cell wall. *Acta Biologica Scientifica Hungarica*, **28**, 49–58.

Fleet, G.H. (1991) Cell walls. In *The Yeasts, 2nd edn, Vol. 4: Yeast Organelles* (eds A.H. Rose and J.S. Harrison), pp. 199–277. Academic Press, London.

Freeman, R.F. and Peberdy, J.F. (1983) Protoplast fusion in yeasts. In *Yeast Genetics. Fundamental and Applied Aspects* (eds J.F.T. Spencer, D.M. Spencer and A.R.W. Smith), pp. 243–253. Springer, New York.

Friis, J. and Ottolenghi, P. (1970) The genetically determined binding of Alcian Blue by a minor fraction of yeast cell walls. *Comptes Rendus des Travaux du Laboratoire Carlsberg*, **37**, 327–341.

Fukuhara, H. (1995) Linear DNA plasmids of yeast. *FEMS Microbiology Letters*, **131**, 1–9.

Futcher, A.B. (1988) The 2 μm circle plasmid of *Saccharomyces cerevisiae*. *Yeast*, **4**, 27–40.

Gardiner, R., Podgorski, C. and Day, A.W. (1982) Serological studies in the fimbriae of yeasts and yeastlike species. *Botanical Gazette*, **143**, 534–541.

Gift, E.A., Park,. H.J., Paradis, G.A., Demain, A.L. and Weaver, J.C. (1996) FACS-based isolation of slowly growing cells: double encapsulation of yeast in gel microdrops. *Nature Biotechnology*, **14**, 884–887.

Glick, B.S. and Pon, L.A. (1995) Isolation of highly purified mitochondria from *Saccharomyces cerevisiae*. *Methods in Enzymology*, **260**, 213–223.

Goffeau, A. *et al.* (15 others) (1996) Life with 6000 genes. *Science*, **274**, 546 567.

Golubev, W.I. (1991) Capsules. In *The Yeasts, 2nd edn*,

Vol. 4: Yeast Organelles (eds A.H. Rose and J.S. Harrison), pp. 175–198. Academic Press, London.

Govindan, B. and Novick, P. (1995) Development of cell polarity in budding yeast. *Journal of Experimental Zoology*, **273**, 401–424.

Griffin, D.H. (1994) *Fungal Physiology*. 2nd edn. Wiley-Liss Inc., New York.

Guérin, B. (1991) Mitochondria. In *The Yeasts, 2nd edn, Vol. 4: Yeast Organelles* (eds A.H. Rose and J.S. Harrison), pp. 541–600. Academic Press, London.

Guinard, J.-X. and Lewis, M.J. (1993) Study of the phenomenon of agglomeration in the yeast *Saccharomyces cerevisiae*. *Journal of the Institute of Brewing*, **99**, 487–503.

Haas, A. and Wickner, W. (1996) Organelle inheritance in a test tube – the yeast vacuole. *Seminars in Cell and Developmental Biology*, **7**, 517–524.

Hammond, J.R.M. (1993) Brewer's yeasts. In *The Yeasts, 2nd edn, Vol. 5: Yeast Technology* (eds A.H. Rose and J.S. Harrison), pp. 7–67 Academic Press, London.

Harsay, E. and Bretscher, A. (1995) Parallel secretory pathways to the cell surface in yeast. *Journal of Cell Biology*, **131**, 297–310.

Hashida, M., Sakai, K. and Kogame, M. (1995) Analytical methods of yeast activity by image processing. European Brewery Convention. Proceedings of the 25th Congress, Brussels, 1995, pp. 3353–360. IRL Press at Oxford University Press, Oxford.

Haugland, R.P. (1996) *Handbook of Fluorescent Probes and Research Chemicals*. 6th edn. Molecular Probes Inc., Eugene, OR, USA.

Hay, J.D. (1993) Novel yeast products from fermentation processes. *Journal of Chemical Technology and Biotechnology*, **58**, 203–205.

Heath, I.B. (1995) The cytoskeleton. In *The Growing Fungus* (eds N.A.R. Gow and G.M. Gadd), pp. 99–134. Chapman & Hall, London.

Henschke, P.A. and Rose, A.H. (1991) Plasma membranes. In *The Yeasts, 2nd edn, Vol. 4: Yeast Organelles* (eds A.H. Rose and J.S. Harrison), pp. 297–435. Academic Press, London.

Herrmann, J.M., Fölsch, H., Neupert, W. and Stuart, R.H. (1994) Isolation of yeast mitochondria and study of mitochondrial protein translation. In *Cell Biology: A Laboratory Handbook* (ed. J.E. Celis), pp. 538–544. Academic Press Inc., New York.

Hilt, W. and Wolf, D.H. (1995) Proteasomes of the yeast *Saccharomyces cerevisiae* – genes, structure and function. *Molecular Biology Reports*, **21**, 3–10.

Hilt, W. and Wolf, D.H. (1996) Proteasomes – destruction as a programme. *Trends in Biochemical Sciences*, **21**, 96–102.

Hirata, A. and Shimoda, C. (1994) Structural modification of spindle pole bodies during meiosis II is essential for the normal formation of ascospores in *Schizosaccharomyces pombe*: ultrastructural analysis of *SPO* mutants. *Yeast*, **10**, 173–183.

Hobden, D., Teevan, C., James, L. and O'Shea, P. (1995) Hydrophobic properties of the cell surface of *Candida albicans*: a role in aggregation. *Microbiology (UK)*, **141**, 1875–1881.

Höfer, M. (1997) Membranes. In *Yeasts in Natural and Artificial Habitats* (eds J.F.T. Spencer and D.M. Spencer), pp. 95–132. Springer-Verlag, Berlin, Heidelberg.

Horazdovsky, B. F., DeWald, D.B. and Emr, S.D. (1995) Protein transport to the yeast vacuole. *Current Opinion in Cell Biology*, **7**, 544–551.

Hough, J.S., Briggs, D.E., Stevens, R. and Young, T.W. (1982) *Malting and Brewing Science. Vol. II. Hopped Wort and Beer*. Chapman & Hall, London.

Huls, P.G. Nanninga, N., Van Spronsen, E.A., Valkenburg, J.A.C., Vischer, N.O.E. and Woldringh, C.L. (1992) A computer-aided measuring system for the characterization of yeast populations combining 2D-image analysis, electronic particle counter and flow cytometry. *Biotechnology and Bioengineering*, **39**, 343–350.

Hurt, E.C. and Van Loon, A.P.G.M. (1986) How proteins find mitochondria and intramitochondrial compartments. *Trends in Biochemical Sciences*, **11**, 204–207.

Hyman, A.A. and Sorger, P.K. (1995) Structure and function of kinetochores in budding yeast. *Annual Review of Cell and Development Biology*, **11**, 471–495.

Jiang, Y., Sacher, M., Singer-Krüger, B., Liam, J.P., Store, S. and Ferro-Novick, S. (1995) Factors mediating the late stages of ER-to-Golgi transport in yeast. *Cold Spring Harbor Symposia on Quantitative Biology*, **60**, 119–125.

Johnston, E.A. and Gil-Hwan, A. (1991) Astaxanthin from microbial sources. *Critical Reviews in Biotechnology*, **11**, 297–326.

Johnston, J.R. (1988) Yeast genetics, molecular aspects. In *Yeast. A Practical Approach* (eds I. Campbell and J.H. Duffus), pp. 107–123. IRL Press, Oxford and Washington.

Johnston, J.R. (1994) Pulse field gel electrophoresis. In *Molecular Genetics of Yeast. A Practical Approach* (ed. J.R. Johnston), pp. 83–96. Oxford University Press, New York.

Jones, L., Hobden, C. and O'Shea, P. (1995) Use of a real-time fluorescent probe to study the electrostatic properties of the cell surface of *Candida albicans*. *Mycological Research*, **99**, 969–976.

Kaback, D.B. (1995) Yeast genome structure. In *The Yeasts, 2nd edn, Vol. 6: Yeast Genetics* (eds A.N. Wheals, A.H. Rose and J.S Harrison), pp. 179–222. Academic Press, London.

Kahana, J.A., Schnapp, B.J. and Silver, P.A. (1995) Kinetics of spindle pole body separation in budding yeast. *Proceedings of the National Academy of Sciences (USA)*, **92**, 9707–9711.

Käppeli, O., Sauer, M. and Fiechter, A. (1982) Convenient procedure for the isolation of highly enriched cytochrome P-450-containing microsomal fraction from *Candida tropicalis*. *Analytical Biochemistry*, **126**, 179–182.

Kawaguchi, J., Yoshida, M. and Yamashita, I. (1992) Nutritional regulation of meiosis-specific gene expression in *Saccharomyces cerevisiae*. *Bioscience, Biotechnology and Biochemistry*, **56**, 289–297.

Kilmartin, J.V. and Adams, A.E.M. (1984) Structural rearrangements of tubulin and actin during the cell cycle of the yeast *Saccharomyces*. *Journal of Cell Biology*, **98**, 922–933.

Kingsman, A.J. and Kingsman, S.M. (1988a) Ty: A retroelement moving forward. *Cell*, **53**, 333–335.

Kingsman, S.M. and Kingsman, A.J. (1988b) Polyvalent recombinant antigens: a new vaccine strategy. *Vaccine*, **6**, 304–307.

Kingsman, A.J., Adams, S.E., Burns, N.R., Martin-Rendon, E., Marfany, G., Hurd, D.W. and Kingsman, S.M. (1994) Virus-like particles: Ty retrotransposons. In *Molecular Genetics of Yeast. A Practical Approach* (ed. J.R. Johnston), pp. 203–216. IRL Press, Oxford and Washington.

Kippert, F. and Lloyd, D. (1995) The aniline blue fluorochrome specifically stains the septum of both live and fixed *Schizosaccharomyces pombe* cells. *FEMS Microbiology Letters*, **132**, 215–219.

Klionsky, D.J., Herman, P.K. and Emr, S.D. (1990) The fungal vacuole: composition, function and biogenesis. *Microbiological Reviews*, **54**, 266–292.

Kocková-Kratochvilová, A. (1990) *Yeasts and Yeast-like Organisms*. VCH Publishers, New York.

Kolaczkowski, M., van der Rest, M., Cybularz-Kolaczkowska, A., Soumillion, J.-P., Konings, W.N. and Goffeau, A. (1996). Anticancer drugs, ionophoric peptides, and steroids as substrates of the yeast multidrug transporter Pdr5p. *Journal of Biological Chemistry*, **271**, 31543–31548.

Koning, A.J., Lum, P.Y., Williams, J.M. and Wright, R. (1993) $DiOC_6$ staining reveals organelle structure and dynamics in living yeast cells. *Cell Motility and Cytoskeleton*, **25**, 111–128.

Kopecka, M. (1977) The use of the antibiotic Lomofungin for demonstration of nuclei and chromosomes in live yeast cells and protoplasts. *Folia Microbiologica*, **21**, 406–408.

Kreutzfeldt, C. and Witt, W. (1991) Structural biochemistry. In *Saccharomyces* (eds M.F. Tuite and S.G. Oliver), pp. 5–58. Plenum Press, New York and London.

Kurtzman, C.P. (1987) Molecular taxonomy of industrial yeasts. In *Biological Research on Industrial Yeasts. Vol. I* (eds G.G. Stewart, I. Russell, R.D. Klein and R.R. Hiebsch), pp. 27–45. CRC Press, Boca Raton, Florida, USA.

Lechner, J. and Ortiz, J. (1996) The *Saccharomyces cerevisiae* kinetochore. *FEBS Letters*, **389**, 70–74.

Lee, J.C. (1991) Ribosomes. In *The Yeasts, 2nd edn, Vol. 4: Yeast Organelles* (eds A.H. Rose and J.S. Harrison), pp. 489–540. Academic Press, London.

Léger-Silvestre, I., Noaillac-Depeyre, J., Faubladier, M. and Gas, N. (1997) Structural and functional analysis of the nucleolus of the fission yeast *Schizosaccharomyces pombe*. *European Journal of Cell Biology*, **72**, 13–23.

Lewis, M.J. and Pelham, H.R.B. (1996) SNARE-mediated retrograde traffic from the Golgi complex to the endoplasmic reticulum. *Cell*, **85**, 205–215.

Lloyd, D. and Cartledge, T.G. (1991) Separation of yeast organelles. In *The Yeasts, 2nd edn, Vol. 4: Yeast Organelles* (eds A.H. Rose and J.S. Harrison), pp. 121–174. Academic Press, London.

Lo, W.-S. and Dranginis, A.M. (1996) *FLO11*, a yeast gene related to the *STA* genes, encodes a novel cell surface flocculin. *Journal of Bacteriology*, **178**, 7144–7151.

Lohr, D. (1988) Isolation of yeast nuclei and chromatin for studies of transcription-related processes. In *Yeast. A Practical Approach* (eds I. Campbell and J.H. Duffus), pp. 125–145. IRL Press, Oxford and Washington.

Louis, E.J. (1995) The chromosome ends of *Saccharomyces cerevisiae*. *Yeast*, **11**, 1553–1573.

Lupashin, V.V., Hamamoto, S. and Schekman, R. (1996) Biochemical requirements for the targeting and fusion of ER-derived transport vesicles with purified yeast Golgi membranes. *Journal of Cell Biology*, **132**, 277–289.

Lyman, S.K. and Schekman, R. (1996) Polypeptide translocation machinery of the yeast endoplasmic reticulum. *Experientia*, **52**, 1042–1049.

Malone, R.E. (1990) Dual regulation of meiosis in yeast. *Cell*, **61**, 375–378.

Mann, K. and Mecke, D. (1982) The isolation of *Saccharomyces cerevisiae* nuclear membranes with nuclease and high-salt treatment. *Biochimica et Biophysica Acta*, **687**, 57–62.

Mann, W. and Jeffery, J. (1986) Yeasts in molecular biology. Spheroplast preparation with *Candida utilis*, *Schizosaccharomyces pombe* and *Saccharomyces cerevisiae*. *Bioscience Reports*, **6**, 597–602.

McDonald, K., O'Toole, E.T., Mastronarde, D.M., Winey, M. and McIntosh, J.R. (1996) Mapping the 3-dimensional organization of microtubules in mitotic spindles of yeast. *Trends in Cell Biology*, **6**, 235–239.

Menendez, A., Larsson, C. and Ugalde, U. (1995)

Purification of functionally sealed cytoplasmic side out plasma membrane vesicles from *Saccharomyces cerevisiae*. *Analytical Biochemistry*, **230**, 308–314.

Middleberg, A.P.J. (1993) Process-scale disruption of microorganisms. *Biotechnology Advances*, **13**, 491–551.

Miki, B.L.A., Poon, N.H., James, A.P. and Seligy, V.L. (1982) Possible mechanism for flocculation interactions governed by gene *FLO1* in *Saccharomyces cerevisiae*. *Journal of Bacteriology*, **150**, 878–889.

Naumov, G.I., Naumova, E.S., Lantto, R.A., Louis, E.J. and Korhola, M. (1992) Genetic homology between *Saccharomyces cerevisiae* and its sibling species *S. paradoxus* and *S. bayanus*: electrophoretic karyotypes. *Yeast*, **8**, 599–612.

Niedenthal, R.K. Riles, L., Johnston, M. and Hegemann, J.H. (1996) Green fluorescent protein as a marker for gene expression and subcellular localization in budding yeast. *Yeast*, **12**, 773–786.

Nishimoto, F. and Yamashoji, S. (1994) Rapid assay of cell activity of yeast cells. *Journal of Fermentation and Bioengineering*, **77**, 107–108.

Novick, P., Garrett, M.D., Brenwald, P., Louring, A., Finger, F.P., Collins, R. and TerBusch, D.R. (1995) Control of exocytosis in yeast. *Cold Spring Harbor Symposia on Quantitative Biology*, **60**, 171–177.

O'Connor-Cox, E.S.C., Lodolo, E.J. and Axcell, B.C. (1996) Mitochondrial relevance to yeast fermentative performance: a review. *Journal of the Institute of Brewing*, **102**, 19–25.

O'Shea, D.G. and Walsh, P.K. (1996) Morphological characterization of the dimorphic yeast *Kluyveromyces marxianus* var. *marxianus* NRRLy 2415 by semi-automated image analysis. *Biotechnology and Bioengineering*, **51**, 679–690.

Osumi, M., Yamada, N., Yaguchi, H., Kobori, H., Takashi, N. and Sato, M. (1995) Ultrahigh-resolution low-voltage SEM reveals ultrastructure of the glucan network formation from fission yeast protoplasts. *Journal of Electron Microscopy*, **44**, 198–206.

Panaretou, B. and Piper, P. (1996) Isolation of yeast plasma membranes. In *Methods in Molecular Biology. Vol. 53: Yeast Protocols* (ed. I.H. Evans), pp. 117–121. Humana Press, Totowa, NJ, USA.

Park, C.M., Lopinski, J.D., Masuda, J., Tzeung, T.H. and Bruenn, J.A. (1996) A second double-stranded RNA virus from yeast. *Virology*, **216**, 451–454.

Parent, S.A. and Bostian, K.A. (1995) Recombinant DNA technology: yeast vectors. In *The Yeasts, 2nd edn, Vol. 6: Yeast Genetics* (eds A.N. Wheals, A.H. Rose and J.S. Harrison), pp. 121–178. Academic Press, London.

Pelham, H.R.B., Banfield, D.K. and Lewis, M.J. (1995) SNAREs involved in traffic through the Golgi complex. *Cold Spring Harbor Symposia on Quantitative Biology*, **60**, 105–111.

Pérez-Ortin, J.E., Matallana, E. and Franco, L. (1989) Chromatin structure of yeast genes. *Yeast*, **5**, 219–238.

Phaff, H.J., Miller, M.W. and Mrak, E.M. (1978) *The Life of Yeasts*. 2nd edn. Harvard University Press, Cambridge (USA) and London.

Planta, R.J., Gonçalves, P.M. and Mager, W.H. (1995) Global regulators of ribosome biosynthesis in yeast. *Biochemical and Cellular Biology*, **73**, 825–834.

Podgornik, A., Koloini, T. and Raspor, P. (1997) Online measurement and analysis of yeast flocculation. *Biotechnology and Bioengineering*, **53**, 179–184.

Pon, L. and Schatz, G. (1991) Biogenesis of yeast mitochondria. In *The Molecular and Cellular Biology of the Yeast* Saccharomyces: *Genome Dynamics, Protein Synthesis and Energetics* (eds J.R Broach, J.R. Pringle and E.W. Jones), pp. 333–406. Cold Spring Harbor Laboratory Press, Cold Spring Harbor, New York.

Pons, M.N., Vivier, H., Rémy, J.F. and Dodds, J.A. (1993) Morphological characterization of yeast by image analysis. *Biotechnology and Bioengineering*, **42**, 1352–1359.

Porro, D. and Srienc, F. (1995) Tracking of individual cell cohorts in asynchronous *Saccharomyces cerevisiae* populations. *Biotechnology Progress*, **11**, 342–347.

Porro, D., Ranzi, B.M., Smeraldi, C., Martegani, E. and Alberghina, L. (1995) A double flow cytometric tag allows tracking of the dynamics of cell cycle progression of newborn *Saccharomyces cerevisiae* cells during balanced exponential growth. *Yeast*, **11**, 1157–1169.

Poulter, R.T. (1995) Genetics of *Candida* species. In *The Yeasts, 2nd edn, Vol. 6: Yeast Genetics* (eds A.N. Wheals, A.H. Rose and J.S. Harrison), pp. 285–308. Academic Press, London.

Pringle, J.R., Preston, R.A., Adams, A.E., Stearns, T., Drubin, D.G., Haarer, B.K. and Jones, E.W. (1989) Fluorescence microscopy methods for yeast. In *Methods in Cell Biology, Vol. 31* (ed. A.M. Tartakoff), pp. 357–435. Academic Press Inc., San Diego, USA.

Pringle, J.R., Adams, A.E., Drubin,D. and Haarer, B.K. (1991) Immunofluorescence methods for yeast. *Methods in Enzymology*, **194**, 565–665.

Pryer, N.K., Wuestehube, L.J. and Schekman, R. (1992) Vesicle-mediated protein sorting. *Annual Review of Biochemistry*, **61**, 471–516.

Querol, A, Huerta, T., Barrio, E. and Ramón, D. (1992) Dry yeast strain for use in fermentation of Alicante wines; selection and DNA patterns. *Journal of Food Science*, **57**, 193–185, 216.

Rallabhandi, P. and Yu, P.-L. (1996) Production of therapeutic proteins in yeasts: a review. *Australasian Biotechnology*, **6**, 230–237.

Rambourg, A., Clermont, Y., Ovtracht, L. and Kepes, F. (1995) 3-Dimensional structure of tubular networks, presumably Golgi in nature, in various yeast strains a comparative study. *Anatomical Record*, **243**, 283–293.

Raymond, C.K., Roberts, C.J., Moore, K.E., Howald, I. and Stevens, T.H. (1992) Biogenesis of the vacuole in *Saccharomyces cerevisiae*. *International Review of Cytology*, **139**, 59–120.

Rickwood, D., Dujon, B. and Darley-Usmar, V.M. (1988) Yeast mitochondria. In *Yeast. A Practical Approach* (eds I. Campbell and J.H. Duffus), pp. 185–254. IRL Press, Oxford and Washington.

Riezman, H. (1993) Yeast endocytosis. *Trends in Cell Biology*, **3**, 273–277.

Robinow, C.F. (1981) The view through the microscope. In *Current Developments in Yeast Research* (eds G.G. Stewart and I. Russell), pp. 623–628. Pergamon Press, Toronto.

Robinow, C.F. and Hyams, J.S. (1989) General cytology of fission yeasts. In *Molecular Biology of the Fission Yeast* (eds A. Nassim, P. Young and B.F. Johnson), pp. 273–330. Academic Press, San Diego.

Robinow, C.F. and Johnson, B.F. (1991) Yeast cytology: an overview. In *The Yeasts, 2nd edn, Vol. 4: Yeast Organelles* (eds A.H. Rose and J.S. Harrison), pp. 7–120. Academic Press, London.

Rose, A.H. (1993) Composition of the envelope layers of *Saccharomyces cerevisiae* in relation to flocculation and ethanol tolerance. *Journal of Applied Bacteriology Symposium Supplement*, **74**, 110S–118S.

Rose, A.H. and Harrison, J.S. (1991) (eds) *The Yeasts. 2nd edn. Vol. 4: Yeast Cytology*. Academic Press, London.

Rose, A.H. and Veazey, K.J.H. (1988) Membranes and lipids of yeasts. In *Yeast. A Practical Approach* (eds I. Campbell and J.H. Duffus), pp. 255–275. IRL Press, Oxford and Washington.

Rose, M.D. (1996) Nuclear fusion in the yeast *Saccharomyces cerevisiae*. *Annual Reviews in Cell and Developmental Biology*, **12**, 663–695.

Rothblatt, J. and Schekman, R. (1989) A hitchhiker's guide to analysis of the secretory pathway in yeast. *Methods in Cell Biology*, **32**, 3–36.

Rothman, J.E. (1994) Mechanism of intracellular membrane fusion. *Nature*, **372**, 55–63.

Rout, M.P. and Kilmartin, J.V. (1994) Preparation of yeast spindle pole bodies. In *Cell Biology: A Laboratory Handbook* (ed. J.E. Celis), pp. 605–612. Academic Press Inc., New York.

Salama, N.R. and Schekman, R. (1995) The role of coat proteins in the biosynthesis of secretory proteins. *Current Opinion in Cell Biology*, **7**, 536–543.

Sanderson, C.M. and Meyer, D.I. (1994) Purification and functional characterization of membranes derived from the rough endoplasmic reticulum of *Saccharomyces cerevisiae*. *Journal of Biological Chemistry*, **266**, 13423–13430.

Sawin, K.E. and Nurse, P. (1996) Identification of fission yeast nuclear markers using random polypeptide fusions with green fluorescent protein. *Proceedings of the National Academy of Sciences (USA)*, **93**, 5146–5151.

Sazer, S. and Sherwood, S.W. (1990) Mitochondrial growth and DNA synthesis occur in the absence of nuclear replication in fission yeast. *Journal of Cell Science*, **97**, 509–516.

Serrano, R. (1991) Transport across yeast vacuolar and plasma membranes. In *The Molecular Biology of the Yeast* Saccharomyces. *Genome Dynamics, Protein Synthesis and Energetics* (eds J.N. Strathern, E.W. Jones and J.R. Broach), pp. 523–585. Cold Spring Harbor Laboratory Press, Cold Spring Harbor, New York.

Schreuder, M.P., Mooren, A.T.A., Toschka, H.U., Verrips, C.T. and Kils, F.M. (1996a) Immobilizing proteins on the surface of yeast cells. *Trends in Biotechnology*, **14**, 115–120.

Schreuder, M.P., Dean, C., Boersma, W.J.A., Pouwels, P.H. and Klis, F.M. (1996b) Yeast expressing hepatitis B virus surface antigen determinants on its surface: implications for a possible oral vaccine. *Vaccine*, **14**, 383–388.

Schwencke, J. (1991) Vacuoles, internal membranous systems and vesicles. In *The Yeasts, 2nd edn, Vol. 4: Yeast Organelles* (eds A.H. Rose and J.S. Harrison), pp. 347–432. Academic Press, London.

Shepherd, M.G. (1987) Cell envelope of *Candida albicans*. *Critical Reviews in Microbiology*, **15**, 7–25.

Silverman, S.J. (1989) Similar and different domains of chitin synthases 1 and 2 of *S. cerevisiae*: two isozymes with different functions. *Yeast*, **5**, 459–467.

Simmons, P., Tobin, J.M. and Singleton, I. (1995) Considerations on the use of commercially available yeast biomass for the treatment of metal-containing effluents. *Journal of Industrial Microbiology*, **14**, 240–246.

Skowronek,, P., Krummeck, G., Haferkamp, O. and Rödel, G. (1990) Flow cytometry as a tool to discriminate respiratory-competent and respiratory-deficient yeast cells. *Current Genetics*, **18**, 265–267.

Slodki, M.E. and Cadmus, M.C. (1978) Production of microbial polysaccharides. *Advances in Applied Microbiology*, **23**, 19–54.

Smart, K.A., Boulton, C.A., Hinchcliffe, E. and Molzahn, S. (1995) Effect of physiological stress on the surface properties of brewing yeasts. *Journal of the American Society of Brewing Chemists*, **53**, 33–38.

Soares, E.V. and Mota, M. (1996) Flocculation onset,

growth phase and genealogical age in *Saccharomyces cerevisiae*. *Canadian Journal of Microbiology*, **42**, 539–547.

Soares, E.V. and Mota, M. (1997) Quantification of yeast flocculation. *Journal of the Institute of Brewing*, **103**, 93–98.

Soares, E.V., Teixeira, J.A. and Mota, M. (1994) Effect of cultural and nutritional conditions on the control of flocculation expression of *Saccharomyces cerevisiae*. *Canadian Journal of Microbiology*, **40**, 851–857.

Sobel, S.G. (1997) Mini review: mitosis and the spindle pole body in *Saccharomyces cerevisiae*. *Journal of Experimental Zoology*, **277**, 120–138.

Speers, R.A., Tung, M.A. Durance, T.D. and Stewart, G.G. (1992) Biochemical aspects of yeast flocculation and its measurement: a review. *Journal of the Institute of Brewing*, **98**, 293–300.

Staleva, L., Waltscheva, L., Golovinski, E. and Venkov, P. (1996) Enhanced cell permeability increases the sensitivity of a yeast test for mutagens. *Mutation Research*, **370**, 81–89.

Stambio-de-Castilla, C., Blobel, G. and Rout, M.P. (1995) Isolation and characterization of nuclear envelopes from the yeast *Saccharomyces*. *Journal of Cell Biology*, **131**, 19–31.

Stearns, T. (1995) The green revolution. *Current Biology*, **5**, 262–264.

Stratford, M. (1992) Yeast flocculation: a new perspective. *Advances in Microbial Physiology*, **33**, 1–71.

Stratford, M. (1994) Another brick in the wall? Recent developments concerning the yeast cell envelope. *Yeast*, **10**, 1741–1752.

Stratford, M. (1996) Induction of flocculation in brewing yeasts by change in pH value. *FEMS Microbiology Letters*, **136**, 13–18.

Stratford, M. and Assinder, S. (1991) Yeast flocculation: Flo1 and New Flo phenotypes and receptor structure. *Yeast*, **7**, 559–574.

Straver, M.H. and Kijne, J.W. (1996) A rapid and selective assay for measuring cell surface hydrophobicity of brewers yeast cells. *Yeast*, **12**, 207–213.

Straver, M.H., Kijne, J.W. and Smit, G. (1993) Cause and control of flocculation in yeast. *Trends in Biotechnology*, **11**, 228–232.

Streiblová, E. (1988) Cytological methods. In *Yeast. A Practical Approach* (eds I. Campbell and J.H. Duffus), pp. 9–49. IRL Press, Oxford and Washington.

Sudbery, P.E. (1994) The non-*Saccharomyces* yeasts. *Yeast*, **10**, 1707–1726.

Suizu, T., Tsutsumi, H., Kawado, A., Suginami, K., Imagyasu, S. and Murata, K. (1995) Calcium ion influx during sporulation in the yeast *Saccharomyces cerevisiae*. *Canadian Journal of Microbiology*, **41**, 1035–1037.

Sutherland, I.W. and Elwood, D.C. (1979) Microbial exopolysaccharides – industrial polymers of current and future potential. In *Microbial Technology: Current State, Future Prospects* (eds A.T. Bull, D.C. Elwood and C. Ratledge), pp. 107–150. Cambridge University Press, Cambridge, UK.

Takita, M.A. and Castilho-Valavicius, B. (1993) Absence of cell wall chitin in *Saccharomyces cerevisiae* leads to resistance to *Kluyveromyces lactis* killer toxin. *Yeast*, **9**, 589–598.

Teunissen, A.W. and Steensma, H.Y. (1995) Review the dominant flocculation genes of *Saccharomyces cerevisiae* constitute a new subtelomeric gene family. *Yeast*, **11**, 1001–1013.

Theis, J.F. and Newlon, C.S. (1996) The replication of yeast chromosomes. In *The Mycota III. Biochemistry and Molecular Biology* (eds R. Brambl and G.A. Marzluf), pp. 3–28. Springer-Verlag, Berlin and Heidelberg.

Thevelein, J.M. (1994) Signal transduction in yeast. *Yeast*, **10**, 1753–1790.

Titorenko, V.I., Waterham, H.R., Cregg, J.H, Harder, W. and Veenhuis, M. (1993) Peroxisome biogenesis in the yeast *Hansenula polymorpha* is controlled by a complex set of interacting gene products. *Proceedings of the National Academy of Sciences* (*USA*), **90**, 7470–7474.

Tkacz, J.S. and Lampen, L.O. (1972) Wall replication in *Saccharomyces* species: use of fluorescein-conjugated Concanavalin A to reveal the site of mannan insertion. *Journal of General Microbiology*, **72**, 243–247.

Trivedi, A., Fantin, D.J. and Tustanofff, E.R. (1985) *Saccharomyces cerevisiae* as a model eukaryote for studies on mitochondriogenesis. *Microbiological Sciences*, **2**, 10–13.

Tschopp, J.F., Hansen, W., Emr, S.D. and Schekman, R. (1987) Plasma membrane assembly in yeast. In *Biological Research on Industrial Yeasts. Vol. III* (eds G.G. Stewart, I. Russell, R.D. Klein and R.R. Hiebsch), pp. 87–104. CRC Press, Boca Raton, Florida, USA.

Tuite, M.F. and Lindquist, S.L. (1996) Maintenance and inheritance of yeast prions. *Trends in Genetics*, **12**, 467–471.

Tzagoloff, A. and Myers, A.M. (1986) Genetics of mitochondrial biogenesis. *Annual Review of Biochemistry*, **55**, 249–285.

Uno, I. (1992) Role of signal transduction systems in cell proliferation in yeast. *International Review of Cytology*, **139**, 309–332.

Ushinsky, S.C. Bussey, H., Ahmed, A.A., Wang, Y., Friesen, J., Williams, B.A. and Storms, R.K. (1997) Histone H1 in *Saccharomyces cerevisiae*. *Yeast*, **13**, 151–161.

Van Dam, K. (1996) Role of glucose signalling in yeast

metabolism. *Biotechnology and Bioengineering*, **52**, 161–165.

Van den Hazel, H.B., Kielland-Brandt, M.C. and Winther, J.R. (1996) Review: Biosynthesis and function of yeast vacuolar proteases. *Yeast*, **12**, 1–16.

Van der Rest, M.E., Kamminga, A.H., Nakanoi, A., Anraku, Y., Poolman, B. and Konings, W.N. (1995) The plasma membrane of *S. cerevisiae*: structure, function and biogenesis. *Microbiological Reviews*, **59**, 304–322.

Veenhuis, M. and Harder, W. (1991) Microbodies. In *The Yeasts, 2nd edn, Vol. 4: Yeast Organelles* (eds A.H. Rose and J.S. Harrison), pp. 601–653. Academic Press, London.

Venkov, P. and Ivanov, V. (1982) Uptake of DNA by fragile mutants of *S. cerevisiae*. *Current Genetics*, **5**, 153–155.

Venkov, P.V., Hadjiolov, A.A., Battaner, E. and Schlessinger, D. (1974) *Saccharomyces cerevisiae* sorbitol dependent fragile mutants. *Biochemical and Biophysical Research Communications*, **56**, 599–604.

Vendramin-Pintar, M., Jernejc, K. and Cimerman, A. (1995) A comparative study of lipid composition of bakers and brewers yeast. *Food Biotechnology*, **9**, 207–215.

Visser, W., Van Sponsen, E.A., Nanninga, N., Pronk, J.T., Kuenen, J.G. and van Dijken, J.P. (1995) Effects of growth conditions on mitochondrial morphology in *Saccharomyces cerevisiae*. *Antonie van Leeuwenhoek*, **67**, 243–253.

Volesky, B. May, H. and Holan, Z. (1993) Cadmium biosorption by *Saccharomyces cerevisiae*. *Biotechnology and Bioengineering*, **41**, 826–829.

Vukovic, R. and Mrsa, V. (1995) Structure of the *Saccharomyces cerevisiae* cell wall. *Croatia Chemica Acta*, **68**, 597–605.

Waddle, J.A., Karpova, T.S., Waterston, R.H. and Cooper, J.A. (1996) Movement of cortical actin patches in yeast. *Journal of Cell Biology*, **132**, 861–870.

Walker, G.M. (1982) Cell cycle specificity of certain antimicrotubular drugs in *Schizosaccharomyces cerevisiae*. *Journal of General Microbiology*, **128**, 61–71.

Walker, G.M., Birch-Andersen, A., Hamburger, K. and Kramhøft, B. (1982) Magnesium-induced mitochondrial polymorphism and changes in respiratory metabolism in the fission yeast *Schizosaccharomyces pombe*. *Carlsberg Research Communications*, **47**, 205–214.

Warner, J.R. (1989) Synthesis of ribosomes in *Saccharomyces cerevisiae*. *Microbiological Reviews*, **53**, 256–271.

Waterham, H.R. and Cregg, J.M. (1997) Peroxisome biogenesis. *BioEssays*, **19**, 57–66.

Wickner, R.B. (1995) Non-mendelian genetic elements *in Saccharomyces cerevisiae* – RNA viruses, 2μm DNA, ψ, [URE 3], 20s RNA and other wonders of nature. In *The Yeasts, 2nd edn, Vol. 6: Yeast Genetics* (eds A.N. Wheals, A.H. Rose and J.S. Harrison), pp. 309–356. Academic Press, London.

Wickner, R.B. (1996) Double-stranded RNA viruses of *Saccharomyces cerevisiae*. *Microbiological Reviews*, **60**, 250–265.

Wickner, R.B., Masison, D.C. and Edskes, H.K. (1996) [URE 3] and [PSI] as prions of *Saccharomyces cerevisiae*: genetic evidence and biochemical properties. *Seminars in Virology*, **7**, 215–223.

Wiemken, A., Schellenberg, M. and Urech, K. (1979) Vacuoles: the sole compartments of digestive enzymes in yeast (*Saccharomyces cerevisiae*). *Archives of Microbiology*, **123**, 2325.

Wilcocks, K.L. and Smart, K.A. (1995) The importance of surface charge and hydrophobicity for the flocculation of chain-forming brewing yeast strains and resistance of these parameters to acid washing. *FEMS Microbiology Letters*, **134**, 293–297.

Wilkie, D. (1993) Early recollections of fungal genetics and the cytoplasmic inheritance controversy. In *The Early Days of Yeast Genetics* (eds M.N. Hall and P. Linder), pp. 259–270. Cold Spring Harbor Laboratory Press, Cold Spring Harbor, New York.

Williamson, D.H. (1991) Nucleus, chromosomes and plasmids. In *The Yeasts, 2nd edn, Vol. 4: Yeast Organelles* (eds A.H. Rose and J.S. Harrison), pp. 433–488. Academic Press, London.

Williamson, D.H. and Fennell, D.J. (1975) The use of fluorescent DNA-binding agent for detecting and separating yeast mitochondrial DNA. In *Methods in Cell Biology. Vol. XII* (ed. D.M. Prescott), pp. 335–351. Academic Press, New York.

Wu, L., Shiozaki, K. Aligue, R. and Russell, P. (1996) Spatial organization of the Nim1-wee1-cdc2 mitotic control network in *Schizosaccharomyces pombe*. *Molecular Biology of the Cell*, **7**, 1749–1758.

Yaffe, M.P., Haranta, D., Verde, F., Eddison, M., Toda, T. and Nurse, P. (1996) Microtubules mediate mitochondrial distribution in fission yeast. *Proceedings of the National Academy of Sciences* (*USA*), **93**, 1664–1668.

Yamamoto, M. (1996) Regulation of meiosis in fission yeast. *Cell Structure and Function*, **21**, 431–436.

Yang, S., Ayscough, K.R. and Drubin, D.G. (1997) A role for the actin cytoskeleton of *Saccharomyces cerevisiae* in bipolar bud-site selection. *Journal of Cell Biology*, **136**, 95–110.

Zakian, V.A. (1996) Telomere functions: lessons from yeast. *Trends in Cell Biology*, **6**, 29–33.

Zalewiski, K. and Buckholz, R. (1996) Morphological analysis of yeast cells using an automated image processing system. *Journal of Biotechnology*, **48**, 43–49.

Zimmerman, M. and Fournier, P. (1996) Electrophoretic karyotyping of yeasts. In *Nonconventional Yeasts in Biotechnology* (ed. K. Wolf), pp. 101–116. Springer-Verlag, Berlin and Heidelberg.

Zimmerman, M. and Sipiczki, M. (1996) Protoplast fusion of yeasts. In *Nonconventional Yeasts in Biotechnology* (ed. K. Wolf), pp. 83–99. Springer-Verlag, Berlin and Heidelberg.

Zinser, E. and Daum, G. (1995) Isolation and biochemical characterization of organelles from the yeast *Saccharomyces cerevisiae*. *Yeast*, **11**, 493–536.

Zinser, E.M., Sperka-Gottleib, C.D., Fasch, E.-V., Kohlwein, S.D., Paltauf, F. and Daum, G. (1991) Phospholipid synthesis and lipid composition of subcellular membranes in the unicellular eukaryote *Saccharomyces cerevisiae*. *Journal of Biotechnology*, **173**, 2026–2034.

Zlotnick, H., Fernandez, M.P., Bowers, B. and Cabib, E. (1984) *Saccharomyces cerevisiae* mannoproteins form in external cell wall layer that determines wall porosity. *Journal of Bacteriology*, **159**, 1018–1026.

3

YEAST NUTRITION

3.1 INTRODUCTION
3.2 YEAST NUTRITIONAL REQUIREMENTS
3.2.1 Yeast Cell Chemistry and Essential Elements
3.2.2 Sources of Utilizable Nutrients for Yeasts
3.2.2.1 Carbon
3.2.2.2 Hydrogen
3.2.2.3 Oxygen
3.2.2.4 Nitrogen
3.2.2.5 Sulphur
3.2.2.6 Phosphorus
3.2.2.7 Mineral Elements
3.2.2.8 Growth Factors
3.2.3 Yeast Cultivation Media
3.3 NUTRIENT ACQUISITION BY YEASTS
3.3.1 General Nutritional Strategies Adopted by Yeasts
3.3.1.1 Physiological Responses to Nutrient Availability
3.3.1.2 Translocation of Nutrients into the Yeast Cell
3.3.1.3 Intracellular Fate of Nutrients in Yeasts
3.3.2 Specific Nutritional Strategies Adopted by Yeasts
3.3.2.1 Transport of Water
3.3.2.2 Transport of Sugars
3.3.2.3 Transport of Alcohols
3.3.2.4 Transport of Hydrocarbons
3.3.2.5 Transport of Organic Acids
3.3.2.6 Transport of Sterols
3.3.2.7 Transport of Fatty Acids
3.3.2.8 Transport of Nitrogenous Compounds
3.3.2.9 Transport of Anions
3.3.2.10 Transport of Monovalent Cations
3.3.2.11 Transport of Divalent Metal Cations
3.3.2.12 Transport of Heavy Metals
3.4 SUMMARY
3.5 REFERENCES

3.1 INTRODUCTION

Yeast nutrition refers to how yeast cells feed. More specifically, it refers to how yeasts translocate water and essential organic and inorganic nutrients from their surrounding growth medium through the cell wall, across the cell membrane and into the intracellular milieu. Membrane transport is the key to selectivity of nutrient uptake and it represents the fundamental way in which yeast cells interact and communicate with their growth environment. The various membrane transport mechanisms adopted by yeasts are discussed further in this chapter. Yeast nutrition also refers to the subsequent utilization of essential food sources for both anabolic and catabolic reactions which ultimately ensure the growth and survival of the cell. Nutrient assimilation in yeast growth and metabolic processes is discussed further in Chapters 4 and 5, respectively.

Understanding yeast nutritional requirements and acquisition strategies, together with the regulation of nutrient transport, is important not only for successful cultivation of yeasts in the laboratory but also for the optimization of industrial fermentation processes. The manner by which yeasts transport nutrients from complex feedstocks is therefore additionally considered in this chapter. Yeast nutrition is also pertinent to environmental and clinical microbiologists evaluating the ecological role of yeasts and their parasitic behaviour as animal and plant pathogens. In various niches, from laboratory culture flasks to industrial fermenters to natural habitats, yeasts employ remarkably diverse ways in which to feed on an equally diverse range of nutrients.

3.2 YEAST NUTRITIONAL REQUIREMENTS

3.2.1 Yeast Cell Chemistry and Essential Elements

Elemental analysis of 100 g of baker's yeast would yield the following 'empirical formula':

$$C_{47}\,H_{6.3}\,O_{33}\,N_{8}\,P_{1.2}\,\mathrm{Salts}_{4.5}\ \text{(Berry, 1988)}$$

Such a formulation is of very limited value since it would vary considerably depending on the yeast strain and growth conditions. Nevertheless, it does give a broad indication as to the nutritional make-up of the yeast cell. Thus, yeast is comprised of the major elemental building blocks (carbon, hydrogen, oxygen, nitrogen, phosphorus, sulphur) of macromolecules (proteins, polysaccharides, nucleic acids, lipids), together with the bulk inorganic ions (potassium, magnesium) and trace elements which play a variety of structural and functional roles in the yeast cell.

Yeasts acquire essential elements from their growth environment from simple food sources which need to be available at the macronutrient level (approx. 10^{-3} M) in the case of C, H, O, N, P, K, Mg and S or at the micronutrient level (approx. 10^{-6} M) in the case of trace elements. Table 3.1 summarizes the main elemental requirements of yeasts.

Many yeast species will grow very well in a simple aqueous medium at pH 5.5 comprising a hexose sugar, ammonium salt, various minerals, trace elements and a few vitamins. Nevertheless, yeasts are extremely diverse in their nutritional requirements and this diversity will now be considered in terms of the chemical sources of yeast nutrients.

3.2.2 Sources of Utilizable Nutrients for Yeasts

3.2.2.1 Carbon

Yeasts are chemoorganotrophic organisms. This means that they obtain carbon and energy from compounds in fixed, organic linkage. These compounds are most commonly sugars of which glucose is the most widely utilized by yeast. Glucose, however, may not be the most effectively metabolized sugar for all yeasts. It should also be remembered that while glucose is routinely added to laboratory culture media for growing yeasts, glucose is not freely available in natural yeast habitats (being polymerized in cellulose, starch and other polysaccharides) or in many industrial fermentation substrates (where maltose, sucrose, fructose, xylose and lactose are the more common sugars). Indeed, glucose generally exhibits a repressive and inhibitory effect on the assimilation of other sugars by yeasts.

Table 3.2 categorizes the carbon sources that various yeasts can assimilate with varying degrees of effectiveness. As for *Saccharomyces cerevisiae*, this yeast has a relatively narrow range of sugars which can be considered as good growth and fermentation substrates;

Table 3.1. Summary of elemental requirements of yeasts.

Element	Common sources	Cellular functions
Carbon	Sugars	Major structural element of yeast cells in combination with hydrogen, oxygen and nitrogen. Catabolism of carbon compounds also provides energy
Hydrogen	Protons from acidic environments	Transmembrane proton-motive force vital for yeast nutrition. Intracellular acidic pH (around 5–6) necessary for yeast metabolism
Oxygen	Air, O_2	Substrate for respiratory and other mixed-function oxidative enzymes. Essential for ergosterol and unsaturated fatty acid synthesis
Nitrogen	NH_4^+ salts, urea, amino acids	Structurally and functionally as organic amino nitrogen in proteins and enzymes
Phosphorus	Phosphates	Energy transduction, nucleic acid and membrane structure
Potassium	K^+ salts	Ionic balance, enzyme activity
Magnesium	Mg^{2+} salts	Enzyme activity, cell and organelle structure
Sulphur	Sulphates, methionine	Sulphydryl amino acids and vitamins
Calcium	Ca^{2+} salts	Possible second messenger in signal transduction
Copper	Cupric salts	Redox pigments
Iron	Ferric salts	Haem-proteins, cytochromes
Manganese	Mn^{2+} salts	Enzyme activity
Zinc	Zn^{2+} salts	Enzyme activity
Nickel	Ni^{2+} salts	Urease activity
Molybdenum	Na_2MoO_4	Nitrate metabolism, vitamin B_{12}

namely, glucose, fructose, mannose, galactose, sucrose and maltose. Other carbon substrates (e.g. ethanol, acetate) can act as respiratory substrates only in *S. cerevisiae*.

Although it has already been stated that yeasts are chemoorganotrophs, a small proportion (approx. 5%) of their carbon requirement may be incorporated from carbon dioxide. Such 'fixation' of CO_2 is necessary in anaplerotic reactions to replace dicarboxylic acids of the tricarboxylic acid cycle employed for amino acid biosynthesis, and in the biosynthesis of fatty acids, purines and pyrimidines (Bull and Bushell, 1976). Carbon dioxide should, therefore, also be considered as a carbon 'substrate' for yeast nutrition.

3.2.2.2 Hydrogen

Elemental hydrogen is present in yeast cellular macromolecules and is available from carbohydrates and other sources. Hydrogen ions (protons) are very important in yeast cell physiology since variations in both extracellular and intracellular pH can have a dramatic influence on growth and metabolism of yeast cells. For example, hydrogen ion concentrations of 1 μM (pH 6.0) and 10 μM (pH 5.0) have been reported for optimal yeast growth and fermentation on glucose, respectively (Jones and Greenfield, 1984). Yeasts generally grow very well when the initial culture medium pH is between 4–6, but many yeasts are capable of growth over quite a wide pH range (2–8). The general ability of yeasts to grow at lower pH than most bacteria helps them to occupy different ecological niches and also to spoil acidic foods. Although yeasts do not generally grow well at alkaline pH, certain marine yeasts are specially adapted to growing in slightly alkaline seawater. Actively growing yeasts acidify their growth medium through a combination of differential ion uptake,

Table 3.2. Various carbon sources for the growth of yeasts.

Carbon source	Typical examples	Comments
Hexose sugars	D-glucose, D-galactose, D-fructose, D-mannose	Glucose metabolized by all yeasts If a yeast does not ferment glucose, it will not ferment other sugars. If a yeast ferments glucose, it will also ferment fructose and mannose, but not necessarily galactose
Pentose sugars	L-arabinose, D-xylose, D-xylulose, L-rhamnose	Some yeasts respire pentoses better than glucose. *S. cerevisiae* can utilize xylulose but not xylose
Disaccharides	Maltose, sucrose, lactose, trehalose, melibiose, cellobiose, melezitose	If a yeast ferments maltose, it does not generally ferment lactose and vice versa. Melibiose utilization used to distinguish ale and lager brewing yeasts. Large number of yeasts utilize disaccharides
Trisaccharides	Raffinose, maltotriose	Raffinose only partially used by *S. cerevisiae*, but completely used by other *Saccharomyces* spp. (*S. carlsbergensis, S. kluyveri*)
Oligosaccharides	Maltotetraose, maltodextrins	Metabolized by amylolytic yeasts, not by brewing strains
Polysaccharides	Starch, inulin	Polysaccharide-fermenting yeasts are rare. *Saccharomycopsis* spp. and *S. diastaticus* can utilize soluble starch; *Kluyveromyces* spp. possess inulinase
Lower aliphatic alcohols	Methanol, ethanol	Respiratory substrates. Several methylotrophic yeasts have industrial potential
Sugar alcohols	Glycerol, glucitol	Can only be respired by yeasts
Organic acids	Acetate, citrate, lactate, malate, pyruvate, succinate	Many yeasts can respire organic acids, but few can ferment them
Fatty acids	Oleate, palmitate	Several species of oleaginous yeasts can assimilate fatty acids as carbon and energy sources
Hydrocarbons	n-Alkanes	Many species grow well on C_{12}–C_{18} *n*-alkanes
Aromatics	Phenol, cresol, quinol, resorcinol, catechol, benzoate	Few yeasts can utilize these compounds. Several *n*-alkane utilizing yeasts use phenol as carbon source via the β-ketoadipate pathway
Miscellaneous	Adenine, uric acid, butylamine, pentylamine, putrescine.	Some yeasts, for example, *Arxula adeninivorans* and *A. terestre* can utilize such compounds as sole source of carbon and nitrogen

proton secretion during nutrient transport, direct secretion of organic acids and CO_2 evolution. Ethanol production appears to be especially sensitive to alterations in medium pH. Intracellular pH, which is regulated to within relatively narrow ranges in growing cells (at pH 5.25 in *S. cerevisiae*; Cimprich *et al.*, 1995) through the action of the plasma membrane proton-pumping ATPase, plays a major role in governing yeast nutrient uptake and subsequent metabolism. Extracellular pH variations have limited impact on yeast cytosolic pH unless the medium comprises organic acids such as acetate or sorbate.

3.2.2.3 Oxygen

Yeasts are unable to grow well in the complete absence of oxygen. This is because, as well as providing a substrate for respiratory enzymes during aerobic growth, oxygen is required for certain growth-maintaining hydroxylations such as those involving the biosynthesis of sterols and unsaturated fatty acids. Specifically, yeasts need molecular oxygen for the mixed-function oxidase mediated cyclization of squalene 2,3-epoxide to form lanosterol and for the synthesis of unsaturated fatty acyl coenzyme-A esters. The ergosterol and oleic acid requirement for *S. cerevisiae* to grow in the absence of oxygen dispels the notion that this yeast can grow truly anaerobically. Oxygen should therefore be regarded as an important yeast *growth factor*. Different yeasts possess different requirements for molecular O_2 and at high pressures, pure oxygen can even be strongly inhibitory to yeast cells. The toxicity of hyperbaric oxygen towards yeasts may be cell cycle-dependent (Bull and Bushell, 1976).

In certain yeast biotechnologies where optimization of respiratory growth is of paramount importance (e.g. baker's yeast and yeast biomass protein industries), sufficient oxygen must be maintained in bioreactors to support rapid yeast growth. Oxygen is sparingly soluble in aqueous solution and various bioengineering practices have evolved in yeast propagations to increase oxygen absorption rate, $K_La c$ (where K_L is the rate of O_2 passage from the atmosphere through the liquid interface and into solution, **a** is the area of the interface and **c** is the oxygen concentration in the medium). In the laboratory, even the presence of baffled indentations in Erlenmeyer vessels for shake-flask culture can dramatically increase $K_La c$ values and improve oxygen transfer and consequentially, yeast growth. Also in brewing fermentations, which are often regarded as solely anaerobic, the initial presence of oxygen and the prior aerobic growth of the pitching yeast is extremely important in dictating the progress of the subsequent fermentation (Boulton and Quain, 1987).

3.2.2.4 Nitrogen

Yeast cells have a nitrogen content of around 10% of their dry weight. Although yeasts cannot fix molecular nitrogen, simple inorganic nitrogen sources such as ammonium salts are widely utilized. Ammonium sulphate is a commonly used nitrogen source in yeast growth media since it also provides a source of assimilable sulphur. Some yeasts can also grow on nitrate as a source of nitrogen, and if able to do so, may also utilize low, sub-toxic concentrations of nitrite. Nitrate assimilation ability has long been used as a physiological discriminator between certain yeast genera. For example, *Hansenula* spp. are nitrate-positive, whereas *Pichia* spp. are classed as nitrate-negative. Similarly, with regard to urea utilization, most basidiomycetous yeasts are urease-positive while most ascomycetous yeasts are classed as urease-negative.

A variety of organic nitrogen compounds: amino acids, peptides, purines, pyrimidines and amines can also provide the nitrogenous requirements of the yeast cell. Glutamine and aspartic acids are readily deaminated by yeasts and therefore act as good nitrogen sources. In industrial fermentation media, for example malt wort, mixtures of amino acids are present and are sequentially assimilated by yeasts (see Table 3.11). For some laboratory yeast culture media, amino acid mixtures also provide growth and may be supplied in the form of protein (e.g. casein) hydrolysates. However, the individual addition of cystine, glycine, histidine, lysine, proline or threonine does not allow good growth of *S. cerevisiae* (Matthews and Webb, 1991). Some yeasts, although not generally *Saccharomyces* spp., are able to utilize amino acids and low molecular mass peptides generated following secretion of extracellular proteases.

3.2.2.5 Sulphur

Yeasts require sulphur principally for the biosynthesis of sulphur-containing amino acids.

Yeast sulphur content represents around 0.3% of cell dry weight. Sulphur sources can be in the form of a variety of sulphur compounds including: sulphate, sulphite, thiosulphate, methionine and glutathione. However, inorganic sulphate and the sulphur amino acid methionine are the two compounds central to the sulphur metabolism of yeast. Methionine is the most effectively used amino acid in yeast nutrition. Virtually all yeasts can synthesize sulphur amino acids from sulphate, the most oxidized form of inorganic sulphur, and it is therefore routinely included in yeast culture media.

3.2.2.6 Phosphorus

Phosphorus is present in nucleic acids and in phospholipids and therefore is essential for all yeasts. A significant contribution to the negative charge of the yeast cytoplasm is due to the presence of inorganic phosphates and the phosphate groups in organic compounds. The phosphate content of yeast cells accounts for around 3–5% of dry weight; the major part of this is in the form of orthophosphate (Aiking and Tempest, 1976; Theobald *et al.*, 1996a). Orthophosphate ($H_2PO_4^-$) and condensed inorganic phosphate are common sources of phosphorus in yeast growth media. Orthophosphate acts as a substrate and effector of many enzymes, including those involved in energy transduction. The levels of cytosolic phosphate are very low in yeast cells, at around 0.2% of cell dry weight. However, these levels do fluctuate depending on the mode of sugar catabolism adopted by yeast. For example, intracellular phosphate has been observed to rise dramatically during a glucose pulse when cells exhibit a short-term Crabtree effect (Theobald *et al.*, 1996b). Yeasts can also effectively store phosphate in organelles and Okorokov *et al.* (1980) have shown that a 110-fold higher concentration of phosphate is found in the vacuole compared with the cytoplasm.

3.2.2.7 Mineral Elements

Yeast requirements for minerals are similar to that of other cells with a supply of potassium, magnesium and several trace elements being necessary for growth. K and Mg are regarded as bulk or macroelements which are required in millimolar concentrations to establish the main metallic cationic environment in the yeast cell. Other minerals, which are generally required in the micromolar or even nanomolar range, are referred to as micro or trace elements and include: Mn, Ca Fe, Zn, Cu, Ni, Co and Mo. A host of other metals (e.g. Ag, As, Ba, Cs, Cd, Hg, Li and Pb) represent toxic minerals since they adversely affect yeast growth at concentrations greater than 100 μM (Rose, 1976).

Concerning potassium, yeasts have an absolute growth requirement for this mineral which is essential as a cofactor for a wide variety of enzymes involved in oxidative phosphorylation, protein biosynthesis and carbohydrate metabolism. It is also involved in the uptake of other nutrients like phosphate, as a non-specific charge-balancer and as a stabilizer of macromolecules and ribosomes. Yeast cellular K content varies according to growth conditions, but generally represents 1–2% of dry weight. Aiking and Tempest (1976) demonstrated in K-limited chemostat cultures of *Candida utilis* that cellular potassium varied as a linear function of yeast growth rate. The K^+ analogue, Rb^+, was the only alkali cation which could substitute for K^+ in terms of controlling growth rate and yields of *C. utilis*. Perkins and Gadd (1996) have found that in addition to Rb^+, NH_4^+ ions were able to substitute partially for K^+ in the activation of pyruvate kinase in *Rhodotorula rubra*.

Although some yeasts grow well in saline environments, there is no evidence to suggest that yeasts require sodium, even at very low concentrations. Halotolerant yeasts such as *Debaryomyces hansenii*, *Pichia miso* and *Zygosaccharomyces rouxii* have adapted special osmoregulatory mechanisms for growth in high concentrations of NaCl. Physiological responses of

Table 3.3. Optimum concentrations of cations stimulating yeast growth.

Cation	Concentration
H^+	1.0 μM (pH 6.0)
K^+	2–4 mM
Mg^{2+}	2–4 mM
Mn^{2+}	2–4 μM
Ca^{2+}	< μM
Cu^{2+}	1.5 μM
Fe^{2+}	1–3 μM
Zn^{2+}	4-8 μM
Ni^{2+}	10–90 μM
Mo^{2+}	1.5 μM
Co^{2+}	0.1 μM
B^+	0.4 μM

Adapted from Jones and Greenfield (1984).

yeasts to osmotic and ionic stress are discussed further in Chapter 4 (section 4.4).

Magnesium is an absolute requirement for yeast growth and is present in cells at around 0.3% of dry weight (representing concentrations in the mM range within cells) where it plays essential structural and metabolic functions (Walker, 1994). For example, transphosphorylation enzymes have a strict dependence on Mg^{2+} ions (e.g. in ATP synthesis) as do enzymes involved in the synthesis, expression and translation of genetic information. Yeast cell Mg^{2+} levels vary with growth conditions (Aiking and Tempest, 1976) and with cell cycle phases (Walker and Duffus, 1980). Other cations cannot replace Mg^{2+} in terms of its influence on yeast cell physiology, although cationic polyamines like spermidine and putrescine may be synthesized in response to Mg^{2+}-limited yeast growth (Gildenhuys and Slaughter, 1983).

With regard to calcium, there are conflicting views on the levels of calcium ions required for yeast growth. Concentrations as low as 10^{-9} M have been reported (Youatt, 1993). High levels of Ca^{2+} may be inhibitory to yeast due to antagonism of essential Mg^{2+}-dependent functions (Walker, 1994).

Jones and Greenfield (1984) and Jones and Gadd (1990) have reviewed yeast ionic nutrition in terms of the role and influence of ions in fermentation. Optimum levels of several cations and anions which promote yeast growth have been calculated using elemental mass balances and programmed search techniques (Table 3.3). It is, however, very difficult to generalize on ionic requirements due to factors such as strain differences, culture media chelation and ionic interactions. With regard to the latter, Chandrasena *et al.* (1997) have developed a statistical modelling process to determine cationic interactions which may be employed to predict yeast fermentation performance.

3.2.2.8 Growth Factors

These are organic compounds required in very low concentrations for specific catalytic or structural roles in yeast, but are not utilized as energy sources. Yeast growth factors include: vitamins (which serve vital metabolic functions as components of coenzymes), purines and pyrimidines, nucleosides and nucleotides, amino acids, fatty acids, sterols and other miscellaneous compounds (e.g. polyamines, choline, *meso*-inositol). When a yeast species is said to have a growth factor requirement this indicates that it cannot synthesize the particular factor, resulting in the curtailment of growth and key metabolic processes without its addition to the culture medium. In some cases, this requirement may not be absolute, as when the addition of a growth factor stimulates the rate of yeast growth. This is known as a relative growth factor requirement. Yeasts do vary widely in their growth factor requirements, as is evident from Table 3.4.

Vitamins commonly required by yeast (at μM levels) include: biotin (which serves as a cofactor in carboxylase-catalysed reactions); pantothenic acid (the functional form of which, coenzyme A, is involved in acetylation reactions); nicotinic acid (in the form of nicotinamide which is involved in redox reactions) and thiamine or vitamin B (in the form of thiamine pyrophosphate is involved in decarboxylation reactions).

Table 3.4. Growth factor requirements of some yeasts.

Yeast	Requirements	Comments
Saccharomyces cerevisiae	Biotin, pantothenic acid, Inositol, thiamine	Required by practically all strains Required by some strains
Saccharomyces carlsbergensis	Biotin, pantothenic acid Inositol, uracil, guanine Thiamine, pantothenic acid, nicotinic acid	Required by all strains Required by some strains Stimulatory for growth of some strains
Schizosaccharomyces pombe	Inositol, biotin, pantothenic acid Nicotinic acid Pyridoxine, thiamine	Required by all strains Required by some strains Stimulatory for growth of some strains
Pichia spp.	Variable	Some strains (e.g. *P. membranefaciens*) have no requirements whereas others require one (e.g. thiamine) or more (e.g. biotin plus pyridoxine)
Hansenula spp.	Variable	Vitamin requirements of this genus have been used in differentiating species. *H. anomala* has no requirements for pre-formed vitamins
Schwanniomyces spp.	Biotin	Required by *S. occidentalis*
Candida spp.	Biotin, nicotinic acid Thiamine Cyanocobalamin	Required by most species Required by *C. lipolytica* and stimulatory to *C. albicans* Required by *C. albicans*. *C. utilis* has no requirements for preformed vitamins
Kluyveromyces spp.	Nicotinic acid	All lactose-fermenting yeasts need this vitamin
Rhodotorula spp.	Thiamine, *p*-aminobenzoic acid	Generally required by this genus

Summarized from information provided in Koser (1968).

3.2.3 Yeast Cultivation Media

Due to their relatively simple nutritional requirements, it is quite easy to grow yeast cells. For general laboratory cultivation of yeasts, various synthetic and complex media are available commercially and formulations of standard media have been provided by Campbell (1988), Boekhout and Kurtzman (1996) and Spencer and Spencer (1996). Specific growth media recipes are available for the cultivation, sporulation and mating of *Schwanniomyces occidentalis* (Dohmen and Hollenberg, 1996); *Kluyveromyces lactis* (Wésolowski-Louvel, Breunig and Fukuhara, 1996); *Pichia pastoris* (Sreekrishna and Kropp, 1996); *Hansenula polymorpha* (Hansen and Hollenberg, 1996); *Yarrowia lipolytica*, (Barth and Gaillardin, 1996); *Candida maltosa* (Mauersberger *et al.*, 1996) and *Trichosporon cutaneum* (Reiser *et al.*, 1996). For vigorous and extensive yeast growth, malt extract broth (or beer wort) is a traditionally used complex medium employed which can also be prepared in solid form with the addition of 1–2% (w/v) agar. Yeast extract, prepared by autolysing or hydrolysing yeast cells, represents a complex mixture of the breakdown products of yeast structural and storage macromolecules, and can be supplemented with the addition of peptone and glucose (as in YEPG) for the routine short-term maintenance of laboratory strains. Other complex yeast growth media commonly employed in the laboratory include Sabouraud's medium (glucose plus mycological peptone) for medically important yeasts, and Wallerstein Laboratories Nutrient for brewing yeasts. Examples of synthetic media include Yeast Nitrogen Base (available from Difco) which was originally formulated by Wickerham (1951) for carbon assimilation and fermentation tests and contains ammonium sulphate and asparagine as nitrogen sources, the principal minerals, a range of vitamins, trace elements and (in specified formulations) the amino acids histidine, methionine and tryptophan which promote the growth of certain fastidious yeast species. The carbon source of choice is added, usually to a final concentration of 1% (w/v). Yeast Carbon Base is also available which requires the supplementation of a nitrogen source. Other synthetic growth media have been developed for laboratory cultivation of yeasts, for example, Edinburgh Minimal Medium was developed for the growth of *Schizosaccharomyces pombe* (Mitchison, 1970). To enrich synthetic media, the addition of yeast extract may be employed. The formulations then become 'semi-synthetic'.

In the preparation of selective growth media, several antibacterial, antifungal and anti-yeast agents can be incorporated into media in order to select against other microorganisms when isolating yeasts from environmental samples (Davenport, 1980). For example, acidified agar (to pH 3.7) can be employed to inhibit soil bacteria when isolating terrestrial yeasts. Crystal violet plus fuchsin–sulphite can be used (in Lin's Medium) selectively to determine wild *Saccharomyces* yeasts in the presence of brewing yeasts (Lin, 1975). Cycloheximide is an antibiotic displaying selective inhibitory action against certain yeasts and can also be employed to detect the presence of wild yeasts in brewing samples (Campbell, 1996). Carefully prepared defined media can also be used in selecting certain yeasts with specific nutritional requirements. For example, media with a high osmotic pressure for selecting *Zygosaccharomyces rouxii* in food samples and lysine (as sole N source) agar for selecting non-*Saccharomyces* wild yeasts in brewery samples.

For continuous cultivation of yeasts in *chemostats*, nutrients are continually fed into a culture vessel which is maintained at a constant volume by the simultaneous removal of culture fluid and actively growing cells. Chemostats have proved very useful for cell physiologists in understanding different aspects of yeast nutrition. For example, in optimizing growth medium constituents, in studying sugar transport and metabolism (Weusthuis *et al.*, 1994), and in determining competition either between different yeast species for the same sugar (Postma *et al.*, 1989) or between sugars in a mixture for a single species (Johnston and Barford, 1991).

Chemostat media are usually designed by ensuring that all nutrients for yeast growth are present in excess, except one (the growth-limiting nutrient). Chemostats can therefore facilitate studies into the influence of a single nutrient on yeast cell physiology, with all other factors being kept constant. Media prepared with increasing concentrations of the growth-limiting nutrient can be used to calculate the specific growth rate of yeast cells. The dependent relationship between the growth rate and the growth-limiting nutrient concentration is defined in the Monod equation:

$$\mu = \mu_m \frac{[s]}{K_s + [s]}$$

where μ is the specific growth rate; μ_m is the maximum specific growth rate; $[s]$ is the limiting nutrient concentration and K_s is the saturation constant equal to the nutrient concentration which limits growth to half its maximum rate. For chemostat operations, the limiting nutrient concentration in the feed reservoir is usually chosen at a value above K_s. For yeast growth, Table 3.5 shows that K_s values for sugars and nitrogen sources are generally of the order of mg/l, for inorganic ions and amino acids µg/l, and much less for vitamins (since the affinities of yeast for vitamins are

Table 3.5. K_s values for some yeast nutrients.

Growth-limiting nutrient	Yeast species	K_s	References
Glucose	*Saccharomyces cerevisiae*	108 mg/l	Bull and Bushell (1976)
Glucose	*Saccharomyces cerevisiae*	110 μM	Postma *et al.* (1989)
Glucose	*Candida utilis*[a]	15 μM	Postma *et al.* (1989)
Glycerol phosphate	*Saccharomyces cerevisiae*	4.6 mg/l	Bull and Bushell (1976)
Phosphate	*Rhodotorula rubra*[b]	1.0 μg/l	Bull and Bushell (1976)
Mg^{2+}	*Saccharomyces cerevisiae*	36 μM	Walker and Maynard (1996)
Mg^{2+}	*Schizosaccharomyces pombe*	20 μM	Walker *et al.* (1990)
Thiamine	*Cryptococcus albidus*	4.3×10^{-13}g/l	Bull and Bushell (1976)

[a] *C. utilis*, with a high-affinity (low K_s) for glucose is a potential threat to the production of baker's yeast which has a relatively low affinity for glucose.
[b] Low value of K_s due to its natural phosphate-depleted marine environment.

several orders of magnitude greater than for macronutrients).

For yeast propagation and fermentation in industry, several types of complex agriculturally based media are employed. The principal ingredients of some industrial media are listed qualitatively in Table 3.6.

Several points which are pertinent to the usage of such media in yeast biotechnology industries should be made. Firstly, a yeast propagation medium may not be suited for yeast fermentation. This is exemplified when comparing the use of molasses for both yeast biomass and ethanol production. In the former case, levels of assimilable sugars must be kept low by controlled nutrient feeding (as in 'fed-batch' fermenters; see Chapter 4) to prevent glucose repression and promote respiratory metabolism; while in the latter case, sugar levels can remain quite high since fermentative metabolism is desired. As well as precise sugar levels in molasses, the availability of other key nutrient needs to be addressed. For example, levels of assimilable nitrogen and phosphate are low in molasses which requires to be supplemented accordingly for proper yeast growth. The concentration of essential cations like Mg^{2+} may also be sub-optimal for yeast growth and metabolism in molasses (Walker *et al.*, 1996). The question of bioavailability of metal ions in microbial growth media has been addressed by Hughes and Poole (1991) who have highlighted chelation, precipitation and binding to inert surfaces as circumstances which can lead to the loss of essential ions. Precautionary measures to counteract such losses, for example, by separate sterilization of mineral salts, may be necessary. Conversely, several industrial yeast media such as molasses and malt wort can effectively chelate potentially toxic metal ions. The use solely of beet molasses requires adjustment of vitamin levels since this feedstock is deficient in biotin.

Some yeast fermentations benefit from the addition of media supplements or 'yeast foods', which are usually based on mixtures of yeast extract, ammonium, phosphate and metal (e.g. Zn^{2+}) ions. For example, although malt wort is generally regarded as a complete growth medium for brewing, the addition of nutrient blends to brewer's wort has been advocated in order to balance the supply of nutrients during yeast fermentations (e.g. Hsu *et al.*, 1980). Low-level supplementations (to 40 mg/l) with a blend of amino acids, lower peptides, proteolipids, minerals and vitamins are proposed to minimize flavour variations in the final product by ensuring more consistent yeast activity.

Some industrial fermentation media may have low levels of certain compounds present which

Table 3.6. Principal ingredients of selected industrial media for yeasts.

Components	Molasses	Malt wort	Wine must	Cheese whey
Carbon Sources	Sucrose Fructose Glucose Raffinose	Maltose Sucrose Fructose Glucose Maltotriose	Glucose Fructose Sucrose (trace)	Lactose
Nitrogen Sources	Nitrogen compounds as unassimilable proteins. Nitrogen sources need to be supplemented	Low molecular α-amino nitrogen compounds, ammonium ions and a range of amino acids	Variable levels of ammonia nitrogen, which may be limiting. Range of amino acids	Unassimilable globulin and albumin proteins. Low levels of ammonium and urea nitrogen
Minerals	Supply of P, K, and S available. High K^+ levels may be inhibitory	Supply of P, K, Mg and S available	Supply of P, K, Mg and S available. High levels of sulphite often present	Supply of P, K, Mg, S.
Vitamins	Small, but generally adequate supplies. Biotin is deficient in beet molasses	Supply of vitamins is usually adequate. High adjunct sugar wort may be deficient in biotin	Vitamin supply generally sufficient	Wide range of vitamins present.
Trace Elements	Range of trace metals present, although Mn^{2+} may be limiting	All supplied, although Zn^{2+} may be limiting	Sufficient quantities available	Fe, Zn, Mn, Ca, Cu present
Other Components	Unfermentable sugars (2-4%), organic acids, waxes, pigments, silica, pesticide residues, caramelized compounds, betaine	Unfermentable maltodextrins, pyrazines, hop compounds	Unfermentable pentoses. Tartaric and malic acids. Decanoic and octanoic acids may be inhibitory. May be deficient in sterols and unsaturated fatty acids	Lipids, NaCl. Lactic and citric acids

are potentially toxic or inhibitory to yeast. Some of these compounds may actually be generated during heat treatments employed to sterilize or pasteurize the medium. For example, excessive heating of molasses may generate undesired caramelization products following the Maillard condensation between sugars and amino nitrogen compounds. Heat may also destroy vitamins and other growth factors essential for yeast growth.

3.3 NUTRIENT ACQUISITION BY YEASTS

3.3.1 General Nutritional Strategies Adopted by Yeasts

3.3.1.1 Physiological Responses to Nutrient Availability

Lachance (1990) has defined yeast communities in terms of their nutritional breadth in natural environments. Thus, **generalistic** yeasts are associated with diverse habitats which are nutritionally complex and dilute, such as the phylloplane or aquatic niches. **Competent** yeasts have a broader nutritional spectrum and include 'polytrophic' genera such as *Rhodotorula* and *Cryptococcus* found in fruit, nectar, etc. Specialized yeast communities have found sugar-rich niches in, for example, decaying fruit and tree exudates. **Oligotrophic** yeasts such as *Kloeckera* and *Pichia* spp. utilize a narrow range of nutrients and possess a competitive advantage over other species capable of sharing the same habitat. Oligotrophic environments are nutrient-poor and include the surface of leaves and non-rhizosphere soil. Jennings (1995) has noted that CO_2 fixation may be part of the metabolic processes which allow the oligotrophic growth of yeasts.

One can describe yeasts as non-motile, saprophytic (and occasionally parasitic), chemoorganotrophic microfungi. However, that is not to say that yeasts lack a dynamic interaction with their nutritional environment. For example, many yeasts (including *S. cerevisiae*) can be regarded as polymorphic in that they are capable of altering cell shape in response to the bioavailability of nutrients. This may be exemplified by the filamentous mode of growth observed at the periphery of yeast colonies growing in agar (Gimeno *et al.*, 1992). Such behaviour is akin to a foraging for nutrients as observed in certain fungi. Metabolic, as well as morphological, dynamism is also evident in yeasts. Although not like some higher fungi which are avid secreters of hydrolytic enzymes, yeasts are nevertheless able to secrete enzymes and other compounds in an effort to degrade otherwise non-utilizable polymers or scavenge limited supplies of certain solutes. For example, yeast species are known which can hydrolyse: polysaccharides (e.g. amylolytic and inulinolytic yeasts), hydrocarbons (e.g. bioemulsifying yeasts), proteins (e.g. proteolytic yeasts), pectins (e.g. polygalacturonidase-secreting yeasts), and lipids (e.g. lipolytic yeasts). Other secretions made by yeasts in response to limited nutrient bioavailability include siderophores which are released by some strains when the levels of iron in their growth environment are very low. Some secretions by yeasts may be regarded as a means of safeguarding particular niches. For example, the secretion of ethanol or acids may prevent the growth of competitor microorganisms. The secretion of killer toxins by some yeast species may be viewed as a rather specialized case of inter-yeast rivalry for nutrients.

Other strategies for conferring a competitive advantage for nutrients over their microbial neighbours, especially other yeasts, relate to rates of nutrient uptake. Yeasts need to translocate organic and inorganic solutes into the cell at a particular rate. This rate will govern the rate at which the cells will grow. Rate-limiting processes occur at the level of the yeast cell membrane and can be described in terms of the nutrient transport affinity constant, K_T (akin to the Michaelis–Menten enzyme kinetic constant, K_M) and the solute transport flux (in moles transported per unit surface area of membrane per unit time). The values of K_T calculated for different solutes and for different yeasts are therefore useful in gaining insight into nutrient transport affinities since low values of K_T (μM range) indicate a high transport affinity, whereas higher values of K_T (mM range) indicate lower transport affinities. Two values obtained for the same yeast growing on the same sugar, for example, mean that the sugar is taken up by two processes and the organism exhibits biphasic kinetics in response to sugar availability. Information on K_T values is useful in understanding sugar uptake during yeast fermentations, especially when mixed substrates or mixed cultures are present. For example, *Candida utilis* is a yeast with a very high affinity (low K_T) for glucose and is a potential growth contaminant in production of *Saccharomyces cerevisiae* biomass for the food industry. This is because the latter yeast has a lower affinity (high K_T) for glucose which is taken up by facilitated diffusion rather than active transport as is the case in *C. utilis* (Postma *et al.*, 1989).

Other physiological responses by yeasts to nutrient availability are evident when cells encounter starvation conditions. Yeast cells generally arrest in the G1 phase and rest in the G0 state of their cell division cycle when deprived of essential nutrients such as carbon, nitrogen or phosphorus sources. In sexual yeasts, depletion of carbon or nitrogen source may lead to sporulation as a survival mechanism. In *S. cerevisiae*, the glucose transport system is inactivated during sporulation (Cid *et al.*, 1987). Schulze *et al.* (1996) have made a detailed study of the physiological effects of nitrogen starvation in *S. cerevisiae*. The cellular contents of RNA, protein, trehalose and glycogen, as well as yeast growth energetics, all change dramatically when cells were deprived of ammonium ions.

3.3.1.2 Translocation of Nutrients into the Yeast Cell

Role of the cell wall

Several barriers to the movement of solutes within and outwith yeast cells exist; namely, the capsular layer, the cell wall, the periplasm, the cell membrane and intracellular (organellar) membranes. Although the yeast cell wall does not represent a selectively permeable barrier to the same extent as the plasma membrane, it cannot be considered as a freely porous structure. Generally, yeast cell walls are considered porous to molecules up to an average molecular mass of around 300 Da, but will retain molecules greater than around 700 Da (Scherrer *et al.*, 1974). However, some macromolecules with relative masses as great as 200 K Da, or even 400 K Da, are often able to diffuse freely through intact yeast cell walls (de Nobel and Barnett, 1991). As with other yeast organelles, the cell wall must be regarded as a dynamic system in terms of its structure. In this regard, de Nobel *et al.* (1991) have revealed that the cell wall of *S. cerevisiae* exhibits variations in porosity during growth and cell division. For example, maximum porosity is observed during bud growth where the wall is in a more plastic, expanded state compared with stationary phase cells (which exhibit relatively low porosity). In addition, when subjected to environmental stress such as heat-shock, the yeast cell wall may become weakened leading to plasma membrane stretching (Kamada *et al.*, 1995). This may, in turn, lead to increased permeability of the cell envelope to large molecules, including exogenous DNA.

Role of the cell membrane

The yeast plasma membrane represents the principal selective barrier which isolates the cell from its environment and which dictates those nutrients entering and those metabolites leaving the cell. Understanding transport mechanisms at the level of the cell membrane is important in yeast physiology since such mechanisms govern the rates at which yeasts metabolize, grow and divide.

Before discussing the various ways in which essential nutrients pass through the cell membrane, it should be noted that yeasts may occasionally transport macromolecules by endocytosis in a manner not too dissimilar to that found in animal cells. For example, *S. cerevisiae* α-factor pheromone may be internalized by budding of the plasma membrane to form endosomes en route to the vacuole (Riezman, 1993; see also Chapter 2). The study of endocytic mutants in yeast may provide new insights into membrane trafficking events in eukaryotic cells.

With regard to the general nutrient transport mechanisms, the following discussion will highlight the different modes of passive and active uptake at the level of the yeast cell membrane. There are four basic mechanisms whereby nutrients are taken up, and metabolites exported, across yeast membranes: free diffusion, facilitated diffusion, diffusion channels (pores), and active transport. These are depicted schematically in Figure 3.1.

Free diffusion is the simplest and slowest mode of nutrient transport in yeast which involves passive penetration of lipid-soluble solutes through the plasma membrane. Solutes move by the law of mass action from a high extracellular concentra-

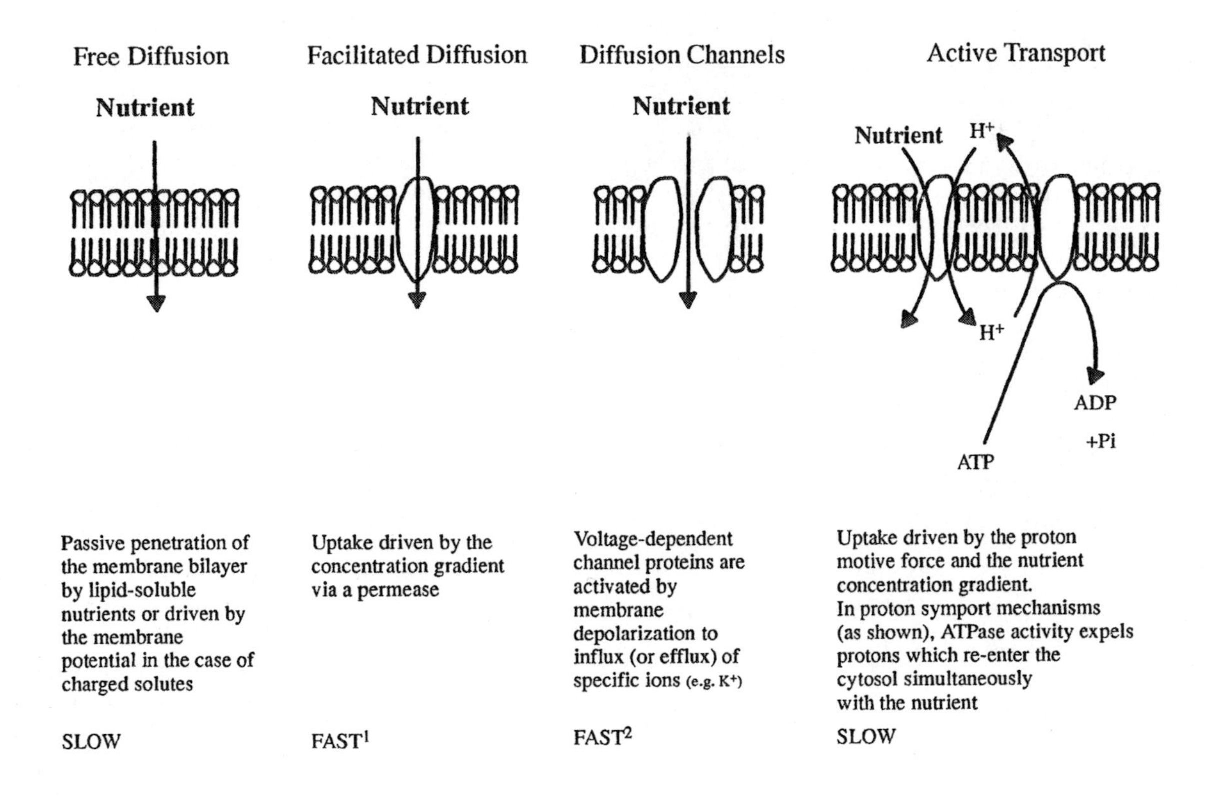

Figure 3.1. Mechanisms of nutrient uptake in yeasts.

tion to a lower intracellular concentration. Passive diffusion may account for the uptake of undissociated organic acids (e.g. benzoate, sorbate), short-chain alkanes and long-chain fatty acids into the yeast cell and the export of ethanol and gaseous compounds from the yeast cell. Charged solutes will only be accumulated if the driving force of an appropriate electrical potential difference is developed across the yeast membrane.

Facilitated diffusion is faster than simple diffusion since solutes are translocated down a transmembrane concentration gradient in an enzyme-mediated fashion. The enzyme is referred to as a permease, or carrier, or facilitator, which traverses the membrane and exhibits stereospecificity for the transported solute. As in passive diffusion, transport of nutrients continues until the intracellular concentration equals that of the extracellular medium. Uptake is saturable with respect to substrate concentration. Certain sugars are transported in this matter, notably glucose in *S. cerevisiae*.

Diffusion channels for certain ions exist in yeast as voltage-dependent 'gates' to move transiently ions down concentration gradients. Such proteinaceous channels are normally closed at the negative membrane potential of resting yeast cells and open when the membrane potential becomes positive. Reid *et al.* (1996) have molecularly characterized the K^+-efflux channel in *S. cerevisiae* which is activated by membrane depolarization. The plasma membrane K^+ channel is the predominant ion channel in yeast, but others exist in cellular, mitochondrial and vacuolar membranes. For example, Gustin *et al.* (1988) have identified a mechanosensitive channel in the yeast cell membrane using the patch–clamp technique. Such a channel conducts both cations and anions and may be involved in yeast salt stress responses and in controlling Ca^{2+} in homeostasis. Plasma membrane diffusion pores in the yeast cell membrane may also exist to permit the passage of water and sugar alcohols (Cartwright *et al.*, 1989; André, 1995). In addition, under conditions of hypoosmotic stress, *S. cerevisiae* facilitates glycerol permeation through a membrane channel, the FPS1 facilitator protein (Luyten *et al.*, 1995).

Active transport is a concentrative, energy-dependent mechanism based on chemiosmotic principles which is responsible for the uptake of the majority of nutrients into yeast cells. The primary inward driving force for nutrients against a concentration gradient during active transport is the membrane potential and the transmembrane electrochemical proton gradient generated by the plasma membrane H^+-ATPase. The latter controls intracellular pH by extruding protons using the free energy of ATP hydrolysis and is an enzyme of central importance in yeast nutrition and growth (Serrano, 1989; Rao and Slayman, 1996). The net efflux of protons per ATP hydrolysed is not a fixed parameter and Venema and Palmgren (1996) have discussed the metabolic regulation of transport coupling in *S. cerevisiae* plasma membrane H^+-ATPase. The proton motive force enables nutrients to either enter with influxed protons, as in 'symport' mechanisms, or against effluxed protons, as in 'antiport' mechanisms. For many yeasts, concentrative sugar uptake occurs by proton symport. It should also be noted that metabolites can be expelled from yeast by such means.

Specific permeases, which may either be constitutive or inducible, facilitate the translocation of solutes with simultaneous movement of protons across the yeast cell membrane. The term 'porter', rather than permease, more accurately describes these carriers since they do not function in the classical enzymatic sense. They do, nevertheless, straddle the cell membrane and catalyse reversible, vectorial translocation of nutrients using the energy derived from proton gradients.

3.3.1.3 Intracellular Fate of Nutrients in Yeasts

Once nutrients are translocated from the extracellular medium, across the plasma membrane and into the intracellular milieu, they can either be rapidly metabolized (catabolically or anabolically) or stored for subsequent usage by the cell. In the case of inorganic ions, they can also be sequestered by intracellular chelators in the cytosol or in the membrane-bound compartments such as the vacuole. In fact, intracellular sequestration and compartmentalization represent the major means whereby yeasts govern their ionic homeostasis. For example, yeast cells control cytoplasmic orthophosphate concentrations by transporting phosphate to the vacuole where it is linearly polymerized to insoluble polyphosphate. The extent of storage of this polymer depends on the availability of phosphate in the growth medium and the metabolic activity of the cells. In *S. cerevisiae* vacuoles, phosphate is concentrated around 100-fold compared with the cytosol. The vacuole is, in fact, a major compartment for nutrient storage in yeast. This organelle has been shown to accumulate divalent metal cations (Mg^{2+}, Mn^{2+}, Ca^{2+} and Zn^{2+}), amino acids (notably arginine, but also several other amino acids) and tricarboxylic acid derivatives (citrate, malate, α-ketoglutarate). The main mechanism of vacuolar membrane uptake appears to be proton-nutrient antiport (Okorokov *et al.*, 1985).

3.3.2 Specific Nutritional Strategies Adopted by Yeasts

3.3.2.1 Transport of Water

Water is essential for yeasts, as it is for all organisms, as a solvent to allow movement of solutes within the cell and for intracellular enzyme activity. When a yeast cell grows, it increases in volume which is partly brought about by the inflow of water. Water permeability is rate-limited by passage through both the yeast cell wall and plasma membrane. The driving force for water inflow into fungal cells is the *turgor potential*, being the difference between the cells osmotic potential and its water potential. The osmotic potential in *S. cerevisiae* is created by the accumulation of solutes from the growth environment. Principally, these solutes are (in order of intracellular concentration magnitude): K^+, amino acids, organic acids, phosphate, Mg^{2+} and Na^+. Meikle *et al.* (1988) have studied the water relations in *S. cerevisiae* and have calculated turgor potential values of 0.81 and 0.61 MPa (mega Pascals) in stationary and exponential phase cells, respectively. Under steady-state conditions, yeasts may be considered as growing in relatively constant water potential. However, there are many situations in natural environments and in both laboratory and industrial culture where cells have to adapt, sometimes very rapidly, to alterations in external water potential. Thus, water 'stressed' cells have to adjust quickly to new osmotic potential conditions by regulating their cell volume or their turgor. For example, if the growth medium water potential decreases (in the presence of high sugar or salt), cellular water will be drawn out and cell volume will decrease, whereas water will be drawn in and cell volume will increase when the converse situation of rising water potential arises. Although this latter situation would be unlikely to cause lysis of a thick-walled yeast cell, the rate of water inflow nevertheless would require to be regulated in some way. The vital roles of *compatible solutes* in osmoregulation and water stress responses in yeast cells are discussed further in Chapter 4 (section 4.4).

Although water would be expected to diffuse freely across the yeast cell membrane lipid bilayer, the high rates of water movements encountered when cells rapidly osmoregulate in changing growth environments cannot be explained by simple passive diffusion. Several members of the MIP (Membrane Intrinsic Protein) family of channel proteins have been implicated in control of cell water and turgor, and in osmoregulation in general. Perhaps the existence of water channel proteins analogous to the 'aquaporins' of plant and animal cells act to regulate water transport in yeast. It is therefore interesting to note that some *S. cerevisiae* membrane protein genes show close sequence similarities to plant aquaporins (André, 1995).

3.3.2.2 Transport of Sugars

A variety of mechanisms exist in yeast cells to facilitate sugar translocation across the plasma membrane. These range from simple net diffusion (a passive or free mechanism) and facilitated (catalysed) diffusion to active (energy-dependent) transport. The general mechanisms of these modes of membrane transport in yeast were discussed in section 3.3.1. The methods adopted and how they are regulated will depend on a number of factors, most notably: the species and strain of yeast in question; the availability of particular sugar molecules in the growth environment; and the physical conditions under which the yeasts are grown.

The following discussion covers not only the fundamental mechanisms and regulation of mono- and disaccharide transport in yeasts but also the biotechnological implications for understanding yeast sugar transport mechanisms.

Monosaccharides

Yeasts can utilize a range of hexose monosaccharides such as glucose, fructose, mannose and galactose as carbon growth substrates. Generally in yeast nutrition, cells transport glucose preferentially. However, yeast cell membranes are not freely permeable to highly polar sugar molecules and various complex mechanisms exist for the translocation of glucose and other saccharides into the cell. In *S. cerevisiae*, 20 hexose transporters are known to exist which differ in their abundance and intrinsic affinities for hexoses (Kruckeberg, 1996). It is possible that at the very high sugar concentrations encountered in industrial fermentation media, free diffusion may account for a very small proportion of uptake into yeast cells (Walsh *et al.*, 1994). Sugar transport in yeast has been studied for many years due to the biotechnological relevance of this topic for the fermentation industries. For example, it is known that the rate of alcohol production by yeast is primarily limited by the rate of sugar uptake. Thus in winemaking, the loss of hexose transport toward the end of fermentation may be responsible for reduced alcohol yields. Several papers in recent years have reviewed the physiology, biochemistry and molecular biology of sugar transport in yeasts (e.g. Fuhrmann and Völker 1992; Bisson *et al.*, 1993; Lagunas, 1993; van Dijken *et al.*, 1993; Weusthuis *et al.*, 1994; André, 1995; Horak, 1997), but there are still several unanswered questions relating to precise mechanisms and regulation of transport. For example, what regulates sugar uptake kinetics? What is the role of phosphorylation? How many transporters are there? Why are there multiple transporters? How is sugar uptake controlled in non-*Saccharomyces* species?

Glucose. Concerning glucose transport, research, particularly with *Saccharomyces*, and to a lesser extent with *Kluyveromyces*, *Candida*, *Rhodotorula* and *Schizosaccharomyces* species, has shown that there is no singular, common way in which yeasts transport glucose. The precise manner will depend on several factors, not least the availability of extracellular glucose, the growth conditions and the yeast strain in question (Deák, 1978; Loureiro-Dias, 1988; Kilian *et al.*, 1991). Nevertheless, the following characteristics of glucose transport in *S. cerevisiae* can be noted:

- Glucose uptake is rapid and down a concentration gradient
- Glucose transporters are stereospecific for certain hexoses and will carry glucose, fructose and mannose
- Glucose uptake reaches equilibrium and is not accumulative
- Glucose carriers facilitate glucose diffusion and several carriers exist
- No metabolic energy is required
- Phosphorylation accompanies glucose entry into the cell

The above refers to the high-affinity system which is absent in cells growing in high [c. 2% (w/v)] levels of glucose. Under these conditions, low-affinity glucose transport systems are operable which are constitutive and independent of phosphorylation. In *S. cerevisiae*, high-affinity glucose transport is summarized as being by specific, energy-independent, carrier-mediated facilitated diffusion. Proton symport is not involved. High-affinity glucose translocation in *S. cerevisiae* is coupled with phosphorylation by glycolytic kinases (two hexokinases, *HXK1* and *HXK2* and a glucokinase, *GLK1*). Phosphorylation may be at the level of glucose itself or an interaction between the kinases and glucose transporters may exist (Romano, 1986; Lagunas, 1993). Glucose uptake exhibits differential kinetics depending on the extracellular glucose availability. The general consensus is that under non-steady-state growth conditions, for example, during batch culture growth (Fuhrmann and Völker, 1992), there is a biphasic (or occasionally multiphasic) catalytic uptake exhibited by high- and low-affinity carriers. The former is characterized by a relatively low transport affinity constant, K_T, of around 1 mM glucose and the latter by a higher K_T of around 20 mM glucose. In comparison with some other yeasts, these K_T values for *S. cerevisiae* are uncharacteristically high (van Urk *et al.*, 1989). The possibility that some glucose, when in abundance, may freely permeate yeast cells via the low-affinity system by passive diffusion (Fuhrmann and Völker, 1992) has recently been discounted by Gamo *et al.* (1995).

Table 3.7 reveals that the situation in yeasts other than *S. cerevisiae* is different. Generally speaking, energy expenditure from a proton electrochemical gradient across the cell membrane drives glucose accumulation against a concentration gradient in non-*Saccharomyces* yeasts. In fact, the energy cost is quite high for active proton symport of glucose; perhaps one ATP per glucose in certain aerobic yeasts. Griffin (1994) has attempted to rationalize the apparent discrepancies in energy requirements for glucose uptake in different yeasts. Thus, *S. cerevisiae* being a strongly fermentative yeast yields only 2 ATP per glucose metabolized. If half of this ATP were required for glucose uptake, this would impose too great an energy demand on the cell. In aerobic yeasts like *Candida utilis* and *Rhodotorula glutinis*, on the other hand, the energy 'slippage' is much less due to a higher ATP yield from respiration.

Table 3.7. Glucose transport systems in some biotechnologically important yeasts.

Yeast[a]	Affinity, K_T[b]	Mechanism[c]	Control[d]	Comments
Candida utilis	High, K_T 0.015 mM	Proton symport	Repressible	Facilitated diffusion apparent in repressed cells
	Low, K_T 0.2–2 mM	Facilitated diffusion		Low-affinity system operative only at high growth rates
Candida shehatae	High, K_T 0.12 mM	Proton symport	Repressible	Also transports D-mannose
	Low, K_T 2 mM	Facilitated diffusion	Non-repressible	Also transports D-xylose
Kluyveromyces marxianus	High, K_T 0.025–0.1 mM	Proton symport	Repressible	Also transports D-galactose
	Low, K_T 1.8–90 mM	Proton symport	Repressible	Also transports D-fructose
Pichia stipitis	High, K_T 0.015–0.5 mM	Proton symport	Non-repressible	No low-affinity system identified
Rhodotorula glutinis	Variable, K_T 0.8–2 mM	Proton symport	?	Also transports D-xylose
Saccharomyces cerevisiae	High, K_T c. 1 mM	Facilitated diffusion	Repressible	Also transports D-fructose and D-mannose
	Low, K_T 10–20 mM	Facilitated diffusion	Constitutive	Others may exist
Schizosaccharomyces pombe	High, K_T 1–3 mM	Both proton symport and facilitated diffusion reported	?	Transports only D-glucose

[a] Yeasts listed alphabetically.
[b] K_T refers to the affinity constant of the transport system and is analogous to the Michaelis–Menten constant in enzyme kinetics.
[c] Facilitated diffusion is an energy-independent system mediated by protein carriers, whereas proton symport refers to active energy-dependent, ATPase-mediated proton-motive force.
[d] Control refers to whether or not the transport system is repressible by glucose or is constitutively expressed.
Information from: van Urk *et al.* (1989); Lagunas (1993); Weusthuis *et al.* (1994) and Jennings (1995).

Fermentative, but not respiratory, yeasts exhibit glucose catabolite repression and proteolytic inactivation of the high-affinity glucose transporter (Does and Bisson, 1989) and this control is associated with fermentative ability. Van Urk *et al.* (1989) have postulated that in 'Crabtree-positive' yeasts like *S. cerevisiae*, which actively ferment glucose under aerobic conditions (see Chapter 5), facilitated diffusion may allow an unrestricted flow of glucose into the cell, whereas proton symport prevents this in aerobic (Crabtree-negative) yeasts. Crabtree-negative yeasts possess proton-symport glucose transporters of very high affinity and unlike *S. cerevisiae*, can concentrate glucose (against a gradient). Such yeasts may not be found in very glucose-rich environments such as grape juice. However, the reasons why *S. cerevisiae* behaves differently in comparison with other fermentative yeasts (Loureiro-Dias, 1988; Van Urk *et al.*, 1989) in terms of its energy-independent glucose transport remains unclear. Many workers regard *S. cerevisiae* as a model eukaryotic cell, but it may not be regarded as a model yeast when it comes to sugar uptake mechanisms.

In both respiratory and fermentative yeasts glucose uptake increases as cells grow faster (up to μ_{max}), perhaps indicating an increasing (altered turnover) number of glucose transport sites. It makes sense that the rate of glucose uptake will govern the rate of yeast growth and in *S. cerevisiae* glucose carriers may represent 'sensors' of glucose availability which enable yeasts rapidly to adjust their growth and metabolic activities to changing environmental conditions. Glucose sensing may itself be regulated by repression and inactivation mechanisms (Thevelein, 1991; Lagunas, 1993). In this way, the high-affinity glucose carrier is repressed or switched off during growth on high concentrations (e.g. 100 mM) of glucose, whereas it is derepressed or switched on when cells experience low (e.g. 5 mM) concentrations of glucose (Bisson and Fraenkel, 1984). Although a gene called *General Glucose Sensor* (*GGSI*) has been implicated in glucose-induced regulation in *S. cerevisiae*, and has been cloned (Van Aelst *et al.*, 1993), the precise mechanism of glucose sensing in yeasts remains to be elucidated (Thevelein, 1994). Nevertheless, various components of the *S. cerevisiae* glucose sensing and signalling system have now been characterized (Thevelein, 1994; Sillјé *et al.*, 1996; Walsh *et al.*, 1996) and key roles for trehalose metabolism and intracellular cAMP implicated (Thevelein and Hohmann, 1995; Hohmann *et al.*, 1996; Wheals, 1996). In addition, Özcan *et al.* (1996) have identified two unusual glucose transporters in *S. cerevisiae* which appear to act as glucose sensors that generate intracellular glucose signals for induction of gene expression.

Catabolite repression or inactivation of the high-affinity transport system appears to be a general phenomenon in yeasts when two transport systems for a given sugar are present. Note that the low-affinity glucose transport system in *S. cerevisiae* is constitutively expressed. In *S. cerevisiae*, the affinity of the glucose carrier for its substrate may not only depend on the availability of glucose, but also on the presence of oxygen (Fuhrmann and Völker, 1992) and the growth rate and energy status of the cell. The involvement of glucose carriers in metabolic regulation, for example as rate-limiting glycolytic steps in expression of the Pasteur effect, is discussed further in Chapter 5.

Other hexoses. Recent work on the molecular biology of glucose (and other sugar) transport in yeasts has identified several genes encoding high- and low-affinity transporters. A large number of putative glucose transporters (e.g. *HXT1–HXT14* and *SNF3*) belonging to a multigene sugar permease family have been characterized in *S. cerevisiae* (André, 1995) and their regulation by glucose induction studied (Özcan and Johnston, 1995). The *HXT* genes examined affect glucose (*HXT2*), galactose (*GAL2*), glucose and mannose (*HXT1*), or glucose, fructose and galactose (*HXT4*) uptake; however, none has been identified in *S. cerevisiae* as specifically affecting fructose. Other genes may be implicated for high- and low-affinity systems (Reifenberger *et al.*, 1995).

Although *S. cerevisiae* transports fructose via facilitated diffusion rather than by active transport, the presence of fructose-proton symporters has been detected in other *Saccharomyces* species, most notably lager brewing strains of *S. carlsbergensis* and in a wine yeast, *S. bayanus*. The presence or absence of active fructose transport mechanisms has been deemed useful in differentiating species of *Saccharomyces sensu stricto* (Rodriguez de Sousa *et al.*, 1995). Thus, *S. bayanus* and *S. pastorianus*, which are fructose proton-symport positive, are delimited from *S. cerevisiae* and *S. paradoxus*, which are fructose proton-symport negative (Tornai-Lehoczki *et al.*, 1996).

Whereas yeast glucose transporters will translocate other hexoses, a singular galactose transporter (encoded by *GAL2* gene) operates only with D-galactose. Nevertheless, both the glucose and galactose transporters in *S. cerevisiae* act by facilitated diffusion. *S. cerevisiae* possesses both an inducible high-affinity (K_T value 4 mM; De Juan and Lagunas, 1986) and a constitutive low-affinity transport system for galactose (Ramos *et al.*, 1989). As in the case with glucose, high-affinity galactose transport in *S. cerevisiae* is associated with the activity of kinases, galactose being phosphorylated by a specific galactokinase. Regulation of galactose transport in *S. cerevisiae* is similar to transport of glucose (De Juan and Lagunas, 1986; Ramos *et al.*, 1989). In non-*Saccharomyces* yeasts, such as *Kluyveromyces marxianus*, galactose transport appears to be by a proton symporter (De Bruijne *et al.*, 1988).

It has been suggested (e.g. Bun-Ya *et al.*, 1991) that some yeast sugar transporters may have additional roles quite unconnected with those in transporting sugar. For example, phosphate uptake has been implicated and further analysis of sugar transport genes may reveal other transport functions.

Pentoses. Considering the transport of pentose sugars (e.g. D-xylose, L-rhamnose, D-arabinose and D-ribose) by yeasts, it is again evident that multiple transport systems exist depending on the particular species of yeast and their nutritional status. Pentoses are generally transported into yeasts by generation of membrane potential due to proton symport, although facilitated diffusion may mediate in certain low-affinity transport systems (Singh and Mishra, 1995; Table 3.8).

Several uptake systems exist in pentose-utilizing yeasts like *Pichia stipitis*, *Candida shehatae* and *Rhodotorula glutinis*. The latter yeast has been used as a model organism in studying pentose transport due to its strict dependence on respiratory energy metabolism. Under aerobic (but not anaerobic) conditions, an electrogenic proton symporter is evident during high-affinity D-xylose uptake in *R. glutinis*, which is selectively derepressed in starving cells. A low-affinity system involving a proton symport mechanism may also exist in this yeast. In *P. stipitis*, xylose transport is a rate-limiting step under aerobic growth conditions and uptake of this sugar is mediated by both low- and high-affinity carriers. In *Candida shehatae*, facilitated diffusion and proton symport mechanisms coexist for D-xylose transport. The former mechanism is operative under conditions of repression but is absent in starving cultures of this yeast. Figure 3.2 represents a schematic illustration of uptake of lignocellulose-related hexoses and pentoses in *C. shehatae*. *S. cerevisiae* may be able to take up xylose under aerobic conditions and a low-affinity system operates in this yeast by facilitated diffusion.

Generally, the presence of glucose will inhibit xylose uptake in pentose-fermenting yeasts, ensuring preferential utilization of glucose in cells grown on complex sugar mixtures like hemicellulosic hydrolysates. This sequential pattern of uptake represents a major factor which limits the biotechnological exploitation of pentose-fermenting yeasts in the bioconversion of wood hydrolysates (Spencer-Martins, 1994).

Disaccharides

Certain disaccharides such as sucrose, maltose, lactose and cellobiose are much more abundant

Table 3.8. D-Xylose transport system in yeasts.

Yeast	Affinity	Mechanism	Comments
Candida shehatae	High, K_T 1.0 mM Low, K_T 100–153 mM	Proton symport Facilitated diffusion	xylose both respired and fermented
Candida utilis	High, K_T mM Low, K_T 68 mM	Proton symport ?	only high-affinity system operates in glucose derepressed cells
Pichia stipitis	High, K_T 0.08–0.9 mM Low, K_T 380 mM	Proton symport	xylose both respired and fermented
Metschnikowia reukaufii	High, K_T 2 mM	Proton symport	xylose only respired
Rhodotorula glutinis	High, K_T 0.6–1.1 mM Low, K_T 15–18 mM	Proton symport Facilitated diffusion	xylose only respired
Saccharomyces cerevisiae	Low, K_T 95–120 mM	Facilitated diffusion	xylose non-metabolizable
Kluyveromyces fragilis (*marxianus*)	Low, K_T 290 mM	Facilitated diffusion	xylose metabolizable
Schizosaccharomyces pombe	Low, K_T 206 mM	Facilitated diffusion	xylose non-metabolizable

Adapted from: Kilian *et al.* (1993); Singh and Mishra (1995); and Spencer-Martins (1994).

in industrial yeast growth substrates than glucose and other monosaccharides. Whereas glucose and galactose are transported in *S. cerevisiae* by facilitated diffusion, disaccharides generally enter by concentrative active proton-symport mechanisms. Disaccharides may either be transported intact or firstly hydrolysed outwith the yeast plasma membrane prior to transport of the component monosaccharides (Barnett, 1981). Molecular biology has revealed that yeast disaccharide transporters (e.g. for maltose and lactose) are structurally and functionally related to one another.

Sucrose. Considering sucrose (β-D-fructofuranosyl α-D-glucopyranoside), this sugar is readily assimilated and fermented by yeasts and is abundant in many natural, complex growth media like molasses and sugar cane juice. Some strains of *S. cerevisiae* have been reported possessing a sucrose proton symporter (Santos *et al.*, 1982), but predominantly in this species sucrose is firstly converted to glucose and fructose. This is accomplished at the yeast cell envelope by a periplasmic invertase. Expression of the invertase gene (*SUC2*) is repressed in glucose-grown cells and derepressed when glucose availability is low. Previously, it was thought that the *SNF3* (for *s*ucrose *n*on-*f*ermenting) gene was involved in regulation of invertase, but it is more generally implicated in the utilization of sugars when present at low concentrations. Grimes and Overvoorde (1996) have identified a unique plasma membrane sucrose-binding protein in *S. cerevisiae* which mediates uptake of sucrose in a non-saturable, linear manner. This uptake system appears to mimic that which occurs in higher plants (Overvoorde *et al.*, 1996).

In *Kluyveromyces marxianus*, sucrose is again initially hydrolysed but in this yeast an extracellular inulinase (β-D-fructofuranosidase) releases and transports glucose and fructose when grown on sucrose.

Maltose. Maltose (4-O-α-D-glucopyranosyl-D-glucopyranose) is the disaccharide which is the most

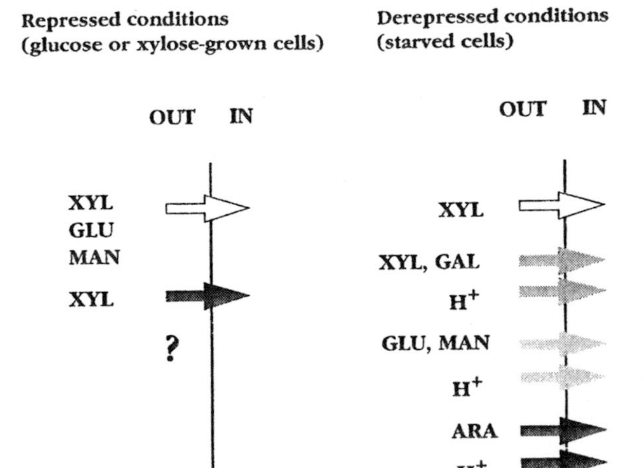

Figure 3.2. Proposed uptake of lignocellulose-related sugars in *Candida shehatae*. Abbreviations: XYL, D-xylose; GLU, D-glucose; MAN, D-mannose; GAL, D-galactose; ARA, L-arabinose. Arrows of different shades represent different transport systems. Adapted from Spencer-Martins (1994): *Bioresource Technology* ©1994, vol. 50, pp. 51–57, with kind permission from Elsevier Science Ltd, The Boulevard, Langford Lane, Kidlington, OX5 1GB, UK.

abundant fermentable sugar present in barley malt extracts for brewer's, distiller's and baker's yeast applications. In fact, maltose uptake may be a rate-limiting determinant in such fermentations. Unlike invertase, maltase (α-glucosidase) is not located in the yeast cell wall. *S. cerevisiae* transports maltose by an energy-dependent proton-symport mechanism and hydrolyses the sugar by an intracellular maltase to two molecules of glucose. One molecule of maltose is co-transported with a proton and electroneutrality is maintained by extrusion of K^+ ions from the cell. Although maltose transport by laboratory strains of *S. cerevisiae* was once thought to consist of an inducible, high-affinity (K_T 4 mM) and a constitutive, low-affinity (K_T 70 mM) component operable under high maltose concentrations, it has now been established that the latter component is an artefact due to non-specific binding of maltose in the yeast periplasm (Benito and Lagunas, 1992). In industrial (brewing) strains of *S. cerevisiae*, however, Crumplen *et al.* (1996) have provided evidence for the existence of a low-affinity maltose transporter. Brewing yeasts clearly differ from laboratory strains with regard to maltose uptake characteristics. The high-affinity maltose transporter in *S. cerevisiae* which is related to the *HXT* family, is induced by maltose, repressed by high (>0.4%, w/v) glucose and inactivated following growth on glucose or by nitrogen source exhaustion. Riballo *et al.* (1995) have shown that rapid irreversible catabolite inactivation of the maltose transporter, which is brought about when glucose is added to maltose-fermenting cells, occurs by vacuolar proteolysis following internalization of the protein by endocytosis. Transfer of cells back to maltose-containing media results in recovery of maltose permease activity. Note that the physiological consequences for a yeast cell of uncontrolled maltose uptake are severe and can lead to substrate-accelerated death caused by glucose 'poisoning' (Entian and Loureiro-Dias, 1990; Postma *et al.*, 1990).

The high-affinity maltose transporter has been characterized both genetically and biochemically. In *S. cerevisiae*, three maltose utilization genes (collectively referred to as *MAL*) are involved: a maltose-specific carrier or permease; maltase; and a transcriptional activator.

The regulation of maltose uptake in *S. cerevisiae* is of great significance to the fermentation industries. In brewing, for example, malt wort comprises a mixture of sugars which are sequentially transported and metabolized by yeast in the approximate order: sucrose, glucose, fructose, maltose and maltotriose. Maltotetraose and branched oligosaccharides known as *dextrins* are not transported and remain unfermented (see Hough *et al.*, 1982). Maltose is present at much higher concentrations than glucose (c. 40%, w/v compared with c. 1%, w/v) in wort, but glucose regulates the rate at which maltose is utilized and thus the overall rate of fermentation. Glucose (>0.4%, w/v) impairs maltose uptake by acting both as catabolite repressor and inactivator of maltose permease. The rate of maltose utilization in brewing fermentations will remain low until glucose falls to a level which no longer represses maltose uptake. Such effects will be even more pronounced when using additional carbohydrate sources (adjuncts) in brewing with high glucose levels. Brewing yeast strains (e.g. derepressed mutants) which do not exhibit repression of maltose uptake will attenuate at greater rates and to greater extents than repressed strains (Russell *et al.*, 1987).

The situation is very similar in baking. In fact, the first microorganism altered by recombinant DNA technology to be given approval for food use in the UK (in 1990) was a baker's yeast which over-rode the normal glucose-regulated maltose uptake (see Chapter 6).

Cellobiose. Certain species of yeast are able to ferment the disaccharide cellobiose (4-O-β-D-glucopyranosyl-D-glucopyranose), which is one of the principal sugars present in wood hydrolysates, and such yeasts are potentially exploitable in large-scale simultaneous saccharification and fermentation (SSF) of cellulose in biotechnology. Two *Candida* species, namely, *C. lusitaniae* and *C. wickerhamii*, transport cellobiose by different mechanisms (Spencer-Martins, 1994). The former species transports cellobiose into the cell where it is hydrolysed. *C. wickerhamii*, on the other hand, firstly hydrolyses cellobiose extracellularly by a membrane-bound β-glucosidase and the resulting glucose is then taken up by an active proton symport mechanism. Glucose represses β-glucosidase synthesis, meaning that *C. wickerhamii* cannot metabolize cellobiose when grown in the presence of glucose.

Lactose. Lactose (4-O-β-D-galactopyranosyl-D-glucopyranose), or milk sugar, is a disaccharide of glucose and galactose found at levels of around 5% (w/v) in cheese whey. Although *S. cerevisiae* cannot transport and metabolize lactose, other yeasts, notably *Kluyveromyces* spp. and certain *Candida* spp., are able to ferment lactose. They are able to do so by transporting lactose via an inducible lactose permease (which catalyses lactose proton-symport) and subsequently hydrolysing it to fermentable glucose and galactose by an intracellular β-galactosidase. Lactose permease in *K. lactis* has an affinity constant, K_T, of 0.7–2.8 mM and the carrier may also transport free galactose. Transport of lactose in *K. lactis* and *K. marxianus* is energy-dependent and involves proton symport. Another transport system may operate by facilitated diffusion when cells are grown anaerobically. Derepression of lactose permease (*LAC12*) genes is accomplished by growth on lactose or galactose. *LAC12* genes have been transformed into *S. cerevisiae* enabling this yeast to grow successfully on lactose (Sreekrishna and Dickson, 1985). This has biotechnological significance in the fermentation of cheese whey by lactose-utilizing *S. cerevisiae* since this yeast is more alcohol tolerant than the currently used *Kluyveromyces* spp.

Trehalose. Trehalose (α-D-glucopyranosyl α-D-glucopyranoside) is a disaccharide more often associated with intracellular regulatory phenomena in yeast such as osmoregulation, glucose metabolism and cell cycle progression (Thevelein, 1996). However, many yeasts can utilize trehalose as a carbon substrate. *S. cerevisiae* possesses trehalase activity but can also absorb this disaccharide without prior hydrolysis to glucose. Several trehalose carriers may be involved in yeast, possibly one being an inducible active proton symporter distinct from the maltose carrier (Crowe *et al.*, 1991). Stambuk *et al.* (1996) have characterized a glucose-repressible, trehalose-H^+ symporter in *S. cerevisiae* which was able to accumulate the disaccharide against a concentration gradient with high affinity (K_T 4 mM). A lower-affinity transport system ($K_T > 100$ mM) was also identified which presumably translocated trehalose via facilitated diffusion. The biotechnological significance of trehalose transport in yeast lies in the role of this disaccharide as a stress-protectant molecule. Accumulation of trehalose on both sides of the plasma membrane is thought to confer stress protection in yeast by stabilizing membrane structure. Eleutherio *et al.* (1993) have shown that protection against dehydration and desiccation was mediated by a specific trehalose carrier. Similarly, the regulation of trehalose carrier synthesis or activity may be involved in protection against ethanol toxicity in yeast (Eleutherio *et al.*, 1993). Due to the multiple roles of trehalose in increasing survival of yeasts exposed to several physical and chemical stresses, trehalose accumulation in yeast cells has important commercial applications, particularly in the baking and brewing industries (Kim *et al.*, 1996).

Trisaccharides

Maltotriose (O-α-D-glucopyranosyl-(1→4)-O-α-D-glucopyranosyl-(1→4)-O-α-D-glucose) is a trisaccharide found in brewer's wort following amylolysis of barley starch. It is respired but poorly fermented by brewing yeast. Nevertheless, *S. cerevisiae* synthesizes a constitutive maltotriose transporter with low affinity (K_T 50 mM) which transports this sugar by facilitated diffusion. Transporters have also been reported in industrial strains of *S. cerevisiae* (Michaljanicova *et al.*, 1982).

Oligosaccharides and polysaccharides

There is little evidence that yeast cells can transport specific higher saccharides intact. Brewer's yeast is a very poor fermenter of tetrasaccharides like maltotetraose and *S. cerevisiae* has a general inability to transport and utilize higher saccharides. This is of consequence to fermentation industries like brewing and distilling since those compounds represent potentially fermentable sugar. Numerous approaches have been undertaken to enable branched glucose substrates (e.g. dextrins) to be metabolized in yeast and these are discussed further in Chapter 6.

3.3.2.3 Transport of Alcohols

Small lipid-soluble molecules such as glycerol, methanol and ethanol are known to be transported into yeast to an appreciable extent by passive diffusion. The rate of transport of most sugars is negligible by this mechanism but acyclic polyols (erythritol, xylitol, ribitol, arabinitol, mannitol, sorbitol and galactitol) are known to be taken up by passive diffusion in *S. cerevisiae.*

Concerning **ethanol**, the possible existence of ethanol concentration gradients in yeast has now been discounted with free diffusion accepted as the means whereby this important compound gets into and out of cells. Ethanol efflux is obviously important to the yeast fermentation industries while ethanol influx is pertinent to industries growing yeast (e.g. *Candida utilis*) on ethanol for biomass production. Ethanol uptake by *S. cerevisiae* is also significant during aerobic growth on glucose since it accounts for the phenomenon of diauxie, as described in Chapter 4. Methanol transport is of interest to biotechnologists studying and exploiting methylotrophic yeast species such as *Pichia pastoris* and *Hansenula polymorpha* (Chapter 6).

Concerning transport of **glycerol**, compared with certain yeasts (like *C. utilis*), the plasma membrane of *S. cerevisiae* has a relatively low permeability for glycerol under 'normal' growth conditions. Nevertheless, controlled permeability for glycerol must be maintained when glycerol is utilized as a sole respiratory carbon source or when glycerol efflux is reduced under osmotic stress. Two components for glycerol translocation are thought to exist: one based on free diffusion, and the other based on facilitated diffusion. No direct evidence has been provided for active glycerol uptake in *S. cerevisiae*. Luyten *et al.* (1995) have identified a protein (Fps1) which facilitates glycerol permeation through a channel in the yeast plasma membrane. The *FPS1* gene is a member of the MIP family of channels identified in many organisms. The existence of such a glycerol channel allows the yeast cell to adjust rapidly its plasma membrane permeability for glycerol depending on the growth conditions. The channel is thought to control both influx and efflux of glycerol and thus plays an important role in osmoregulation when yeast cells synthesize glycerol as a compatible solute in times of osmotic stress (see Chapter 4).

In contrast to *S. cerevisiae*, osmotolerant yeasts such as *Zygosaccharomyces rouxii* and *Debaryomyces hansenii* accumulate glycerol by an active uptake system (Lucas *et al.*, 1990; van Zyl *et al.*, 1990) apparently unrelated to the FPS1 channel.

3.3.2.4 Transport of Hydrocarbons

Assimilation of *n*-alkanes by yeast is via a passive uptake of the apolar substrate facilitated by special hydrophobic lipopolysaccharide cell wall structures which adhere microemulsions of alkanes (Mauersberger *et al.*, 1996). Several species of yeast, particularly *Candida* spp., are able to grow on short-chain alkanes as carbon and energy sources in the production of yeast biomass protein from petrochemical residues. Such yeasts synthesize surface-active compounds such as lipopeptides and sophorose lipids to emulsify the hydrocarbons into micellar droplets which surround aggregates of yeast cells. Other oleaginous yeasts, such as *Yarrowia lipolytica*, secrete an emulsifier called liposan which aids internalization of small droplets of *n*-alkanes (Barth and Gaillardin, 1996). Such droplets then diffuse into the yeast cell in the form of microbodies (microsomes) in which the alkanes are then metabolized, for example by the cytochrome P450 mono-oxygenase system (Chapter 5).

3.3.2.5 Transport of Organic Acids

There are several reasons why an understanding of organic acid transport in yeasts is important in biotechnology: the ability of yeasts to use certain carboxylic acids for growth is useful for taxonomic purposes; the metabolism of the predominant organic acids in wine (namely, malic and tartaric acids) is important in wine fermentations (Subden and Osothsilp, 1987); the production of citric acid by yeasts is commercially important (Mattey, 1992); and some food spoilage yeasts (e.g. *Zygosaccharomyces bailii*) can grow in the presence of weak acid preservatives such as benzoic, sorbic and acetic acids. Transport of organic acids is important not only when yeasts utilize such compounds as carbon sources for growth but also in control of intracellular pH and in contributing to the intracellular charge balance by enhancing K^+ ion uptake (Jennings, 1995). Transport mechanisms can either involve proton–organic acid anion symport or simple diffusion of the undissociated acid. The precise mechanism depends on the yeast species and the acid in question (Table 3.9).

In winemaking, the use of grapes grown in cool climates can often lead to excess levels of malic acid in wine, leaving the product with a sour taste. Controlled malo-lactic fermentation with lactic acid bacteria can be used to reduce malate levels in high acid grape musts but genetic selection of *Saccharomyces* wine strains capable of complete malic acid reduction during fermentation would be of distinct interest to winemakers. Malate is thought to enter *S. cerevisiae* by simple diffusion, but other yeasts (*Z. bailii*, *Kloeck. javanica*, *Sch. pombe*, *K. lactis* and *Z. rouxii*) possess malic acid permease transport proteins (Subden and Osothsilp, 1987).

Concerning lactic acid, *Candida utilis* accumulates lactate by an electrochemical H^+-lactate symport in a 1 : 1 stoichiometry in an inducible, glucose-repressible system which also transports pyruvate, propionate and acetate (Leão and van Uden, 1986).

3.3.2.6 Transport of Sterols

The principal yeast sterol, ergosterol, is integrated into the yeast plasma membrane and accounts for around 50% of total sterols in yeast cells. Cholesterol uptake and internalization by *S. cerevisiae* has been studied by Hapala and Hunakova (1996). The other sterols are generally esterified in lipid storage granules and can be hydrolysed to release sterols and fatty acids when cells commence active membrane biogenesis at the onset of growth (Kessler *et al.*, 1992). Under aerobic conditions, *S. cerevisiae* synthesizes

Table 3.9. Yeast organic acid transport systems.

Acid	Yeast	Affinity constant (K_T mM)	Mechanism	Comments
L-Lactic	*Saccharomyces cerevisiae*	0.13	Electroneutral proton/anion symport	Also transports: D-lactate, acetate, pyruvate, propionate
L-Lactic	*Candida utilis*	0.06	Proton/anion symport	Also transports acetate, D-lactate, propionate, pyruvate
L-Lactic	*Kluyveromyces marxianus*	0.42	Uniport	
L-Malic	*Saccharomyces cerevisiae*		Simple diffusion uptake	Active efflux
L-Malic	*Kluyveromyces marxianus*	0.1	Proton/anion symport	Also transports succinate, fumarate, 2-oxoglutarate, D-malate
L-Malic	*Schizosaccharomyces pombe*	3.0	Constitutive	Not repressed by glucose
L-Malic	*Hansenula anomala*	0.076	Proton/anion symport	Also transports succinate, fumarate, 2-oxoglutarate, D-malate, oxaloacetate
Succinic	*Kluyveromyces lactis*	0.018		Also transports malate, fumarate, 2-oxoglutarate
Citric	*Candida utilis*	0.056 (High aff.)	Proton/anion symport	Also transports isocitrate
		0.59 (Low aff.)	Facilitated diffusion	Also transports isocitrate, L-lactate fumarate

Adapted from Jennings (1995).

sterols in the endoplasmic reticulum and subsequently transports them to other cellular membranes. Since molecular oxygen is involved in *de novo* sterol biosynthesis in yeast, aerobically grown cells do not take up sterols. However, under anaerobic conditions, *S. cerevisiae* translocates ergosterol via a specific sterol transporter, encoded by the *SUT* (sterol uptake) gene which has been characterized by Bourdot and Karst (1995).

3.3.2.7 *Transport of Fatty Acids*

Some oleaginous yeasts accumulate very large amounts of fatty acids of which a significant proportion exist as saturated triglycerides. For example, approximately 40% of the dry weight of *Candida antarctica* is comprised of this material. In *Yarrowia lipolytica*, two fatty acid carrier systems have been detected (Kohlwein and Paltauf, 1983). One is specific for C_{12} and C_{14}

fatty acids, and the other for C_{16} and C_{18} saturated or unsaturated fatty acids.

3.3.2.8 *Transport of Nitrogenous Compounds*

Yeasts cannot fix atmospheric nitrogen but are capable of transporting and subsequently utilizing a few inorganic and several organic nitrogen compounds. Yeast nitrogen sources generally serve anabolic roles for the biosynthesis of structural proteins and functional enzymes whereas some sources, notably amino acids, may be catabolized immediately on entry into the cell.

The favoured sources of nitrogen for maximizing yeast growth are ammonium ions, glutamate and glutamine. Ammonium salts commonly provide the necessary nitrogen in yeast media; however, many natural and industrial yeast fermentation substrates contain a wide spectrum of amino acids and peptides. Knowledge of yeast nitrogen transport systems is therefore relevant in biotechnology and the following section not only summarizes the basic uptake mechanisms and their regulation, but also the biotechnological significance of such processes.

Ammonium ions

All yeasts are able to use ammonia efficiently as a source of nitrogen and ammonium salts are routinely incorporated into yeast growth media for this reason. Although there is no convenient method for directly studying ammonium ion uptake in yeast due to the lack of a convenient radioisotope of N, uptake can be studied indirectly using ^{14}C-methylammonium. Both ammonium and methylammonium ions are thought to be transported by the same system and studies in *S. cerevisiae* have shown that a carrier-mediated, active transport system exists with the following characteristics:

- High-affinity (K_T 0.2 mM), concentrative system following saturation kinetics
- Requires a fermentable or respiratory substrate
- Non-competitively inhibited by amino acids
- Precise driving force for NH_4^+ translocation still unclear, but may be driven by the membrane potential

In addition to this high-affinity system, a low-affinity system (K_T 2 mM) also exists in *S. cerevisiae* (Cartwright *et al.*, 1989). In nitrogen-limited chemostat cultures of *S. cerevisiae*, this dual-affinity uptake system for ammonium ions is not apparent and in such circumstances regulation of uptake may depend more on intracellular, rather than extracellular, ammonia availability (Schulze *et al.*, 1996).

In *Candida utilis*, a very high-affinity system (K_T 1 μM) has been identified in ammonium-limited chemostat cultures, indicating that an efficient nitrogen scavenging process may operate in some yeasts, as it does in filamentous fungi.

The role of ammonium ions in regulation of nitrogen utilization in yeasts through 'nitrogen catabolite control' is an important phenomenon in yeast nutrition since it may serve to prioritize assimilation of nitrogenous compounds. This aspect is considered below in relation to amino acid transport.

Nitrate and nitrite ions

Some yeasts can utilize other inorganic nitrogen compounds; namely, nitrate and nitrite. The ability of yeasts to utilize nitrate as a sole source of nitrogen is an important taxonomic criterion and Hipkin (1989) has grouped yeasts generally on the basis of nitrate (and nitrite) utilization. For example, *Brettanomyces*, *Candida*, *Hansenula*, *Pachysolen* and *Rhodotorula* are generally included as genera that assimilate nitrate; *Debaryomyces* can assimilate nitrite but not nitrate; and *Kluyveromyces*, *Pichia*, *Saccharomyces* and *Schizosaccharomyces* are generally noted for their inability to assimilate nitrate. Thus, *S. cerevisiae* is not able to utilize nitrate or nitrite as nitrogen sources and this is one reason why nitrate uptake has not been studied in great detail in yeasts.

Nevertheless, some information on nitrate transport in yeasts has emanated from studies in *Candida* spp. such as *C. utilis* and *C. nitratophila.* The former species takes up nitrate in a carrier-mediated process under aerobic conditions while the latter possesses an electrogenic nitrate-proton symporter. The proton inflow which accompanies nitrate uptake in *C. utilis* results in the cell maintaining charge balance by expelling K^+ ions (Eddy and Hopkins, 1985). Note that membrane transport of nitrate appears to be the rate-limiting step in nitrate assimilation in yeast which is discussed further in Chapter 5.

While probably not an important natural source of nitrogen, nitrite is nevertheless transported and subsequently reduced (to NH_4^+ by nitrite reductase) by *C. nitratophila* in an apparently energy-dependent mechanism (Hipkin, 1989).

Urea

Urea is widely utilized by yeasts. Two uptake systems have been reported in *S. cerevisiae* (Cartwright *et al.*, 1989). One is a high-affinity (K_T 14 μM), nitrogen-repressible system that can actively concentrate urea 200-fold in cells and the other is a constitutive, low-affinity (K_T 2.5 mM) system that functions by facilitated diffusion when excess ($>$0.5 mM) urea is present.

Amino acids

Amino acids transported across the yeast cell membrane can either be incorporated intact into proteins or they can be intracellularly catabolized to serve as nitrogen or carbon sources. Yeasts also contain relatively large pools of endogenous amino acids, most notably arginine in the vacuole (Eddy, 1980). Although the growth of yeasts is often better with ammonium salts than when any single amino acid is employed, growth of some yeasts is more rapid in mixtures of amino acids. This is the case with brewer's yeast, where in fact the presence of ammonium ions may inhibit amino acid uptake.

A general succession of events is involved in yeast amino acid transport (Stoppani and Ramos, 1978):

- *Binding:* this is accomplished by recognition of the amino acid at a specific receptor.
- *Translocation:* this is mediated by permeases in the plasma membrane and results in concentrative accumulation of amino acids.
- *Coupling:* uptake is driven by a spontaneous influx of protons coupled to ion efflux.
- *Release:* amino acids are released inside the cell and may be compartmentalized into vacuoles which accounts for the uptake irreversibility.

Two classes of amino acid uptake system are synthesized in yeast (Cooper, 1982). One is broadly specific and effects the uptake of all naturally occurring amino acids including citrulline. This is referred to as the *general amino acid permease*, or GAP. The other system displays specificity for one or a small number of related amino acids. Several of these singular transporters have been characterized in *S. cerevisiae* and are summarized in Table 3.10.

Amino acid transport (both general and specific) in *S. cerevisiae* and *C. albicans* has been shown to be active and dependent on proton-symport mechanisms. Thus, the transmembrane pH gradient provides energy and proton influx is counterbalanced by expulsion of K^+ ions in an antiport mechanism. In this way, chemiosmotic coupling between an electrochemical proton gradient and solute transport accounts for amino acid uptake in yeast.

Considering regulatory aspects of yeast amino acid uptake, the GAP may be viewed as a scavenger whose expression strongly depends on the nitrogen source availability. The GAP is not synthesized when yeasts are grown in nitrogen-rich media, for example, containing ammonium ions. While the GAP is subject to ammonium catabolic repression, the specific amino acid transport systems in *S. cerevisiae* are not thought to be subject to metabolically regulated repression. Therefore, nitrogen catabolite repression in yeasts enables cells to adapt to the

Table 3.10. Specific amino acid transporters in *S. cerevisiae*.

Amino acid	Affinity (K_T)	Comments
L-Arginine	High, K_T 10 μM Low, K_T 1 mM	System seems to facilitate uptake of all basic amino acids
L-Lysine	High, K_T 25–78 μM Low, K_T 0.2 mM	
L-Histidine	High, K_T 20 μM	Histidine permease was first yeast transporter to be molecularly characterized
	Low, K_T 0.5 mM	Diffusion may account for this
L-Methionine	High, K_T 3–12 μM Low, K_T 0.6–0.8 mM	One high-affinity and two low-affinity methionine permeases have been identified
S-adenosyl-L-methionine	High, K_T 1.6–3.3 μM	
L-Cysteine	Low, K_T 0.25 mM	Precise identity unclear. *S. cerevisiae* apparently unable to transport cystine
L-Serine	Low, K_T 0.58 mM	Single, but not necessarily identical, transport systems
L-Threonine	Low, K_T 0.21 mM	
L-Leucine	High, K_T 30 μM Low, K_T 0.5–4.5 mM	Transport of leucine and other branched-chain amino acids (isoleucine and valine) mediated by a specific gene, *BAP2*, which is regulated by leucine availability
L-Glutamate	High, K_T 20 μM Low, K_T 3.3 mM	Three transport systems may exist, the other is the GAP
L-Asparagine	Low, K_T 0.35 mM	Also transports glutamine, histidine, threonine and tryptophan
L-Proline	High, K_T 25 μM	Low-affinity system may also exist
L-Alanine	High	GAP is responsible of high-affinity uptake
L-Glycine	Low	Role of this system uncertain

Summarised from information provided in Cartwright *et al.* (1989), Jennings (1995), Didion *et al.* (1996) and Isnard *et al.* (1996).

changing availability of amino acids (Ter Schure *et al.*, 1995).

The biotechnological significance of control of amino acid transport in yeast is especially apparent in brewing fermentations where mixtures of amino acids provide the bulk of the yeasts' nitrogen nutrition. In key research conducted by Jones and Pierce (1964), it was determined that the uptake of amino acids from brewer's wort by yeast occurs in an ordered sequence as indicated in Table 3.11.

The orderly uptake was considered due to the involvement of particular classes of amino acid permeases: one generally specific for Group A and C amino acids (since Group A amino acids compete for transport with those of Group C) and another for Group B amino acids. Perhaps in the initial stages of brewing fermentations only amino acids with specific permeases would be transported due to ammonium ion inhibition of the GAP.

Thus, the relative availability of different amino acids and the concentration of ammonium ions may dictate the progress of brewing fermentations based on controls operating at the level of yeast membrane transport of amino acids.

Table 3.11. Sequential uptake of amino acids by brewer's yeast.

Group	Amino acid	Comments
A	Glutamate, glutamine, aspartate, asparagine, serine, threonine, lysine, arginine	Immediately transported and assimilated during the first 20 h of fermentation
B	Histidine, valine, methionine, leucine, isoleucine	Transported gradually during the course of fermentation
C	Glycine, phenylalanine, tryptophan, tyrosine, alanine, ammonia	Amino acids transported after an appreciable lag time. Ammonium ions may inhibit amino acid uptake during logarithmic growth.
D	Proline	Negligible net transport

Adapted from Jones and Pierce (1964).

Peptides

Peptides are often made available in laboratory yeast growth media through additions of 'peptones' (protein hydrolysates of animal or microbial origin) or in industrial media following controlled proteolysis of plant protein (such as barley hordein in the brewing process). Such complex sources will provide varying mixtures of di-, tri-, oligo- and polypeptides not all of which will be available for yeast nutrition. In *S. cerevisiae*, a single, broadly specific, constitutive transport system exists which is capable of taking up di- and tripeptides without prior hydrolysis in a carrier-mediated, active manner (Becker and Naider, 1980). This transport system is distinct from amino acid transporters and has a preference for small hydrophobic peptides but anything larger than a tripeptide is unlikely to be transported intact by *S. cerevisiae*. Peptide uptake, as is the case with amino acid uptake, will be influenced greatly by the availability of different nitrogen sources in yeast growth media. For example, growth in the presence of a poor nitrogen source like proline, or in very low concentrations of amino acids, may stimulate a general peptide transporter. Conversely, peptide transport may be subject to ammonium ion repression in a similar manner to the general amino acid permease.

Oligopeptides require extracellular hydrolysis to smaller peptides and free amino acids by secreted peptidases, an ability not commonly attributed to *S. cerevisiae*. Such peptides generated in this manner do not, therefore, serve as growth substrates for this yeast.

In *C. albicans*, two peptide transporters may exist, one of which is thought to transport dipeptides and the other oligopeptides. Knowledge of such systems is pertinent in the design of anti-*Candida* drugs since it is possible to introduce normally impermeant compounds by coupling to a peptide in a process referred to as 'illicit transport' (Becker and Naider, 1980).

Miscellaneous nitrogenous compounds: purines, pyrimidines and vitamins

Concerning the transport of purines and pyrimidines by yeast, both proton symport and facilitated diffusion systems have been reported to operate in *S. cerevisiae* (Cartwright *et al.*, 1989). Uptake of adenine, adenosine, cytosine, guanine and hypoxanthine is driven by proton-motive force. Uptake of adenosine is rapid and of very high transport affinity (K_T 0.12 μM). Uracil and uridine uptake may be mediated by facilitated diffusion.

Some information on yeast vitamin uptake has been provided by Cartwright *et al.* (1989) and Jennings (1995). Generally, vitamins appear to be accumulated in yeast cells in very high-affinity, active, carrier-mediated processes. For example, biotin and thiamine, which are frequently required by yeasts, have transport affinity constants (K_T) of 0.32 μM and 0.18 μM, respectively, but uptake is strongly dependent on levels of exogenous vitamins. Many yeasts have active transport systems for pyridoxine (Yagi *et al.*, 1996).

3.3.2.9 Transport of Anions

In the following section, the transport of phosphate, sulphate and chloride ions into yeast cells is overviewed and salient aspects of the physiological regulation of anion uptake by yeast discussed. The transport of nitrate by yeast has already been covered in the discussion of transport of nitrogenous compounds.

Phosphate

Phosphorus is essential to yeast cells for incorporation into structural molecules (e.g. phospholipids), nucleic acids and phosphorylated metabolites. Yeasts may also store phosphorus in the vacuole in the form of polymeric orthophosphate. Phosphorus is commonly available to yeasts in the form of inorganic orthophosphate ($H_2PO_4^-$) which is metabolized very rapidly to nucleoside triphosphate on entry into yeast cells.

Orthophosphate transport in yeast occurs against a concentration gradient and is therefore active. Generally, transport is highly dependent on both the intra- and extracellular pH, and on Mg^{2+}, K^+ and phosphate concentrations in the growth medium (Borst-Pauwels, 1981). It is also dependent on the presence of an exogenous fermentable substrate like glucose but respirable substrates like acetate or ethanol do not support phosphate uptake. In *S. cerevisiae*, at least three systems are thought to translocate orthophosphate (Borst-Pauwels and Peters, 1987) and these are summarized in Table 3.12.

In *C. tropicalis*, a high-affinity (K_T 4.5 μM) phosphate transporter has been demonstrated under phosphate-starvation conditions. In *S. cerevisiae*, it is conceivable that both low- and high-affinity systems may operate simultaneously depending on phosphate availability. In fact, phosphate uptake may be controlled by the concentration of intracellular orthophosphate. When this is high, no net phosphate transport occurs, but when it declines (as during fermentation or respiration) the rate of phosphate uptake increases (Borst-Pauwels, 1981).

In addition to yeast cell membrane transport of phosphate, other membrane transporters are

Table 3.12. Phosphate transport systems in *S. cerevisiae*.

System	Transport affinity (K_T)	Comments
High-affinity system	c. around 10 μM at pH 4.5	An inducible, specific carrier, encoded by the *PHO84* gene, operates as a proton symporter. Proton influx is counterbalanced by K^+ efflux.
Low-affinity system	c. around 1 mM at pH 4.5	Operates under high phosphate concentrations as a constitutive proton symporter.
Sodium-phosphate transporter	c. around 1 μM at pH 7.2	A very high-affinity, inducible system operating primarily under non-physiological high Na^+ concentrations.

Adapted from Cartwright *et al.* (1989); Jennings (1995); and André (1995).

known to operate. For example, in yeast mitochondria, which exhibit both high- and low-affinity transport, and in vacuoles where the formation of insoluble polyphosphate may contribute to the way in which yeast controls cytosolic phosphate levels.

Molecular biological studies of phosphate uptake in *S. cerevisiae* have revealed that the gene *PHO84* (and possibly also *PHO86*, *PHO87* and *PHO88*) encodes a high-affinity phosphate transporter, with a K_T value of 8.2 μM (Bun-Ya *et al.*, 1996; Yompakdee *et al.*, 1996). The *PHO84* protein shares amino acid sequence homology with the sugar transporter superfamily referred to earlier.

Sulphate and sulphite

Inorganic sulphur, in the form of sulphate anions, is transported by yeast for assimilation into sulphur-containing amino acids like methionine and peptides like glutathione. Yeast sulphate uptake is an active process. The mechanism involves an inducible carrier which is energized by proton-motive force. Sulphate-proton symport is counterbalanced by K^+ efflux.

Although several molecular genetic studies have been carried out into yeast sulphate uptake (e.g. Jin *et al.*, 1995; Smith *et al.*, 1995), no singular sulphate transporter gene has been identified in yeasts at present (André, 1995). Several transport systems are thought to exist in *S. cerevisiae* which possess different transport affinities for exogenous sulphate depending on the availability of sulphate and the pH. Generally, K_T values for sulphate transport are in the μM range. A study into ^{35}S-labelled sulphate uptake kinetics in *S. cerevisiae* (Breton and Surdin-Kerjan, 1977) revealed a biphasic Lineweaver–Burke relationship which indicated the presence of both high-affinity (K_T 5 μM) and low-affinity (K_T 350 μM) sulphate transporters. The existence of two (high- and low-affinity) independent sulphate transporter proteins in yeast has been confirmed by Smith *et al.* (1995).

In the presence of excess sulphate, yeast can store sulphur intracellularly in the form of glutathione, which can account for as much as 1% of total dry weight of the cells (Elskens *et al.*, 1991). Thus, under conditions of sulphate limitation or starvation, glutathione may act as an endogenous sulphur source for the biosynthesis of sulphur amino acids. It is important to note that glutathione also plays other significant roles in yeast cell physiology, such as a primary scavenger of oxygen free radicals and in conferring protection in yeast to oxidative stress (see Chapter 4).

Another inorganic form of sulphur, namely sulphite, is of interest in yeast biotechnology because of the long-standing usage of sulphiting agents (sulphites, SO_2) to preserve grape must before wine fermentations. Membrane transport of sulphite in *S. cerevisiae*, which is comparatively resistant to sulphite levels employed in must, is by simple diffusion of liberated SO_2 rather than being carrier-mediated. Only molecular SO_2 enters yeast cells where it converts to SO_3^{2-} and HSO_3^- at physiological pH (5–6). A K_T value of 0.1 mM at pH 4.0 has been reported (Rose, 1989). Evidence exists for sulphite transport processes in other yeasts. For example, the spoilage yeasts *Saccharomycodes ludwigii* and *Zygosaccharomyces bailii* take up both sulphate and sulphite in an energy-dependent manner.

A biotechnologically significant aspect of inorganic sulphur transport by yeast lies in the inhibition of sulphate uptake by sulphite, the latter being preferentially utilized. Stratford and Rose (1985) reported that H_2S production during yeast fermentation, which can give rise to noxious odours in alcoholic beverages, may be greater when sulphite rather than sulphate is the sole sulphur source and this reflects differences in the regulation of sulphite and sulphate transport.

Chloride

Although there is no evidence of an active chloride uptake system in the yeast plasma membrane, it is has been suggested that Cl^- transport occurs via a proton–chloride or sodium–chloride symport mechanism (André, 1995) which may be involved in regulation of yeast cell water content.

3.3.2.10 Transport of Cations

Yeast cells take up inorganic cations for several reasons. These may involve: regulation of intracellular pH homeostasis and generation of proton motive force (in the case of H^+ transport); osmoregulation and charge balancing (in the case of K^+); enzyme cofactor functions (in the case of Mg^{2+} and Mn^{2+}); metalloenzyme structural functions (in the case of micronutrient divalent cations such as Fe^{2+}, Zn^{2+}, and Ni^{2+}) and signal transduction second messenger functions (in the case of Ca^{2+}). In fact, Silver (1995) has stated with regard to microbial cation transport: 'There are specific physiological and biochemical roles for each cation formed from an element from atomic number 1 (hydrogen) to 30 (zinc)'. Although advances are being made into the elucidation of cation transport mechanisms in yeast using molecular biological approaches (e.g. Gaber, 1992), relatively few cell physiological studies have been reported in recent years. Nevertheless, the following section describes the transport and regulation of several monovalent and divalent cations in yeast (depicted schematically in Figure 3.3) and discusses the significance of this with respect to yeast biotechnology.

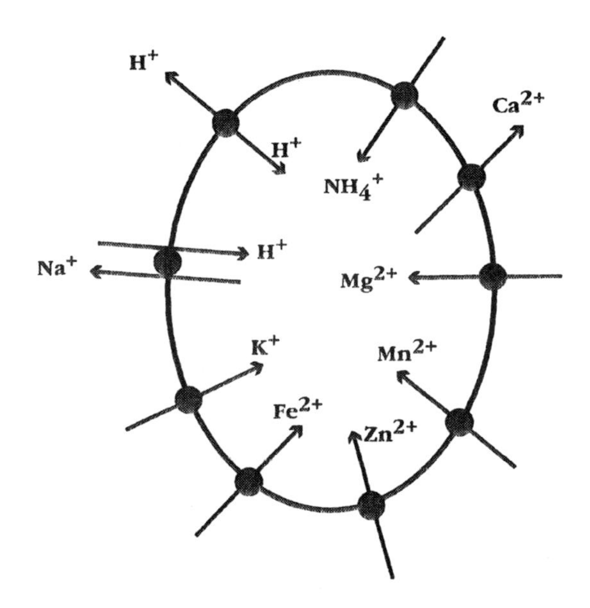

Figure 3.3. Schematic representation of some key cation transport systems in yeast.

Hydrogen ions

Yeast cell membranes are not freely permeable to hydrogen ions and transmembrane proton gradients are established by active proton-pumping mechanisms. The electrochemical transmembrane proton gradient is generated by H^+-translocating ATPase enzymes which provide the driving force for the transport of many yeast nutrients. The yeast cell membrane H^+-ATPase is a major constituent of the plasma membrane (comprising as much as 50% of total membrane protein in *S. cerevisiae*) and has been described by Serrano (1989) as a 'master enzyme' in fungal physiology. This is because it controls cell pH, nutrient and ion transport and overall cell growth. With regard to the latter aspect, yeast mutants of the plasma membrane ATPase gene display reduced growth rates, altered budding patterns and elongated cell morphology. In addition, the activity of the enzyme declines significantly as *S. cerevisiae* cells enter the stationary phase of growth.

With regard to pH homeostasis in yeast, the H^+-ATPase is instrumental in modulating both intra- and extracellular pH. Intracellular pH of *S. cerevisiae* remains relatively constant (to within around 0.4 pH units) at about pH 5.2, even when the extracellular pH fluctuates (Cimprich *et al.*, 1995). This constancy is maintained primarily through the activities of the cell membrane H^+-ATPase. Extracellular acidification and concomitant intracellular alkalinization are important yeast growth responses. The plasma membrane H^+-ATPase activity is therefore inextricably linked with yeast growth and has the capability of generating a 10 000-fold difference between the concentration of protons on either side of the membrane. The magnitude of the gradient in yeast depends on the presence of other cations, notably K^+ which is exchanged for H^+ in a 1:1 stoichiometry. Control of proton exchanges in growing yeast cells is relevant to some aspects of yeast biotechnology (Castrillo *et al.*, 1995). For example, the acidification response of yeasts to addition of a carbon substrate can be exploited in assessing the metabolic competence of industrial yeast cultures (Susta *et al.*, 1984). So-called 'acidification power' tests for yeast membrane proton efflux capacity are useful in distinguishing vitality from viability (see Chapter 4).

Extracellular acidification by yeasts may also be significant in inhibiting the growth of competitor microorganisms in natural environments, since this can sometimes reach values as low as pH 1.5. Intracellular acidification, on the other hand, is significant in food microbiology because this may be responsible for the effect of weak acid preservatives (e.g. benzoic, propionic and sorbic acids) in inhibiting growth of spoilage yeasts. It should be noted that although the electrogenic proton-pumping ATPase in the yeast plasma membrane creates most of the acidity in culture media, it is not the sole contributor. In metabolically active yeast cells, excreted acidity is also caused by the hydration and dissociation reaction of produced CO_2 which yields HCO_3^- and H^+. This CO_2-derived acidity accounts for around 3% of the overall acidity of yeast culture media (Sigler and Höfer, 1991). Other contributory factors are the production of organic acids by yeast; most notably succinate, with smaller amounts of malate, lactate, butyrate, formate and acetate. In *Dekkera* and *Brettanomyces* spp., major acetate excretion may occur in traditional Belgian brewing (gueuze and lambic beer).

Molecular biological knowledge of yeast ATPases has advanced significantly in recent years. The yeast plasma membrane H^+-ATPase belongs to the P-type cation-translocating ATPases of higher eukaryotic cells. The main gene encoding the *S. cerevisiae* and *Sch. pombe* enzyme is *PMA1*. Gaber (1992) and de Kerchove d'Exaerde *et al.* (1996) have reviewed the structural, functional and regulatory aspects of H^+-translocating ATPases in *S. cerevisiae* and other yeasts.

Potassium ions

Potassium is the most prevalent cation in the yeast cytoplasm (at around 150 mM) and several systems for translocating K^+ exist. These include: active transport via carrier permeases, counter-transport with protons via the H^+-ATPase and passive transport via K^+ ion channels.

Active K^+ transport in yeast requires a fermentable or respirable substrate. It occurs against a considerable concentration gradient (5000:1) and exhibits substrate saturation kinetics. The K^+ carriers also transport other monovalent (e.g. Rb^+, Cs^+, Li^+, NH_4^+) and divalent (Ca^{2+}, Mg^{2+}) cations, albeit with much lower affinity than K^+ (Borst-Pauwels, 1981). In *S. cerevisiae*, several components of the monovalent cation transport system have been elucidated. One site is a transporter for alkali metal cations and also possibly Mg^{2+} when this is present in high concentrations. Another (modifier) site exists where other ions (both monovalent and divalent) bind in a non-competitive fashion. The third site, referred to as the activation site, may be equivalent to the high-affinity K^+ carrier (with a K_T around 20 µM) which is only expressed in cells grown in low K^+ ion concentrations. Molecular genetic analysis of K^+ transport (e.g. Ramos *et al.*, 1990) has verified the existence of independent high- and low-affinity carriers. Such parallel K^+ uptake systems are necessary to maintain high intracellular K^+ levels, even when the extracellular K^+ availability fluctuates widely. High- and low-affinity K^+ carriers in yeast are encoded by the *TRK1* and *TRK2* genes, respectively, and these have been cloned (Gaber, 1992). The TRK1 protein displays sequence homology to K^+-H^+ symporters of higher plants.

The multi-component K^+ transporter is also implicated in potassium efflux from yeast, and net translocation into cells is dependent upon the balance between uptake and efflux. In fermenting yeast cells, the net K^+ uptake is fast. Resting cells leak K^+ slowly in the absence of an energy source. The genetic basis of energy-dependent K^+ efflux in yeast is not known (Gaber, 1992).

K^+ uptake in active cultures of *S. cerevisiae* is much higher than can be accounted for if membrane potential is the sole driving force. Normally, outward proton pumping mediated by the electrogenic plasma membrane ATPase is accompanied by K^+ uptake. This indicates a key involvement of a K^+-H^+ antiport mechanism to influx K^+, efflux H^+ and to maintain electrical neutrality. Net accumulation of K^+ into yeast cells, therefore, is energized not directly by ATP but by a combination of the membrane potential and the proton-motive force.

Another facet of K^+ transport by yeast is the presence of plasma membrane channels which act as voltage-gated diffusion pores (Gustin *et al.*, 1986). These are highly selective for K^+ (over Na^+) and permit K^+ to be very rapidly transported down electrochemical gradients. Such channels, termed YPK1, also operate outwardly in transient effluxing of K^+. They play roles in charge balancing during H^+-symport of solutes and in osmoregulation during osmotic stress. K^+ efflux through membrane channels is also thought to occur when glucose is added to anaerobic cultures (at pH 7) of *S. cerevisiae*. Reid *et al.* (1996) have identified and characterized a novel outwardly rectifying, voltage-dependent potassium channel in *S. cerevisiae* which is encoded by a single gene, *DUK1* (duplicate-pore K channel). Various physical and chemical factors can induce K^+ efflux from yeast cells and a study of these has practical significance in our understanding of antifungal agent action and drug resistance. For example, cationic dyes, uncouplers, heavy metals, detergents, nystatin, polymyxin B, and yeast killer toxins all elicit K^+ leakage from yeast cells. This phenomenon may also be used to assess the efficacy of anti-yeast agents using K^+-selective electrodes or atomic absorption spectrophotometry to correlate alterations in cell membrane permeability with cell viability.

Sodium ions

Yeast cells do not accumulate Na^+ ions under normal growth conditions. Conversely, yeasts continuously excrete Na^+ to maintain very low cytosolic concentrations of this cation (Rodriguez-Navarro and Ortega, 1982). This is accomplished via a Na^+-H^+ antiport mechanism which has been the subject of molecular genetic analysis in *S. cerevisiae*, *Sch. pombe* and *Zygosaccharomyces rouxii* (André, 1995; Watanabe *et al.*, 1995; Prior *et al.*, 1996). In *S. cerevisiae*, Na^+ efflux is mediated by the P-type ATPase in the cell membrane and is encoded by the *ENA1/PMR2* gene. Additional genes, including *NHA1*, are implicated in Na^+ transport in *S. cerevisiae* (Prior *et al.*, 1996). There is a similarity between Li^+ and Na^+ efflux in *S. cerevisiae*, both ions are expelled via an electrogenic Li^+ or Na^+-H^+ antiporter. *S. cerevisiae* cells can be artificially 'loaded' with Na^+ which, under such non-physiological conditions, probably enters via a low-affinity K^+ transporter and perhaps also Na^+-substrate symporters. Na^+ toxicity in *S. cerevisiae* may be due to antagonism of essential K^+-functions. Some yeast species, such as *Debaryomyces hansenii*, *Pichia sorbitophila* and *Zygosaccharomyces rouxii*, can survive in high Na^+ environments by maintaining intracellular Na^+ levels low and K^+ levels sufficiently high. In the presence of high salt concentrations, yeast cells also osmoregulate by producing intracellular compatible solutes such as glycerol and arabinitol. Ferrando *et al.* (1995) have isolated halotolerance *HAL3* genes in *S. cerevisiae*, which, when overexpressed from plasmids, enable this species to grow in otherwise toxic concentrations of NaCl. It was proposed that HAL3p activity regulates physiological responses to salt stress in yeast by increasing cell K^+ while decreasing cell Na^+. A very unusual halotolerant mutant of *S. cerevisiae* has been studied by Gaxiola *et al.* (1996) which is capable of growth in 1–2 M NaCl. Such cells have increased K^+ but greatly reduced intracellular Na^+ levels compared with halosensitive wild-type cells.

Table 3.13. Specific very high-affinity divalent metal transport systems in yeasts.

Yeast	Metal	K_T	Comments
Candida utilis	Zn^{2+}	1.3 μM	Inhibited by lack of glucose
	Mn^{2+}	16.4 nM	Inhibited by other divalent cations
	Cu^{2+}	3.1 μM	Inhibited by high internal K^+
Saccharomyces cerevisiae	Cu^{2+}	1–13 μM	Inhibited by equimolar Zn^{2+}, Mn^{2+}, Mg^{2+} and Ca^{2+}
	Zn^{2+}	3.7 μM	Lower-affinity (K_T 10 μM) system also observable in Zn-replete cells
	Mn^{2+}	0.3 μM	A lower-affinity (K_T 62 μM) system is also present over 5–200 μM Mn^{2+}
	Fe^{2+}	0.15 μM	Lower-affinity (K_T 30 μM) system is also present

Data from: Parkin and Ross (1986); White and Gadd (1987); Jennings (1995); Zhao and Eide (1996a); Gadd and Laurence (1996) and Askwith, De Silva and Kaplan (1996).

3.3.2.11 Transport of Divalent Metal Cations

Although we still have much to learn about divalent (and trivalent) cation uptake in yeasts, several general statements concerning transport mechanisms can be made. Uptake is biphasic, involving firstly non-specific, cell-surface binding of cations followed by a more regulated, carrier-mediated translocation across the plasma membrane. This secondary phase involves energy-dependent transport driven by the electrochemical membrane gradients generated by proton and potassium ion pumps. However, it is the transmembrane potential which is the **primary** driving force for divalent cation uptake (Jones and Gadd, 1990). The extracellular concentrations of glucose, phosphate and potassium greatly influence divalent cation uptake. Once transported, certain cations are subject to intracellular compartmentalization, most notably in the yeast vacuole. Some cation carriers may have a very high affinity and be singularly specific for certain ions (Table 3.13), whereas others may possess broader specificities and be capable of transporting a multitude of divalent cations. Controlled efflux of certain cations (e.g. Ca^{2+}, Cu^{2+}) also exists and this is important to maintain intracellular levels at very low, subtoxic levels.

The following sections deal with aspects relating to the transport of particular divalent metal ions in yeast and discusses why the accumulation of such cations is important both in fundamental aspects of yeast physiology and in practical aspects of yeast biotechnology.

Magnesium ions

Magnesium is the most abundant intracellular divalent cation in yeast cells where it acts primarily as an enzyme cofactor. Although still far from being fully understood, uptake of Mg^{2+} ions in yeast is thought to be driven by both the proton and potassium ion transmembrane gradients. Mg^{2+} uptake through the low-affinity K^+ transporter is thought to be of minor significance in yeast. It is not known how many Mg^{2+} carriers exist, but a general divalent cation transport was described around 30 years ago in *S. cerevisiae* (Fuhrmann and Rothstein, 1968), with the following order of affinity:

$$Mg^{2+} > Co^{2+} > Zn^{2+} Mn^{2+} > Ni^{2+} > Ca^{2+} > Sr^{2+}$$

Mg^{2+} transport occurs with simultaneous uptake of phosphate and reserves of polymeric Mg-orthophosphate are found in the yeast vacuole (Okorokov *et al.*, 1974). This intracellular sequestration of Mg^{2+} indicates that vacuolar transport mechanisms are involved in regulating free Mg^{2+} ion concentrations (in the mM range) in the yeast cytosol. Beeler *et al.* (1997) have shown that cytosolic Mg^{2+} levels in *S. cerevisiae* are maintained at 0.1–1.0 mM by the regulation of Mg^{2+} fluxes across both the vacuolar and plasma membranes.

The molecular biology of Mg^{2+} transport in bacteria is much more advanced than in yeast (Tao *et al.*, 1995). In *Salmonella typhimurium*, for example, the involvement of several P-type ATPases which are primary active Mg^{2+} cation membrane pumps, has been established (Smith and Maguire, 1993). These enzymes display homology with eukaryotic ATPases and it is conceivable that similar such transporters exist in yeast, although no Mg^{2+} transport mutants have yet been isolated to facilitate molecular genetic studies.

The biotechnological significance of Mg^{2+} transport in yeast lies in the central importance of this metal cation in governing several aspects of yeast growth and metabolism (Walker, 1994). With regard to growth, cell Mg^{2+} has been shown to fluctuate during the cell cycle in budding and fission yeasts (Walker and Duffus, 1980) and has been postulated to coordinate cell growth and division by regulating key events during mitosis (Walker, 1986). With regard to yeast fermentative metabolism, Walker *et al.* (1996) and Walker and Maynard (1997) have shown that there is a correlation between cellular Mg^{2+} uptake and alcoholic fermentation in industrial strains of *S. cerevisiae*. Exogenous Mg^{2+} may also exert a protective effect on yeast cells subjected to a variety of physical and chemical stresses, as discussed further in Chapter 4.

Manganese ions

Manganese is essential for yeast growth and metabolism in trace (μM) levels and may also act as an intracellular regulator of key enzymes (Auling, 1994). Mn^{2+} ions are accumulated to a greater extent than Ca^{2+} in yeast cells, but to a much lesser extent than Mg^{2+}. Although Mn^{2+} can substitute for Mg^{2+} as an enzyme cofactor *in vitro*, this is unlikely to be of any physiological significance due to the different transport magnitudes and resulting intracellular concentration differences between Mn^{2+} and Mg^{2+} in yeast cells (μM versus mM, respectively). The possibility that Mn^{2+} may substitute for Ca^{2+} ions in regulating cell division cycle progression in yeast is discussed further below. Mn^{2+} uptake in yeast, which is strongly inhibited by Mg^{2+}, is maximal during exponential growth and decreases greatly on entry into the stationary phase. Like Mg^{2+}, Mn^{2+} is accumulated in the yeast vacuole (Kihn *et al.*, 1988). Energy-dependent transport of Mn^{2+}, which is optimal at pH 5, is counterbalanced by K^{+} efflux to maintain cellular electroneutrality. In *S. cerevisiae*, Blackwell *et al.* (1997) have shown that Mn^{2+} uptake and toxicity is strongly influenced by the intracellular levels of Mg^{2+} ions.

A dual transport system for Mn^{2+} ions has been identified in *S. cerevisiae* by Gadd and Laurence (1996). Thus, at low Mn^{2+} concentrations (< 1 μM), a high-affinity (but low specificity) Mn^{2+} transporter exists with a K_T of 0.3 μM. At higher concentrations of Mn^{2+} (5–200 μM), a lower-affinity (K_T 62 μM) system is operational. Gadd and Laurence (1996) also found that the competitive inhibition of Mn^{2+} uptake caused by Mg^{2+} was not absolute, indicating that yeast cells retain an ability to acquire trace elements like Mn^{2+} even in the presence of much higher concentrations of other divalent cations.

Similarly with regard to Mg^{2+} transport mechanisms, little is known of the molecular biology of Mn^{2+} uptake in yeast. Nevertheless, Supek *et al.* (1996) have identified a gene, *SMF1*, which encodes a high-affinity Mn^{2+} transporter in *S. cerevisiae*. Furthermore, *CCC1*, a gene known to function in calcium metabolism in yeast, has now been shown to control cellular homeostasis of Mn^{2+} in *S. cerevisiae* by specifically sequestering this metal ion in the Golgi apparatus (Lapinskas *et al.*, 1996).

Calcium ions

Yeast cells maintain cytosolic Ca^{2+} at very low levels, around 10^{-8} M (Youatt, 1993). This is accomplished by means of efflux and compartmentalization via plasma membrane and tonoplast Ca^{2+} transporters and through sequestration with specific Ca^{2+}-binding proteins like calmodulin. Eilam (1982) demonstrated the presence of Ca^{2+}-H^+ antiporter activity in the yeast cell membrane. Such a carrier also exists in the vacuolar membrane, indicating that energy-dependent uptake of Ca^{2+} into the yeast vacuole may be involved in regulating Ca^{2+} homeostasis in yeast. In *S. cerevisiae*, components of the secretory pathway (e.g. ER, Golgi) all possess Ca^{2+} uptake capability (Okorokov *et al.*, 1996 and L. Okorokov, personal communication). The isolation of Ca^{2+}-dependent mutants (*cal* mutants) and the cloning of fluorescent Ca^{2+} indicators such as aequorin into *S. cerevisiae* (Nakajima-Shimada *et al.*, 1991) has provided further insight into the role of Ca^{2+} uptake in yeast cell physiology (Gaber, 1992). Several genes which encode P-type Ca^{2+}-ATPase enzymes have now been identified in both *S. cerevisiae* and *Sch. pombe*.

The physiological and biotechnological significance of Ca^{2+} uptake in yeast lies in the multifunctional roles of this cation as a second messenger in modulation of growth and metabolic responses of cells to external stimuli. For example, in *Rhodosporidium toruloides* and *S. cerevisiae*, one of the responses of cells to mating pheromones is a rapid and transient uptake of Ca^{2+}. This suggests that Ca^{2+} mobilization may be involved in mating signal transduction mechanisms in yeasts (Iida *et al.*, 1990). In relation to *S. cerevisiae* cell division, Ca^{2+} ions have been linked to cell cycle regulation (Iida *et al.*, 1990; Anraku *et al.*, 1991) and have been implicated in the transition from lag phase to exponential phase in batch cultures of this yeast (Friis *et al.*, 1994). Controversy surrounding culture media requirements for Ca^{2+} in yeast growth and division (Youatt, 1993) has been highlighted recently by the finding that Mn^{2+} can effectively replace Ca^{2+} in modulating events leading to cell cycle progression in *S. cerevisiae* (Loukin and Kung, 1995).

Zinc ions

Trace levels of Zn^{2+} (in the pM range) are essential for yeast growth. For example, Zn^{2+} deprivation in *S. cerevisiae* prevents budding and arrests cells in the G_1 phase of the cell cycle, while in *Candida utilis*, Zn^{2+} maximally accumulates in the exponential phase of growth. Zn^{2+} requirements for the growth of yeasts cannot be met by other metal ions. Concerning roles in metabolism, Zn^{2+} is essential for the structure and function of many enzymes. For example, the important terminal step enzyme in yeast alcoholic fermentation, namely, alcohol dehydrogenase is a zinc-metalloenzyme. The biotechnological significance of this lies in the phenomenon of 'stuck' fermentations, occasionally encountered in the brewing industry and which may be ameliorated following appropriate supplementation of zinc salts.

Zn^{2+} uptake by yeast is energy-dependent and is driven by both the proton and potassium gradients across the cell membrane. Zhao and Eide (1996a,b) have demonstrated both high- and low-affinity Zn^{2+} transporters in *S. cerevisiae* which operate in Zn^{2+}-limiting and Zn^{2+}-replete conditions, respectively. The high-affinity Zn^{2+} transporter gene, *ZRT1*, has been molecularly characterized in *S. cerevisiae*, the first such characterization for any organism (Zhao and Eide, 1996a). *ZRT2* has been shown to encode a separate low-affinity Zn^{2+} transporter (K_T 10 μM) which is active in zinc-replete cells of *S. cerevisiae* (Zhao and Eide, 1996b). In addition to inwardly-directed Zn^{2+} transport, separate efflux carriers exist which play a role in intracellular Zn^{2+} detoxification mechanisms.

Copper and iron ions

Both copper and iron are essential nutrients for yeasts which act as cofactors in several enzymes, including the redox pigments of the respiratory chain. The assimilation of these two metals and their subsequent metabolism is closely interconnected in yeasts, as in other organisms. The study of yeast copper and iron transport is relevant to our fundamental understanding of several human disorders (e.g. Wilson's and Menkes' diseases) and to yeast pathogenesis (e.g. iron assimilation by *Candida albicans*).

Knowledge of strategies adopted by yeast to facilitate the uptake of insoluble ferric and cupric salts into the cell, together with subsequent maintenance of low cytosolic concentrations of ferrous and cuprous ions, has expanded dramatically in recent years. The various ways in which yeast cells regulate their transport of copper and iron will now be described and the significance of such regulation discussed in terms of more practical aspects of yeast physiology.

Copper is an essential micronutrient at low concentrations, but is toxic at high concentrations. Copper toxicity towards yeast cells involves intracellular interaction between copper and nucleic acids and enzymes. However, the major mode of action is disruption of plasma membrane integrity (Avery *et al.*, 1996). Copper ion homeostasis in yeast is controlled by several uptake, efflux and chelation strategies depending on the external bioavailability of copper. One mechanism relates to sequestration of copper by a copper–metallothionein protein, encoded by the *CUP1* gene. Such low molecular mass proteins are generally synthesized as a protective response to high levels of potentially toxic metal ions (Mehra and Winge, 1991). Up to 60% of cellular copper in *S. cerevisiae* can be in the form of copper–metallothionein (Butt and Ecker, 1987), and this protein therefore plays an important role in copper resistance in this yeast. The biotechnological significance of the *CUP1* gene in yeast is that it has a very efficient inducible promoter which has proved extremely useful in heterologous gene expression (see Chapter 6). In addition, transformation of normally genetically-intractable brewing yeasts with *CUP1* and selection in increasing concentrations of copper ions has proved very successful. Another aspect of yeast metallothioneins which has great potential in biotechnology is in microbial 'mining', especially in intracellular accumulation and subsequent recovery of precious metals. For example, gold may accumulate intracellularly in yeast by binding to synthesized metallothionein (Butt and Ecker, 1987).

In addition to the role of copper-inducible metallothioneins in chelating excess Cu^{2+}, other mechanisms of cellular export or intracellular (vacuolar) compartmentalization contribute to Cu^{2+} detoxification in yeast. Another way in which yeast cells prevent copper-induced toxicity is by facilitating copper biomineralization – a process which leads to accumulation of CuS at the cell surface and which turns the cells brown as a consequence. Yu *et al.* (1996) have now identified a gene in *S. cerevisiae* (*SLF1*) which controls this process. Copper accumulation by yeast through cell wall biosorption has potential in environmental biotechnology in bioremediation and biorecovery of copper from industrial wastewater (Brady and Duncan, 1994).

Concerning inwardly directed copper transport mechanisms, yeasts possess a plasma membrane protein, encoded on the *CTR 1* gene, which is required for high-affinity copper uptake (Dancis *et al.*, 1994), and which is only expressed when cells are grown in copper levels below 10 μM. This transporter may also be involved in Fe^{2+} uptake in yeast (Chang and Fink, 1994). A low-affinity copper transporter is also thought to exist in *S. cerevisiae*. Research into the molecular biology of copper transporters in yeast is pertinent to studies of certain human hereditary diseases associated with abnormal cellular copper homeostasis. For example, Wilson's disease results from defective copper efflux from the liver leading to cirrhosis, and Menkes' disease, a neurological degenerative disorder, leads to cellular deficiency of copper. Both diseases are characterized by genes defective in P-type ATPase-mediated copper transport. Biochemical and molecular genetic knowledge of copper transport being gained in yeast may therefore provide insight into the underlying regulation of copper homeostasis in humans.

With regard to iron uptake, some yeasts (but not *S. cerevisiae*) may secrete specific iron-chelating compounds called **siderophores** in response to growth under iron-limitation. These so-called ferri-siderophores may be reduced by extracellular reductases to yield the ferrous ion, which is then transported. For example, *Rhodotorula toruloides* secretes a hydroxamate siderophore called rhodotorulic acid to sequester iron and transfer it to the cell surface for subsequent translocation into the cell.

Yeasts have adopted additional strategies for converting insoluble ferric (Fe^{3+}) into biologically active and soluble ferrous (Fe^{2+}) ions. In *S. cerevisiae*, *Sch. pombe* and *Candida albicans*, this is accomplished by extracellular reduction by plasma membrane ferric reductase activity. In *S. cerevisiae*, this enzyme is encoded by *FRE1* (Dancis *et al.*, 1992) which is repressed by both iron and copper. *C. albicans* has also been shown to possess iron- and copper-regulated ferric reductase activity (Morrissey *et al.*, 1996). This may account for the way in which this opportunistic pathogen assimilates iron when systemically infecting the human body.

In *S. cerevisiae*, it is now recognized that several iron transporters exist in the cell membrane (Figure 3.4). A tightly regulated and specific high-affinity system, encoded by the *FET3* gene, and a low-affinity system, encoded by *FET4* gene exist in *S. cerevisiae* (Askwith *et al.*, 1996). The latter system is relatively non-specific for iron and will also transport cobalt, cadmium and nickel.

It is now thought that copper and iron exhibit regulatory interdependence with respect to their transport into yeast cells. The high-affinity copper transport protein, Ctr1, is not, however, a common transporter of both copper and iron. Ctr1 supplies Cu^{2+} to enable iron uptake via Fet3, a copper-dependent ferro-oxidase. Uptake of Fe^{2+} is not only dependent on extracellular reduction of ferric iron, but is also coupled to intracellular re-oxidation of ferrous iron. A model for copper-dependent iron uptake in yeast (Figure 3.5) has been proposed by Chang and Fink (1994).

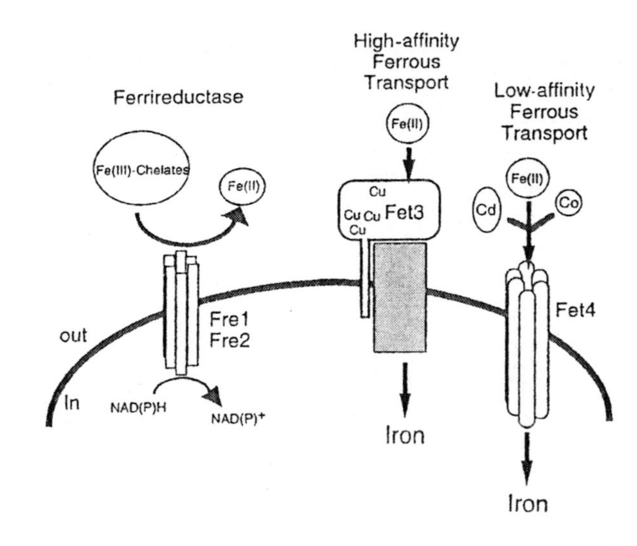

Figure 3.4. Elemental iron transport in *S. cerevisiae*. Fre1 or Fre2 are ferric reductases which convert ferric iron (usually present as ferric chelates) to ferrous iron. The *FRE1* and *FRE2* gene products convert NAD(P)H to NAD(P)+ in the process of reducing Fe(III) to Fe(II). The ferrous iron is then recognized by the elemental transport systems. The high-affinity transport system requires the action of the copper-containing protein Fet3 and probably a second molecule (the transporter). The low-affinity system is mediated by the *FET4* gene product which also may transport cadmium and cobalt. The manner in which the proteins are displayed in the figure is not meant to indicate actual physiological structure. Reproduced with permission from Askwith *et al.* (1996) and Blackwell Science Ltd.

Once inside the yeast cell, little is known of the distribution of iron, although storage in the vacuole and the synthesis of molecules akin to haemoglobin (Oshino *et al.*, 1973) and ferritin (Raguzzi *et al.*, 1988) have been reported in various yeasts. In addition, a yeast mitochondrial ferrochelatase is known which catalyses the insertion of ferrous iron into haemoproteins of the cell (Gora *et al.*, 1996).

3.3.2.12 Transport of Heavy Metals

Heavy metals are more commonly regarded as toxic to yeast (at concentrations >100 μM),

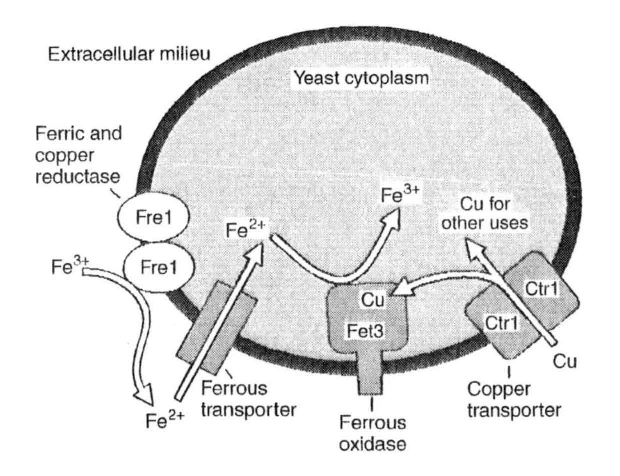

Figure 3.5. Model of copper and iron uptake in yeast. Reproduced with permission from Chang and Fink (1994) and Current Biology Ltd.

rather than acting as sources of nutrients. Nevertheless, *S. cerevisiae* has been shown, principally through passive cell wall biosorption, to bind heavy metals including: cadmium, caesium, cobalt, gold, lead, mercury, nickel, strontium and uranium. This yeast has also been shown to accumulate titanium intracellularly (Hegoczki *et al.*, 1995). Brewer's yeast is also a rich source of an organic form of chromium, although little is known about chromium utilization by yeast. Several yeasts have also been shown to synthesize cytoplasmic Cd^{2+}-binding metallothionein complexes which function in cadmium resistance (Inouhe *et al.*, 1996). Yeast cells may, therefore, through both surface absorption and intracellular sequestration, assist in removal of potentially toxic metals, including radionuclides, from contaminated aqueous process streams in industry (Gadd, 1993; Brady and Duncan, 1994; Blackwell *et al.*, 1995). Immobilization of yeasts in calcium alginate or polyacrylamide gels has distinct advantages over free cell systems in such bioremediation technology (Brady and Duncan, 1994).

3.4 SUMMARY

In general, yeasts have quite straightforward, minimal nutritional needs. Most species can grow perfectly well in the presence of simple carbon and nitrogen backbone compounds together with inorganic ions and a few growth factors. Yeasts absorb these nutrients through the cell membrane as low molecular weight compounds dissolved in water. Entry of water, which is essential for growth, is controlled at the level of the cell envelope.

Sugars are translocated into yeast cells by active transport or facilitated diffusion mechanisms which are controlled at the plasma membrane and which depend on the sugar, yeast species and growth conditions. For example, *S. cerevisiae* translocates glucose by facilitated diffusion and maltose by active transport. Active transport means that plasma membrane ATPases act as directional proton pumps in accordance with Peter Mitchell's chemiosmotic hypothesis (Mitchell, 1963). pH gradients thus drive nutrient transport either via proton symporters (as is the case with certain sugars and amino acids), or via proton antiporters (as is the case with certain inorganic ions, such as potassium). There is great diversity with regard to the physiological processes whereby yeasts transport sugars, even within a particular species (e.g. Kilian *et al.*, 1991). Yeast cells may also employ two different transport systems for the same sugar, depending on its availability. There is still a great deal to learn about precise mechanisms of sugar transport, not only in *S. cerevisiae*, but also in other exploitable yeasts.

Understanding the regulation of how yeasts acquire sugar and other nutrient solutes is important in biotechnology, since efficient transport relates to efficient yeast growth and metabolism. A coordination of physiological, biochemical and molecular genetic approaches to the study of yeast nutrition and yeast nutrient transport systems will undoubtedly yield the most meaningful information for yeast scientists and technologists.

3.5 REFERENCES

Aiking, H. and Tempest, D.W. (1976) Growth and physiology of *Candida utilis* NCYC 321 in potassium-limited chemostat culture. *Archives of Microbiology*, **108**, 117–124.

André, B. (1995) An overview of membrane transport proteins in *Saccharomyces cerevisiae*. *Yeast*, **11**, 1575–1611.

Anraku, Y., Ohya, Y. and Iida, H. (1991) Cell cycle control by calcium and calmodulin in *Saccharomyces cerevisiae*. *Biochimica et Biophysica Acta*, **1093**, 169–177.

Askwith, C.C., de Silva, D. and Kaplan, J. (1996) Molecular biology of iron acquisition in *Saccharomyces cerevisiae*. *Molecular Microbiology*, **20**, 27–34.

Auling, G (1994) Manganese: function and transport in fungi. In *Metal Ions in Fungi* (eds G. Winkleman and D. Winge), pp. 215–236. Marcel Dekker, New York.

Avery, S.V., Howlett, N.G. and Radice, S. (1996) Copper toxicity towards *Saccharomyces cerevisiae*: dependence on plasma membrane fatty acid composition. *Applied and Environmental Microbiology*, **62**, 3960–3966.

Barnett, J.A. (1981) The utilization of disaccharides and some other sugars by yeasts. *Advances in Carbohydrate Chemistry and Biochemistry*, **39**, 347–404.

Barth, G. and Gaillardin, C. (1996) *Yarrowia lipolytica*. In *Nonconventional Yeasts in Biotechnology. A Handbook* (ed. K. Wolf), pp. 313–358. Springer-Verlag, Berlin and Heidelberg.

Becker, J.M. and Naider, F. (1980) Transport and utilization of peptides by yeast. In *Microorganisms and Nitrogen Sources* (ed. J.W. Payne), pp. 257–379. J. Wiley, Chichester.

Beeler, T., Bruce, K. and Dunn, T. (1997) Regulation of cellular Mg^{2+} by *Saccharomyces cerevisiae*. *Biochimica et Biophysica Acta*, **1323**, 310–318.

Benito, B. and Lagunas, R. (1992) The low-affinity component of *Saccharomyces cerevisiae* maltose transport is an artefact. *Journal of Bacteriology*, **174**, 3065–3069.

Berry, D.R. (1988) Products of primary metabolic pathways. In *Physiology of Industrial Fungi* (ed. D.R. Berry), pp. 130–160. Blackwell Scientific Publications, Oxford.

Bisson, L.F. and Fraenkel, D.G. (1984) Expression of kinase-dependent glucose uptake in *Saccharomyces cerevisiae*. *Journal of Bacteriology*, **159**, 1013–1017.

Bisson, L.F., Coons, D.M., Kruckeburg, A.L. and Lewis, D.A. (1993) Yeast sugars transporters. *Critical Reviews in Biochemistry and Molecular Biology*, **28**, 259–308.

Blackwell, K.J., Singleton, I. and Tobin, J.M. (1995) Metal cation uptake by yeast: a review. *Applied Microbiology and Biotechnology*, **43**, 579–584.

Blackwell, K.J., Tobin, J.M. and Avery, S.V. (1997) Manganese uptake and toxicity in magnesium-supplemented and unsupplemented *Saccharomyces cerevisiae*. *Applied Microbiology and Biotechnology*, **47**, 180–184.

Boekhout, T. and Kurtzman, C.P. (1996) Principles and methods used in yeast classification and an overview of currently accepted yeast genera. In *Nonconventional Yeasts in Biotechnology. A Handbook* (ed. K. Wolf), pp. 1–81. Springer-Verlag, Berlin, Heidelberg and New York.

Borst-Pauwels, G.W.F.H. (1981) Ion transport in yeast. *Biochimica et Biophysica Acta*, **650**, 88–127.

Borst-Pauwels, G.W.F.H. and Peters, P.H.J. (1987) Phosphate uptake in *Saccharomyces cerevisiae*. In *Phosphate Metabolism and Cellular Regulation in Microorganisms* (eds A. Torriani-Gorini, F.G. Rothman, S. Silver, A. Wright and E. Yagil), pp. 205–209. American Society of Microbiology, Washington.

Boulton, C.A. and Quain, D.E. (1987) Yeast, oxygen and the control of brewery fermentations. European Brewery Convention. Proceedings of the 21st Congress, Madrid 1987, pp. 401–408. IRL Press at Oxford University Press, Oxford, UK.

Bourdot, S. and Karst, F. (1995) Isolation and characterization of the *Saccharomyces cerevisiae SUT1* gene involved in sterol uptake. *Gene*, **165**, 97–102.

Brady, D. and Duncan, J.R. (1994) Bioaccumulation of metal cations by *Saccharomyces cerevisiae*. *Applied Microbiology and Biotechnology*, **41**, 149–154.

Breton, A. and Surdin-Kerjan, Y. (1977) Sulphate uptake in *S. cerevisiae*: biochemical and genetic study. *Journal of Bacteriology*, **132**, 224–232.

Bull, A.T. and Bushell, M.E. (1976) Environmental control of fungal growth. In *The Filamentous Fungi, Vol. 2* (eds J.E. Smith and D.R. Berry), pp. 1–31. J Wiley, New York.

Bun-Ya, M., Nishimura, M., Harashima, S. and Oshima, Y (1991) The *PHO84* gene of *Saccharomyces cerevisiae* encodes an inorganic phosphate transporter. *Molecular and Cellular Biology*, **11**, 3229–3238.

Bun-Ya, M., Shikata, K., Nakade, S., Yompakdee, C., Harashimi, S. and Oshima, Y. (1996) Two new genes, *PHO86* and *PHO87*, involved in inorganic phosphate uptake in *Saccharomyces cerevisiae*. *Current Genetics*, **29**, 344–351.

Butt, T.R. and Ecker, D.J. (1987) Yeast metallothionein and its applications in biotechnology. *Microbiological Reviews*, **15**, 351–364.

Campbell, I. (1988) Standard media for cultivation of

yeasts. In *Yeast. A Practical Approach* (eds I. Campbell and J.H. Duffus), pp. 277–279. IRL Press, Oxford and Washington.
Campbell, I. (1996) Wild yeasts in brewing and distilling. In *Brewing Microbiology*, 2nd edn (eds F.G. Priest and I. Campbell), pp. 193–203. Chapman & Hall, London.
Cartwright, C.P., Rose, A.H., Calderbank, J. and Keenan, M.H.J. (1989) Solute Transport. In *The Yeasts, Vol. 3*, 2nd edn (eds A.H. Rose and J.S. Harrison), pp. 5–56. Academic Press, London.
Castrillo, J.I., De Miguel, I. and Ugalde, U.O. (1995) Proton production and consumption pathways in yeast metabolism. *Yeast*, **11**, 1353–1365.
Chandrasena, G., Walker, G.M. and Staines, H. (1997) Use of response surfaces to investigate metal ion interactions in yeast fermentations. *Journal of the American Society of Brewing Chemists*, **55**, 24–28.
Chang, A. and Fink, G.R. (1994) The copper-ion connection. *Current Biology*, **4**, 532–533.
Cid, A., Gancedo, C. and Lagunas, R. (1987) Inactivation of the glucose transporting system in sporulating yeast. *FEMS Microbiology Letters*, **41**, 59–61.
Cimprich, P., Slavik, J. and Kotyk, A. (1995) Distribution of individual pH values in a population of the yeast *Saccharomyces cerevisiae*. *FEMS Microbiology Letters*, **130**, 245–252.
Cooper, T.G. (1982) Transport in *Saccharomyces cerevisiae*. In *The Molecular Biology of the Yeast Saccharomyces. Metabolism and Gene Expression* (eds J.N. Strathern, E.W. Jones and J.R. Broach), pp. 399–461. Cold Spring Harbor Press, Cold Spring Harbor, New York.
Crowe, J.H., Panek, A.D., Crowe, L.M., Panek, A.M. and Soares-de Araujo, P. (1991) Trehalose transport in yeast cells. *Biochemistry International*, **24**, 721–730.
Crumplen, R.M., Slaughter, J.C. and Stewart, G.G. (1996) Characteristics of maltose transporter activity in an ale and lager strain of the yeast *Saccharomyces cerevisiae*. *Letters in Applied Microbiology*, **23**, 448–452.
Dancis, A., Roman, D.G., Anderson, G.J., Hinnebusch, A.G. and Klausner, R.D. (1992) Ferric reductase of *Saccharomyces cerevisiae*: molecular characterization, role in iron uptake and transcriptional control by iron. *Proceedings of the National Academy of Sciences* (*USA*), **89**, 3869–3873.
Dancis, A., Yuan, D.S., Haile, D., Askwith, C., Eide, D., Moehle, C., Kaplan, J. and Klausner, R. (1994) Molecular characterization of a copper transport protein in *S. cerevisiae*: an unexpected role for copper in iron transport. *Cell*, **76**, 393–402.
Davenport, R.R. (1980) An outline guide to media and methods for studying yeasts and yeast-like organisms. In *Biology and Activities of Yeasts* (eds F.A. Skinner, S.M. Passmore and R.R. Davenport), pp. 261–278. Academic Press, London.
Deák, T. (1978) On the existence of H^+- symport in yeasts. *Archives of Microbiology*, **1978**, 2005–2011.
De Bruijne, A.W., Schuddemat, J., Van den Broek, P.J.A. and Van Steveninck, J. (1988) Regulation of sugar transport systems of *Kluyveromyces marxianus*: the role of carbohydrates and their catabolism. *Biochimica et Biophysica Acta*, **939**, 569–576.
De Kerchove d'Exaerde, A., Supply, P. and Goffeau, A. (1996) Review: subcellular traffic of the plasma membrane H^+- ATPase in *Saccharomyces cerevisiae*. *Yeast*, **12**, 907–916.
De Juan, C. and Lagunas, R. (1986) Inactivation of the galactose transport system in *Saccharomyces cerevisiae*. *FEBS Letters*, **207**, 258–261.
De Nobel, J.G. and Barnett, J.A. (1991) Passage of molecules through yeast cell walls. A brief essay review. *Yeast*, **7**, 313–324.
De Nobel, J.G. Klis, F.M., Ram, A., Van Unen, H., Priem, J., Munnik, T. and Van den Ende, H. (1991) Cyclic variations in the permeability of the cell wall of *Saccharomyces cerevisiae*. *Yeast*, **7**, 589–598.
Didion, T., Grauslund, M., Kielland-Brandt, M.C. and Anderson, H.A. (1996) Amino acids induce expression of *BAP2*, a branched-chain amino acid permease gene in *Saccharomyces cerevisiae*. *Journal of Bacteriology*, **178**, 2025–2029.
Does, A.L. and Bisson, L.F. (1989) Comparison of glucose uptake kinetics in different yeasts. *Journal of Bacteriology*, **171**, 1303–1308.
Dohmen, R.J. and Hollenberg, C.P. (1996) *Schwanniomyces occidentalis*. In *Nonconventional Yeasts in Biotechnology. A Handbook* (ed. K. Wolf), pp. 117–137. Springer-Verlag, Berlin, Heidelberg.
Eddy, A.A. (1980) Some aspects of amino acid transport in yeast. In *Microorganisms and Nitrogen Sources* (ed. J.W. Payne), pp. 35–62. J. Wiley and Sons Ltd., Chichester.
Eddy, A.A. and Hopkins, P.G. (1985) The putative electrogenic nitrate-proton symport of the yeast *Candida utilis*. *Biochemical Journal*, **231**, 291–297.
Eilam, Y. (1982) Studies on calcium efflux in the yeast *Saccharomyces cerevisiae*. *Microbios*, **35**, 99–110.
Eleutherio, E.C.A., Araujo, P.S. and Panek, A.D. (1993) Role of trehalose carrier in dehydration resistance of *Saccharomyces cerevisiae*. *Biochimica et Biophysica Acta*, **1156**, 263–266.
Elskens, M.T., Jaspers, C.J. and Penninckx, M.J. (1991) Glutathione as an endogenous sulphur source in the yeast *Saccharomyces cerevisiae*. *Journal of General Microbiology*, **137**, 637–644.
Entian, K.D. and Loureiro-Dias, M.C. (1990) Misregulation of maltose uptake in a glucose defective mutant of *Saccharomyces cerevisiae* leads to glucose

poisoning. *Journal of General Microbiology*, **36**, 855–860.

Ferrando, A., Kron, S.J., Rios, G., Fink, G.R. and Serrano, R. (1995) Regulation of cation transport in *Saccharomyces cerevisiae* by the salt tolerance gene *HAL3*. *Molecular and Cellular Biology*, **15**, 5470–5481.

Friis, J., Szablewski, L., Christensen, S.T., Schousboe, P. and Rasmussen, L. (1994) Physiological studies on the effect of Ca^{2+} on the duration of the lag phase of *Saccharomyces cerevisiae*. *FEMS Microbiology Letters*, **123**, 33–36.

Fuhrmann, G.F. and Rothstein, A. (1968) The active transport of Zn^{2+}, Co^{2+} and Ni^{2+} into yeast cells. *Biochimica et Biophysica Acta*, **163**, 325–330.

Fuhrmann, G. and Völker, B. (1992) Regulation of glucose transport in *Saccharomyces cerevisiae*. *Journal of Biotechnology*, **27**, 1–15.

Gaber, R.F. (1992) Molecular genetics of yeast ion transport. *International Review of Cytology*, **137**, 299–353.

Gadd, G.M. (1993) Interactions of fungi with toxic metals. *New Phytologist*, **124**, 25–60

Gadd, G.M. and Laurence, O.S. (1996) Demonstration of high affinity Mn^{2+} uptake in *Saccharomyces cerevisiae* – specificity and kinetics. *Microbiology (UK)*, **142**, 1159–1167.

Gamo, F.J., Moreno, E. and Lagunas, R. (1995) The low-affinity component of the glucose transport system in *Saccharomyces cerevisiae* is not due to passive diffusion. *Yeast*, **11**, 1393–1398.

Gaxiola, R., Corona, M. and Zinker, S. (1996) A halotolerant mutant of *Saccharomyces cerevisiae*. *Journal of Bacteriology*, **178**, 2978–2981.

Gildenhuys, P.T. and Slaughter, J.C. (1983) The metabolism of putrescine, spermidine and spermine by yeast in relation to the availability of magnesium. *Journal of the Institute of Brewing*, **89**, 333–340.

Gimeno, C.J., Ljungdahl, P.O., Styles, C.A. and Fink, G.R. (1992) Unipolar cell divisions in the yeast *S. cerevisiae* lead to filamentous growth: regulation by starvation and RAS. *Cell*, **68**, 1077–1090.

Gora, M., Grybowska, E., Rytka, J. and Labbe-Bois, R. (1996) Probing the active site residues in *Saccharomyces cerevisiae* ferrochelatase by directed mutagenesis. *In vivo* and *in vitro* analyses. *Journal of Biological Chemistry*, **271**, 1810–1816.

Griffin, D.H. (1994) *Fungal Physiology*, 2nd edn. Wiley-Liss Inc., New York.

Grimes, H.D. and Overvoorde, P.J. (1996) Functional characterization of sucrose binding protein-mediated sucrose uptake in yeast. *Journal of Experimental Botany*, **47**, 1217–1222.

Gustin, M.C., Martinac, B., Saimi, Y., Culbertson, M.R. and Kung, C. (1986) Ion channels in yeast. *Science*, **233**, 1995–1197.

Gustin, M.C. Zhou, X.l., Martinac, B. and Kung, C. (1988) A mechanosensitive ion channel in the yeast plasma membrane. *Science*, **242**, 762–765.

Hansen, H. and Hollenberg, C.P. (1996) *Hansenula polymorpha* (*Pichia angusta*). In *Nonconventional Yeasts in Biotechnology. A Handbook* (ed. K. Wolf), pp. 293–311. Springer-Verlag, Berlin and Heidelberg.

Hapala, I. and Hunakova, A. (1996) Mechanism of sterol uptake in yeast. *Folia Microbiologia*, **41**, 95–96.

Hegoczki, J. Janzso, B. and Suhajda, A. (1995) Preparation of titanium enriched *Saccharomyces cerevisiae*. *Acta Alimentaria*, **24**, 181–190.

Hipkin, C.R. (1989) Nitrate assimilation in yeast. In *Molecular and Genetic Aspects of Nitrate Assimilation* (eds J.L. Wray and J.R. Kinghorn), pp. 51–68. Oxford Science Publications, Oxford.

Hohmann, S., Bell, W., Neves, M.J., Valckx, D. and Thevelein, J.M. (1996) Evidence for trehalose-6-phosphate-dependent and -independent mechanisms in the control of sugar influx into yeast glycolysis. *Molecular Microbiology*, **20**, 981–991.

Horak, J. (1997) Yeast nutrient transporters. *Biochimica et Biophysica Acta*, **1331**, 41–79.

Hough, J.S., Briggs, D.E, Stevens, R and Young, T.W. (1982) *Malting and Brewing Science. Vol. II Hopped Wort and Beer*. Chapman & Hall, London.

Hsu, N.P., Vogt, A. and Bernstein, L. (1980) Yeast nutrients and beer quality. *Master Brewers Association of the Americas Technical Quarterly*, **17**, 85–88.

Hughes, M.N. and Poole, R.K. (1991) Metal speciation and microbial growth – the hard (and soft) facts. *Journal of General Microbiology*, **137**, 725–734.

Iida, H., Yagawa, Y. and Anraku, Y. (1990) Essential role for induced Ca^{2+} influx followed by $[Ca^{2+}]_i$ rise in maintaining viability of yeast cells later in the mating pheromone response pathway. *Journal of Biological Chemistry*, **265**, 13391–1399.

Iida, H., Sakaguchi, S., Yagawa, Y. and Anraku, Y. (1990) Cell cycle control by Ca^{2+} in *Saccharomyces cerevisiae*. *Journal of Biological Chemistry*, **265**, 21216–21222.

Inouhe, M., Sumiyoshi, M. Tohoyama, H. and Joho, M. (1996) Resistance to cadmium ions and formation of a cadmium-binding complex in various wild-type yeast cells. *Plant Cell Physiology*, **37**, 341–346.

Isnard, A-D., Thomas, D. and Surdin-Kerjan, Y. (1996) The study of methionine uptake in *Saccharomyces cerevisiae* reveals a new family of amino acid permeases. *Journal of Molecular Biology*, **262**, 473–484.

Jennings, D.M. (1995) *The Physiology of Fungal Nutrition*. Cambridge University Press, Cambridge.

Jin, Y.H., Jang, Y.K., Kim, M.J., Rad, M.R., Kirchrath, L., Seong, R.H., Hong, S.H., Hollenberg, C.P. and Park, S.D. (1995). Characterization of *SFP2*, a putative sulphate permease gene of *S. cerevisiae*. *Biochemical and Biophysical Research Communications*, **214**, 709–715.

Johnston, J.H. and Barford, J.P. (1991) Continuous growth of *Saccharomyces cerevisiae* on a mixture of glucose and fructose. *Journal of General and Applied Microbiology*, **37**, 133–140.

Jones, M. and Pierce, J.S. (1964) Absorption of amino acids from wort by yeasts. *Journal of the Institute of Brewing*, **70**, 307–315.

Jones, R.P. and Gadd, G.M. (1990) Ionic nutrition of yeast – physiological mechanisms involved and implications for biotechnology. *Enzyme and Microbial Technology*, **12**, 402–418.

Jones, R.P. and Greenfield, P.F. (1984) A review of yeast ionic nutrition. I. Growth and fermentation requirements. *Process Biochemistry*, **April**, 48–60.

Kamada, Y., Jung, U.S., Piotrowski, J. and Levin, D.E. (1995) The protein kinase C-activated MAP kinase pathway of *Saccharomyces cerevisiae* mediates a novel aspect of the heat shock response. *Genes and Development*, **9**, 1559–1571.

Kessler, G.A., Laster, S.M. and Parks, L.W. (1992) A defect in the sterol : steryl ester interconversion in a mutant of the yeast, *Saccharomyces cerevisiae*. *Biochimica et Biophysica Acta*, **1123**, 127–132.

Kihn, J.C., Dassargues, C.M. and Mestdagh, M.M. (1988) Preliminary ESR study of Mn (II) retention by the yeast *Saccharomyces*. *Canadian Journal of Microbiology*, **34**, 1230–1234.

Kilian, S.G., van Deemter, A., Kock, J.L.F. and du Preez, J.C. (1991) Occurrence and taxonomic aspects of proton movements coupled to sugar transport in the yeast genus *Kluyveromyces*. *Antonie van Leeuwenhoek*, **59**, 199–206.

Kilian, S.G., Prior, B.A. and du Preez, J.C. (1993) The kinetics and regulation of D-xylose transport in *Candida utilis*. *World Journal of Microbiology and Biotechnology*, **9**, 356–360.

Kim, J. Alizadeh, P., Harding, T. Hefner-Gravink, A. and Klionsky, D.J. (1996) Disruption of the yeast *ATH1* gene confers better survival after dehydration, freezing and ethanol shock – potential commercial applications. *Applied and Environmental Microbiology*, **62**, 1563–1569.

Kohlwein, S.D. and Paltauf, F. (1983) Uptake of fatty acids by the yeasts, *Saccharomyces uvarum* and *Saccharomycopsis lipolytica*. *Biochimica et Biophysica Acta*, **792**, 310–317.

Koser, S. A. (1968) *Vitamin Requirements of Bacteria and Yeasts*. Charles C. Thomas, Springfield, Illinois, USA.

Kruckeberg, A.L. (1996) The hexose transporter family of *Saccharomyces cerevisiae*. *Archives of Microbiology*, **166**, 283–292.

Lachance, M.-A. (1990) Yeast selection in nature. In *Yeast Strain Selection* (ed. C. Panchal), pp. 24–41. Marcel Dekker Inc., New York.

Lagunas, R. (1993) Sugar transport in *Saccharomyces cerevisiae*. *FEMS Microbiology Reviews*, **104**, 229–242.

Lapinskas, S.J., Lin, S-J. and Culotta, V.C. (1996) The role of the *Saccharomyces cerevisiae CCC1* gene in the homeostasis of manganese ions. *Molecular Microbiology*, **21**, 519–528.

Leão, C. and Van Uden, N. (1986) Transport of lactate and other short-chain monocarboxylates in the yeast *Candida utilis*. *Applied Microbiology and Biotechnology*, **23**, 389–393.

Lin, Y. (1975) Detection of wild yeasts in the brewery. Efficiency of differential media. *Journal of the Institute of Brewing*, **81**, 410–417.

Loukin, S. and Kung, C. (1995) Manganese effectively supports yeast cell cycle progression in place of calcium. *Journal of Cell Biology*, **131**, 1025–1037.

Loureiro-Dias, M.C. (1988) Movement of protons to glucose transport in yeasts. A comparative study among 248 yeast strains. *Antonie van Leeuwenhoek*, **54**, 331–343.

Lucas, C., da Costa, M. and Van Uden, N. (1990) Osmoregulatory active sodium-glycerol co-transport in the halotolerant yeast *Debaryomyces hansenii*. *Yeast*, **6**, 187–191.

Luyten, K., Albertyn, J., Skibbe, W.F., Prior, B.A., Ramos, J., Thevelein, J.M. and Hohmann, S. (1995) *Fps1*, a yeast member of the *MIP* family of channel proteins is a facilitator of glycerol uptake and efflux and is inactive under osmotic stress. *EMBO Journal*, **14**, 1360–1371.

Mattey, M. (1992) The production of organic acids. *Critical Reviews in Biotechnology*, **12**, 87–132.

Matthews, T.M. and Webb, C. (1991) Culture systems. In *Saccharomyces* (eds M.F. Tuite and S.G. Oliver), pp. 249–282. Plenum Press, New York.

Mauersberger, S., Ohkuma, M. Schunck, W.H. and Takagi, M. (1996) *Candida maltosa*. In *Nonconventional Yeasts in Biotechnology. A Handbook* (ed. K. Wolf), pp. 411–580. Springer-Verlag, Berlin and Heidelberg.

Mehra, P.K. and Winge, D.R. (1991) Metal ion resistance in fungi – molecular mechanisms and their regulated expression. *Journal of Cellular Biochemistry*, **45**, 30–40.

Meikle, A.J., Reed, R.J. and Gadd, G.M. (1988) Osmotic adjustment and the accumulation of organic solutes in whole cells and protoplasts of *Saccharomyces cerevisiae*. *Journal of General Microbiology*, **134**, 3049–3060.

Michaljanicova, D., Hodan, J. and Kotyk, A. (1982) Maltotriose transport and utilization in baker's and brewer's yeast. *Folia Microbiologia*, **27**, 217–221.

Mitchell, P. (1963) Molecule, group and electron translocation through natural membranes. *Biochemical Society Symposia*, **22**, 142–169.

Mitchison, J.M. (1970) Physiological and cytological methods for *Schizosaccharomyces pombe*. *Methods in Cell Physiology*, **4**, 131–165.

Morrissey, J.E., Williams, P.H. and Cashmore, A.M. (1996) *Candida albicans* has a cell-associated ferric reductase activity which is regulated in response to lack of iron and copper. *Microbiology (UK)*, **142**, 485–492.

Nakajima-Shimada, J., Iida, H., Tsuji, F.I. and Anraku, Y. (1991) Monitoring of intracellular calcium in *Saccharomyces cerevisiae* with an apoaequorin cDNA expression system. *Proceedings of the National Academy of Sciences (USA)*, **88**, 6878–6884.

Okorokov, L.A., Lichko, L.P., Kadomtseva, V.M., Kholodenko, V.P. and Kulaev, I.S. (1974) Metabolism and physicochemical state of Mg^{2+} ions in fungi. *Microbiologiya*, **43**, 410–416.

Okorokov, L.A., Lichko, L.P. and Kulaev, I.S. (1980) Vacuoles: main compartments of potassium, magnesium and phosphate ions in *Saccharomyces carlsbergensis* cells. *Journal of Bacteriology*, **144**, 661–665.

Okorokov, L. A., Kulakovskaya, T.V., Lichko, L.P. and Polorotova, E.V. (1985) H^+/ion antiport as the principal mechanism of transport systems in the vacuolar membrane of the yeast *Saccharomyces cerevisiae*. *FEBS Letters*, **192**, 303–306.

Okorokov, L.A., Kuranov, A.J., Kuranova, E.V. and dos Santos Silva, R. (1996) Ca^{2+}-transporting ATPase(s) of the reticulum type in intracellular membranes of *Saccharomyces cerevisiae*: biochemical identification. *FEMS Microbiology Letters*, **146**, 39–46.

Oshino, R. Oshino, N., Chance, B. and Hagihara, B. (1973) Studies on yeast hemoglobin. The properties of yeast hemoglobin and its physiological function in the cell. *European Journal of Biochemistry*, **35**, 23–33.

Overvoorde, P.J., Frommer, W.B. and Grimes, H.D. (1996) A soybean sucrose binding protein independently mediates nonsaturable sucrose uptake in yeast. *The Plant Cell*, **8**, 271–280.

Özcan, S. and Johnston, M. (1995) Three different regulatory mechanisms enable yeast hexose transport (*HXT*) genes to be induced by different levels of glucose. *Molecular and Cellular Biology*, **15**, 1564–1572.

Özcan, S., Dover, J., Rosenwold, A.G., Wolfl, S. and Johnston, M. (1996) Two glucose transporters in *Saccharomyces cerevisiae* are glucose sensors that generate a signal for induction of gene expression. *Proceedings of the National Academy of Sciences (USA)*, **93**, 12428–12432.

Parkin, M.J. and Ross, I.S. (1986) The specific uptake of manganese in the yeast *Candida utilis*. *Journal of General Microbiology*, **132**, 2155–2160.

Perkins, J. and Gadd, G.M. (1996) Interactions of Cs^+ and other monovalent cations (Li^+, Na^+, K^+, Rb^+, NH_4^+) with K^+-dependent pyruvate kinase and malate dehydrogenase from the yeasts *Rhodotorula rubra* and *Saccharomyces cerevisiae*. *Mycological Research*, **100**, 449–454.

Postma, E., Kuiper, A., Tomasouw, W.F., Scheffers, W.A. and Van Dijken, J.P. (1989) Competition for glucose between the yeasts *Saccharomyces cerevisiae* and *Candida utilis*. *Applied and Environmental Microbiology*, **55**, 3214–3220.

Postma, E., Verduyn, C., Kuiper, A., Scheffers, W.A. and Van Dijken, J.P. (1990) Substrate-accelerated death of *Saccharomyces cerevisiae* CBS 8066 under maltose stress. *Yeast*, **6**, 149–158.

Prior, C., Potier, S., Souciet, J-L. and Sychrova, H. (1996) Characterization of the *NHA1* gene encoding a Na^+/H^+-antiporter of the yeast *Saccharomyces cerevisiae*. *FEBS Letters*, **387**, 89–93.

Raguzzi, F., Lesuisse, E. and Crichton, R.R. (1988) Iron storage in *Saccharomyces cerevisiae FEBS Letters*, **231**, 253–258.

Ramos, J., Szkutnicka, K. and Cirillo, V.P. (1989) Characteristics of galactose transport in *Saccharomyces cerevisiae* cells and reconstituted lipid vesicles. *Journal of Bacteriology*, **171**, 3539–3544.

Ramos, J., Haro, R. and Rodriguez-Navarro, A. (1990) Regulation of potassium fluxes in *Saccharomyces cerevisiae*. *Biochimica et Biophysica Acta*, **1029**, 211–217.

Rao, R. and Slayman, C.W. (1996) Plasma-membrane and related ATPases. In *The Mycota III. Biochemistry and Molecular Biology* (eds. R. Brambl and G.A. Marzluf), pp. 3–28. Springer-Verlag, Berlin and Heidelberg.

Reid, J.D., Lukas, W., Shafaatian, R., Bertl, A., Scheurmann-Kettner, C., Guy, H.R. and North, R.A. (1996) The *Saccharomyces cerevisiae* outwardly-rectifying potassium channel (*DUK1*) identifies a new family of channels with duplicated pore domains. *Receptors and Channels*, **4**, 51–62.

Reifenberger, E., Freidel, K. and Ciriacy, M. (1995) Identification of novel *HXT* genes in *Saccharomyces cerevisiae* reveals the impact of individual hexose transporters on glycolytic flux. *Molecular Microbiology*, **16**, 157–167.

Reiser, T., Ochsner, U.A., Kalin, M., Glumoff, V. and Fiechter, A. (1996) *Trichosporon*. In *Nonconventional Yeasts in Biotechnology. A Handbook* (ed. K. Wolf), pp. 581–606. Springer-Verlag, Berlin and Heidelberg.

Riballo, E., Herweijer, M., Wolf, D.H. and Lagunas, R. (1995) Catabolite inactivation of the yeast maltose transporter occurs in the vacuole after internalization by endocytosis. *Journal of Bacteriology*, **177**, 5622–5627.

Riezman, H. (1993) Yeast endocytosis. *Trends in Cell Biology*, **3**, 273–277.

Rodriguez de Sousa, H., Madeira-Lopes, A. and Spencer-Martins, I. (1995) The significance of active fructose transport and maximum temperature for growth in the taxonomy of *Saccharomyces sensu stricto*. *Systematic and Applied Microbiology*, **18**, 44–51.

Rodriguez-Navarro, A. and Ortega, M.D. (1982) The mechanisms of sodium efflux in yeast. *FEBS Letters*, **138**, 205–208.

Romano, A.H. (1986) Sugar transport systems of bakers' yeast and filamentous fungi. In *Carbohydrate Metabolism in Cultured Cells* (ed. M.J. Morgan), pp. 225–244. Plenum Publishing Corp., New York.

Rose, A.H. (1976) *Chemical Microbiology*. 3rd edn. Plenum Press, New York.

Rose, A.H. (1989) Transport and metabolism of sulphur dioxide in yeasts and filamentous fungi. In *Nitrogen, Phosphorus and Sulphur Utilization by Fungi* (eds. L. Boddy, R. Marchant and D.J. Reed), pp. 59–70. Cambridge University Press, Cambridge.

Russell, I., Jones, R and Stewart, G.G. (1987) Yeast – the primary industrial microorganism. In *Biological Research on Industrial Yeasts* (eds G.G. Stewart, I. Russell, R.D. Klein and R.R. Hiebsch), pp. 1–20. CRC Press, Boca Raton.

Santos, E., Rodriguez, L., Elorza, M.V. and Sentandreu, R. (1982) Uptake of sucrose by *Saccharomyces cerevisiae*. *Archives of Biochemistry and Biophysics*, **216**, 652–660.

Scherrer, R., Louden, L. and Gerhardt, P. (1974) Porosity of the yeast cell wall and membrane. *Journal of Bacteriology*, **118**, 534–540.

Schulze, V. Liden, G. and Villadsen, J (1996) Dynamics of ammonia uptake in nitrogen-limited anaerobic cultures of *Saccharomyces cerevisiae*. *Journal of Biotechnology*, **46**, 33–42.

Schulze, V., Liden, G., Nielson, J. and Villadsen, J. (1996) Physiological effects of nitrogen starvation in an anaerobic batch culture of *Saccharomyces cerevisiae*. *Microbiology* (*UK*), **142**, 2299–2310.

Serrano, R. (1989) Structure and function of the plasma membrane ATPase. *Annual Reviews of Plant Physiology and Plant Molecular Biology*, **40**, 61–94.

Sigler, K. and Höfer, M. (1991) Mechanism of acid extrusion in yeast. *Biochimica et Biophysica Acta*, **1071**, 375–391.

Silljé, H.H.W., ter Schure, E.G., Verkleij, A.J. Boonstra, J. and Verrips, C.T. (1996) The Cdc25 protein of *Saccharomyces cerevisiae* is required for normal glucose transport. *Microbiology* (*UK*), **142**, 1765–1773.

Silver, S. (1995) Transport of inorganic cations. In *Escherichia coli* and *Salmonella typhimurium*: Cellular and Molecular Biology. 2nd edn (ed. F.C. Neidhardt), Chapter 72, pp. 1091–1102. American Society for Microbiology, Washington, DC.

Singh, A. and Mishra, P. (1995) Microbial pentose utilization. Current applications in biotechnology. *Progress in Industrial Microbiology*. Vol. **33**. Elsevier, Amsterdam.

Smith, D.L. and Maguire, M.E. (1993) Molecular aspects of Mg^{2+} transport systems. *Mineral Electrolyte Metabolism*, **19**, 266–276.

Smith, F.W., Hawkesford, M.J., Prosser, I.M. and Clarkson, D.T. (1995) Isolation of cDNA from *S. cerevisiae* that encodes a high-affinity sulfate transporter at the plasma membrane. *Molecular and General Genetics*, **247**, 709–715.

Spencer-Martins, I. (1994) Transport of sugars in yeasts: implications in the fermentation of lignocellulosic materials. *Bioresource Technology*, **50**, 51–57.

Spencer, J.F.T. and Spencer, D.M. (1996) Maintenance and culture of yeasts. In *Methods in Molecular Biology. Vol. 53: Yeast Protocols* (ed. I. H. Evans), pp. 5–16. Humana Press, Totowa, NJ, USA.

Sreekrishna, K. and Dickson, R.C. (1985) Construction of strains of *Saccharomyces cerevisiae* that grow on lactose. *Proceedings of the National Academy of Sciences* (*USA*), **82**, 7909–7913.

Sreekrishna, K. and Kropp, K.E. (1996) *Pichia pastoris*. In *Nonconventional Yeasts in Biotechnology. A Handbook* (ed. K. Wolf), pp. 203–253. Springer-Verlag, Berlin and Heidelberg.

Stambuk, B.U., de Araujo, P.S., Panek, A.D. and Serrano, R. (1996) Kinetics and energetics of trehalose transport in *Saccharomyces cerevisiae*. *European Journal of Biochemistry*, **237**, 876–881.

Stoppani, A.O.M and Ramos, E.J. (1978) Amino acid transport in yeasts. In *Biochemistry and Genetics of Yeasts* (eds M. Bacilla, B.L. Horecker and A.O.M. Stoppani), pp. 171–196 Academic Press, New York.

Stratford, M. and Rose, A.H. (1985) Hydrogen sulphide formation from sulphide by *S. cerevisiae*. *Journal of General Microbiology*, **131**, 1417–1424.

Subden, R.E. and Osothsilp, C. (1987) Malic acid metabolism in wine yeasts. In *Biological Research on Industrial Yeasts, Vol. II* (eds G.G. Stewart, I. Russell, R.D. Klein and R.R. Hiebsch), pp. 67–76. CRC Press Inc., Boca Raton, Florida.

Supek, F., Supekova, L. Nelson, H. and Nelson, N. (1996) A yeast manganese transporter related to the macrophage protein involved in conferring resistance to mycobacteria. *Proceedings of the National Academy of Sciences* (*USA*), **93**, 5105–5110.

Susta, J., Hodan, J., Opekarova, M. and Sigler, K. (1984) A simple method for determining the metabolic activity of brewer's yeast during the brewing process. *Food Microbiology*, **1**, 169–171.

Tao, T., Snavely, M.D., Farr, S.G. and Maguire, M.E. (1995) Magnesium transport in *Salmonella typhimurium*: *mgtA* encodes a P-type ATPase and is regulated by Mg^{2+} in a manner similar to that of the *mgtB* P-type ATPase. *Journal of Bacteriology*, **177**, 2654–2662.

Ter Schure, E.G., Silljé, H.H.W., Verkleij, A.J., Boonstra, J. and Verrips, C.T. (1995) The concentration of ammonia regulates nitrogen metabolism in *Saccharomyces cerevisiae*. *Journal of Bacteriology*, **177**, 6672–6675.

Theobald, U., Mohns, J. and Rizzi, M. (1996a) Dynamics of orthophosphate in yeast cytoplasm. *Biotechnology Letters*, **18**, 461–466.

Theobald, U., Mohns, J. and Rizzi, M. (1996b) Determination of *in vivo* cytoplasmic orthophosphate concentration in yeast. *Biotechnology Techniques*, **10**, 297–302.

Thevelein, J.M. (1991) Fermentable sugars and intracellular acidification as specific activators of the RAS-adenylate cyclase signalling pathway in yeast: the relationships to nutrient-induced cell cycle control. *Molecular Microbiology*, **5**, 1301–1307.

Thevelein, J.M. (1994) Signal transduction in yeast. *Yeast*, **10**, 1753–1790.

Thevelein, J.M. (1996) Regulation of trehalose metabolism and its relevance to cell growth and function. In *The Mycota III* (eds R. Brambl and G.A. Marzluf), pp. 395–420. Springer-Verlag, Berlin and Heidelberg.

Thevelein, J.M. and Hohmann, S. (1995) Trehalose synthase: guard to the gate of glycolysis in yeast? *Trends in Biochemical Sciences*, **20**, 3–10.

Tornai-Lehoczki, J., Peter, G., Dlauchy, D. and Deák, T. (1996) Some remarks on a taxonomic key for the genus *Saccharomyces* (Vaughan-Martini and Martini 1993). *Antonie van Leeuwenhoek*, **69**, 229–233.

Van Aelst, L. *et al.* (17 others) (1993) Molecular cloning of a gene involved in glucose sensing in the yeast *Saccharomyces cerevisiae*. *Molecular Microbiology*, **8**, 927–943.

Van Dijken, J.P., Weusthuis, R.A. and Pronk, J.T. (1993) Kinetics of growth and sugar consumption in yeasts. *Antonie van Leeuwenhoek*, **63**, 343–352.

Van Urk, H., Postma, E., Scheffers, W.A. and Van Dijken, J.P. (1989) Glucose transport in Crabtree-positive and Crabtree-negative yeasts. *Journal of General Microbiology*, **134**, 2399–2406.

Van Zyl, P.J., Kilian, S.G. and Prior, B.A. (1990) The role of an active transport mechanisms in glycerol accumulation during osmoregulation by *Zygosaccharomyces rouxii*. *Applied Microbiology and Biotechnology*, **34**, 231–235.

Venema, K. and Palmgren, M.G. (1996) Metabolic modulation of transport coupling ratio in yeast plasma membrane H^+-ATPase. *Journal of Biological Chemistry*, **270**, 19659–19667.

Walker, G.M. (1986) Magnesium and cell cycle control: an update. *Magnesium*, **5**, 9–23.

Walker, G.M. (1994) The roles of magnesium in biotechnology. *Critical Reviews in Biotechnology*, **14**, 311–354.

Walker, G.M. and Duffus, J.H. (1980) Magnesium ions and the control of the cell cycle in yeast. *Journal of Cell Science*, **42**, 329–356.

Walker, G.M. and Maynard, A.I. (1996) Magnesium-limited growth of *Saccharomyces cerevisiae*. *Enzyme and Microbial Technology*, **18**, 455–459.

Walker, G.M. and Maynard, A.I. (1997) Accumulation of magnesium ions during fermentative metabolism in *Saccharomyces cerevisiae*. *Journal of Industrial Microbiology and Biotechnology*, **18**, 1–3.

Walker, G.M., Maynard, A.I. and Johns, C.G.W. (1990) The importance of magnesium ions in yeast biotechnology. In *Fermentation Technologies. Industrial Applications* (ed. P.-L. Yu), pp. 233–240. Elsevier Applied Science, London and New York.

Walker, G.M., Birch, R.M., Chandrasena, G. and Maynard, A.I. (1996) Magnesium, calcium and fermentative metabolism in industrial yeasts. *Journal of the American Society of Brewing Chemists*, **54**, 13–18.

Walsh, M.C., Smits, H.P., Scholte, M. and Van Dam, K (1994) Affinity of glucose transport in *Saccharomyces cerevisiae* is modulated during growth on glucose. *Journal of Bacteriology*, **176**, 953–958.

Walsh, M.C., Scholte, M., Valkier, J., Smits, H.P. and van Dam, K. (1996) Glucose sensing and signalling properties in *Saccharomyces cerevisiae* require the presence of at least two members of the glucose transporter family. *Journal of Bacteriology*, **178**, 2593–2597.

Watanabe, Y., Miwa, S. and Tamai, Y. (1995) Characterization of Na^+/H^+-antiporter gene closely related to the salt tolerance gene of yeast *Zygosaccharomyces rouxii*. *Yeast*, **11**, 829–838.

Wésolowski-Louvel, M., Breunig, K.D. and Fukuhara, H. (1996) *Kluyveromyces lactis*. In *Nonconventional Yeasts in Biotechnology. A Handbook* (ed. K. Wolf), pp. 139–201. Springer-Verlag, Berlin and Heidelberg.

Weusthuis, R.A., Pronk, J.T., van den Broek, P.J.A. and van Dijken, J.P. (1994) Chemostat cultivation as a tool for studies on sugar transport in yeasts. *Microbiological Reviews*, **8**, 616–630.

Wheals, A.E. (1996) Anthony H. Rose memorial lecture. The ins and outs of the yeast plasma membrane. *Journal of the Institute of Brewing*, **102**, 291–294.

White, C. and Gadd, G.M. (1987) The uptake and cellular distribution of zinc in *Saccharomyces cerevisiae*. *Journal of General Microbiology*, **133**, 727–737.

Wickerham, L.J. (1951) *Taxonomy of yeasts*. US Department of Agriculture Technical Bulletin, Washington. pp. 1–56.

Yagi, T., Kim, Y., Hiraoka, Y., Tanouchi, A., Yamamoto, T. and Yamamoto, S. (1996) Active transport activities of free B-6 vitamins in various yeast strains. *Bioscience Biotechnology and Biochemistry*, **60**, 893–897.

Yompakdee, C., Ogawa, N., Harashima, S. and Oshima, Y. (1996) A putative membrane protein, Pho88p, involved in inorganic phosphate transport in *Saccharomyces cerevisiae*. *Molecular and General Genetics*, **251**, 580–590.

Youatt, J. (1993) Calcium and microorganisms. *Critical Reviews in Microbiology*, **19**, 83–97.

Yu, W., Farrell, R.A., Stillman, D.J. and Winge, D.R. (1996) Identification of *SLF1* as a new copper homeostasis gene involved in copper sulphide mineralization in *Saccharomyces cerevisiae*. *Molecular and Cellular Biology*, **16**, 2464–2472.

Zhao, J. and Eide, D. (1996a) The yeast *ZRT1* gene encodes the zinc transporter protein of a high-affinity uptake system induced by zinc limitation. *Proceedings of the National Academy of Sciences (USA)*, **93**, 2454–2458.

Zhao, J. and Eide, D. (1996b) The *ZRT2* gene encodes the low-affinity zinc transporter in *Saccharomyces cerevisiae*. *Journal of Biological Chemistry*, **271**, 23203–23210.

4

YEAST GROWTH

4.1 INTRODUCTION
4.2 CELLULAR GROWTH OF YEASTS
4.2.1 Vegetative Reproduction in Yeasts
4.2.1.1 Budding
4.2.1.2 Fission
4.2.1.3 Filamentation
4.2.2 Control of Yeast Growth and Cell Division
4.2.2.1 The Cell Cycle in Yeasts
4.2.2.2 Molecular Aspects of Cell Cycle Control in Yeasts
4.2.3 Sexual Reproduction in Yeasts
4.3 POPULATION GROWTH OF YEASTS
4.3.1 Colonial Yeast Growth
4.3.2 Population Yeast Growth
4.3.2.1 Batch Growth of Yeasts
4.3.2.2 Continuous Growth of Yeasts
4.3.2.3 Synchronous Yeast Growth
4.3.3 Cultural Yeast Growth
4.3.3.1 Yeast Pure Culturing and Maintenance Strategies
4.3.3.2 Assessments of Yeast Viability and Vitality
4.3.3.3 Cultivation Strategies in Yeast Biotechnology
4.4 THE PHYSICOCHEMICAL ENVIRONMENT AND YEAST GROWTH
4.4.1 Physical Requirements for Yeast Growth
4.4.1.1 Temperature
4.4.1.2 Water
4.4.1.3 Media pH and pO_2
4.4.2 Effects of Physical Stresses on Yeast Growth
4.4.2.1 Temperature Stress
4.4.2.2 Water Stress
4.4.2.3 Other Physical Stresses
4.4.3 Effects of Chemical Stresses on Yeast Growth
4.5 BIOTIC FACTORS INFLUENCING YEAST GROWTH
4.5.1 Yeast–Plant Interactions
4.5.2 Yeast–Animal Interactions
4.5.3 Yeast–Microbe Interactions
4.6 YEAST CELL DEATH
4.6.1 Physical Parameters and Yeast Cell Death
4.6.2 Chemical Factors and Yeast Cell Death
4.6.3 Biological Factors Influencing Yeast Cell Death
4.7 SUMMARY
4.8 REFERENCES

4.1 INTRODUCTION

Yeast growth is concerned with how yeasts transport and assimilate nutrients (which were discussed in Chapter 3) and then integrate numerous component functions in the cell in order to increase in mass and eventually divide. The growth, reproduction and, ultimately, the death of yeast cells is considered in the present chapter from the fundamental cell physiological viewpoint and with regard to practical implications for yeast biotechnology.

In the section on cellular growth of yeast, the rudiments of budding, fission and filamentation in yeasts will be described with emphasis on changes in gross cellular morphology, cell wall growth, cytoskeletal rearrangements and spindle dynamics. Major advances in our understanding of the growth of individual yeast cells, in particular the molecular mechanisms which govern DNA synthesis and mitosis, have been made in recent years. The mechanics of the cell cycle 'engine' in yeast will be described with reference to studies conducted in two model eukaryotes, the budding yeast, *Saccharomyces cerevisiae*, and the fission yeast, *Schizosaccharomyces pombe*. Such studies are emphasized since they enhance our fundamental knowledge of cell proliferation in higher eukaryotes. In addition to describing the physiology of vegetative reproduction and its control in budding and fission yeasts, sexual mating in *S. cerevisiae* and the pheromone response signal transduction pathway will be introduced.

The control of growth of yeast cell populations in liquid culture is crucial to the performance of industrial processes which exploit yeasts. Population yeast growth on solid surfaces also has significant practical relevance in industry and medicine. Descriptions of colonial and cultural yeast growth kinetics will therefore be presented before discussing population dynamics of yeast cells propagated in batch, fed-batch, continuous, phased and immobilized culture systems. The practicalities of growing yeasts in large-scale industrial bioreactors will be discussed with regard to preservation, inoculum development and cultivation strategies. The design, monitoring and control of yeast bioreactors will also be briefly covered.

Since the growth physiology of yeasts in biotechnology is strongly influenced by physical, chemical and biological factors in the growth environment, these will be explained (in sections 4.4 and 4.5) with regard to their beneficial and detrimental influences on yeast cells. Concerning the latter, stress effects in yeast and the physiological and molecular adaptive responses of cells will be discussed with particular reference to stresses caused by temperature extremes, reduced water availability, ethanol shock and oxidative stress. These environmental insults would be expected to be encountered by industrial yeasts during the course of fermentation and so understanding the physiological basis of stress in yeast is of practical significance.

In the final section in this chapter, factors which cause the death of yeast cells will be discussed. An understanding of these physical, chemical and biological factors is important since they directly impinge on the efficiency of yeast bioreactions and are also relevant in the eradication of undesirable yeasts in industrial and clinical situations.

4.2 CELLULAR GROWTH OF YEASTS

4.2.1 Vegetative Reproduction in Yeasts

4.2.1.1 Budding

Budding is the most common mode of vegetative reproduction in yeasts (Table 4.1) and multilateral budding is a typical reproductive characteristic of ascomycetous yeasts.

Generally speaking, yeast buds are initiated when mother cells attain a critical cell size at a time coinciding with the onset of DNA synthesis. This is followed by localized weakening of the cell wall and this, together with tension exerted by turgor pressure, allows extrusion of cytoplasm in an area bounded by new cell wall material which is synthesized by enzymes such as glucan and chitin synthetases. The regulation of cell wall synthetic enzymes, together with fusion of actin-directed secretory vesicles with specific bud plasma membrane receptors, are key events in the emergence of a new bud (Tschopp *et al.*, 1987; Mulholland *et al.*, 1994; Harold, 1995). Chitin, a polymer of *N*-acetylglucosamine, forms a ring at the junction between the mother cell and the bud (Figure 4.1). This chitin ring will eventually form the characteristic *bud scar* after cell division.

Table 4.1. Modes of vegetative reproduction in yeasts.

Mode	Description	Representative yeast genera
Multilateral budding	Buds may arise at any point on the mother cell surface, but never again at the same site. Branched chaining may occasionally follow multilateral budding when buds fail to separate.	*Saccharomyces**, *Zygosaccharomyces, Torulaspora, Pichia, Pachysolen, Kluyveromyces, Williopsis, Debaryomyces, Yarrowia, Saccharomycopsis, Lipomyces*
Bipolar budding	Budding restricted to poles of elongated cells (apiculate or lemon-shaped) along their longitudinal axis.	*Nadsonia, Saccharomycodes, Hanseniaspora, Wickerhamia, Kloeckera*
Unipolar budding	Budding repeated at same site on mother cell surface. In *Trigonopsis*, buds are restricted to the three apices of triangular cells.	*Pityrosporum, Trigonopsis*
Monopolar budding	Buds originate at only one pole of the mother cell.	*Malassezia*
Binary fission	A cell septum (cell plate or cross-wall) is laid down within cells after lengthways growth and which cleaves cells into two	*Schizosaccharomyces*
Bud fission	Broad cross-wall at base of bud forms which separates bud from mother.	Occasionally found in: *Saccharomycodes, Nadsonia* and *Pityrosporum*
Budding from stalks	Buds formed on short denticles or long stalks.	*Sterigmatomyces*
Ballistoconidiogenesis	Ballistoconidia are actively discharged from tapering outgrowths on the cell.	*Bullera, Sporobolomyces*
Pseudomycelia	Cells fail to separate after budding or fission to produce a single filament. Pseudomycelial morphology is quite diverse and the extent of differentiation variable depending on yeast species and growth conditions.	Several yeast species may exhibit 'dimorphism', e.g. *Candida albicans, Saccharomycopsis fibuligera*. Even *S. cerevisiae* exhibits pseudohyphal growth depending on conditions.

* A fuller description of budding patterns in *S. cerevisiae* is provided in the text.

Once a new daughter bud has been initiated, cell surface growth during the remainder of the cell division cycle is restricted to the bud. In other words, the mother cell wall does not grow very much during budding. The mother and daughter bud cell walls are contiguous during bud development. Once mitosis is complete and the bud nucleus and other organelles (e.g. mitochondria) have migrated into the bud, cytokinesis ensues and a septum is formed in the isthmus between mother and daughter. A ring of proteins called **septins** are involved in positioning cell division in *S. cerevisiae* in that they define the cleavage plane which bisects the spindle axis at cytokinesis (reviewed by Chant, 1996a). These septins encircle the neck between mother and daughter for the duration of the cell cycle.

In *S. cerevisiae*, cell size at division is asymmetrical with buds being smaller than mother cells when they separate. Since daughter buds require time (in G1) to attain a *critical cell size* before they in turn give birth, cell division cycle times of mother and daughter are also asymmetrical (Wheals, 1987). Mother and daughter budding

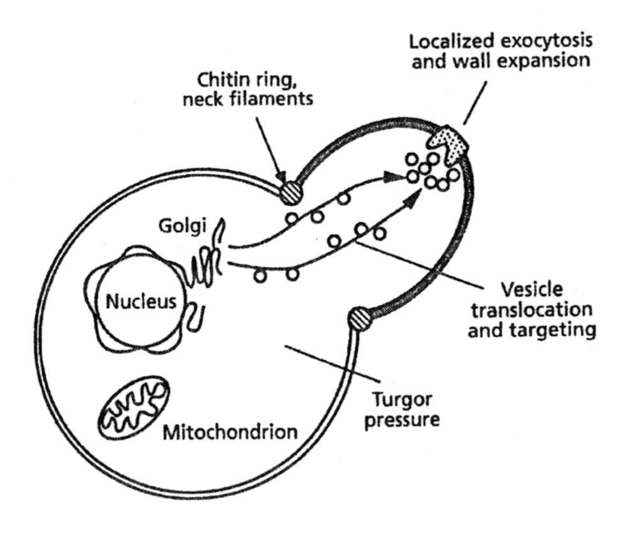

Figure 4.1. Physiology of bud growth in yeast. The diagram identifies processes known or surmised to contribute to the enlargement of the bud's surface. Reproduced with permission from Harold (1995), and the Society for General Microbiology.

cycles in *S. cerevisiae* may thus also be referred to as asynchronous (Kron and Gow, 1995). Occasionally, *S. cerevisiae* divides symmetrically. This behaviour is prevalent during pseudohyphal growth (Kron *et al.*, 1994) and in aged populations when mother cells enlarge as they get older giving rise to similarly enlarged daughters (Kennedy *et al.*, 1994).

The percentage of budding yeast cells in a population can be assessed quite easily using phase-contrast microscopy and this is useful in characterizing the cell cycle state of a culture (Futcher, 1995). For example, rapidly dividing populations have a budding index (percentage fraction of budded cells) of around 50%, while specific cell cycle blockage before septation (by mutation or drugs) would be reflected in, for example, bud indices of $>80\%$, and, if arrest occurs at or before 'Start' (see later), $<20\%$.

When the bud eventually detaches, cell surface scar regions appear on both mother and daughter cells. Mother cells are left with a **bud scar** which is visible in scanning electron micrographs as a concave structure surrounded by a raised ring of cell wall material (see Chapter 2, Figure 2.7). This ring is rich in chitin (Bulawa, 1993) and can be stained with dyes such as Calcofluor White and visualized by fluorescence microscopy. Conversely, the daughter bud cells are left with a low chitin-containing, weakly-fluorescent convex **birth scar** following cell separation, usually at one end of the long axis of ovidal cells. Note that in *S. cerevisiae*, bud sites may occasionally overlap with the birth scar, but *never* with bud scars (Streiblova and Beran, 1963; Chant and Pringle, 1995). The number of bud scars produced on the cell surface of multilateral budding yeasts is a useful determinant of cellular age (see section 4.6). This is because bud scars represent a completed cell division cycle; successive buds never emanate from the same surface site; and the number of buds produced is finite. In *S. cerevisiae*, although as many as 100 (Bartholomew and Mittwer, 1953) or 50 (Cook, 1963) bud scars have been reported on the surface of cells, a typical stationary phase population would consist mainly of cells with very few bud scars (Phaff, Miller and Mrak, 1978). In unipolar budding yeasts, very few bud scars are observable at the cell surface because bud sites are superimposed on each other. This gives rise instead to scar tissue characterized by ringed ridges (e.g. *Saccharomycodes*) or concentric collars (e.g. *Rhodotorula*) on cell surfaces.

Although *Saccharomyces* is listed in Table 4.1 as an example of a multilateral budding yeast genus, this requires further explanation for *S. cerevisiae* since budding in this species is not a randomized, uncontrolled process (Beran, 1968). Cellular geometry is undoubtedly important in localizing budding sites in yeast. Generally, in ellipsoidal cells such as *S. cerevisiae*, buds arise at loci of maximum cell curvature or, at the 'end of the rugby ball-shaped cell' (Kron and Gow, 1995). General cell morphology and cytoskeletal organization in *S. cerevisiae* dictates that budding is not random. Alterations of the yeast cytoskeleton, for example by microgravity, have been shown the randomize budding patterns in *S. cerevisiae* (Walther *et al.*, 1996). In spherical yeasts, however, such as *Debaryomyces* spp.,

buds apparently arise randomly on the cell surface and lack polarized growth as typified by *S. cerevisiae*.

In *S. cerevisiae*, numerous studies have endeavoured to explain at the cellular and molecular levels how polarized cell growth is regulated and how the site of emerging buds is chosen (Chant, 1995; Chant and Pringle, 1995; Herskowitz *et al.*, 1995; Lew and Reed, 1995; Pringle *et al.*, 1995; Chant, 1996b). *S. cerevisiae* provides an ideal model in which to study detailed mechanisms of asymmetric cellular organization pertinent to development in multicellular organisms. Bud-site selection in this yeast depends on several physiological and genetic factors. With regard to the latter, cell mating type is important and **a** and α haploid cells are known to exhibit an **axial** budding pattern, whereas **a**α diploid cells exhibit a **bipolar** budding pattern (Figure 4.2).

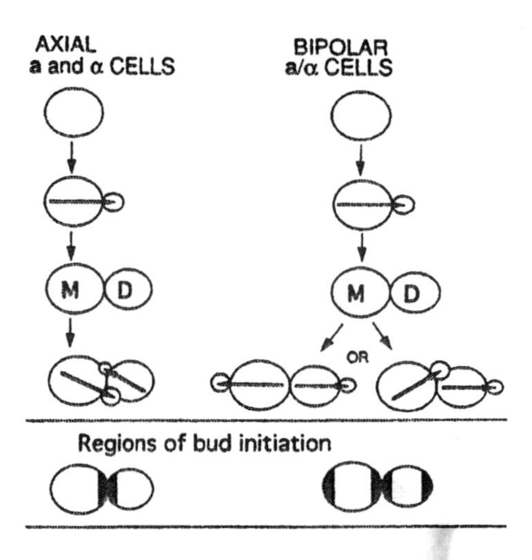

Figure 4.2. Bud site selection patterns in *S. cerevisiae*. Left: in axial budding, the mother cell (M) buds immediately adjacent to its last daughter; the daughter cell (D) buds toward its mother. Right: in bipolar budding, the mother cell can bud at or near either of its poles; the daughter cell buds away from its mother. The arrows within the cells in the figure indicate the axis of polarity and cell division. The bud produced by the mother cell is drawn longer than the bud from its daughter cell because mother cells initiate budding before daughter cells. Reproduced with permission from Chant and Herskowitz (1991) © Cell Press.

Axial budding is when mother and daughter cells bud towards each other with mother cells forming a bud near the preceding bud scar and daughter cells forming a bud near the preceding birth scar. Nutrient starvation may induce haploid **a** and α cells to bud non-axially at sites adjacent to that of the most recent bud (Chant and Pringle, 1995). Bipolar budding in *S. cerevisiae* is when daughter cells bud firstly away from their mother (near the pole distal to the birth scar) while mother cells either bud away from or toward daughter cells (either near the preceding bud site or near the opposite pole). It is thought that such polarized growth in *S. cerevisiae* (see Figure 4.3) may, in the case of the axial pattern, facilitate mating between cells of the opposite mating type whereas in the bipolar pattern, it may allow cells to grow away from each other and avoid competition for nutrients (Herskowitz *et al.*, 1995).

The establishment of cell polarity in *S. cerevisiae* is now thought to be governed by a 'morphogenetic hierarchy' involving the interplay between various genes which dictate the orientation of cytoskeletal elements (Figure 4.4). Thus, genes for bud-site selection (*BUD* genes) are involved in determining the orientation of actin fibres and genes for bud formation (e.g. *CDC24*, *CDC42*, *BEM1*) direct cell surface growth to the developing bud (Herskowitz *et al.*, 1995). GTPase activity is required for numerous aspects of yeast budding including the orientation of the actin cytoskeleton toward the bud site (Chant and Stowers, 1995). Cyclin-dependent kinases and cyclins (see section 4.2.2) also play important roles in actin assembly and in localizing and timing bud emergence (Kron and Gow, 1995). In *S. cerevisiae*, cytoskeletal actin fibres thus play crucial roles in establishing cell polarity and the precise siting of new bud growth (Chant, 1995; Lew and Reed, 1995; Yang *et al.*, 1997). Chant and Pringle (1995) have presented models to explain budding patterns in *S. cerevisiae* based on 'positional signals' which direct the selection of bud sites and Zahner *et al.*, (1996) have

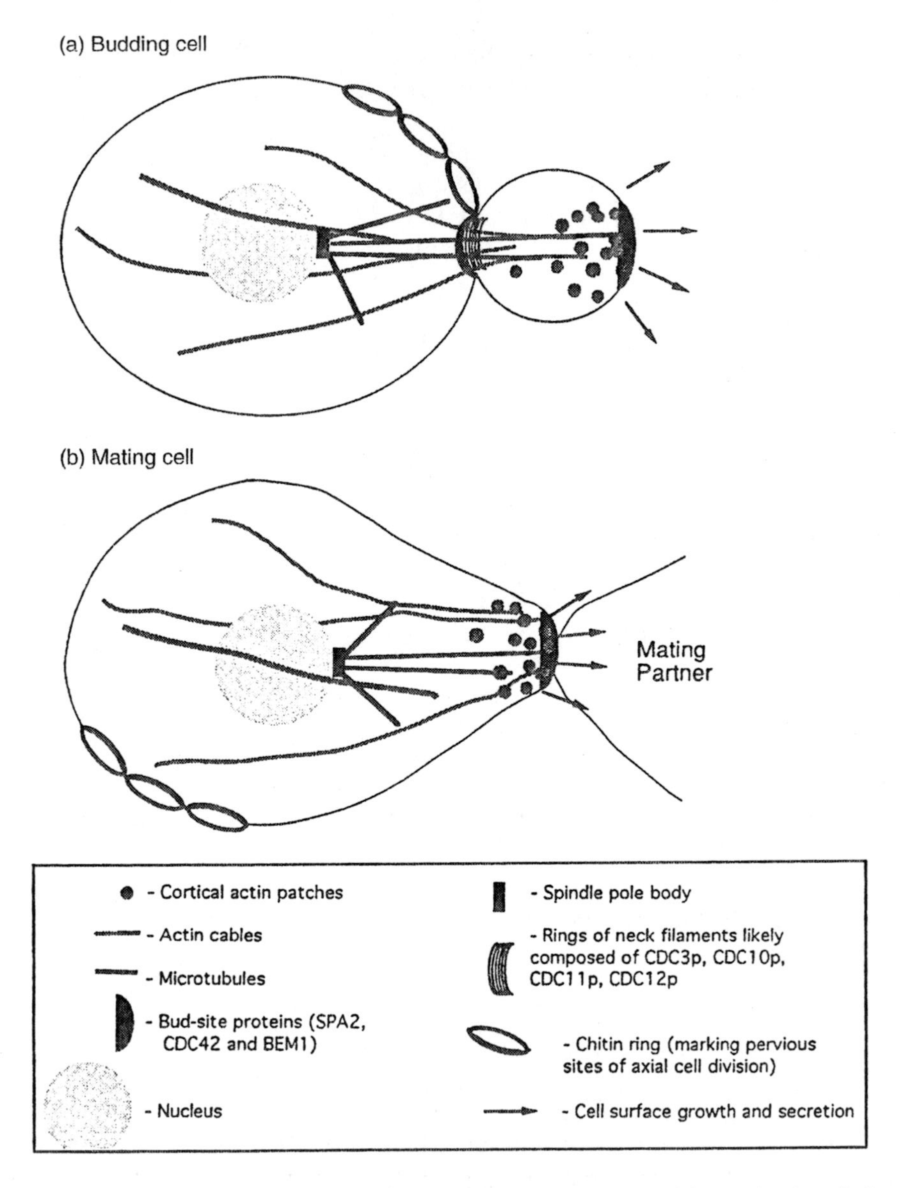

Figure 4.3. The polarized yeast cell. (a) Polarization during budding in *S. cerevisiae*. (b) Polarization during mating in *S. cerevisiae*. Reproduced with permission from Chant (1995) and Academic Press.

genetically analysed positional signalling in this yeast. Roemer *et al.*, (1996) have reviewed the roles of key molecular components, including products of bud-site selection genes, polarity-establishment genes and septin-ring genes, in governing polarized cell growth and cell division in *S. cerevisiae*.

4.2.1.2 Fission

Fission yeasts (e.g. *Schizosaccharomyces* spp.) divide exclusively by forming a cell septum, analogous to the mammalian cell cleavage furrow, which constricts the cell into two equal-sized daughters. Pioneering studies of the model eukar-

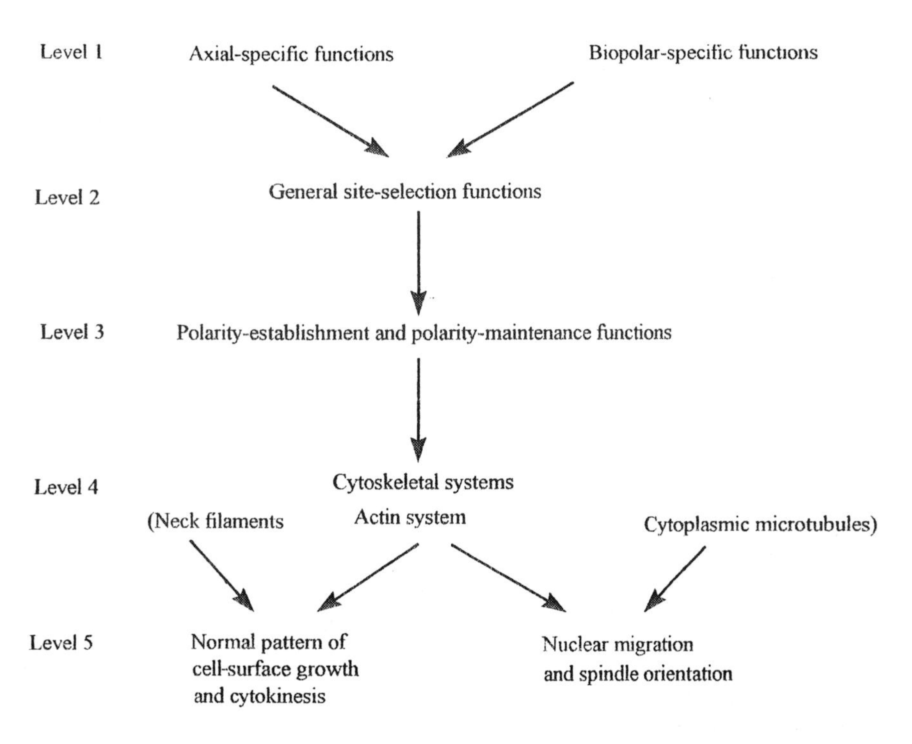

Figure 4.4. Morphogenetic hierarchy in the *S. cerevisiae* cell cycle. According to this model (modified from Pringle *et al.*, 1995), cortical signals specific to the axial or bipolar budding patterns (Level 1) localize the action of the general site-selection functions (Level 2), which in turn localize the action of the polarity-establishment functions (Level 3), which in turn direct the polarization of the cytoskeletal elements (Level 4), which carry out the actual morphogenetic events (Level 5). The pattern of cell-surface growth is determined primarily by the actin system but is influenced by the neck filaments, which are also essential for cytokinesis. Nuclear migration and spindle orientation are determined primarily by cytoplasmic microtubules, but are also influenced by the actin system.

yote, *Schizosaccharomyces pombe*, by Mitchison (e.g. Mitchison, 1957) laid the foundations for our understanding of fission yeast growth and cell division. The cytology of *Sch. pombe* growth, division and cellular morphogenesis has been extensively reviewed (Mitchison, 1970, 1971; Streiblova, 1981; Johnson *et al.*, 1982; Hirano and Yanagida, 1989; Johnson *et al.*, 1989; Robinow and Hyams, 1989). Here, only the salient features of cell wall growth and cytoskeletal rearrangements during fission will be summarized.

In *Sch. pombe*, newly divided daughter cells (about 8 μm long) grow lengthways in a monopolar fashion from the old-end cell tip for about one-third of their new cell cycle. Cells then switch to bipolar growth from the new-end cell tip for about three-quarters of the cell cycle until mitosis is initiated at the constant cell length stage (about 14 μm). Mitosis proceeds and daughter nuclei move to opposite ends of the dividing cell before renewing their positions about one-quarter cell length from the ends. Cell septum formation and cytokinesis in *Sch. pombe* is associated with extensive cytoskeletal element (actin fibres and cytoplasmic microtubules) rearrangement, together with dramatic cell wall changes at the cell equatorial plane.

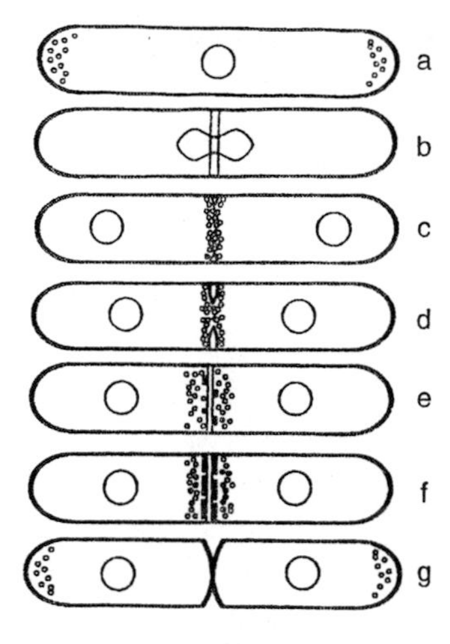

Figure 4.5. Septum formation and cytokinesis in *Sch. pombe.* During interphase, fission yeast cells grow primarily by elongation at their tips. When mitosis is initiated cells stop elongating and enter the so-called constant-volume phase of the cell cycle. Upon completion of mitosis they form a centrally placed division septum and then divide by binary fission. There is a strong temporal and spatial association between elements of the cytoskeleton and the events that occur during mitosis and cytokinesis. In interphase cells, actin is located at the growing ends of the cell (**a**) and a basket of microtubules run from end to end in the cell. At the onset of mitosis, there are extensive rearrangements of both the actin and tubulin elements of the cytoskeleton. The cytoplasmic microtubule array disappears and is replaced by a short intranuclear spindle, which elongates as mitosis proceeds. Actin relocates from the ends of the cell to form a hollow equatorial ring overlying the nucleus (**b**). The position of this ring anticipates the site of septum formation. As mitosis is completed, the primary septum begins to grow centripetally, arising from the periplasmic space just outside the plasma membrane. As the synthesis of the septum proceeds, the appearance of actin staining changes from a filamentous ring to a cluster of dots (**c-e**). The sites of deposition of the septum material correlate well with the location of actin vesicles or 'dots'. Secondary septa are formed on either side of the primary one (**e-f**), which is then 'dissolved' away to bring about cell separation. When the cell completes the septum and undergoes cytokinesis, the actin relocates to the old (pre-existing) end of the cell (**g**), from where growth will resume. The figure is a schematic representation of the events of cytokinesis in fission yeast. The cell wall and secondary division septa are shown in black, the primary septa in white and the distribution of actin vesicles or 'dots' by grey circles. For sake of clarity, microtubules have been omitted from the diagrams. Reproduced with permission from Fankhauser and Simanis (1994) and Elsevier Science Ltd.

A summary of these latter activities is depicted in Figure 4.5.

Hyams and his colleagues (e.g. Hagan and Hyams, 1996) have made extensive studies of fission yeast mitosis and have proposed that astral microtubules (radiating into the cytoplasm from spindle pole bodies) exert pushing and pulling forces on the mitotic spindle interacting with motor proteins (e.g. dynein) on the plasma membrane. Such forces enable microtubules to move in a direction which depends on their polarity (Figure 4.6).

Pichová *et al.*, (1995) have also proposed a key role for cytoplasmic microtubule arrays and equatorial tubulin rings during cell division in *Sch. pombe* and their model is presented in Figure 4.7.

The *Sch. pombe* septum forms by lateral growth of the inner cell wall (the primary septum) and proceeds inwardly followed by deposition of secondary septa (on both sides of the primary septum). Medial fission is completed in a manner resembling the closure of an iris diaphragm. During transverse cleavage, *fission scars* (or scar plugs) are formed which, after subsequent repeated cell divisions, are retained as scar rings on the surface of the original mother cell (see Figure 4.8). These scars therefore represent permanent records of the division history of fission yeast and are thus analogous to the bud scars of *S. cerevisiae*. Knowing the number of fission scars it is possible to deduce the most likely progeny (and progenitor) of a fission yeast cell. It appears that the maximum number of fission scars in *Sch. pombe* is very few, around six (Calleja *et al.*, 1980). Thus, fission yeast does not apparently grow as 'old' as the budding yeast, *S. cerevisiae*, which may possess many more bud scars (see section 4.6).

It is important to note that, with regard to cell wall deposition during growth, in *S. cerevisiae* there is a switch at 'Start' from general growth over the daughter cell to restricted growth at the bud site, whereas in *S. pombe* there is growth along one cylindrical end until *new end take off* (NETO) when it grows at both ends. Equatorial septation and symmetrical binary fission, as

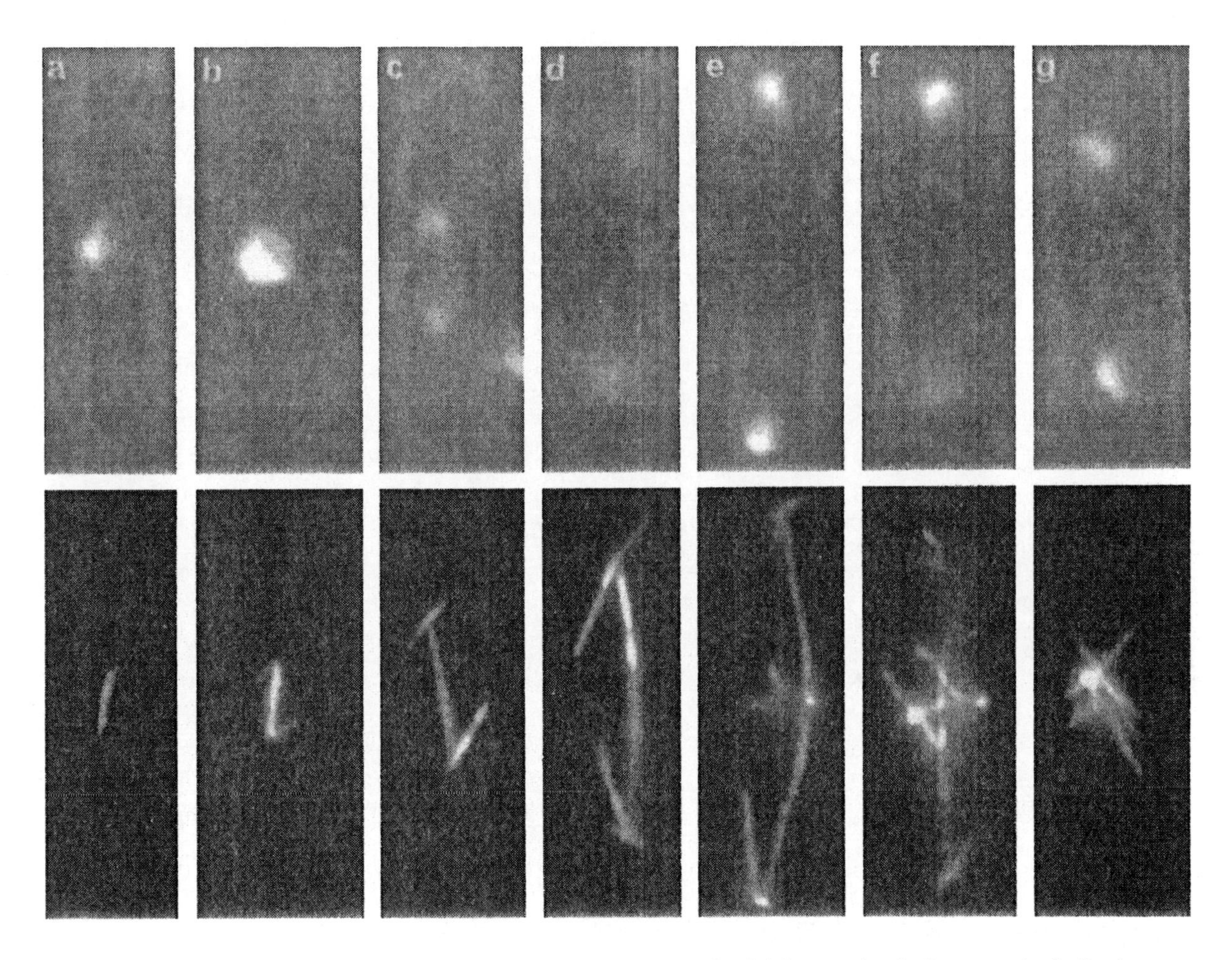

Figure 4.6. Microtubular dynamics during the *Sch. pombe* cell division cycle. Indirect anti-tubulin immunofluorescence microscopy of wild-type fission yeast cells. Each figure is split into two panels, the upper being a DAPI/phase-contrast image to show the location of the nucleus and the lower an anti-tubulin immunofluorescence image to show microtubules. (**a**) Prometaphase/metaphase; kinetochore microtubules are still present and the spindle has a dumbbell appearance. (**b**) Early anaphase spindle, astral microtubules in the parallel configuration. (**c**) Mid-anaphase spindle, astral microtubules in the parallel configuration. (**d**) Late anaphase spindle, astral microtubules in the convergent configuration. (**e**) Late anaphase spindle with newly active equatorial microtubule organizing centres (EMTOCs) giving rise to cytoplasmic microtubules at the cell equator. (**f**) Spindle dissolution a slightly later stage than (**e**) as signified by the more mature post anaphase array emanating from the EMTOCs. (**g**) The EMTOCs appear to coalesce immediately before cytokinesis. Note the bowed spindles and short stubs in addition to the main astral bundles in (df). Magnification, ×5000. Reproduced with permission from Hagan and Hyams (1996) © Wiley-Liss Inc.

exemplified by division of *Sch. pombe*, is much more common in eukaryotic cells compared with the asymmetrical budding process as exemplified by division of *S. cerevisiae*. In addition, *Sch. pombe* chromosomes condense to a greater extent than those of *S. cerevisiae* during mitosis. With respect to these (and other) aspects, *Sch. pombe* is regarded as a much more typical eukaryotic cell than its distant cousin, *S. cerevisiae* (Forsburg and Nurse, 1991; Sipiczki, 1995).

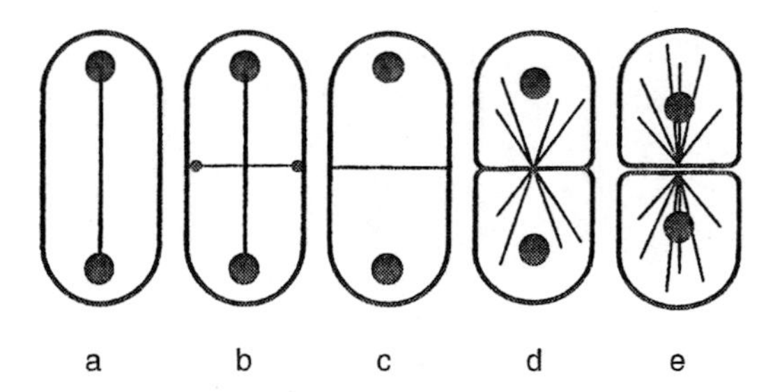

Figure 4.7. Cytoplasmic microtubule arrays during *Sch. pombe* cell division. A proposed scheme of the dynamic changes in microtubule distribution and nucleus positioning during anaphase and telophase of the *Sch. pombe* cell cycle. (**a**) A long spindle and separated nuclear regions. (**b**) Coexistence of a long spindle and a tubulin equatorial ring at the end of anaphase. (**c**) Disappearance of the spindle. (**d**) A persisting tubulin equatorial ring with astral microtubules converging at the septum ingrowth. (**e**) Disappearance of the tubulin equatorial ring and formation of a septum with converging astral microtubules still remaining. Reproduced with permission from Pichová *et al.*, (1995) and Springer-Verlag.

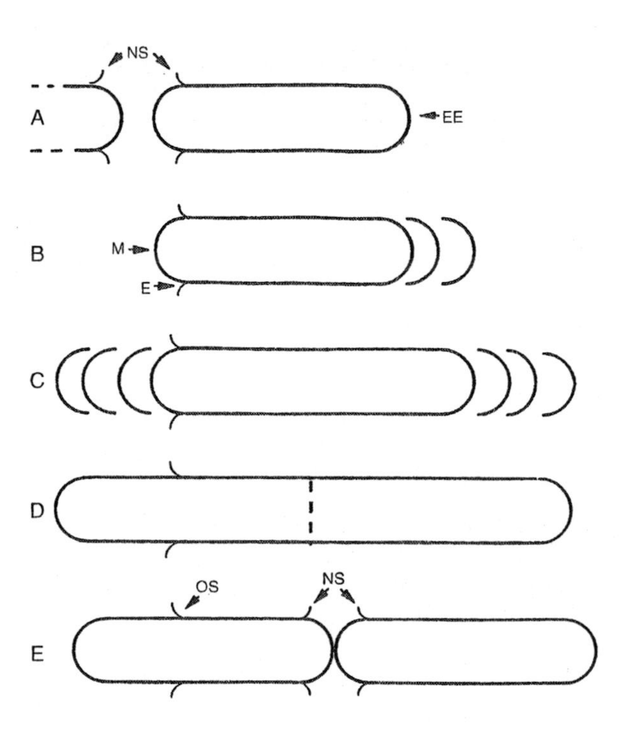

Figure 4.8. Simplified morphogenesis in *Sch. pombe.* (A) Typical fission yeast cell, immediately after cell division. NS, new fission scar; EE, extensile end. (B) Cell from (A) has been extending at older, previously extensile, unscarred (Mitchison's Rule) end. Defines extensile end; NS. [see (A)] defines non-extensile end. M, middle of non-extensile end; E, edge of non-extensile end. (C) Extensile end continues to grow; non-extensile end has become extensile. NS becomes old scar, once it is bounded by a cylindrical wall on both sides. (D) Septum is elaborated (vertical dashed line). (E) Septum has fissioned. OS, old scar. Reproduced with permission from Johnson, Yoo and Calleja (1995) and Chapman & Hall.

4.2.1.3 Filamentation

Filamentous growth occurs in numerous yeast species and may be regarded as an alternative mode of vegetative growth to budding or fission. Different forms of filamentous yeast growth have been described and these are summarized in Table 4.2. Kurtzman (1987) has discussed the presence or absence of pseudohyphae and true hyphae of yeast as taxonomic criteria. In the latter, the ultrastructure of hyphal septa is deemed useful in discriminating between certain ascomycetous yeasts.

Some yeasts exhibit a propensity to grow with true hyphae initiated from **germ tubes** (e.g. *Candida albicans*), but others (including *S. cerevisiae*) may grow in a pseudohyphal fashion when induced to do so by unfavourable conditions. Hyphal and pseudohyphal growth represents different developmental pathways in yeast (Kron and Gow, 1995; Gow, 1996). Such filamentation in yeast is generally reversible in that cells can revert to yeast unicellular growth upon return to more conducive growth conditions. It is conceivable, therefore, that a filamentous mode of growth represents an adaptation by yeast to foraging when nutrients are scarce.

Verde *et al.*, (1995) have conducted genetic analysis of morphogenetic transitions in *Schizosaccharomyces pombe* and have identified mutants with cytoskeletal defects which exhibit branched or bent morphologies. With regard to physiological aspects of filamentous yeast

Table 4.2. Diversity of yeast cell morphologies.

Cell morphology	Description
Yeast form	Spherical or ellipsoidal unicells.
Elongated yeast	Elongated ellipsoidal unicells.
Dimorphic	Yeasts that grow vegetatively in either yeast or filamentous (hyphal or pseudohyphal) form.
Pseudohyphae	Chains of budding yeast cells which have elongated without detachment. Pseudohyphal morphology is intermediate between a chain of yeast cells and a hypha.
Hyphae	Branched or unbranched filamentous cells which form from germ tubes. Septa may be laid down by the continuously extending hyphal tip. Hyphae may give rise to lateral budding cells called blastospores.

growth, it is known that various changes in cultural or environmental conditions will induce dimorphism. Odds (1985) and Gow (1996) have reviewed environmental signals which trigger yeast-to-hyphal transitions in *Candida albicans*. In this yeast, increases in external pH and temperature together with certain nutritional 'germ tube inducers' (e.g. serum, *N*-acetylglucosamine, etc.) are important determinants of dimorphism. Bioavailability of certain metal ions (e.g. Mg^{2+}) is also involved (Walker *et al.*, 1984). In budding yeasts such as *S. cerevisiae* and *Kluyveromyces marxianus*, limitation of the nitrogen source (Gimeno *et al.*, 1992) and the presence or absence of oxygen (Walker and O'Neill, 1990) are known to influence dimorphism. Figure 4.9 summarizes the effect of nutrient stress on filamentous growth of yeasts.

With regard to cultural conditions, propagation of yeasts in nutrient-limited chemostats (Kuriyama and Slaughter, 1995) immobilization of yeasts in alginate beads and invasive colonial

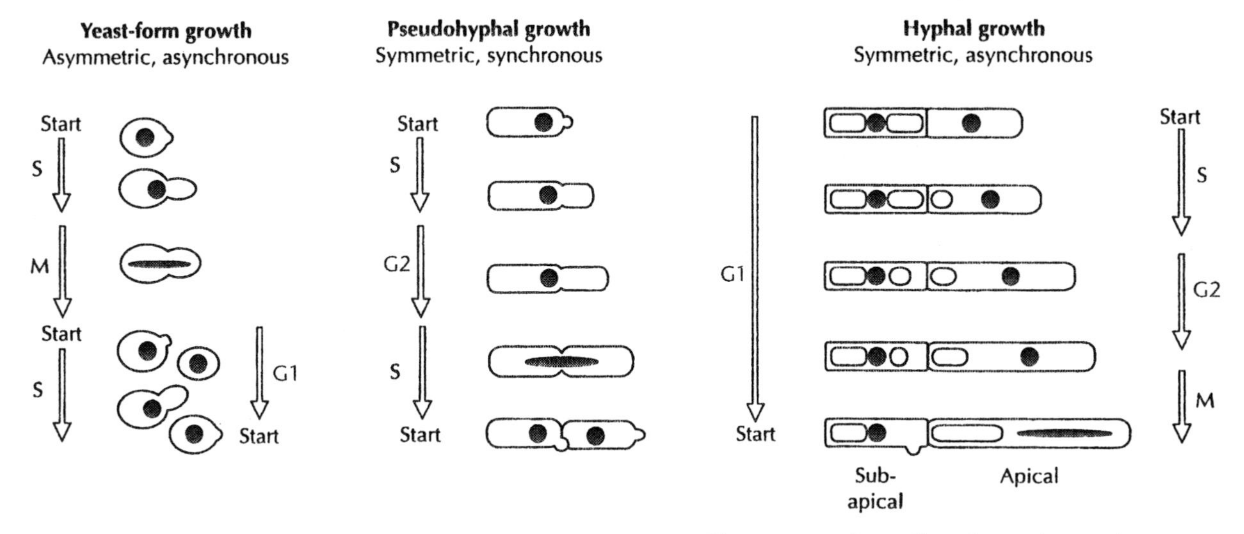

Figure 4.9. Vegetative cell cycles in yeast under nutrient stress. Three vegetative cell cycles under nutrient stress are shown. In yeast-form growth (left), nutrient stress promotes asymmetric cell division before the bud reaches the size of the mother. The nascent daughter grows in G_1 before passing Start, causing asynchrony in cell cycles. In pseudohyphal growth (centre), the elongated budding period leads to symmetric cell division, no G_1 delay and synchronous re-entry to the budding cycle. In hyphal growth (right), asymmetric distribution of vacuole to the apical and subapical cells leads to a G_1 period of growth in the subapical cell before branch emergence whereas the apical cell begins its next cell cycle immediately. Reproduced from Kron and Gow (1995), with permission from Current Biology Ltd.

growth of yeasts on agar can all lead to filamentous cell growth.

Kron and Gow (1995) have discussed the various controls over cytoskeletal organization and cell cycle progression which mediate differentiation into filamentous growth of both *S. cerevisiae* and *C. albicans*. Genetic analysis of dimorphism in these two yeasts has been discussed, respectively by Blacketer *et al.*, (1995) and by Gow *et al.* (1995). Molecular signals (e.g. cyclins and cAMP) are known to be involved in promoting pseudohyphal growth of *S. cerevisiae* and hyphal growth of *C. albicans*. In the former species, dimorphism appears to be controlled primarily by the *Ras*/cAMP pathway and by elements of the mitogen-activated protein kinase (MAPK) cascade of the mating pheromone signal transduction pathway (Gimeno *et al.*, 1992; Liu *et al.*, 1993; Wittenberg and Reed, 1996). In *C. albicans*, specific induction of a hyphally regulated gene, *HYR1*, appears to be crucial in dictating the yeast-to-hypha transition in this organism (Bailey *et al.*, 1996).

Kron *et al.*, (1994) have made interesting comparisons of the division patterns of *S. cerevisiae* pseudohyphae and fission yeast cells of *Sch. pombe*. Thus, in both situations division is symmetrical, daughter cell cycles are initiated without much delay, DNA synthesis is completed early in the cell cycle, cells reside mostly in G2 during the cell cycle and finally, mitosis occurs after cells double in size. Although the 'discovery' of dimorphic laboratory diploid strains of *S. cerevisiae* by Gimeno *et al.* (1992) has stimulated a great deal of interest from yeast molecular geneticists, it should be stressed that dimorphism in industrial strains of *Saccharomyces* has been known for over 100 years (Hansen, 1886).

Dimorphic transitions between unicellular yeast cells to filamentous (hyphal and pseudohyphal) forms (and *vice versa*) are exhibited by many clinically and industrially important yeasts, some of which are described in Table 4.3.

It is evident, therefore, that there are several important practical implications of filamentous yeast growth. For example, many dimorphic yeasts are animal and plant pathogens and knowledge of mechanisms controlling morphogenesis is important in combating yeast pathogenesis. In the opportunistic human pathogen, *C. albicans*, yeast-hyphal dimorphism is one of several virulence traits. This is because hyphae of this yeast are well adapted to penetrate epithelial cell envelopes (Sherwood *et al.*, 1992; Gow *et al.*, 1995). In industrial fermentations, the morphology of yeasts may have important influences on rheological properties of the growth medium, oxygen transfer and nutrient uptake and these in turn may affect the outcome of metabolic reactions in yeast. In industrially relevant yeasts like *S. cerevisiae* and *Kluyveromyces marxianus*, dimorphism resulting in long, thin filamentous cells with a large surface area may be better suited to immobilization compared with ellipsoidal unicellular yeast forms with a low surface area. Finally, filamentous growth of yeasts may be associated with hydrolytic enzyme secretion during nutrient foraging and it may be possible to exploit this in production of carbohydrases and proteases.

4.2.2 Control of Yeast Growth and Cell Division

4.2.2.1 The Cell Cycle in Yeasts

The cell cycle can be defined as the period between division of a mother cell and subsequent division of its daughter progeny. It is a fundamental life process for free-growing unicells like yeast (and also for tissue cells) since it defines the proliferative continuity of a growing cell population and ensures that DNA replication, mitosis and cell division are coordinated with cell growth. The regulatory mechanisms that order and coordinate the progress of the eukaryotic cell cycle have been intensively studied in recent years. Yeasts have proved invaluable in unravelling the major control elements of the eukaryotic cell cycle and research with two species in particular has significantly advanced our understanding of cell cycle regulation: the budding yeast,

Table 4.3. Some pathogenic and biotechnologically important yeasts exhibiting dimorphism.

	Yeast	Comments	References
Human pathogens:	*Candida albicans*	Opportunistic pathogen exhibiting hyphal growth *in vivo*	Odds (1994); Gow (1994)
	Malassezia furfur	Hyphae (in association with yeast cells) are seen in cutaneous infections (e.g. *pityriasis versicolor*)	Guillot *et al.* (1995)
	Histoplasma capsulatum	Virulence associated with mycelial–yeast transition	Maresca and Kobayashi (1989)
	Sporothrix schenckii	Dimorphic pathogen sensitive to yeast killer toxins	Polonelli *et al.* (1989)
Plant pathogen:	*Ophiostoma novo-ulmi*	Yeast–mycelial transitions associated with pathogenesis of Dutch elm disease	Webber (1993)
Biotechnologically important yeasts:	*Saccharomyces cerevisiae*	Nutritional and other factors can induce pseudohyphal growth	Morris and Hough (1956); Hill and Robinson (1988); Gimeno *et al.* (1993)
	Schizosaccharomyces pombe	Fission yeast cells can be physiologically induced (e.g. C or N limitation) to form branched hyphae	Johnson and McDonald (1983)
	Kluyveromyces marxianus	Some strains of this lactose-fermenting yeast exhibit oxygen- and growth phase-dependent dimorphism	Walker and O'Neill (1990); O'Shea and Walsh (1996)
	Saccharomycopsis fibuligera	Cells are yeast-like under anaerobic conditions, but filamentous under aerobic conditions	Svoboda and Masa (1970)
	Yarrowia lipolytica	Cells are mostly mycelial in presence of *N*-acetylglucosamine and yeast-like in glucose	Rodriguez and Dominguez (1984); Barth and Gaillardin (1997)

Saccharomyces cerevisiae and the fission yeast, *Schizosaccharomyces pombe*. Such understanding is important in the field of human cancer, which is fundamentally a disease of the cell cycle. That is, cancerous cells grow autonomously having circumvented 'normal' cell cycle controls.

The eukaryotic cell cycle involves both continuous events (cell growth) and periodic events (DNA synthesis and mitosis). The culmination of such events is cell division. Periodic events can be divided into four phases: DNA synthesis (the S phase), a post-synthetic gap (the G2 phase), mitosis (the M phase) and a pre-synthetic gap (the G1 phase). Figures 4.10 and 4.11 depict in schematic form the periodic cell cycle phases of, respectively, *S. cerevisiae* and *Sch. pombe*. Such yeasts have particular advantages over higher eukaryotic cells in cell cycle studies. For example, they grow rapidly on simple media, they can be readily synchronized into division, numerous cell division cycle (*cdc*) mutants are available, molecular genetics is well advanced and key cell cycle 'landmarks' can be easily visualized. Considering the latter point, the following can be observed in

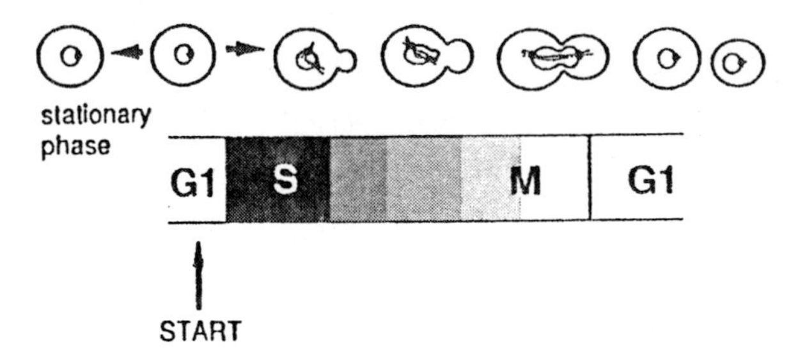

Figure 4.10. Schematic representation of the *S. cerevisiae* cell cycle phases. The position of the haploid cell relative to phases of the cell cycle are depicted. Note that in *S. cerevisiae*, the mitotic spindle forms very early and the S and M phases overlap (resulting in an indistinct G2 phase). Cells can bud only after they have passed Start and can divide only after they have inactivated the cyclin-dependent kinase-cyclin complex known as MPF, making it easy to monitor these cell cycle transitions in living cells. There is, however, no simple morphological or biochemical marker for entry into mitosis. Adapted with permission from Forsburg and Nurse (1991) and *Annual Review of Cell Biology* © 1991, vol. 7, by Annual Reviews, Inc.

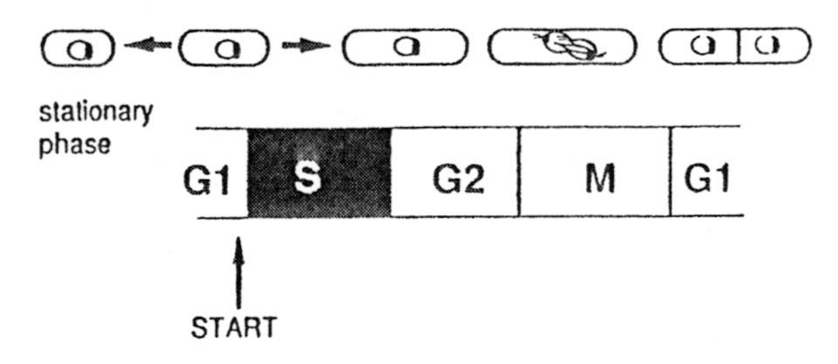

Figure 4.11. Schematic representation of the *Sch. pombe* cell cycle phases. The position of the haploid cell relative to phases of the cell cycle are depicted. Note that in *Sch. pombe*, S and M phases are distinct. That is, there is a clear end of DNA replication and onset of spindle assembly. In addition, *Sch. pombe* (unlike *S. cerevisiae*) lacks a morphological marker for Start. Adapted with permission from Forsburg and Nurse (1991) and *Annual Review of Cell Biology* © 1991, vol. 7, by Annual Reviews, Inc.

the light microscope: bud formation (indication of onset of S phase) and bud size (indication of cell cycle position) in *S. cerevisiae* and cell septum formation (indication of cytokinesis) and cell length (indication of cell cycle position) in *Sch. pombe*. In addition, using video time-lapse phase-contrast microscopy the yeast nucleus, mitosis and septation may be visualized.

It should be stressed, however, that the two yeasts are genetically unrelated and have distinctly different morphologies, life cycles and controls over cell division. Nevertheless, in their own individualistic ways, they have both provided insights into mechanisms which control asymmetric and symmetric cell division in eukaryotes. Table 4.4 summarizes some salient differences in the cell cycles of these two yeasts.

S. cerevisiae and *Sch. pombe* may initiate mitosis at different times during their respective cell cycles but they do share similarities in the way mitosis is conducted. Considering firstly the mitotic apparatus in these yeasts, spindle microtubules are important in governing nuclear migration and eventual chromosome segregation to daughter cells. These intranuclear structures are known to exhibit dynamic changes in morphology and orientation during yeast mitosis. At least six stages of mitosis in *S. cerevisiae* have been distinguished by Yeh *et al.* (1995) using high-resolution time-lapse digital and video-enhanced differential interference contrast microscopy (summarized in Table 4.5). In addition to the polymerization and depolymerization of tubulin (microtubular protein), cytoplasmic dynein is also known to play a key role in yeast spindle dynamics (Yeh *et al.*, 1995). Dynein is a mechanochemical enzyme or *motor protein* which drives microtubule motility in yeast (Streiblova and Bonaly, 1995). Dynein is thought to bind to the plasma membrane at the bud neck and pull the spindle into the bud. Actin filaments, either as cytoskeletal cables or as cortical membrane patches, also play a key role in yeast growth and division and Waddle *et al.* (1996) have described dynamic changes in actin patches during the *S. cerevisiae* cell cycle.

Cytoplasmic as well as nuclear microtubules are also very important in the mechanics of yeast mitosis. In budding yeasts, these cytoskeletal structures emanate from the spindle pole bodies, SPBs (also called microtubule organizing centres,

Table 4.4. Some differences between the cell cycles of *S. cerevisiae* and *Sch. pombe*.

Saccharomyces cerevisiae	*Schizosaccharomyces pombe*
Cell division is asymmetrical	Cell division is symmetrical
Cell growth during the cell cycle is restricted to the developing bud	Mother cells grow only lengthways during the cell cycle
No clear definition between S, G2 and M phases since they overlap	Distinct S and M phases and a clear G2 phase between DNA replication and spindle assembly
In G1, a size-dependent restriction point (Start) limits the onset of S phase	G1 phase is usually very short and remains 'cryptic' due to overlap with previous cytokinesis
Buds form coincidentally with S phase and this provides a morphological marker for Start	Cytokinesis is usually completed in S phase (two haploid nuclei in G1) with no morphological marker for Start
Nutrient starvation blocks passage through Start (in G1)	Start and entry into M phase are equally sensitive to nutrient starvation
Main size-dependent cell cycle control exerted at G1–S transition (Start)	Main size-dependent cell cycle control exerted at G2–M transition
Chromosomes (16) do not show significant condensation at mitosis	Chromosomes (3) condense to some extent at mitosis
Mitotic spindle forms early (in late G1/early S) and no simple morphological marker exists for entry into M phase	Mitotic spindle appears immediately before chromosome condensation
Cells can undergo a diploid mitotic cycle	Unstable diploid unlikely to undergo a mitotic cycle

MTOC, or spindle plaques) toward the new bud and orientate the nucleus and intranuclear spindle at mitosis. In *Sch. pombe*, Hagan and Hyams (1996) have described the dynamics of both intranuclear and cytoplasmic microtubules during mitosis (see Figure 4.6).

The nuclear membrane in both *S. cerevisiae* and *Sch. pombe* remains intact throughout mitosis with the mitotic spindle forming intranuclearly between two SPBs embedded in the nuclear envelope. In *S. cerevisiae*, the mitotic apparatus develops early in the cell cycle with septin ring formation in G1 (Chant, 1996b) and spindles appearing at the G1/S phase. Once the genome replicates, the spindle aligns parallel to the mother-bud axis and elongates eventually to provide each cell with one nucleus. In *Sch. pombe*, spindles assemble immediately before and disassemble shortly after chromosome segregation. Karyokinesis (nuclear division) is accomplished by median constriction of the intact nuclear membrane and nucleoplasm.

Coordination of yeast growth and cell division

Mitchison (1971) indicated that the cell cycle could be dissociated into continuous processes of growth and discontinuous events leading to cell division. There are termed, respectively, the *growth cycle* and the *DNA-division cycle*. In normally dividing cells these 'cycles' are coupled but they can be dissociated by specific inhibitors or mutations. With regard to yeast growth, work with *Sch. pombe* by Mitchison and colleagues (e.g. Mitchison, 1984; Sveiczer *et al.*, 1996) showed that processes associated with cell growth may not be exponentially continuous. For example, cell elongation, total dry mass, total

Table 4.5. Spindle dynamics in *S. cerevisiae* mitosis.

Stage	Description
I	 During pre-anaphase, one spindle pole body (SPB) remains on the mother-bud neck, while the newly synthesized SPB traverses the nuclear envelope destined for the bud
II	 The spindle remains constant in length or elongates very slowly
III	 Spindle rapidly elongates
IV	 'See-saw' movements of the spindle within the mother-bud neck, perhaps governed by cytoplasmic microtubules
V	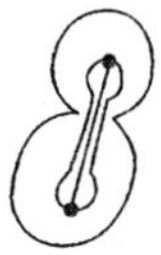 Slower spindle elongation coinciding with transition from 'sausage-shaped' to 'hour-glass'-shaped nuclear morphology
VI	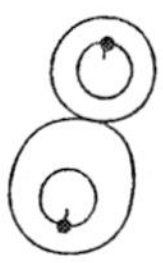 Nucleus divides shortly before cytokinesis

Adapted from information in Yeh *et al.* (1995).

protein synthesis and CO_2 production show increases which are predominantly linear followed by rate change points (RCPs) at characteristic stages of the cell cycle. By analysing cell length and cycle time in synchronized cultures of *Sch. pombe* by time-lapse microscopy, Sveiczer *et al.*, (1996) were able to reveal an important RCP in the G2 phase where the linear rate of lengthways growth increased by around 30%. Periodicity in bulk growth processes in fission yeast may therefore occur with regard to rate changes.

Yeast growth, which is modulated by physiological factors such as nutrient availability is now known to exert a regulatory influence over the DNA-division cycle since cells need to attain a **critical cell size** by growing before progressing through the cell cycle. In other words, growth is rate-limiting for cell cycle progress. Analysis of the asymmetric budding patterns in *S. cerevisiae* has revealed the way in which growth is coordinated with cell division (Figure 4.12).

In *S. cerevisiae*, daughter cell cycles are longer than those of their mothers and thus daughters bud later than mother cells. By altering growth conditions to produce different sized daughters it was shown by Hartwell and his colleagues (Hartwell *et al.*, 1974) that *S. cerevisiae* cells must achieve a threshold or *critical* size at a key transition point called **Start** before they initiate DNA synthesis. Start is the transition in the cell cycle that initiates processes (S phase, budding and spindle pole body duplication) that lead ultimately to cell division (Figure 4.13). Once cells reach a critical size and pass Start, they are irrevocably committed to replicating their DNA and progressing through the cell cycle. Start thus coordinates the cell cycle with cell growth in *S. cerevisiae*.

If *S. cerevisiae* cells are larger than the critical size needed to pass Start, they shorten their cell cycles and divide before doubling in mass. The critical cell size for traversal of Start in *S. cerevisiae* varies with nutrient availability and the cells' growth rate. Nutrient starvation in *S. cerevisiae* blocks passage through Start (at a specialized resting state called G0), but *Sch. pombe*

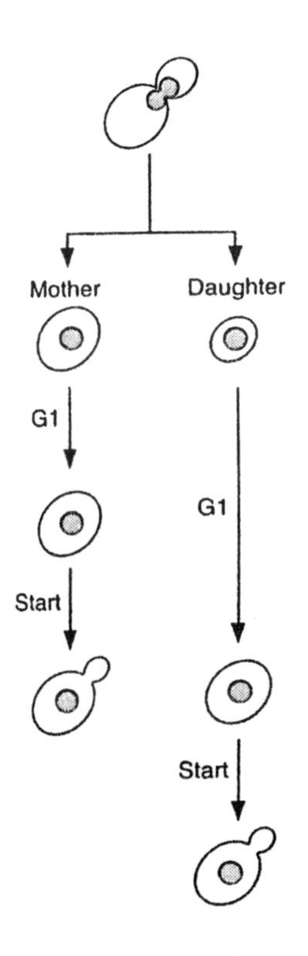

Figure 4.12. Critical cell size for start in *S. cerevisiae*. Daughter cells are smaller than their mothers at cell division and must grow more than their mothers to reach a critical size in order to pass Start (Hartwell and Unger, 1977). Therefore, the time between birth and Start is thus much longer in daughter cells compared with mother cells. From *The Cell Cycle: an Introduction* by A. Murray and T. Hunt. © Copyright 1993 by A. Murray and T. Hunt. Used by permission of Oxford University Press, Inc.

passage through Start and entry into mitosis are equally sensitive to nutrient starvation. In both species, nutrient starvation arrests cells at **checkpoints** in the cell cycle to avoid DNA damage or cell death due to events occurring out of order. Checkpoints are major control points which occur at the G1–S and G2–M boundaries and which can be considered as internal regulatory systems that arrest the cell cycle if prerequisites for progression are not met (Murray, 1995).

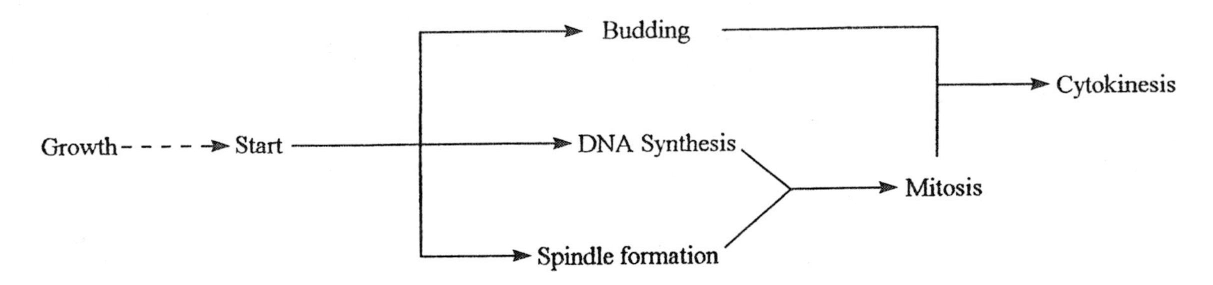

Figure 4.13. Logical map of the *S. cerevisiae* cell cycle. After Start, *S. cerevisiae* cells bud, enter S phase and duplicate their spindle pole bodies. Adapted from Murray and Hunt (1993).

Cell cycle arrest at checkpoints in *S. cerevisiae* due to nutrient starvation may be thought of as a survival mechanism under adverse conditions and this induces a response that includes cessation of growth, decreased protein synthesis and increased resistance to stress. Entry of *S. cerevisiae* cells into the G0/stationary phase (see section 4.3.2) when starved of nutrients represents only one of four alternative developmental fates in this yeast following the end of mitosis. That is, cells can either:

- Pass through Start and complete the cell cycle (if nutrients are present),
- Enter G0 in a state of proliferative rest (if starved of nutrients),
- Undergo conjugation (if different haploid mating types are present), or
- Undergo meiosis and sporulation (if diploid and nitrogen-starved).

Therefore, passage through Start (to culminate in cell division) may be viewed as a process in which *S. cerevisiae* cells ask themselves: are there nutrients present? are mating pheromones absent? am I big enough? (Wheals, 1987).

In addition to a cell size control at Start, *S. cerevisiae* also has a critical size for entry into mitosis. This is usually cryptic (invisible) in wild-type cells but is revealed in certain mutants which possess reduced critical cell size for passage through Start. The period between Start and mitosis then increases since cells must grow to reach the size needed for entry into mitosis.

In *Sch. pombe*, analysis of a particular mutant, the *wee* mutant, by Nurse (Nurse, 1975) revealed that in fission yeast a critical cell size controlled entry into mitosis. Although *wee* mutants grow normally at 25°C, at a restrictive temperature of 35–37°C they shorten their cell cycle and divide at only half the size of wild-type cells (Figure 4.14). Therefore, the minimum size required to pass the cell cycle checkpoint at mitosis is greatly reduced.

The product of the *wee* 1^+ gene thus restrains the cell cycle until cells reach a critical size. The wee1 protein has now been characterized (see below) and acts as part of a homeostatic mechanism to maintain *Sch. pombe* cells at a constant size at birth. This mechanism ensures that cells born smaller than normal have long cell cycles and, conversely, cells born larger than normal have short cell cycles. As in *S. cerevisiae*, the mean size of *Sch. pombe* varies with nutrient availability and cell growth rate (Fantes and Nurse, 1977). Subsequent analysis of *wee* mutants in recent years has revealed that two critical cell size checkpoints exist in the *Sch. pombe* cell cycle: one at Start to initiate DNA replication; and the other at the onset of mitosis. It is important to note that as *S. cerevisiae* has a cryptic size control over mitosis, *Sch. pombe* similarly has a cryptic size control over Start since wild-type cells are born large enough to initiate DNA replication.

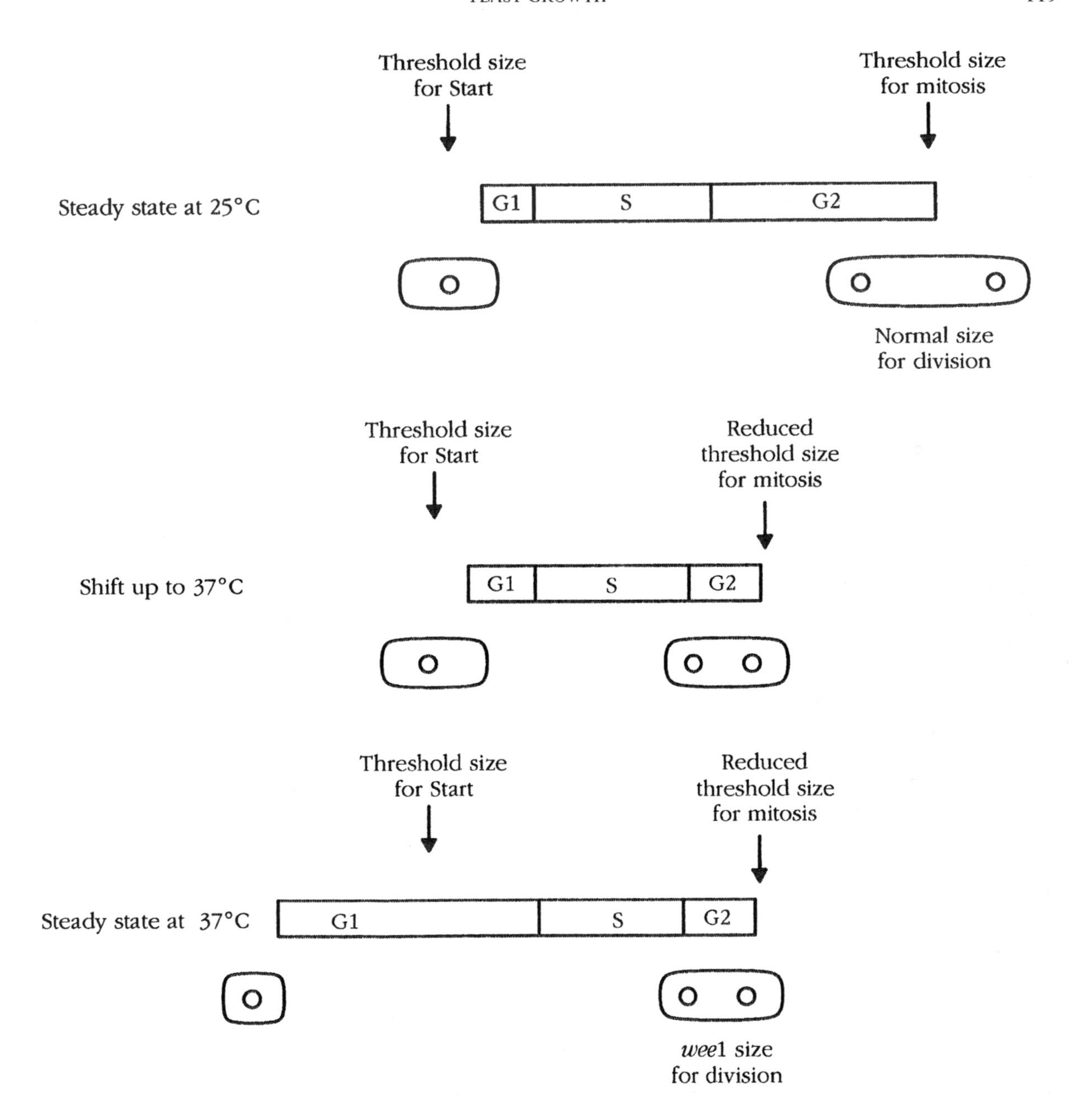

Figure 4.14. Summary of the cell cycle characteristics of the *wee* mutant of *Sch. pombe.* When the *wee* mutant is shifted to the restricted temperature, the minimum size for mitosis is reduced as is the cell cycle length since cells do not have to double in size in order to divide. However, subsequent cycles at the restrictive temperature are of normal duration, since the cells, now smaller at birth, must again double in size before they can divide (Nurse, 1975). Adapted from Murray and Hunt (1993).

4.2.2.2 Molecular Aspects of Cell Cycle Control in Yeasts

It is evident from the above that research with yeasts has revealed that the cell cycle is regulated at Start and at entry into mitosis. These so-called checkpoints are primarily influenced by cell size and nutrient availability. Recent understanding of the molecular biology of the cell cycle checkpoints at G1–S and G2–M has come mainly from studies, respectively, with *S. cerevisiae* and *Sch. pombe*.

G1–S control

Entry into S phase of the cell cycle commits yeast to completing one round of cell division and this entry is governed at Start. Work with *S. cerevisiae* has elucidated some of the molecular mechanisms of the Start cell size-dependent G1 checkpoint. For example, when *S. cerevisiae* reaches a certain size the level of proteins known as **cyclins** (Clns) dramatically increases. Cyclins are proteins, originally discovered in marine invertebrate cells, which are synthesized and degraded at specific points in the cell cycle of eukaryotes. Yeast cyclins are periodically expressed and different cyclins are now known (e.g. at least 11 in *S. cerevisiae*) to be involved in control at G1 (G1 cyclins), G2 (B-type cyclins) and DNA synthesis (S-phase cyclins) (see Futcher, 1996; Küntzel *et al.*, 1996). G1 cyclins in *S. cerevisiae* are transcriptionally regulated and Breeden (1996) has reviewed the role of transcription initiation factors involved in regulating G1–S transition in this yeast. A particular G1 cyclin, Cln3, has been identified as a possible sensor of cell size in *S. cerevisiae* which acts by modulating levels of other cyclins (e.g. Cln1 and 2) which decline once cells pass Start (Tyers *et al.*, 1993). Futcher (1996) has noted that it is difficult to assign specific roles to the various cyclins of *S. cerevisiae* due to their partly overlapping functions.

In addition to cyclin accumulation in G1, the activity of a **cyclin-dependent kinase** (CDK) is involved as an effector of Start and the initiation of budding in *S. cerevisiae*. The particular CDK which plays a crucial role in the induction of Start in *S. cerevisiae* is encoded by the *CDC28* gene. Cdc28 kinase couples growth with cell division through its association with the unstable G1 cyclins. Thus, G1 cyclins activate the Cdc28 kinase to promote transition from G1 to S phase of the cell cycle. The CDK–cyclin complex has been characterized (Nasmyth, 1993) and has been referred to as S phase-promoting factor or SPF. *CDC28* of *S. cerevisiae* shows sequence and functional homology to *Cdc2* of *Sch. pombe* and is now referred to as CDK1. The *CDC28/Cdc2* gene encodes a 34 kDa protein kinase which, as well as being essential for S phase, is also important in controlling entry into mitosis (Nasmyth, 1993; Diffley, 1995; Lew and Reed, 1995). It has therefore been described as 'the linchpin of cell cycle control' (Forsburg and Nurse, 1991). Küntzel *et al.*, (1996) have reviewed the association of the following cyclins with the Cdc28 protein kinase during *S. cerevisiae* cell cycle progression: Cln1–3 (at G1), Clb5,6 (at S), Clb3,4 (at S/G2) and Clb1,2 (at M). Alternation of cell cycle phases appears to be due to mechanisms that ensure that one cyclin family succeeds another. In this regard, Blondel and Mann (1996) have shown that G2 cyclins are necessary for proteolytic degradation of G1 cyclins in *S. cerevisiae*. Thus, G2 cyclin synthesis is coupled with removal of G1 cyclins.

In addition to CDK activation, the presence of CDK inhibitor proteins known as CKIs play an important role in blocking CDK activity in the G1 phase of the cell cycle. CKIs represent a key mechanism by which the onset of DNA replication is regulated in eukaryotic cells and loss of CKIs are thought to contribute to unregulated cell proliferation characteristic of tumour cells. Examples of CKIs in *S. cerevisiae* are *SIC1* (Schwob *et al.*, 1994) and in *Sch. pombe*, *rum 1* (Labib and Moreno, 1996). Futcher (1996), Huberman (1996) and Nasmyth (1996) have discussed the importance of Sic1 proteolysis in triggering the G1–S transition in budding yeast.

Although many gene products are involved in

G1 progression in yeasts, other factors play important roles in influencing such progression. In terms of physiological effectors which regulate Start, mating pheromones (see section 4.2.3) and nutrient availability are important and the former may be incorporated into a scheme of G1–S regulation in *S. cerevisiae* as shown in Figure 4.15.

Nutrient availability in yeast is known to play key roles in regulating the G1–S cell cycle checkpoint. For example, nutrient levels regulate the intracellular concentration of cAMP via a small **G protein** called *Ras*. The so-called Ras/cAMP pathway is a well-documented nutrient regulated signalling pathway in yeasts (Thevelein, 1992; Wittenberg and Reed, 1996). Ras proteins are involved in many aspects of eukaryotic cell physiology in addition to signal transduction. In *S. cerevisiae*, Ras proteins bind GTP in response to glucose and other carbon sources and increase cAMP levels by activating adenylate cyclase. Cellular levels of cAMP and the activities of cAMP-dependent kinases (e.g. trehalase, acid phosphatase) are implicated in progression through the G1 phase of the cell cycle (Figure 4.16). Thevelein (1994) has suggested that only a basal level of cAMP-dependent protein kinase activity may be necessary for progression through the cell cycle, although activity levels may be much higher than this threshold.

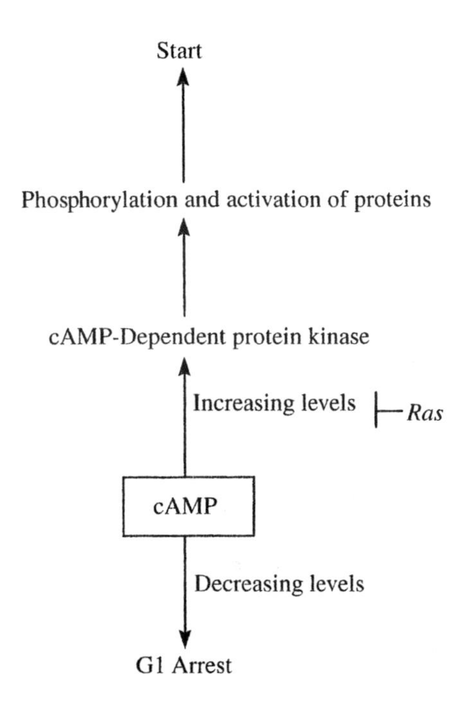

Figure 4.16. The high-go, low-stop model of cAMP-mediated control of G1 progression in *S. cerevisiae*. Adapted from Wheals (1987).

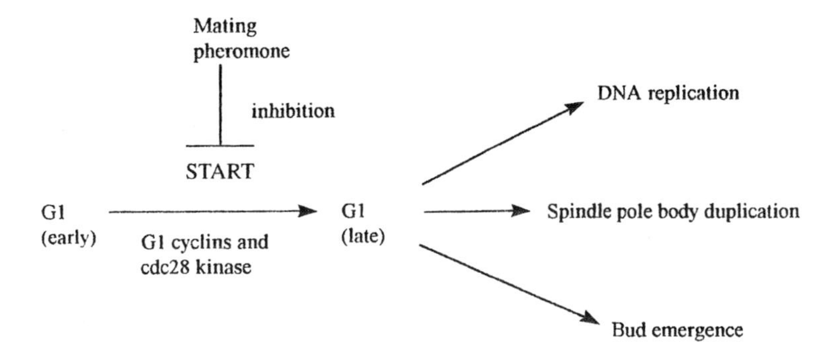

Figure 4.15. Generalized scheme of G1S regulation in *S. cerevisiae*. Adapted from Nasmyth (1993).

Wheals (1987) has discussed the roles of cAMP in cell cycle regulation in yeast and has stated that 'cAMP acts as master switch capable of preventing Start in the absence of favourable conditions'.

Recent research on the role of the Ras/cAMP pathway has provided a link between the availability of nutrients and cell cycle progression at the G1 phase. For example, different effects of cAMP on the synthesis of G1 cyclins have been observed and these can have either positive or negative effects on passage through the G1 checkpoint (Wittenberg and Reed, 1996). It now appears that nutrient-sensing mechanisms involving cAMP influence the levels of G1 cyclins to enable *S. cerevisiae* cells to control their cell cycle in response to external stimuli.

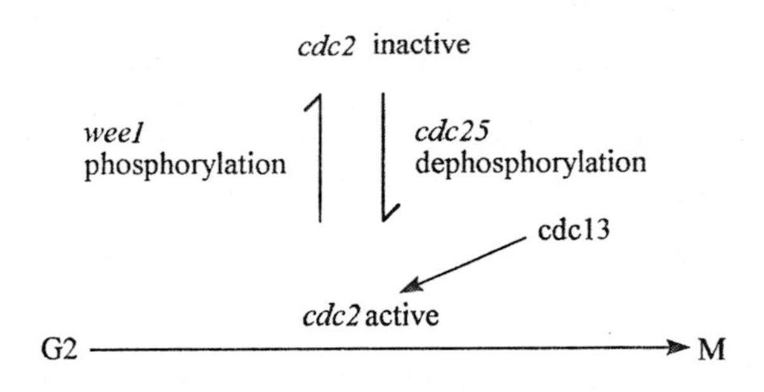

Figure 4.17. Control of mitosis on *Sch. pombe.* Both *cdc2* (which encodes a p34 protein kinase) and *cdc13* (which encodes p56 B-type cyclin) are required for entry into mitosis. $p34^{cdc2}$ is stimulated by a *cdc25*-encoded phosphatase and inhibited by *wee1*-encoded kinase.

G2–M control

Studies with the fission yeast, *Sch. pombe* have played a very important role in our molecular understanding of the control of mitosis in eukaryotic cells (Nurse, 1990; Sheldrick and Carr, 1993; Fanhauser and Simanis, 1994). Key to such control is the 34 kDa protein kinase, designated as **$p34^{cdc2}$** (to signify that the protein is encoded by the $cdc2^{+}$ gene) and its interaction with **B-type cyclins** (encoded by *cdc13*, *cig1* and *cig2*). The cyclin-dependent kinase (CDK)–cyclin complex at mitosis is known as MPF [Mitosis (or Maturation) Promoting Factor]. Activation of the $p34^{cdc2}$ kinase (CDK1) by complexing with *cdc 13*-encoded cyclin brings about the onset of mitosis, whereas degradation of *cdc 13* leads to inactivation of the protein kinase which is a prerequsite for mitotic exit. Although the identity of the substrates for activated $p34^{cdc2}$ are unclear, there is evidence in *Sch. pombe* that phosphorylation of nuclear lamins is involved (Enoch *et al.*, 1991).

Numerous genes and gene products interact with CDK 1 at mitosis. In *Sch. pombe*, the CDK is primarily regulated at the G2/M cell-size checkpoint by gene products encoded by *wee 1* and *cdc 25* (Figure 4.17). Thus, the *wee1* product (a 107 kDa kinase) inhibits mitosis by phosphorylating the $p34^{cdc2}$ kinase, whereas the *cdc25* product (an 80 kDa phosphatase) induces mitosis by dephosphorylating $p34^{cdc2}$ (Feilotter *et al.*, 1992). In actively dividing *Sch. pombe* cells, $p34^{cdc2}$ remains phosphorylated until a critical cell size is reached in G2. The protein kinase is then activated by dephosphorylation, and following interaction with the *cdc13* cyclin, this results in driving cells into mitosis. Such regulation is reflected in the morphological phenotypes of the respective *Sch. pombe* mutants. That is, $wee1^{+}$ mutants have a diminished G2 phase (but extended G1) and divide at a small size; whereas mutants lacking $cdc25^{+}$ elongate and arrest in G2. It is known that the phosphorylation of a single amino acid residue on $p34^{cdc2}$, namely tyrosine 15, regulates activity of MPF (CDK–cyclin complex) and this is crucial for inhibiting G2–M transition (Gould and Nurse, 1989).

$p34^{cdc2}$ exists in more than one state and Broek *et al.* (1991) showed that an S-form of the kinase was appropriate for entry into S phase and an M-form for entry into mitosis. Stern and Nurse (1996) have recently proposed a quantitative model for the *Sch. pombe* cell cycle in which different levels of CDK *activity* control both entry into mitosis and initiation of DNA replication. Thus, a moderate level of kinase activity is sufficient to trigger S phase, whereas a higher level is required for entry into mitosis (Figure 4.18). This ensures that a cell at the beginning of a new cell cycle initiates DNA replication before mitosis. This then avoids potentially catastrophic improper chromosome segregation when the cell divides.

The p34^{cdc2} kinase originally studied fission yeast is also essential for mitosis in *S. cerevisiae* and is now known to be a central component of mitotic control in eukaryotes from yeast to humans. In fact, all eukaryotic cells have a functional homologue of *cdc2* and a human *cdc2*$^{+}$ gene can even complement the *cdc2* mutation in *Sch. pombe* (Lee and Nurse, 1987; Igarashi *et al.*, 1991). In addition to investigating fundamental mechanisms of control in the G2–M transition of

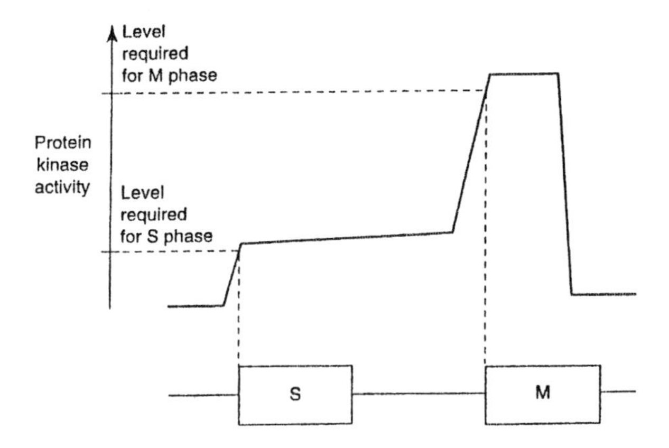

Figure 4.18. Quantitative model for the control of S phase and mitosis in *Sch. pombe.* Based on observations that a single B-type cyclin (*cdc13*) can promote both S phase and mitosis on *Sch. pombe*, Stern and Nurse (1996) have challenged the widely held view that G1- and G2-specific cyclins define the role of *cdc2* at the onset of DNA replication and mitosis. The model proposes that there are only quantitative differences between the *cdc2–cdc13* complexes in G1 and G2. A moderate level of kinase activity is sufficient to trigger S phase and a high level is required for mitosis, thus ensuring that a cell at the beginning of a new cell cycle initiates S phase before mitosis. Reproduced with permission from Stern and Nurse (1996) and Elsevier Science Ltd.

the cell cycle, the p34^{cdc2} kinase in purified form (Beaudette *et al.*, 1993) may also assist in identifying potential anti-mitotic drugs in the treatment of cancer. Indeed, several biotechnology companies are now focusing on several *cdc* gene products for novel anti-cancer drug design targets, including the activator of p34^{cdc2} kinase, *cdc25*-encoded phosphatase (Sedlak, 1997).

In addition to molecular genetic aspects of G2–M control in fission yeast discussed above, several environmental factors have been shown to influence and possibly regulate the onset of mitosis during the cell division cycle in *Sch. pombe.* Considering firstly the availability of major nutrients, a nutrient-modulated size control occurs in *Sch. pombe* and this acts to coordinate growth and cell division (as discussed previously). Thus, cells wait to attain a critical size before entering mitosis and this depends on the level of available nutrients, especially the nitrogen source (Fantes, 1984). However, although it is known that cells monitor how big they are, the way by which they actually sense cell size is not known. Futcher (1996) has discussed a size sensor model (proposed by K. Nasmyth) based on critical levels of the G1 cyclin, Cln3, in the yeast nucleus. This may represent a *putative unstable effector* molecule which varies during the *S. cerevisiae* cell cycle, previously proposed by Wheals (1987) to be involved in cell size monitoring. Walker and his colleagues (Walker and Duffus, 1980; 1983; Walker, 1986) have proposed that cellular magnesium ion concentration acts as the transducer for cell size that coordinates growth with mitosis in *Sch. pombe* (and perhaps in other eukaryotic cells; see Walker, 1986). Other physiological effectors, including: changes in the homeostasis of other metal ions such as calcium (Duffus and Patterson, 1974; Anraku *et al.*, 1994); variation in cell pH (e.g. Imai and Ohno, 1995); cAMP levels (Watson and Berry, 1977; Wheals, 1987; Baroni *et al.*, 1994; Tokiwa *et al.*, 1994); and responses to fluctuations in physical growth conditions such as temperature (Kramhøft *et al.*, 1978; Agar and Bailey, 1982) may also be involved in the complex web of mechanisms which control cell division in *S. cerevisiae*, *Sch. pombe* and other yeasts.

4.2.3 Sexual Reproduction in Yeasts

Many yeasts have the ability to reproduce sexually, but the processes involved are best understood in *S. cerevisiae* and *Sch. pombe*. Although both species have the ability to mate, undergo meiosis and produce spores, they possess distinct sexual life cycles as depicted in Figures 4.19 and 4.20.

In *S. cerevisiae*, mating involves the conjugation of two haploid cells of the opposite mating types, designated **a** and α. These cells synchronize each others' cell cycles at Start in response to peptide mating pheromones known as **a factor** and α **factor.** Once cells progress through Start they are unable to mate until the next cell cycle. Conjugation occurs by the surface contact of specialized projections ('schmoo' formation) on mating cells followed by plasma membrane fusion to form a common cytoplasm. Karyogamy (nuclear fusion) then results in a diploid nucleus. The stable diploid zygote resulting from haploid cell fusion will continue mitotic cell cycles in rich growth media. Once starved of nutrients or if grown on a non-fermentable carbon source such as ethanol or acetate, the *S. cerevisiae* cells are induced to undergo meiosis and sporulation results in the formation of four haploid spores (2**a** and 2α). These spores can in turn germinate in rich media and mate once again to form diploids. After germination, haploid cells of *S. cerevisiae* have the capability of undergoing **mating type switching** which maximizes the chances of diploidy. Note that wild-type strains of *S. cerevisiae* are homothallic and **a**/α diploids represent the usual vegetative state of this yeast. Nevertheless, it should also be stressed that industrial (e.g. brewing) strains of *S. cerevisiae* are generally polyploid and do not undergo a sexual life cycle as described above for 'scientific yeasts'.

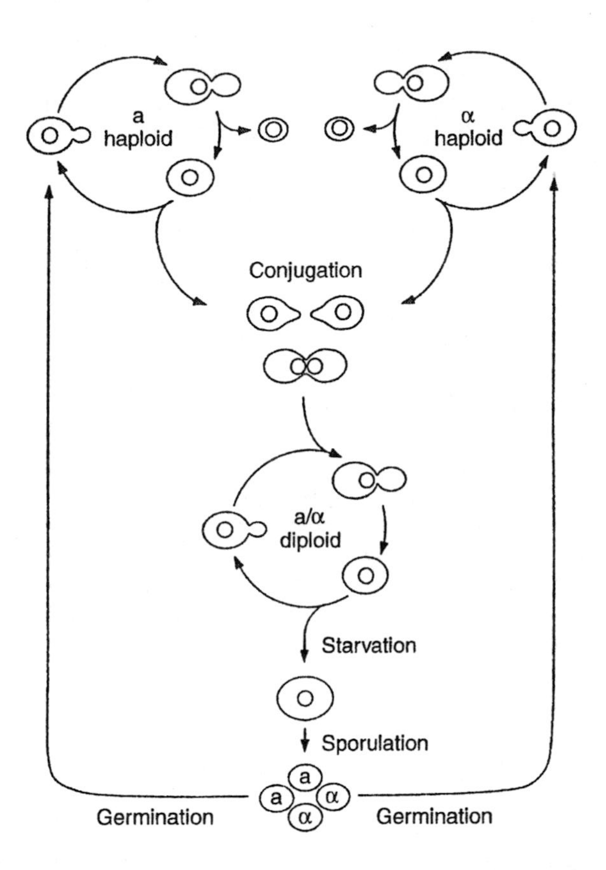

Figure 4.19. Life cycle of *Saccharomyces cerevisiae. S. cerevisiae* can reproduce asexually by budding, as either haploid or diploid cells. Haploid cells exist in two mating types a and α and they can mate to form diploid cells. Cells of opposite mating types secrete mating factors (pheromones) that bind to receptors on each others' surface. This chemical signalling induces cell cycle arrest and the expression of genes that carry out the physical process of mating. If starved of nitrogen, the diploid cells will sporulate to yield four haploid spores. These germinate to form haploid budding cells that can mate with each other to restore the diploid state. From *The Cell Cycle: an Introduction* by A. Murray and T. Hunt. Copyright © 1993 by A. Murray and T. Hunt. Used by permission of Oxford University Press, Inc.

In *Sch. pombe*, haploid cells of the opposite mating types designated h^+ and h^- (or *P* and *M*), secrete mating pheromones and when starved of nutrients (especially a nitrogen source) undergo conjugation to form a diploid. In *Sch. pombe*, however, such diploidization is transient under starvation conditions and cells soon enter

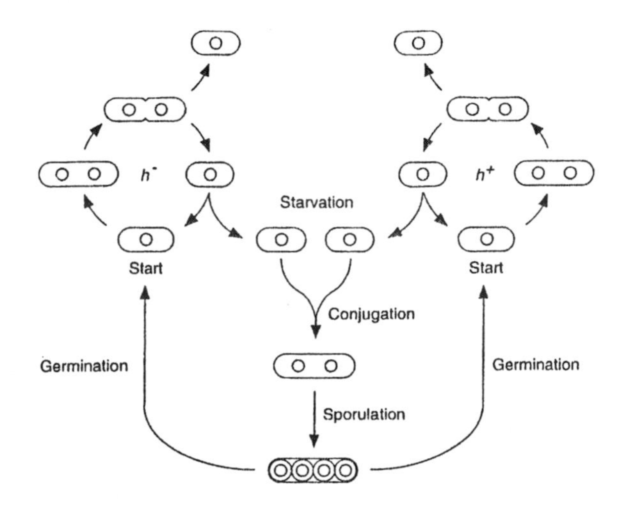

Figure 4.20. Life cycle of *Schizosaccharomyces pombe.* Unlike budding yeast which can divide as either a haploid or diploid, *Sch. pombe* can divide only as a haploid. The haploid exists as one of two mating types h^+ and h^- Cells of the opposite mating types can mate with each other when starved of nitrogen, and immediately after mating they enter meiotic cell cycle and produce four haploid spores. From *The Cell Cycle: an Introduction* by A Murray and T. Hunt. Copyright © 1993 by A. Murray and T. Hunt. Used by permission of Oxford University Press, Inc.

meiosis and sporulate to produce four haploid spores. (Haploidy is the usual vegetative state of wild-type homothallic *Sch. pombe*.) This behaviour of fission yeast has been described as an emergency response to nutrient limitation (Forsburg and Nurse, 1991).

In recent years, our understanding has progressed significantly with regard to the physiology and molecular biology of sexual reproduction in yeasts. For example, a great deal is now known about the regulatory mechanisms of mating-type switching, the mating pheromone response and meiosis in both *S. cerevisiae* and *Sch. pombe* (e.g. Johnson, 1995; Roeder, 1995; Schultz *et al.*, 1995; Sprague, 1995; Amon, 1996; Wittenberg and Reed, 1996; Yamamoto, 1996). Sprague (1995) has reviewed similarities and differences in the sexual reproduction mechanisms between *S. cerevisiae* and *Sch. pombe*.

With regard to mating in *S. cerevisiae*, cells of opposite mating type sense each others' presence by chemical signalling and choose a mating partner according to the 'strength' of pheromonal signals (Jackson and Hartwell, 1990). Thus, α cells communicate with **a** cells by secreting a (13 amino acid) peptide, α-factor pheromone, which binds to **a** cell plasma membrane receptor proteins. Conversely, **a** cells secrete a (12 amino acid) farnesylated lipopeptide, **a**-factor pheromone, which binds likewise to α cell receptors. Although each mating type produces a unique pheromone and pheromone-receptor, the two pheromones elicit common changes in cell physiology and gene expression which eventually leads to the conjugation of two mating cells. These changes may be summarized (according to Sprague, 1995) as:

- increased gene transcription whose products catalyse mating;
- arrest of the cell cycle in the G1 phase; and
- alteration of cell polarity and morphology to locate mating partners.

Some of the key mechanisms of the molecular responses of *S. cerevisiae* cells to mating pheromones will now be considered. Both **a** and α factors in this yeast activate a complex signal transduction pathway in mating cells which culminates in the arrest of cells at Start and the onset of the physical process of mating. The molecular genetics of the pathway has been extensively studied in budding yeast and resembles the intracellular signals elicited when mammalian cells respond to growth factors.

Yeast pheromones and their corresponding receptor proteins interact initially with a heterotrimeric *G protein*. G proteins possess GTPase activity and are involved in regulating intracellular signalling pathways in yeast (Chant and Stowers, 1995). One subunit (α) of this protein binds GDP and in response to mating pheromone exchanges GDP for GTP before dissociating from the other subunits (β and γ). These latter subunits of the G protein initiate the signalling pathway by activating a protein kinase

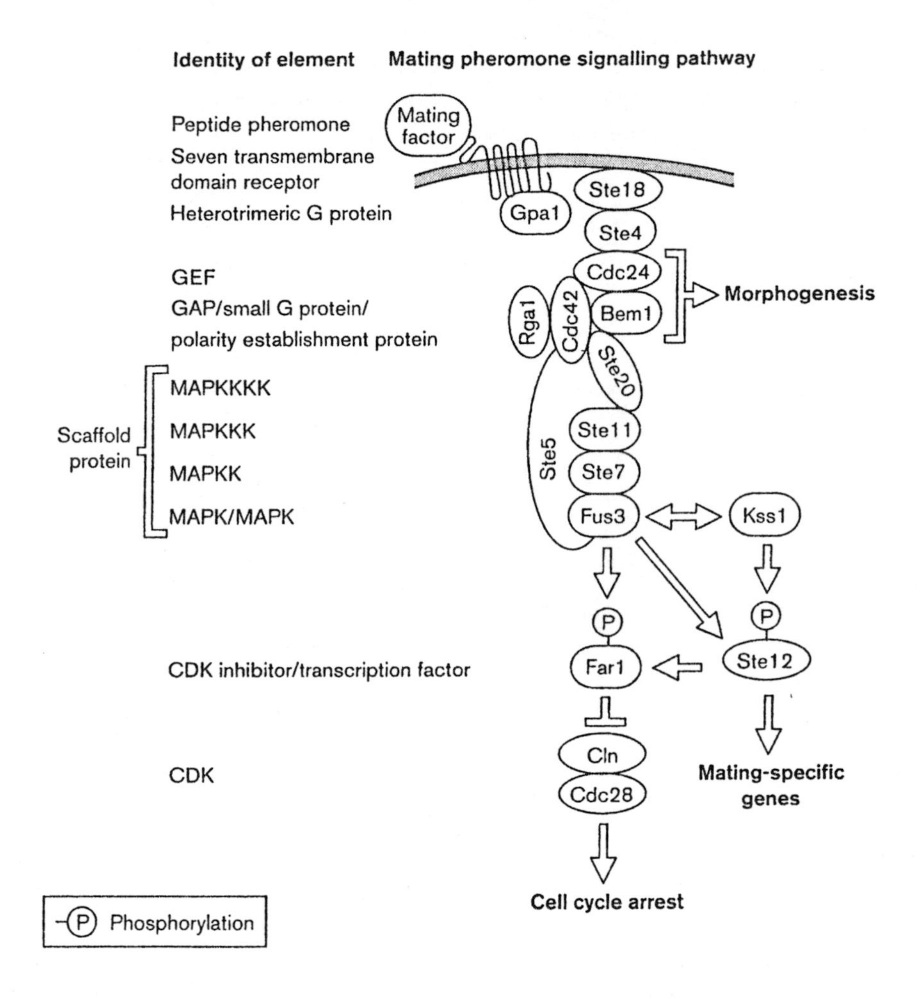

Figure 4.21. The pheromone response pathway of *S. cerevisiae*. The identities of the elements as indicated either by their homology with known proteins or their demonstrated functions are shown (left). Not all associations depicted have been directly demonstrated. MAPKK, MAPK kinase; MAPKKK, MAPKK kinase; MAPKKKK, MAPKKK kinase. Reproduced with permission from Current Biology Ltd. and Wittenberg and Reed (1996) who have described the model in further detail.

cascade. This eventually leads to cell cycle arrest and sexual conjugation. The protein kinases in question are homologous to the MAPKs of higher eukaryotes and their activity is a prerequisite for transmission of the mating pheromone signals. As shown in Figure 4.21, the upstream linkage in the *S. cerevisiae* pathway between the G protein and the protein kinase cascade has now been established. This involves the *CDC42* gene (Simon *et al.*, 1995) whose product is known to modulate the polarity of the actin cytoskeleton during *S. cerevisiae* budding. It has been proposed that *CDC42* has a dual function in coordinating cellular morphogenesis with other mating pheromone responses (Wittenberg and Reed, 1996). Other similarities between control of budding and mating in *S. cerevisiae* are apparent when considering the degradation of cyclins and inhibition of cyclin-dependent kinase (Cdc28) which cause G1 arrest at the end of the pheromone response pathway. Leberer *et al.*, (1997) have reviewed recent developments in the regulation of the *S. cerevisiae* pheromone response pathway.

With regard to the *Sch. pombe* pheromone response pathway, there are strong similarities with the events which occur in *S. cerevisiae*. Nevertheless, interesting differences do exist and these have been discussed by Sprague (1995). For example, in the *Sch. pombe* G protein it is the α subunit (rather than the $\beta\gamma$ subunit in *S. cerevisiae*) which is the positive transducer of the mating signal. In addition, in *Sch. pombe* the *ras1* protein is necessary to propagate the signal and this dependence on *ras* in fission yeast shows parallels with mammalian signal-transduction pathways.

4.3 POPULATION GROWTH OF YEASTS

4.3.1 Colonial Yeast Growth

Morphological and growth kinetic characteristics of yeast colonies growing on solid media such as agar shall firstly be considered. The morphological characteristics of agar-grown yeasts are very diverse with variations observed in the colour, texture and geometry (peripheries, contours) of giant colonies (Figure 4.22).

Individual cells within a yeast colony also show heterogeneity with respect to age, growth kinetics, morphology and metabolism. For example, yeast species such as *S. cerevisiae* and *K. marxianus* which normally grow as budding cells in liquid media, may produce pseudomycelia when grown on agar (Morris and Hough, 1956; Walker and O'Neill, 1990; Gimeno *et al.*, 1992). Such filamentous growth forms are often evident both at the peripheral rough margin of the colony and on its undersurface where pseudomycelia penetrate into the agar. Colonial dimorphism of yeasts is thought to represent nutrient exhausted cells extending the leading edge of the colony in the search for nutrients. In fact, depletion of nutrients (glucose) and oxygen together with an accumulation of toxic yeast metabolites (ethanol) are thought to be primarily responsible for dictating the growth characteristics of individual cells in a yeast colony (Gray and Kirwan, 1974; Jones and Gray, 1978).

In addition to dimorphic growth changes on agar, some yeasts are also known occasionally to develop papilliae or 'tumours' on colonies resulting from adaptive mutations. For example, Rojas and Vondrejs (1995) have described aggressively growing papillae in *rad6* mutants of *S. cerevisiae* which have the capability of growing over neighbouring colonies.

Kamath and Bungay (1988) have made a detailed analysis of growth kinetics in a yeast colony. Both the radial growth rate and the rate of increase in height of colonies of *Candida utilis* were shown to exhibit linear, rather than exponential, increases. Assuming growth is restricted to the periphery, linear radial growth rate of a yeast colony may be expressed as (Hough *et al.*, 1982):

$$rt = K_{rt} + r_o$$

where K_r is the radial growth constant, r is the colony radius at time t and r_o is the radius at zero time.

Figure 4.23 depicts growing zones in a yeast colony and four phases have been distinguished by Kamath and Bungay (1988), depending on the colony height (**h**):

Phase 1 (h ≤ d): when the colony height is less than the oxygen penetration depth (**d**), the entire colony grows exponentially since glucose and oxygen are available throughout the colony (Figure 4.23A).

Phase 2 (d < h ≤ hg): when the colony height increases beyond **d** and until it reaches **hg**, glucose is available throughout the colony, but oxygen is available only on the surface resulting in an annular region of constant thickness which is the growth zone in this phase (Figure 4.23B).

Phase 3 (hg < h < hg + d): when the colony height increases beyond **hg**, there are regions in the

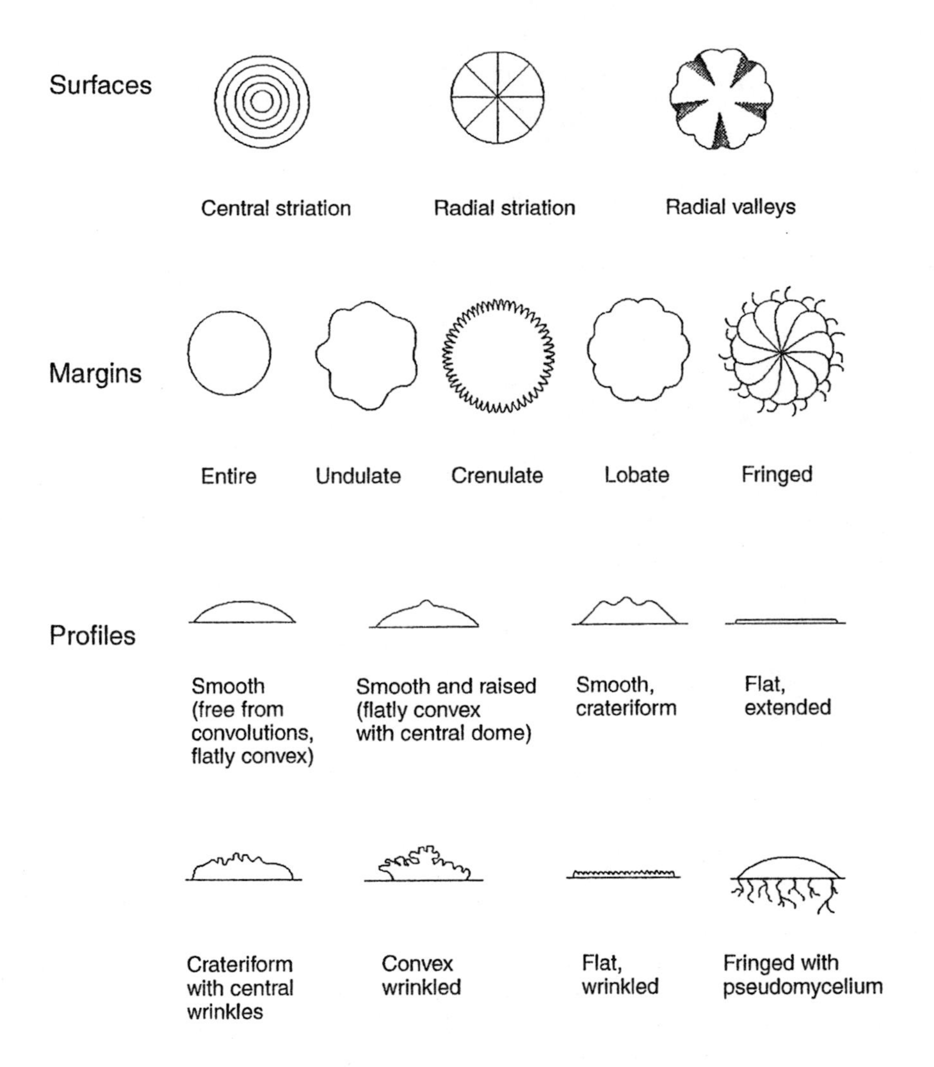

Figure 4.22. Diversity of giant colony morphology of yeasts.

colony (at the top) which are deficient in glucose as well as regions where oxygen cannot diffuse. This phase leads to a growth zone in the form of a truncated annular spherical segment (Figure 4.23C).

Phase 4 (h = hg + d): In this phase, the colony height remains constant at the value **(hg + d)**. Here, only the outer peripheral region of constant thickness **d** and constant height **hg** is the growth zone (Figure 4.23D).

Among these phase relationships, **d** = depth of oxygen penetration into the colony; **h** = height of the colony; and **hg** = height to which glucose from the solid medium can diffuse into the colony.

A rather specialized form of yeast growth on solid media has been observed with the dimorphic pathogenic yeast, *Candida albicans*. Gow and colleagues (Sherwood *et al*., 1992; Gow *et al*., 1994) have observed that when cells of this yeast were grown on membrane filters placed on

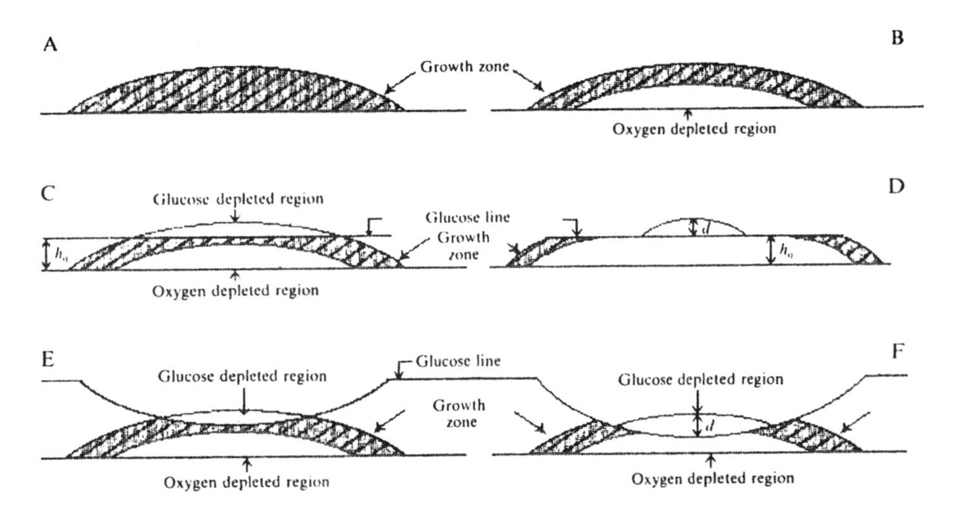

Figure 4.23. Growth phases in a yeast colony. A = growth throughout colony, no limitation; B = no growth in regions remote from air; C = growth in top portion limited by glucose, lower portion limited by oxygen; D = growth only at periphery; overlap of limitations; E = more severe glucose depletion when not supplied continuously; F = glucose not supplied continuously; overlap of limitations. Reproduced with permission from Kamath and Bungay (1988) and the Society for General Microbiology.

top of serum-containing agar, hyphae grew over the membrane surface and through the pores of the membrane (Figure 4.24).

Hyphae of *C. albicans* have also been shown to follow grooves and ridges of inert surfaces (Gow *et al.*, 1994) and such growth responses are due to the phenomenon of **thigmotropism** (contact-mediated sensing and contour guidance) associated with trophic movement of the cells in relation to changes in substratum topography. Chemotropism was ruled out due to the fact that hyphae on the underside of the membrane filters grew into pores *away* from nutrient agar. *C. albicans* thigmotropism, together with characteristic helical or 'cork-screwing' surface growth of this yeast (Sherwood-Higham *et al.*, 1994), may facilitate hyphal penetration of weakened epithelial cell surfaces during pathogenesis of host tissue.

The colonization of yeasts on solid support materials also occurs when cells are immobilized. The immobilization of yeasts onto inert carriers has many advantages over free cell suspension culture in industrial processes and biotechnological applications are discussed further below. Yeast cells may be successfully immobilized either by entrapment, aggregation, containment, attachment or deposition. Table 4.6 summarizes some materials used to immobilize yeast cells using these methods.

Some of the important physiological and metabolic changes which occur in yeast when they are immobilized shall now be briefly considered. Knowledge of the physiological features of immobilized yeasts is considered vital for the successful biotechnological applications of immobilized bioreactors. Considering firstly growth characteristics of immobilized yeast, several studies have shown that cell morphology, budding patterns and growth rates are all affected by immobilization (reviewed by Walsh and Malone, 1995). For example, Masschelein *et al.*, (1994) have discussed growth patterns of gel-entrapped *S. cerevisiae* cells which appear quite different from growth in free cell suspensions. Immobilized yeast cells exhibits four characteristic growth phases:

- Logarithmic phase
- Linear phase I
- Linear phase II
- Pseudostationary phase

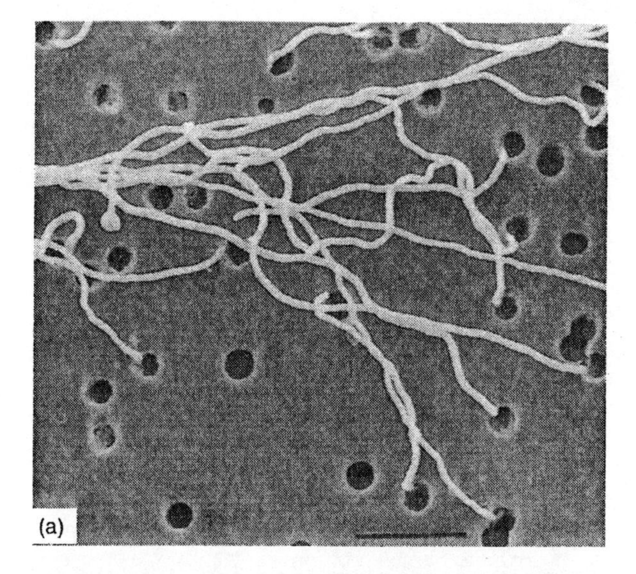

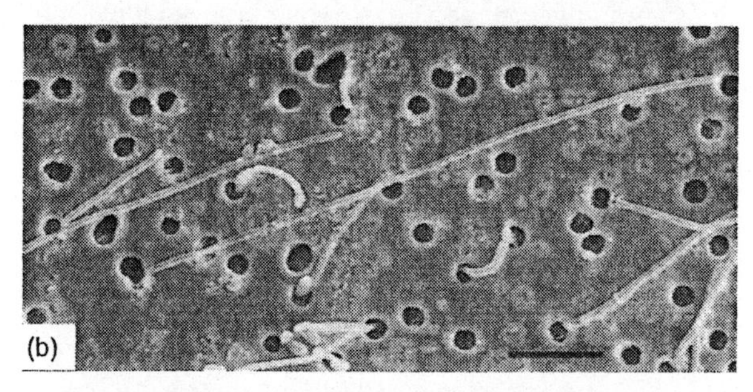

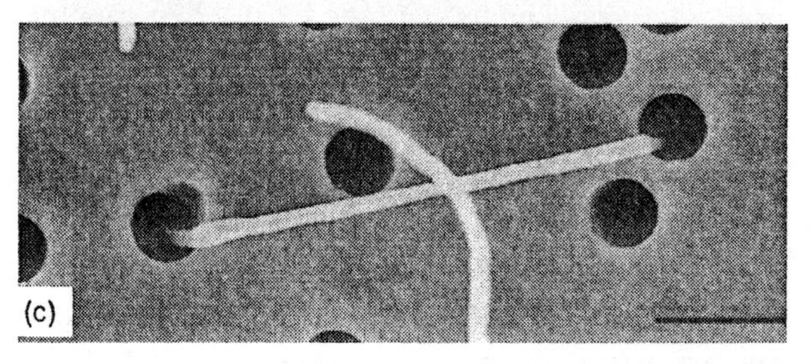

Figure 4.24. Thigmotropism of *Candida albicans*. Scanning electron micrographs of hyphae of *C. albicans* growing into the pores of a nucleopore membrane placed on serum-containing agar. (a) Upper surface of the membrane with several hyphae entering pores. (b) Underside of the membrane with adhering fragments of agar debris; hyphae emerging on the underside, growing along the interface between the membrane and the agar surface. (c) Upper side of the membrane showing a hypha emerging from the lower surface and then re-entering a pore. The hypha is stretched taut during preparation. Bars = 20 μm (a and b); and 10 μm (c). Reproduced from Gow *et al.* (1994), with permission from Scanning Microscopy International.

In the first exponential phase, the specific growth rate is similar to that of freely suspended cells but the linear phases (I, rapid; II, slower) and characterized by decreased diffusion of nutrients into the gel matrix. The duration of the final pseudostationary phase depends on the immobilization matrix which dictates the level of yeast biomass developing in the immobilizate. Growth of gel-entrapped yeasts has been studied by Parascandola and de Alteriis (1996) and shown to be limited by nutrient diffusion which results in biomass gradients within gel beads. That is, most yeast growth occurs in subsuperficial regions of the gel resulting in an adherent biofilm on the gel surface. Doran and Bailey (1986) have also noted growth-related changes in gel-entrapped *S. cerevisiae*, particularly with regard to cell division cycle progression. Results suggested that immobilization retarded budding of daughter cells but permitted the continuation of DNA replication and cell wall polysaccharide biosynthesis.

With regard to metabolic activities following yeast immobilization, Norton and D'Amore (1994) have reviewed alterations in nutrient transport, energy metabolism, storage and structural polysaccharide synthesis and ploidy changes in immobilized cells of *S. cerevisiae*.

Yeast **biofilms** represent a natural form of cell immobilization resulting from yeast attaching to solid support materials (Characklis and Marshall, 1990). Biofilms of yeast have several practical applications in fermentation biotechnology and are also medically important in the colonization of human tissue in the case of pathogenic yeasts. Considering the former, Kunduru and Pometto (1996) have evaluated the fermentation performance of yeasts attached to plastic supports in biofilm reactors. For continuous ethanol production, biofilm fermentations significantly outperformed suspension-culture reactors. In more traditional fermentations, very thick yeast biofilms may develop when hydrophobic strains of *S. cerevisiae* rise to the top of fermenters and accumulate in the foaming head. Such behaviour is characteristic of top-fermenting yeasts used in ale brewing. Characteristics of yeast cell flotation and correlation with surface hydrophobicity has

Table 4.6. Materials and methods of yeast immobilization.

Method of immobilization	Immobilization material	References
Cell–cell aggregation	Yeast biomass itself	Hasal *et al.* (1992) for *S. cerevisiae;* Hsiao *et al.* (1983) for *Sch. pombe*
Entrapment within polymers	Calcium alginate	Nagashima *et al.* (1987)
	Carrageenan	Godia *et al.* (1987)
	Cross-linked gelatin	Parascandola *et al.* (1990)
Covalent attachment	Hydroxyalkyl methacrylate gels	Jirku and Turkova (1987)
Attachment to and adsorption within preformed carriers	Porous sintered glass and ceramic microspheres	Cashin (1996)
	Granular DEAE–cellulose	Cashin (1996)
	Synthetic sponge	Scott and O'Reilly (1995)
	Stainless steel wire	Black *et al.* (1984)
Containment behind a barrier	Ultrafiltration or microporous membranes (sheets or hollow fibres)	Shuler (1987)
Hydrodynamic deposition	Ceramic (Kaolinite) microspheres	Salter *et al.* (1990)

been studied in detail by Palmieri *et al.*, (1996). *S. cerevisiae* strains prone to flotation do not form cell–cell aggregates, unlike flocculant strains. Figure 4.25 shows flotation strains of *S. cerevisiae* attached to a small air bubble.

With regard to pathogenic yeast biofilms, *Candida albicans* has been shown by Douglas and her colleagues to adhere to surgical devices such as heart pacemakers and catheters, human epithelial cells and dental acrylic (Hawser and Douglas, 1994; Douglas, 1995). Figure 4.26 shows the development of a *C. albicans* biofilm on PVC catheter discs.

Such adherence poses particular problems for the clinical mycologist especially since *C. albicans* biofilms are resistant to antifungal agents (Hawser and Douglas, 1995).

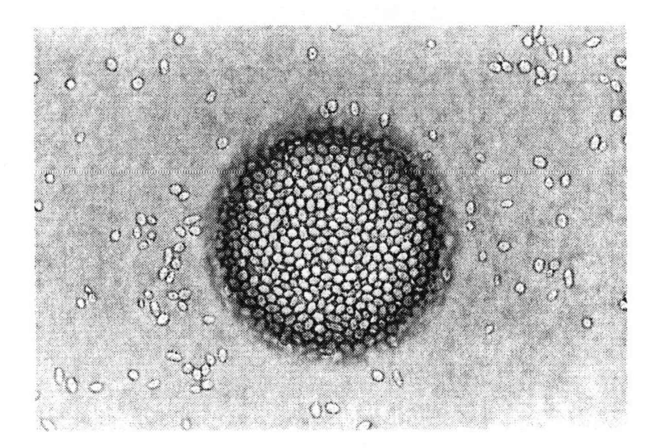

Figure 4.25. Flotation behaviour of *S. cerevisiae* cells. The micrograph shows the affinity of yeast cells for an air bubble. One drop of foam was loaded on a slide in such a manner that the small bubbles were arrested (phase micrography, $\times 600$). No cell–cell aggregation is apparent among the free cells or cells detached from the air bubble. Micrograph kindly provided by Dr Cecilia Laluce, Universidade Paulista, Sao Paulo, Brazil.

4.3.2 Population Yeast Growth

4.3.2.1 Batch Growth of Yeasts

When yeast cells are inoculated into a suitable liquid nutrient medium and incubated under optimal physical growth conditions, a typical **batch growth curve** will result when the viable population of cells is plotted against time (Figure 4.27).

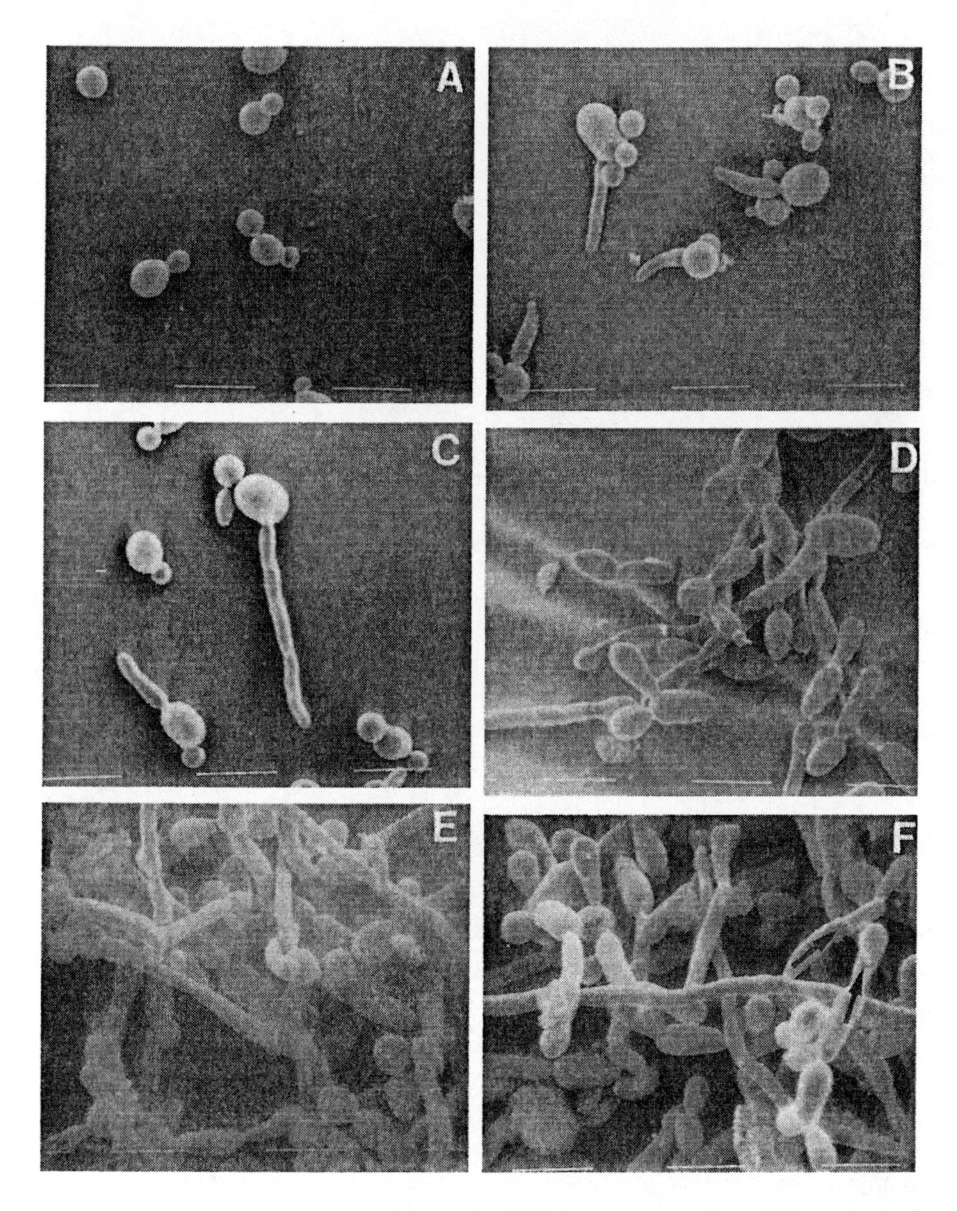

Figure 4.26. Biofilm formation by *Candida albicans.* The figure shows scanning electron micrographs of biofilm formation by *C. albicans* GDH2346 on PVC catheter discs in medium containing 500 mM galactose at 1 (A), 3 (B), 6 (C), 18 (D), 24 (E) and 48 (F) h. Arrows indicate the presence of extracellular polymeric material. Bar, 10 μm. Reproduced with permission from Hawser and Douglas (1994) and the American Society for Microbiology.

The yeast growth curve obtained from batch culture is comprised of characteristic lag, exponential and stationary phases. The **lag phase** represents a period of zero growth (specific growth rate, $\mu = 0$) and is exhibited when inoculum cells experience a change of nutritional status or alterations in physical growth conditions (e.g. temperature, osmolarity). The precise

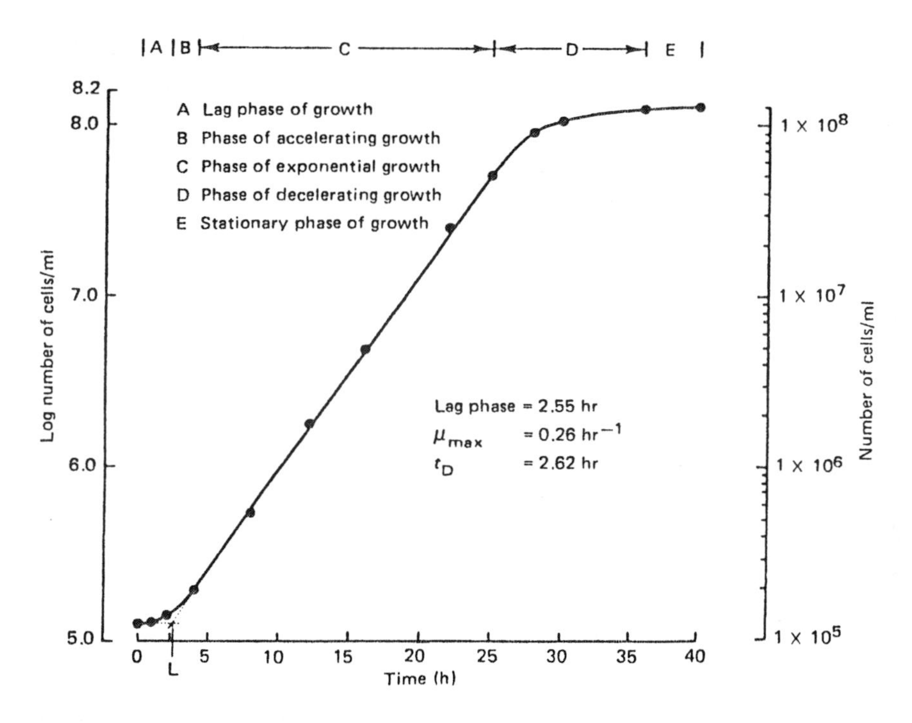

Figure 4.27. Typical yeast growth curve. The figure shows the phases of growth of a yeast in batch culture. μ_{max} is the maximum specific growth rate and t_D is the doubling time. Reproduced from Hough (1985), with permission from Cambridge University Press.

duration of the lag phase is dependent not only on growth conditions but also on inoculation density and the growth 'history' of the inoculum. The lag phase reflects the time required for inoculated yeast cells to adapt to their new physical and chemical growth environment by synthesizing ribosomes and enzymes needed to establish growth at a higher rate. Once cells transit from lag phase and commence active cell division, they enter an **acceleration phase** before exponential growth. The rate of increase (dx/dt) in yeast biomass (x) with time (t) during this phase is expressed as

$$\frac{dx}{dt} = \mu x$$

with the value of the specific growth rate (μ) varying between 0 (lag phase) and μ_{max} (exponential phase). Note that this transition is not a sharp increment. Nevertheless, the point when lag phase ends can be estimated by extrapolating to the abscissa the intercept of the exponential phase line with the initial lag phase cell number (denoted by 'L' on Figure 4.27).

The **exponential phase** represents a period of logarithmic cell doublings and constant, maximum specific growth rate (μ_{max}). The precise value of μ_{max} (in dimensions of reciprocal time, h^{-1}) depends on the yeast species and the prevailing growth conditions. If growth is optimal and cells double logarithmically, then

$$\frac{dx}{dt} = \mu_{max} x$$

when integrated, this yields

$$\ln x - \ln x_0 = \mu_{max} t$$

(where x_0 is the initial cell mass)

or

$$x = x_0 e^{(\mu_{max} t)}$$

which is the fundamental equation for exponential batch growth. According to these kinetic expressions a plot of ln$x$ versus time is *linear* with the slope being μ_{max}. It is often convenient in yeast biotechnology to calculate the doubling time (*td*) of a culture from knowledge of μ_{max}. This can be achieved from

$$\text{td} = \frac{\ln 2}{\mu_{max}} = \frac{0.693}{\mu_{max}}$$

Of course, exponential yeast growth is finite, but one means of prolonging the exponential phase is to employ **fed-batch** culture. This involves incremental feed of nutrients in step with yeast growth and is commonly employed in the industrial propagation of baker's yeast (see section 4.3.3).

Even in conventional batch cultivation, yeasts may occasionally enter a second phase of exponential growth called **diauxie**. This phenomenon may occur when yeasts are exposed to two carbon growth substrates which are utilized sequentially. For example, diauxic growth of *S. cerevisiae* is observed when aerobically grown cells exhaust glucose. Thereafter, many enzymes are derepressed to provide advantageous use of a second substrate, namely ethanol. Diauxic growth in this case is therefore due to firstly incomplete glucose oxidation and secondly to ethanol oxidation and utilization (as described further in Chapter 5).

In batch culture, the exponential phase is of relatively short duration due to depletion of essential nutrients, accumulation of growth inhibitory metabolites or excessive cell flocculation. Following the exponential phase, cell growth is retarded in a **deceleration phase** before cells enter a period of zero growth rate (the stationary phase). As growth becomes limited during the deceleration phase, the increase in cells per unit time will be defined by a lower specific growth rate than μ_{max}. As discussed in Chapter 3, when yeast growth is limited by the concentration of one substrate (*S*) (as in the decelerating phase), the relationship between μ and *S* can be defined by the Monod equation:

$$\mu = \mu_{max} \frac{[S]}{[S] + K_s}$$

where K_s (the saturation constant) is the value of *S* which limits growth rate to ½ μ_{max}.

In the **stationary phase**, the accumulated yeast biomass remains relatively constant and the specific growth rate (μ) returns to zero. After prolonged periods in stationary phase, yeasts may die and autolyse. This in turn may influence the continued growth and survival of the residual viable cells. Death is exponential and may be expressed as

$$-\frac{dx}{dt} = kx$$

where k represents the *death rate*. Section 4.6 discusses in more detail the physical, chemical and biological factors influencing yeast cell death.

In recent years, the physiological nature of stationary phase cells of *S. cerevisiae* has been extensively studied (reviewed by Werner-Washburne *et al.*, 1993, 1996). Table 4.7 summarizes some of the physiological characteristics of stationary phase yeast cells.

The stationary phase has been defined as the ability of yeasts to survive for prolonged periods (i.e. months) without added nutrients (Werner-Washburne *et al.*, 1993). The type of nutrient depletion is important in dictating entry into stationary phase. For example, when *S. cerevisiae* is starved of carbon it enters stationary phase, but when nitrogen-starved, cells may undergo meiosis and sporulation (if diploid) or may develop pseudomycelia (see section 4.2.1). The shift into stationary phase is regulated by the *Ras*/cAMP signal transduction pathway which may monitor the availability of glucose. Once glucose becomes limiting, a cascade system is initiated (analogous to the pheromone response pathway) which triggers cells to enter a state of proliferative rest. Thevelein and Hohmann (1995) have introduced the term *fermentable growth medium induced*

Table 4.7. Physiological characteristics of stationary phase yeast cells.

Characteristics	References
No cell division with cells syncrhonized in a state of proliferative rest (G0)	Werner-Washburne *et al.* (1996)
Long-term (months) survival of cells without nutrients	Werner-Washburne *et al.* (1993)
Slow metabolic rate with very low rates of protein synthesis	Fuge *et al.* (1994)
Increased thermal resistance with increased transcription of heat-shock genes	Mager and Moradas-Ferreira (1993); Mager and DeKruijff (1995)
Increased resistance to oxidants with increased superoxide dismutase activity	Jamieson (1995); Longo *et al.* (1996)
Increased resistance to high-voltage pulses	Gaskova *et al.* (1996)
Cells refractile with thick cell walls resistant to lytic enzymes	Elliot and Futcher (1993)
Higher cell turgor pressure compared with exponential cells (0.2 MPa cf. 0.05 MPa in *S. cerevisiae*)	Martinez de Marañon *et al.* (1996)
Cellular accumulation of glycogen and trehalose	Lillie and Pringle (1980)
Signal transduction pathway initiated	Thevelein (1994)
Unique expression of 'stationary phase genes' (e.g. *SNZ1*)	Werner-Washburne *et al.* (1996)

pathway to identify the fact that a complete growth medium containing a fermentable carbon source is required to exert regulatory effects on entry to and exit form stationary phase in *S. cerevisiae*. Stationary phase yeast cells are often described as being in the G0 phase of the cell cycle. However, Wheals (1987) and Forsburg and Nurse (1991) have discussed the validity of using the G0 designation for stationary phase yeast cells, particularly when compared with the quiescent G0 phase of mammalian cells (which are deprived of growth factors rather than major nutrients as is the case with yeast cells). Furthermore, Werner-Washburne *et al.* (1996) have proposed that there is a subtle nutrient-dependent distinction between G0 arrest and stationary phase in *S. cerevisiae* (Figure 4.28).

In addition to nutrient limitation, other physiological causes may promote entry of yeast cells into stationary phase. These include: toxic metabolites (notably ethanol), low pH, high CO_2, variable O_2 and high temperature. Several of these aspects influence yeast growth in industrial fermentations and so there is distinct practical relevance for studies into the role of external factors in controlling entry to, residence in, and exit from the stationary phase in yeast cells.

4.3.2.2 Continuous Growth of Yeasts

Yeast cells can be propagated in **continuous culture** for prolonged periods of exponential growth without lag or stationary phases. A typical system is described in Figure 4.29. The change in yeast biomass per unit time in continuous culture vessels is described as the difference between the rate of growth and the rate of removal of cells;

$$\frac{dx}{dt} = \mu x - Dx$$

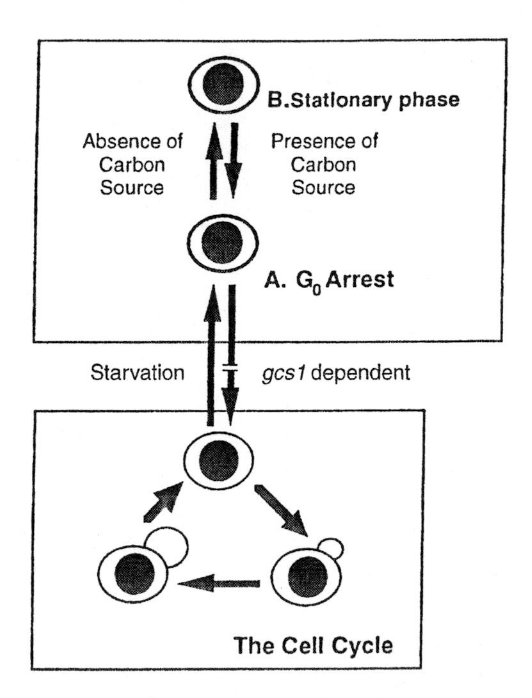

Figure 4.28. Model for G0 arrest and stationary phase in *S. cerevisiae*. The figure represents the proposed model of Werner-Washburne *et al.* (1996) for G0 arrest and acquisition of long-term survival capability of *S. cerevisiae* cells. Cells limited for essential nutrients, such as nitrogen, sulphur, phosphorus or carbon, arrest at G0 as defined by the *gcs1* phenotype. Cells limited for nutrients other than carbon are unable to remain viable in this arrested state (Granot and Snyder, 1993). Cells limited for carbon undergo physiological changes, which include a reduction in metabolism and acquisition of the ability to survive for long periods of time without added nutrients. Reproliferation from this state is, therefore, a two-stage process. During the first stage, metabolic activity is restored and cells prepare for active growth. The second stage is exit from G0 arrest and a return to the mitotic cell cycle. Reproduced with permission from Werner-Washburne *et al.* (1996) and Blackwell Science Ltd.

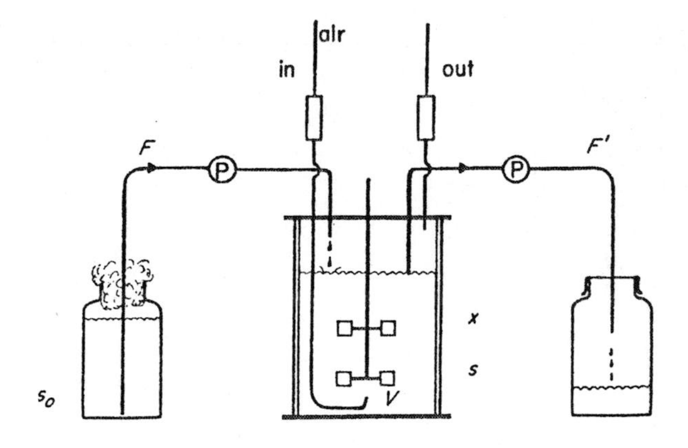

Figure 4.29. Continuous culture of yeasts. Fresh medium with a certain substrate concentration (s_0) is fed by a metering pump (P) to the bioreactor at a constant rate (F). The liquid volume (V) contains cells at the concentration (x) and residual substrate (s), as well as the gas phase. Culture liquid is removed at a rate F', which equals the medium feed F under steady-state conditions. The important variable of the system is the dilution rate, which is defined as the medium flow per unit volume ($D = F/V$). Reproduced with permission from Fiechter, Käppeli and Meussdoerffer (1987) and Academic Press.

where D represents the *dilution rate*, or flow rate per unit volume (h^{-1}). Once continuous growth is established the yeast growth rate remains constant and a steady state cell concentration is maintained such that $dx/dt = 0$. Therefore, $\mu = D$ and so yeast growth rate can be governed by the dilution rate. In practical terms, this is accomplished by altering the rate at which the feed pump delivers nutrients into the continuous culture vessel. If the dilution rate increases beyond μ_{max}, cells will be removed from the vessel faster than they are produced by growth, the population size steady-state is disrupted and all cells will eventually be 'washed out'. Additional theoretical considerations of continuous culture of yeasts have been described by Fiechter *et al.*, (1987).

Continuous cultures which are based on the controlled feeding of a sole growth-limiting substrate are called **chemostats**. In yeast physiology, chemostats have proved extremely valuable in studying the growth and metabolic effects of a single variable nutrient, with all other factors kept constant. They have also proved very useful in studying the stability of plasmids in recombinant yeasts (O'Kennedy *et al.*, 1995; Syamsu *et al.*, 1996) and in optimizng enzyme production in baker's yeast processes (Gregory *et al.*, 1996). Yeast chemostats may be established based on limitation of sources of carbon (e.g. glucose), nitrogen (e.g. ammonium ions), sulphur (e.g. sulphate ions), phosphorus (e.g. phosphate ions), potassium (K^+ ions) and magnesium (Mg^{2+} ions) (Dawson, 1972; Meyer *et al.*, 1985; Fiechter *et al.*, 1987; Walker and Maynard, 1996).

Another way in which to grow yeast cells continuously, besides the chemostat, is in the **turbidostat**. In this system, yeasts are not nutrient-limited and their maximum growth rate is maintained by constantly monitoring the level of bio-mass by optical density. Only when yeast biomass exceeds a set point is fresh medium introduced by a feedback control. Such self-stabilizing continuous cultures are important in biotechnology for selection of improved yeast strains (e.g. Brown and Oliver, 1982; Aarnio *et al.*, 1991). Davey *et al.* (1996) have described a continuous yeast culture system in which the biomass is set by monitoring the radio-frequency electrical capacitance of cells in suspension. Because the biomass is maintained constant by a feed-back based on dielectric permittivity of the culture, the system is called a **permittistat**.

4.3.2.3 Synchronous Yeast Growth

In conventional batch and continuous yeast cultivation systems, cell populations are randomized with respect to their cell division cycles. Non-random or **synchronous** cultures of yeast which are characterized by cells in the population dividing more or less in unison may be established to facilitate studies of the cell cycle. Synchronous cultures may also be employed in biotechnology perhaps to optimize the production of yeast metabolites and cell cycle-modulated proteins (see Dickinson, 1991). Synchronization of yeast cultures may be obtained either by *induction* or *selection* methods (Table 4.8).

Induction of synchrony may be achieved by treating asynchronously dividing (i.e. random) cell populations by physical or chemical agents which inhibit cells at the same stage of their cell cycles, then releasing cells from the inhibition to induce synchrony. Certain temperature sensitive *cdc* mutants of both *S. cerevisiae* and *Sch. pombe* can also be used to elicit synchrony once cells are returned to their permissive growth temperature. In selection synchrony, yeast cells in a growing culture which are at a particular stage of the cell cycle are selected off physically then grown up separately as a synchronous culture. Pons *et al.* (1997) have recently described a rapid field-flow fractionation (FFF) method for eluting and separating *S. cerevisiae* cells on the basis of their individual sizes. Selection methods are preferable in cell cycle studies since they do not perturb normal events of growth and cell division. Alternatives to using synchronized cultures for yeast

Table 4.8. Methods of synchronizing yeast growth

Induction synchrony	Selection synchrony
Periodic nutrient feeding in chemostat culture or feeding–starving key nutrients in batch culture	Velocity separation by density gradient (e.g. lactose, sorbitol) centrifugation
Heat-shock cycles, spaced a cell generation apart	Centrifugal elutriation
Block and release using S phase or M phase inhibitors	Continuous-flow centrifugation
Mating pheromone synchronization by inducing a G1 block	Zonal centrifugation
Releasing *cdc* mutants from their restrictive growth temperature	Equilibrium separation by high-speed ($\sim$45 000 g) centrifugation in dextrin gradients

Information from Mitchison (1988); Creanor and Toyne (1993) and Alfa *et al.* (1993).

cell cycle studies include the use of time-lapse observations in single cells and flow cytometric analysis of fluorescently-labelled asynchronous populations (Porro *et al.*, 1995).

Assessment of the degree of synchrony achieved by induction or selection can be made using a variety of quantitative synchronization indices (e.g. Bakke and Pettersen, 1976) but it should be borne in mind that perfect synchrony is never achieved through human manipulation due to the inherent variability in the duration of mitotic cycles of individual cells.

In batch synchronous cultures of yeast, synchrony of cell division may only be maintained for two or three cycles at best. In addition, the cultures are still unbalanced with respect to continual changes in nutrient status and cell growth rate. One way of achieving more balanced yeast growth is, of course, to employ the steady-state conditions which pertain in a chemostat. However, cells grown in traditional chemostats are generally assumed to be asynchronous. Therefore, the establishment of *continuous–synchronous* cultures would provide yeast cells which are: balanced with respect to nutrients, in steady state with respect to growth rate and synchronous with respect to cell division (Table 4.9). In theory, such cultivation systems should enable the physiologist to study yeast populations as if they were behaving as an individual cell.

Several workers have employed continuous–synchronous cultures in an effort to study the influence of the physiological environment on yeast cell cycle progression. Although the overall growth rate of yeasts in continuous culture is constant, oscillations in the biomass content of chemostat-grown yeast have been reported. For example, Münch *et al.*, (1992a,b) have studied 'forced' synchronization of *S. cerevisiae* in glucose-limited chemostats. Meyer and Beyeler (1984), Abel *et al.* (1989), Bellgardt (1994), Boiteaux (1995) and Duboc *et al.*, (1996) have analysed the growth behaviour and metabolism of self-synchronized cultures of this yeast grown in continuous culture. Duboc *et al.* (1996) have demonstrated in continuous–synchronous cultures of *S. cerevisiae* that growth rate was low ($0.075\ h^{-1}$) during S phase and high ($0.125\ h^{-1}$) during the G2, M and G1 phases at a constant dilution rate of $0.10\ h^{-1}$. The phenomenon of *spontaneous synchrony* or sustained oscillatory growth has been explained by Münch, Sonnleitner and Fiechter (1992a) to be due to periodic fluctuations in carbon supply to different subpopulations of budding cells. These carbon fluctuations may result from the periodicity of glucose feed rate and Rothen *et al.* (1996) have suggested that faint oscillations in glucose concentration may be responsible for synchronizing *S. cerevisiae* cells in continuous culture. Entrained growth oscillations in this yeast may also result from cyclic liberation of ethanol from budding cells and several workers have observed cell cycle-dependent fluctuations in ethanol production by *S. cerevisiae* (e.g. Meyer and Beyeler, 1984; Käppeli *et al.*, 1985; Duboc *et al.*, 1996). In particular, Duboc *et al.* (1996) have demonstrated very dramatic increases in cellular ethanol production coincident with the onset of S phase in *S. cerevisiae*, followed by rapid net consumption of ethanol as cells enter G2 (see Chapter 5, Figure 5.13). The studies of Duboc *et al.* (1996) appear to support previous proposals of Parulekar *et al.* (1986) and Martegani *et al.* (1990) that ethanol acts as a signal between yeast cells to trigger and sustain growth oscillations.

Dawson (1972, 1985) developed and refined a novel system of continuously synchronizing yeasts known as *phased culture*. In this technique, instead of supplying fresh medium as a steady stream of nutrient rations to randomly dividing cells in a chemostat, a single total addition of nutrients is made at doubling time intervals. In this way, cells adjust their growth rate to the nutrient dosing interval and become synchronized to it. By changing the time interval, the doubling time can also be changed. This system delivers a key growth-limiting nutrient at a concentration sufficient to initiate one cell cycle. Over the next cell cycle time, therefore, the population will be exactly doubled (assuming 100% viability). This 'phasing' procedure is repeated continually to maintain synchrony. The mechanics of the system have been described by

Table 4.9. Summary of culture methods used to study yeast cell populations.

Method	Growth pattern	Advantages	Disadvantages
Asynchronous batch culture (closed system)	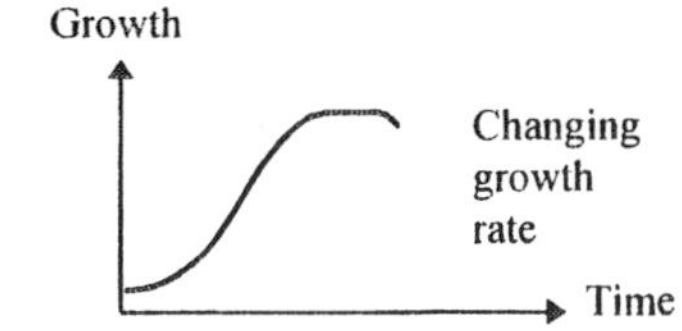 Growth; Changing growth rate; Time	Easy to set up. Simulate traditional yeast fermenters.	Unbalanced with respect to nutrients and yeast biomass and random with respect to cell division.
Asynchronous continuous culture (open system)	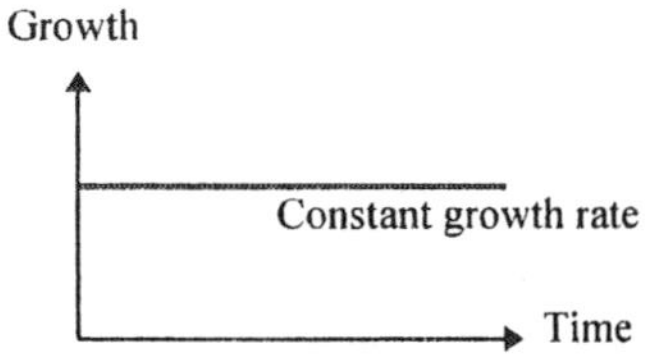 Growth; Constant growth rate; Time	Steady – (exponential) state growth rates easy to manipulate. Balanced with respect to nutrients.	Random with respect to cell division. Low cellular productivity at low dilution rates.
Synchronous batch culture (closed system)	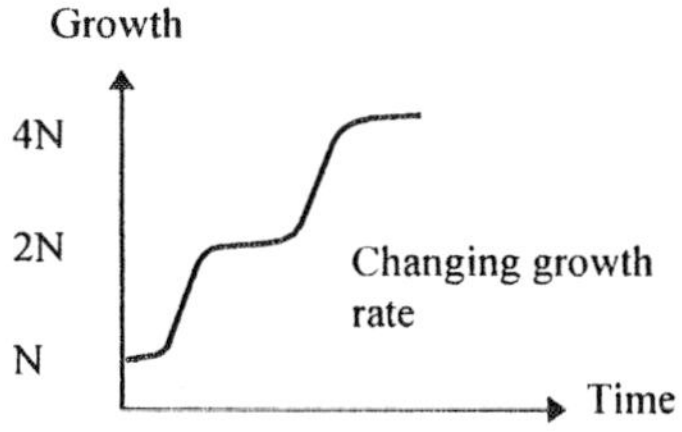 Growth; 4N; 2N; N; Changing growth rate; Time	Choice of synchronization methods. Non-random (briefly) with respect to cell division.	Unbalanced with respect to nutrients. Only one or two synchronous generations possible, which may be 'perturbed'.
Synchronous continuous culture (open system)	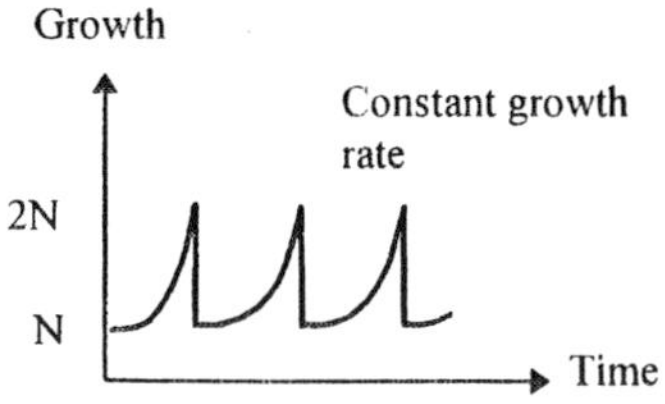 Growth; Constant growth rate; 2N; N; Time	Balanced with respect to nutrients and non-random (indefinitely) with respect to cell division.	Cells are alternately exposed to excess nutrients then periodically starved to hold back cell division.

Dawson (1972). It might be argued that in such nutrient-phased systems yeast cells are periodically starved, but this is no more valid than for the chemostat. Controlled growth limitation is more likely to apply, especially since phased cultures of yeast (*Candida utilis*) can be operated for extended periods (equating to around 3000 generations; Dr Peter Dawson, personal communication, 1986).

4.3.3 Cultural Yeast Growth

Having considered the growth behaviour of yeasts in solid and liquid media, more practical aspects of yeast growth in biotechnology will now be discussed. Laboratory management of yeast cultures based on strategies for pure culturing and maintenance of strains will be considered first. Assessment methods for yeast viability and growth will then be described before discussing cultivation strategies for industrial yeast propagations and fermentations.

4.3.3.1 Yeast Pure Culturing and Maintenance Strategies

With regard to pure culturing of yeast, one has to acknowledge the pioneering strategies developed last century by Emil Christian Hansen (Hansen, 1886). He developed methods for isolating single cells of brewing yeasts and for studying the fermentation performance of selected pure 'starter' cultures in successively larger fermenters. Strict microbiological procedures developed from laboratory to industrial plant ensured consistent product (i.e. Carlsberg beer) quality. Hansen's techniques enabled selection of suitable yeast clones and their indefinite maintenance free from contaminating microorganisms. Clearly, the practices advocated by Hansen for the handling, storage and propagation of yeast cultures are still very pertinent for today's yeast scientists and technologists.

The successful maintenance of pure cultures of yeast necessitates that cells should be kept in a state of high viability and without risk of deleterious genetic changes. Table 4.10 summarizes some of the common strategies for storage and maintenance of yeast cultures. Further discussion of the advantages and disadvantages of the various methods of preserving yeast stock cultures including recombinant strains, can be found

Table 4.10. Methods for preserving and maintaining yeast cultures.

Method	Description	Comments
Serial transfer	Subculturing on nutrient agar slopes and storage at 0–4°C. Overlaying with mineral oil creates anaerobiosis to maintain lower growth rates.	Commonly used method of maintaining growing cultures but unsuitable for some species (e.g. *Brettanomyces*). Mutants may also be selected and there is increased chances of cells sporulating.
Desiccation	Drying of cultures onto filter paper, silica gel, etc.	Simple method in which genetic stability appears to be maintained.
L-Drying	*In vacuo* drying removes water directly from liquid phase in cell suspensions.	Most yeasts can be successfully preserved by this method, except filamentous, zerotolerant or psychrophilic species which are more sensitive.
Lyophilization	Freeze drying by vacuum sublimation of frozen cells. A suitable cryoprotectant is essential.	Commonly used for long-term storage. Some yeasts exhibit high mortality and cell survival is generally low. Genetic damage possible with mutations likely to be induced by the drying process.
Cryopreservation	Deep-freezing cells in electrical freezer at −20°C to −80°C.	Metabolic activity restricted but viability compromised. Freeze–thawing may cause genetic damage. Risk of losses due to mechanical failure.
	Ultra-freezing cells in liquid N_2 at −196°C. Cryoprotectant (e.g. 5–20% glycerol) essential to minimize ice-crystal damage.	High viability of cultures maintained and no chance of genetic changes. Initial capital outlay is high and evaporation of liquefied N_2 requires regular replenishment.

in Kirsop and Henry (1984), Schu and Reith (1995) and Hunter-Cevera and Belt (1996).

4.3.3.2 Assessments of Yeast Viability and Vitality

In yeast biotechnology, the physiological and genetic stability of stock cultures of seed yeast is of vital importance for the success of subsequent fermentations. Physiological stability relates to the maintenance of cell viability and vitality during long-term storage and these are parameters that can be assessed by a variety of methods. Such assessments of the physiological 'fitness' of yeast cells are not just applicable to preservation protocols but are also useful for predicting and controlling yeast activity during industrial processes. Tables 4.11 and 4.12 summarize some of the methods employed to ascertain quantitatively yeast viability and vitality.

Viability refers to measurements of living yeast cells, while **vitality** is a measure of yeast metabolic activity. Note that yeast cells are generally considered to be dead when they irreversibly lose their ability to reproduce. However, cells incapable of division may still be capable of active

Table 4.11. Methods for assessing yeast viability.

Method	Description/Comments	References
Plate count	Colonial growth enumerated on nutrient agar plates. Inaccuracies arise due to flocculant/adherent yeasts.	American Society of Brewing Chemists (1980)
Slide count	Microcolonies counted after 18 h on a film of nutrient agar on a microscope slide. More reliable than stains at low cell viabilities.	Pierce (1970)
Vital stains	Based on membrane permeability and intracellular modification of two types of stains:	
	• Bright-field stains	
	e.g. Methylene blue	
	Crystal Violet	Pierce (1970)
	• Fluorochrome stains	
	e.g. Primuline yellow	Evans and Cleary (1985)
	Rhodamine 123	Trevors *et al.* (1983)
	Carboxyfluorescein (and flow cytometry)	Dinsdale *et al.* (1995) Breeuwer *et al.* (1994)
	Mg-ANS (1-anilino-8-naphthalene sulphonic acid)	McCaig (1990)
	Tetrazolium dyes (e.g. MTT; XTT)	Hodgson *et al.* (1994); Tellier *et al.* (1992)
Metabolic activity	ATP bioluminescence NADH fluorescence, glycolytic flux rate	Jones (1987)
Cellular compounds	Glycogen, trehalose	Slaughter and Nomura (1992)
Intracellular pH	Spectrofluorophotometry	Imai and Ohno (1995)
Dielectric permittivity	Capacitance probe	Molzahn (1992); Davey *et al.* (1996)

metabolism and so another view of yeast death would be the complete cessation of metabolic activity. Discrepancies inevitably arise in viability assessments based on plate counting and vital staining because the criteria for death differs for each method. Since there is no absolute method for measuring the viability of a population of yeast cells, assessments should always make reference to the particular method used. Yeast viability can be determined directly by measuring loss of cell reproduction (e.g. plate count) and indirectly by assessing cellular damage (e.g. vital stains) or loss of metabolic activity (e.g. ATP, NADH). The most accurate measures of yeast viability still remain the time-consuming plate and slide culture methods. Staining procedures are more accurate but generally imprecise, with the exception of the use of certain fluorochromes. Electrosensors such as capacitance probes are finding increasing use especially since they can be incorporated in industrial plants (e.g. breweries) for automatic in-line monitoring of yeast cell viability. Other recent developments include the use of fluorescent probes coupled with flow cytometry which can rapidly analyse cell viability as well as other aspects of yeast physiology, including responses to stress (Edwards, Porter and West, 1996).

Yeast vitality is more difficult to define but generally relates to the fitness or health or vigour of a culture. It can be indirectly assessed by measuring metabolic/fermentative activity, cellular storage molecules, intracellular/extracellular pH and gaseous exchange coefficients like RQ (see Table 4.12). These techniques generally show varying degrees of correlation with fermentative performance and no single technique exists to predict accurately the physiological activity of a yeast sample. In the yeast fermentation industries, the only true indicator of yeast vitality is possibly obtained from the time-consuming operation of a pilot-scale bioreactor which simulates production plant conditions.

Table 4.12. Methods for assessing yeast vitality.

Method	Description	References
Metabolic activity	ATP content by bioluminescence and the ratio of adenosine nucleotides (adenylate energy charge)	Hysert and Morrison (1977)
	Redox indicator (colorimetry)	Pfaller and Barry (1994)
	Redox indicator (chemiluminescence)	Nishimoto and Yamashoji (1994)
	NADH concentration by fluorimetry	Beyeler *et al.* (1981)
Cellular components	Glycogen and sterol content	Quain (1988)
	Protein distributions by FITC and flow cytometry	Porro and Alberghina (1996)
Magnesium ion release	Mg release (assayed colorimetrically) can predict fermentative performance	Mochaba *et al.* (1997)
Fermentation capacity	CO_2 evolution, mini-scale fermentations	Manson and Slaughter (1986)
Acidification power	Extracellular pH	Opekarova and Sigler (1982)
(Membrane potential?)	Intracellular pH	Imai *et al.* (1994)
Oxygen uptake	Respiratory quotients (RQ)	Kara *et al.* (1987)
	Oxygen uptake rate	
Cell size	Cell morphology (by image analysis) correlates with acidification power.	Hashida *et al.* (1995)

4.3.3.3 Cultivation Strategies in Yeast Biotechnology

Yeast inoculum development

Large-scale yeast inocula for industrial processes must initially be prepared by growing sufficient quantities of viable cells from preserved stock cultures in successively larger volumes of nutrient media. Generally speaking, the first few consecutive increases in volume are restricted to ten-fold or less. For example, the contents of a lyophilized ampoule inoculated into 10 ml broth, thence into 100 ml broth, and so on. The performance of fermentation processes will be significantly influenced by the size and quality (physiological vitality) of the yeast inoculum. Regarding inoculum size, it is known that if very small yeast inocula are introduced into large volumes of medium, growth is unlikely to be successful. This is possibly due to the insufficiency of inter-cellular signals to activate cell multiplication at low cell densities (Rasmussen *et al.*, 1996). An analogous situation, known as *quorum sensing*, occurs in bacterial cell populations which sense culture densities and regulate growth by monitoring the accumulation of small, freely diffusible signal molecules (such as *N*-acyl homoserine lactones, or AHLs; see Huisman and Kolter, 1994).

In brewing yeast inoculum development strategies it would be common for 500 ml of active cell culture propagated from an agar slope to be grown in a 'Cornelius' vessel (13 litres) to around 80×10^6 cells/ml before pilot-scale (e.g. 1000-litre) propagation under aseptic, aerobic conditions. Yeast biomass generated in this way can then be employed in production fermenters (e.g. 300 hl). In many brewing operations, yeast will be replaced by recourse to laboratory pure culture stocks after five to ten successive fermentations. Most brewery fermentations are therefore conducted with yeast drawn from a previous one and not from a propagator (Lewis and Young, 1995; Stewart, 1996). The initial inoculum cell density in production fermentations, called the 'pitching rate' in breweries, should be around 10^7 viable yeast cells per ml. Generally, for yeast alcohol fermentations, an optimum inoculation size would be around 1% (v/v) (Birol and Özergen-Ulgen, 1995). In brewing fermentations, the amount of yeast added to wort greatly influences not only the speed of sugar conversion to alcohol but also the amount of yeast growth produced at the end of fermentation. Brewers are often interested in comparing the growth and fermentation kinetics of different yeast strains. The number of biomass doublings achieved at the end of fermentation can be determined from

$$n = 3.32 \log_{10} \left(\frac{x}{x_0} \right)$$

where n is the number of doublings and x_0 and x represent the initial and final yeast biomass levels. Commercial brewery fermentations generally result in two to three doublings of the yeast biomass population (Hough *et al.*, 1982). This, of course, may be regarded as a wasteful diversion of sugar to biomass, instead of to ethanol.

Alternative quantitation of yeast biomass produced at the end of an industrial fermentation can be obtained by calculating growth yields. The growth yield, Y, of a yeast culture may be defined as the amount of biomass, x, produced following consumption of an amount of substrate, S. Or,

$$Y = \frac{dx}{dS}$$

However, the relationship is of limited value when comparing different yeasts since nutrients will be variably metabolized into ethanol, CO_2, numerous minor metabolites, intracellular reserves (e.g. glycogen, trehalose) as well as yeast biomass itself.

Yeast growth monitoring

Once a suitable yeast inoculum has been prepared and the production fermentation initiated, cell growth can then be monitored and controlled using a number of strategies (reviewed by Meyer *et al.*, 1985). Quantification of yeast growth and metabolic activity is required in order to optimize yields and rates of fermentations. Table 4.13 summarizes the principal methods of monitoring yeast growth.

In modern yeast bioreactors, electronic sensors can be used to monitor and control cell growth and metabolic activity (Fiechter *et al.*, 1987). Other aspects of yeast growth during fermentation may be monitored including cell morphology and flocculation properties. Morphological changes during fermentation may be assessed using image analysis (e.g. O'Shea and Walsh, 1996; also Chapter 2) and numerous methods exist for quantifying the degrees of flocculence in yeast populations (e.g. Speers and Ritcey, 1995; Podgornik *et al.*, 1997). In addition, Shimizu *et al.* (1996) have employed a 'fuzzy interference' system to estimate and control yeast growth during fed-batch cultivation.

Industrial cultivation strategies for yeast

Yeast fermenters have evolved from the classical open-batch vessels used in traditional brewing to sophisticated computer-controlled bioreactors used in the production of recombinant pharmaceuticals (see Winkler, 1991). Table 4.14 summarizes the diversity of industrial cultivation systems for yeasts and some of their biotechnological applications.

Numerous advantages can be obtained in food and fermentation industries by immobilizing yeast cells. For example, productivity in continuous bioreactors may be enhanced, cells can be re-used, and very high cell densities can be achieved in smaller-scale facilities with reduced capital expenditure. The principles and practice of yeast cell immobilization for food and beverage fermentations has been reviewed by Groboillot *et al.* (1994); Divies *et al.* (1994); Masschelein *et al.* (1994); Norton and D'Amore (1994) and Norton and Vuillemand (1994). In recombinant DNA technology, Zhang *et al.* (1996) have discussed the advantages of immobilizing *S. cerevisiae* cells in improving plasmid retention. Table 4.15 summarizes some of the biotechnological applications of immobilized yeasts.

4.4 THE PHYSICOCHEMICAL ENVIRONMENT AND YEAST GROWTH

4.4.1 Physical Requirements for Yeast Growth

Most yeasts grow well in warm, moist, sugary, acidic and aerobic environments. Those few species which prefer exceptional physical or chemical conditions are, nonetheless, very important in industry, often as spoilage organisms.

Table 4.13. Methods for assessing yeast growth.

Direct methods	
Cell number	By haemocytometer, electronic (e.g. Coulter) counter, laser-flow cytometer
Cell mass	By wet and dry weights, optical density
Cell volume	By centrifugation (packed cell volume) and electronic size analyser (mean cell volume)
Indirect methods	
Physical and chemical parameters	e.g. Structural (lipids, protein) and storage (polyphosphate, glycogen) cellular components
Metabolic parameters	Carbon source and oxygen consumption, acid and CO_2 production, NADH fluorescence, heat evolution, redox potential, culture fluid viscosity

Table 4.14. Industrial cultivation strategies for yeasts.

Culture system		Description	Applications in biotechnology
Batch	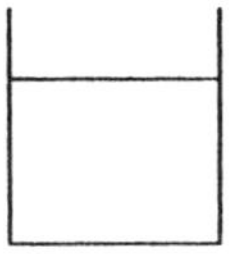	Classical static batch fermenter	Traditional brewing
	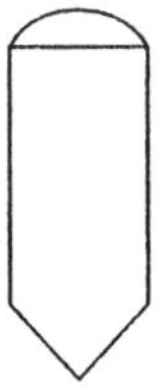	Cylindroconical tower fermenter (turbulent)	Modern brewing
Fed-batch	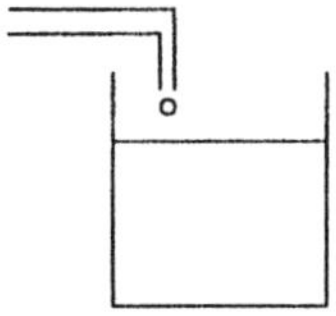	Incremental nutrient feed to batch fermenter	Baker's yeast Recombinant proteins
Continuous	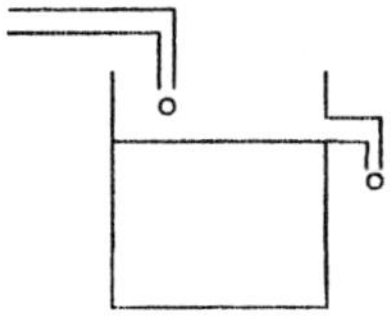	Open system to maintain steady state. Modifications include: chemostat, gradostat, turbidostat	Ethanol Single cell protein
Immobilized	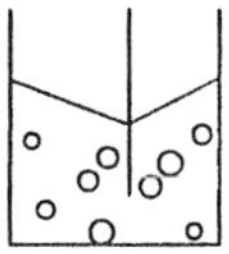	Continuous stirred tank reactor	Beer maturation Bioconversions
		Packed bed	
	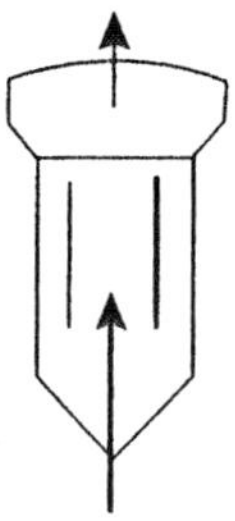	Air lift	
	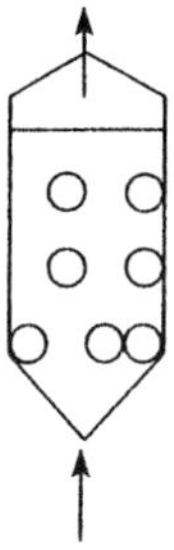	Fluidized bed	

Table 4.15. Examples of products from immobilized yeast systems.

Yeast	Immobilizate	Product	References
Saccharomyces cerevisiae	Various materials (e.g. DEAE–cellulose)	Beer, including low-alcohol beer	Masschelein *et al.* (1994); Norton and D'Amore (1994)
Saccharomyces cerevisiae	Cross-linked gelatin	Wine	Parascandola *et al.* (1992)
Saccharomyces cerevisiae	Coated alginate	Sparkling wine	Godia *et al.* (1991)
Saccharomyces cerevisiae	Coated alginate	Ethanol	Ruggeri *et al.* (1991)
	Alginate	Ethanol	Roukas (1996)
	Porous ceramic	Ethanol	Demuyakor and Ota (1992)
	Membrane filter	Ethanol	Kyung and Gerhardt (1984)
	Preformed cellulose	Ethanol	Szajani *et al.* (1996)
Saccharomyces cerevisiae (recombinant)	Agarose	L-Malate	Neufeld *et al.* (1991)
Saccharomyces cerevisiae (recombinant)	Alginate	β-Glucanase	Cahill *et al.* (1990)
Saccharomyces carlsbergensis	Alginate	Ethanol	Tzeng *et al.* (1991)
Saccharomyces bayanus	Carrageenan	Biomass, ethanol	Taipa *et al.* (1993)
Schizosaccharomyces pombe	Alginate	Deacidified wine	Magyar and Panyik (1989)
Kluyveromyces lactis	Glass wool	Whey lactose hydrolysis	Champluvier *et al.* (1988)
Kluyveromyces bulgaricus	Alginate	Whey lactose hydrolysis	Champluvier *et al.* (1988)
Kluyveromyces marxianus	Alginate	Inulin hydrolysis	Bajpai and Margaritis (1986)
Trichosporon pullulans	Alginate	Cellobiose hydrolysis	Adami *et al.* (1988)
Zygosaccharomyces rouxii	Alginate	Soy sauce fermentation	Hamada *et al.* (1989)

4.4.1.1 Temperature

Considering firstly temperature, this is one of the most important physical parameters which influence yeast growth. Most laboratory and industrial yeasts generally grow best between 20–30°C. Notable exceptions to this range are found when studying yeasts in natural habitats. For example, some species associated with warm-blooded animals will not grow well below 24–30°C, while some psychrophilic yeasts grow optimally between 12–15°C (Phaff *et al.*, 1978). Like all microorganisms, yeasts exhibit characteristic, or 'cardinal', minimum, optimum and maximum growth temperatures (T_{min}, T_{opt} and T_{max}, respectively). The maximum temperature for growth is relatively constant within a species (van Uden, 1984). For *S. cerevisiae* strains, T_{max} values range from 35–43°C (which is similar to *S. paradoxus* strains), whereas strains of *S. bayanus* and *S. pastorianus* fail to grow above 35°C. This has been deemed a taxonomic criterion for differentiating subgroups within *Saccharomyces sensu stricto* (Vaughan-Martini and Martini, 1993; Rodriguez de Sousa *et al.*, 1995).

The lowest T_{max} values for yeasts are around 20°C, while the highest are around 50°C (Slapack *et al.*, 1987), although some thermoduric strains of *Kluyveromyces marxianus* can grow at 52°C (Banat *et al.*, 1992). Actual values of T_{max} are not only species-dependent, but also growth condition-dependent. For example, the influences of carbon source (Gross and Watson, 1996), oxygen availability (Davidson *et al.*, 1996), media water potential (Blomberg and Adler, 1992) and the presence of ethanol (van Uden, 1984) and growth factors (Phaff *et al.*, 1978) have been shown to play a role in dictating T_{max} values.

Yeasts can be grouped, albeit rather awkwardly in comparison with other microbes, as psychrophiles, mesophiles and thermophiles according to their thermal domains for growth (Table 4.16). Most yeasts exploited in biotechnology are mesophilic. Some obligately psychrophilic yeasts have been found in Arctic environments (e.g. *Trichosporon scotii* grows in the range −10 to +10°C). Such adaptation to low temperatures means that certain yeasts are very important spoilage organisms in frozen foods. Although truly thermophilic yeasts do not exist when compared with bacterial high temperature growth, Watson (1987) has described species of yeast 'thermophiles' with T_{min} values at or above 20°C. Some examples are listed in Table 4.16.

4.4.1.2 Water

With regard to water requirements, yeasts (like all organisms) need water in high concentrations for growth and metabolism. Substrates and enzymes are all in aqueous solution or colloidal suspension and no enzymic activity can occur in the absence of water.

The term **water potential** (ψ_w, expressed in megapascals, MPa) refers to the potential energy of water and is used to quantitate the availability of water in the presence of dissolved solutes. It therefore closely relates to the osmotic pressure of yeast growth media. Pure water has a ψ_w of zero, while impure water will have a lower, negative ψ_w value. For example, seawater has a water potential of about −2.5 MPa. As is the case with temperature, yeasts have cardinal water potentials for growth; namely, ψ_{min}, ψ_{opt}, and ψ_{max} (see Figure 4.30). Yeasts able to withstand conditions of low water potential (i.e. high sugar or salt concentrations) are referred to as *osmotolerant* or *zerotolerant*. Some examples are shown in Table 4.17 and the influence of water stress on yeast physiology is discussed further in section 4.42. Note that the terms osmophilic and osmoduric have also been used to describe yeasts

Table 4.16. Yeast growth temperature limits.

Thermal domain	Broad definition	Examples of yeasts
Psychrophile	A yeast capable of growing between 5–18°C. Obligate psychrophiles have an upper growth limit at or below 20°C	*Leucosporidium* spp. (e.g. *L. frigidum*). *Torulopsis* spp. (e.g. *T. psychrophila*)
Mesophile	Yeasts with growth limits at 0°C and up to 48°C	Vast majority of yeast species
Thermophile	Minimum temperature for growth at or above 20°C	*Candida slooffii* *Cyniclomyces guttulatus* *Saccharomyces telluris* *Torulopsis bovina*

Information from Watson (1987).

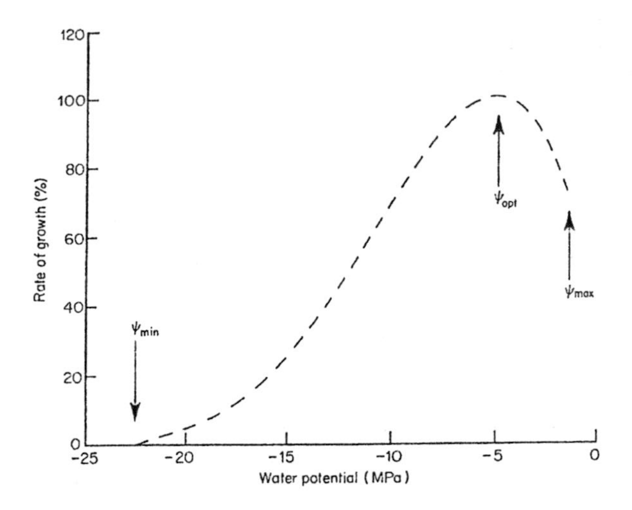

Figure 4.30. Generalized graph of the relative rate of yeast growth in relation to the water potential of the growth medium. Indicated are the cardinal water potentials of growth, namely ψ_{max}, ψ_{opt} and ψ_{min}. Reproduced with permission from Blomberg and Adler (1992) and Academic Press.

Table 4.17. Tolerance of yeasts to low water potential.

Lowest value of ψ_w allowing growth (MPa)	Examples of yeasts
-5 to -20	*Saccharomyces cerevisiae* (non-osmotolerant)
-20 to -40	*Debaryomyces hansenii* (in salt)
Below -40	*Zygosaccharomyces rouxii* (in sugars)

Information from Jennings (1995).

which, respectively, grow well or withstand conditions of low water potential (Phaff *et al.*, 1978). The growth of osmotolerant yeasts (e.g. *Zygosaccharomyces rouxii*) are generally unaffected by ψ_w values between −1.0 and −5.6 MPa, whereas a non-osmotolerant species like *S. cerevisiae* exhibits a sharp ψ_{opt} value around −1.5 MPa. The majority of yeasts display a moderate tolerance to growth at low water potentials, but some of those with the capacity to grow well at very low water potentials (i.e. the osmotolerants) are very important economically as food spoilage yeasts. Examples of osmotolerant yeasts include: *Candida mogii, Debaryomyces hansenii, Metschnikowia bicuspidata, Schizosaccharomyces octosporus* and *Zygosaccharomyces rouxii.*

4.4.1.3 Media pH and pO_2

Other physical growth requirements for yeasts relate to media pH and pO_2. With regard to pH, most yeasts grow very well between pH 4.5 and 6.5, but nearly all species are able to grow, albeit to a lesser extent, in more acidic or alkaline media (around pH 3 or pH 8, respectively). Media acidified with organic acids (e.g. acetic, lactic acids) are more inhibitory to yeast growth compared with those acidified with mineral acids (e.g. hydrochloric, phosphoric acids). This is because undissociated organic acids can lower intracellular pH following translocation across the yeast plasma membrane. This forms the basis of action of weak acid preservatives in inhibiting food spoilage yeast growth.

With regard to oxygen requirements of yeasts in general, it should be stated from the outset that most yeasts are aerobic organisms. The frequently used description of brewing yeasts as being facultatively anaerobic is not strictly speaking correct since brewer's yeast has an *absolute requirement* for oxygen. Lagunas (1986) has outlined the reasons why *S. cerevisiae* is not a facultative anaerobe and these are summarized in Table 4.18.

Yeasts need oxygen not just as the terminal electron acceptor in respiratory growth, but also as an essential *growth factor* for membrane fatty acid and sterol biosynthesis (see Chapter 3). Thus, *S. cerevisiae* is auxotrophic for oleic acid and ergosterol under anaerobic conditions. The influence of oxygen and sugar availability on yeast carbohydrate metabolism has been categorized under various regulatory phenomena: the Pasteur, Crabtree, Custers and Kluyver effects (see Chapter 5). Van Dijken and Scheffers (1986) have discussed the crucial role of redox balances

Table 4.18. Reasons why *S. cerevisiae* is not a facultative anaerobe.

Facultative anaerobe characteristics	*S. cerevisiae* characteristics
Grow aerobically or anaerobically	Anaerobic growth only for a few generations
Preferentially use oxygen	Fermentation is preferred catabolic route
Lower glycolytic rate in presence of oxygen (Pasteur effect)	Glycolytic rate is similar in presence and absence of oxygen (no noticeable Pasteur effect)

Adapted from Lagunas (1981).

in yeast sugar metabolism. Yeasts may be also categorized into different groups with respect to their fermentative properties and growth responses to oxygen availability (Table 4.19).

4.4.2 Effects of Physical Stresses on Yeast Growth

4.4.2.1 Temperature Stress

Growth and metabolic activity of yeasts at various temperatures are functions not only of the genetic background of the cell but also of the growth medium composition and other physical growth parameters. The accumulation of yeast metabolites, both extracellularly and intracellularly, may also influence the temperature profiles of yeast. Watson (1987) has reviewed the influence of suboptimal and supraoptimal growth temperatures on various morphological features (bud scars, cell surface topology, cell size, etc.) of psychrophilic, mesophilic and thermophilic yeasts. With regard to **high temperature stress**, Table 4.20 summarizes some of the general adverse influences of heat on yeast cell physiology.

Thermal damage to yeast cells results from disruption of hydrogen bonding and hydrophobic interactions leading to general denaturation of

Table 4.19. Classification of yeasts based on fermentative capacity.

Class	Examples	Comments
Obligately fermentative	*Candida pintolopesii* (*Saccharomyces telluris*)	Naturally occurring respiratory-deficient yeasts. Only ferment, even in presence of oxygen.
Facultatively fermentative		
Crabtree-positive	*Saccharomyces cerevisiae*	Such yeasts predominantly ferment high sugar-containing media in the presence of oxygen.
Crabtree-negative	*Candida utilis*	Such yeasts do not form ethanol under aerobic conditions and cannot grow anaerobically.
Non-fermentative	*Rhodotorula rubra*	Such yeasts do not produce ethanol, either in the presence or absence of oxygen.

Adapted from information in van Dijken and Scheffers (1986) and Scheffers (1987).

Table 4.20. General effects of high temperature on yeast cell physiology.

Physiological function	Comments
Cell viability	At the highest growth temperature of many yeasts, there is also appreciable cell death. At supramaximal growth temperatures, thermal death rate is exponential.
General cell morphology	Atypical budding, irregular cell wall growth and increased cell size.
Cell division and growth	Growth of non-thermotolerant yeasts inhibited at temperatures >40°C. Actively-dividing cells in S-phase are more thermosensitive compared with resting cells. Heat shock transiently arrests cells in G1 phase of the cell cycle.
Plasma membrane structure/ function	Increased fluidity and reduced permeability to essential nutrients. Ergosterol is known to increase thermotolerance. Decrease in unsaturated membrane fatty acids. Stimulation of ATPase and RAS-adenylate cyclase activity. Decline in intracellular pH.
Cytoskeletal integrity	Extensive disruption of filaments and microtubular network.
Mitochondrial structure/ function	Decrease in respiratory activity and induction of respiratory-deficient petite mutants. Aberrant mitochondrial morphology.
Intermediary metabolism	Inhibition of respiration and fermentation above T_{max}. Immediate increase in cell trehalose and MnSOD following heat shock.
Protein synthesis	Repression of synthesis of many proteins, but specific induction of certain heat-shock proteins. Mitochondrial protein synthesis more thermolabile than cytoplasmic.
Chromosomal structure/ function	Increased frequency of mutation of mitotic cross-over and gene conversion. Increased mitotic chromosomal non-disjunction. Inefficient repair of heat-damaged DNA.

proteins and nucleic acids. Of course, yeasts have no means of regulating their internal temperature and the higher the temperature, the greater the cellular damage, meaning that yeast cell viability will rapidly decline when temperatures are increased beyond growth optimal levels (see section 4.6). One study, however, has suggested that the *in situ* temperatures of *Sch. pombe* interiors may be reduced following heat shock (Komatsu *et al.*, 1996). Temperature optima vary for different yeast species (see Table 4.16) with those termed 'thermotolerant' possessing T_{opt} values above 40°C. Thermotolerance may be defined as the transient ability of cells subjected to high temperatures to survive subsequent lethal exposures to elevated temperatures (Laszlo, 1988). *Intrinsic thermotolerance* in yeast cells is observed following a sudden heat shock (e.g. to 50°C), whereas *induced thermotolerance* occurs when cells are pre-conditioned by exposure to a mild heat shock (e.g. 30 min at 37°C) before a more severe heat shock. Several factors, besides a mild heat shock, are known to influence yeast thermotolerance (Piper, 1993). For example, certain chemicals, osmotic dehydration, low external pH, nutritional status and growth phase are known to play a role.

Concerning pH, yeast thermotolerance increases to a maximum when the external pH declines to 4.0 and Coote *et al.* (1991) have provided convincing evidence which implicates alterations in intracellular pH as the trigger for acquisition of thermotolerance in *S. cerevisiae*. With regard to yeast growth, there is a general correlation between growth rate and stress sensitivity such that cells growing quickly in a glucose-rich medium are more sensitive to heat and other stresses compared with stationary phase cells. This may be due to the fact that glucose itself may exhibit a repressive effect on the synthesis of stress defence proteins in yeast.

Yeast cells exhibit a rapid molecular response when exposed to elevated temperature. This is called the **heat-shock response** and is a ubiquitous regulatory phenomenon in all living cells. Sublethal heat-shock treatment of yeast leads to the induction of synthesis of a specific set of proteins, the highly conserved 'heat-shock proteins' (Hsps). The known functions of the major classes of Hsps in *S. cerevisiae* are summarized in Table 4.21.

Several Hsps have been shown to perform molecular 'chaperoning' functions in the yeast cell, while others are implicated in conferment of thermotolerance, glycolysis and polyubiquitination of proteins (Parsell and Lindquist, 1994). Hsps functioning as chaperons (e.g. Hsp60, chaperonin) prevent protein aggregation and the accumulation of aberrant proteins. Hsps may also assist in the degradation of stress-damaged proteins by enhancing the flow of substrates through proteolytic pathways. Recently, eukaryotic chaperon function has been studied genetically in yeast and *S. cerevisiae* chaperonin Cct (*c*haperonin-*c*ontaining-*t*ailless complex polypeptide) was shown to be required for the *in vivo* assembly of cytoskeletal elements (Stoldt *et al.*, 1996). The heat-shock response in *S. cerevisiae* is one of the best molecularly characterized responses of eukaryotic cells and has been widely reviewed (e.g. Watson, 1990; Piper, 1993; Parsell and Lindquist, 1994; Mager and De Kruijff, 1995; Ruis and Schüller, 1995). Hsp gene expression basically involves increased tran-

Table 4.21. Major heat-shock proteins of *S. cerevisiae*.

Heat-shock protein	Proposed physiological function
Hsp104	Acquisition of stress tolerance. Constitutively expressed in respiring, not fermenting cells and on entry into stationary phase.
Hsp83	Chaperone(s) function.
Hsp70 family	Interact with denatured, aggregated proteins and assists in solubilizing them with simultaneous refolding (i.e. chaperone(s) function). Also involved in post-translational import pathways.
Hsp60	Similar to Hsp70. This chaperonin family facilitate post-translational assembly of proteins.
Small Hsps	
Hsp30 Hsp26 Hsp12	Cellular role still elusive, but may be involved in entry into stationary phase and the induction of sporulation. Hsp30 may regulate plasma membrane ATPase.
Others	
Ubiquitin	Responsible for much of the turnover of stress-damaged proteins.
Some glycolytic enzymes	Enolase (Hsp48), glyceraldehyde 3-phosphate dehydrogenase (Hsp35) and phosphoglycerate kinase.
Catalase	Antioxidant defence.
GP400 and P150	Secretory heat-shock proteins (unknown function)

Information from: Watson (1990); Mager and Moradas-Ferreira (1993); Parsell and Lindquist (1994); Mager and DeKruijff (1995); Tsiomenko and Tuymetova (1995).

scription of genes containing the promoters of the heat shock element (HSE) which occurs in the presence of heat due to the activation of the heat shock transcription factor (HSF). In *S. cerevisiae*, the HSE is unresponsive to other stresses (osmotic, oxidative, DNA damage, glucose repression, etc.) and is exclusively induced by a sub-lethal heat shock. The HSF in yeast, which is an essential protein involved in normal growth, is not required for induced tolerance against severe stress, but its activation by heat shock is required for growth at high temperatures (Ruis and Schüller, 1995). Although several authors have implicated Hsps in yeast thermotolerance (and ethanol tolerance), the functional significance of these proteins in stress tolerance is still unresolved. Indeed, an obligatory role for Hsps in thermotolerance of *S. cerevisiae* has recently been questioned by Gross and Watson (1996) who showed that non-fermenting (acetate-grown) cells acquired thermotolerance in the absence of protein synthesis (and without accumulation of trehalose). Nevertheless, a few specific Hsps are recognized to exert a protective role against thermal stress in *S. cerevisiae* (Piper, 1993). For example, Lindquist and her co-workers have provided compelling evidence that the synthesis of one particular Hsp, namely Hsp104, is required to confer thermotolerance in respiratory (but not fermentative) cultures of *S. cerevisiae* (Lindquist and Kim, 1996).

Other environmental insults (e.g. ethanol and heavy-metal ion stress) also elicit the synthesis of Hsps, indicating the general importance of this response in the stress physiology of yeasts (Watson, 1990). In addition to the induction of Hsps following heat shock, yeast cells also respond by accumulating other putative protective compounds such as trehalose (van Laere, 1989; Wiemken, 1990; Neves and Francois, 1992), glycerol (Omori *et al.*, 1996) and enzymes such as catalase and mitochondrial superoxide dismutase (Costa *et al.*, 1993). Trehalose is thought to act as a thermoprotectant (and cryoprotectant) by stabilizing cell membranes and increasing the temperature stability of yeast cellular proteins by replacing water and forming a hydration shell around proteins (Iwahashi *et al.*, 1995). In *Candida albicans*, Argüelles (1997) has shown that trehalose accumulates markedly in cells exposed to a non-lethal heat shock and has provided additional evidence to link trehalose with acquired thermotolerance in this yeast. Interestingly, Elliot *et al.* (1996) have shown that trehalose, together with Hsp104, act synergistically to confer thermoprotection in *S. cerevisiae*. Stimulation of antioxidant enzymes by heat shock may permit trapping of superoxide radicals thus preventing oxidative damage to cells which would be enhanced at elevated temperatures. Work from Tabor's laboratory (e.g. Balasundaram, Tabor and Tabor, 1996) has shown that polyamines (spermine, spermidine) also play important roles in thermal protection of *S. cerevisiae* cells. Polyamines may be acting in a similar protective fashion to Mg^{2+} ions (Birch and Walker, 1996) in stabilizing the structural integrity of yeast membranes during stress. Increase in cell Ca^{2+}, on the other hand, may activate thermal cell killing due to its role in regulatory degradative enzymes (Jozwichk and Leyko, 1992).

Piper (1993) has noted that there is some degree of functional overlap between thermal and oxidative stress responses of yeast, but not necessarily between thermal and osmotic stress responses. The mechanisms required by yeast to counteract osmotic dehydration (see section on water stress below) are different from those needed to counteract thermal denaturation.

In addition to the utility of yeast as a model organism in basic studies of the molecular mechanisms of thermal stress and responses to it in eukaryotic cells, the topic of thermotolerance has distinct significance in more practical aspects of yeast biotechnology. For example, a greater understanding of how yeast cells precisely acquire thermotolerance may lead to manipulation of strains with greater robustness for industrial fermentations. This is of particular pertinence in the production of bioethanol in tropical countries. Thermal stress is also relevant in yeast recombinant DNA technology. For example, Piper and Kirk (1991) and Cheng and

Table 4.22. Influences of low temperature on yeast cell physiology.

Summary of general observations

- Yeast cells shrink uniformly during freezing
- Low temperatures result in an increase in polyunsaturated membrane fatty acids resulting in decreased solute transport
- Psychrophilic yeasts have membranes rich in polyunsaturated fatty acids
- Sterol synthesis is reduced at 15°C in *S. cerevisiae*
- Stationary-phase cells are relatively resistant to freeze–thaw stress
- Membrane fatty acids/sterols undergo a phase transition from fluid to 'gel' states as the cell membrane cools (membrane integrity is compromised such that it leaks)
- Vacuolar membrane damage and splitting of vacuoles
- Sub-lethal growth arrest during cell division following a cold shock
- Cold shock induces specific protein biosynthesis

Yang (1996) have discussed the utility of temperature-directed expression of heterologous genes in *S. cerevisiae* (see also Chapter 6).

Concerning **low temperature stress**, it is still unclear exactly how yeast cells die at low temperatures. An understanding of aspects of cold stress in yeast is important due to its relevance in food storage (e.g. psychrophilic yeast spoilage) and culture maintenance (e.g. cryopreservation/lyophilization of stock strains). Table 4.22 summarizes some of the physiological effects of low temperatures on yeast.

Survival of frozen yeast cells depends on several genetic, physiological and environmental factors. For example, successful cryopreservation of yeasts in liquid nitrogen (at −196°C) mainly depends on: the rates of freezing and subsequent thawing, the presence or absence of cryoprotectants (e.g. glycerol or trehalose may alleviate freeze–thaw stress) and the culture growth phase (stationary phase cells being much more freeze resistant compared with logarithmic cells). Freeze-drying (lyophilization) of yeast is frequently used for the long-term maintenance of stock cultures, but the procedure does damage cells and may induce respiratory deficient variants (Kirsop and Henry, 1984). There is evidence that phase transitions in membrane lipids are involved in yeast cell death by freeze–thawing (Kruuv *et al.*, 1978). Trehalose, which is known to protect yeast membranes, has been studied by Coutinho *et al.* (1988) in connection with its involvement in cold-shock sensitivity of yeast. Exogenous trehalose was shown to act as a powerful cryoprotectant for yeast cells frozen in water. Furthermore, cellular trehalose was shown by Hino *et al.* (1990) to be associated with freeze tolerance of yeast while Meric *et al.* (1995) demonstrated that baker's yeast with a 4–5% trehalose level was protected from freezing injury and could be used in the successful production of frozen bread doughs. With regard to genetic manipulation of yeast to provide cells with improved resistance to cryo-damage, Driedonks *et al.* (1995) have successfully expressed anti-freeze peptides from polar fish into *S. cerevisiae*.

With regard to the physiological responses to cold stress by yeast, Fargher and Smith (1995) have shown that cold (4°C)-shocking brewing yeast strains inhibits budding and induces vacuolar re-arrangements. Treatment of *S. cerevisiae* to a 10°C cold-shock resulted in the specific induction of a novel cold-shock protein (of 33 kDa) which was distinct from Hsps (Kaul *et al.*, 1992). Komatsu *et al.* (1990) demonstrated that prior heat shock (e.g. 43°C for 30 min) of *S. cerevisiae* significantly increased cell viability following subsequent freezing in liquid nitrogen and thawing. It was proposed that heat shock proteins protected frozen cells by stabilizing macromolecules and increasing hydrophobic interactions within the yeast cell.

Low temperature stress is important in several aspects of yeast biotechnology. Freeze tolerance in baker's yeast in relation to trehalose has been previously mentioned, and there are similar correlations noted in brewing yeast. For example, cold storage of seed yeast in breweries for later pitching (inoculation) is a widespread practice, but one which can adversely affect subsequent fermentation performance (Boulton *et al.*, 1989).

D'Amore *et al.* (1991) have shown that brewer's yeast with elevated levels of trehalose was better able to survive storage conditions (4°C for 30 days in beer at 5%, v/v, ethanol), indicating that trehalose may be an important determinant in fermentation performance and in the maintenance of brewing yeast cell viability.

4.4.2.2 Water Stress

Since water is absolutely essential for yeast enzymic activity, any external conditions which result in reduced water availability to cells (i.e. 'water stress') will impair yeast growth and metabolism (see Gervais *et al.*, 1996). Note that the most severe form of water stress is experienced by yeast cells when water is removed physically by dehydration. This **anhydrobiosis** is commonly encountered in the industrial drying (by spray, drum or fluidized-bed driers) of yeast biomass (e.g. dried baker's yeast). Rapoport and his colleagues have studied yeast anhydrobiosis in terms of changes in the structure and function of vacuolar, nuclear and cell membranes in *S. cerevisiae* (e.g. Beker and Rapoport, 1987; Rapoport *et al.*, 1995). Dramatic (but reversible) changes in membrane structural arrangements and permeability properties occur in this yeast under conditions of severe water stress. Figure 4.31 depicts gross morphological and ultrastructural changes which occur in *S. cerevisiae* cells when they are dehydrated and rehydrated and Figure 4.32 summarizes schematically the structural transformations taking place. Mild water stress occurs in yeasts during osmotic dehydration when cells are placed in a medium with low water potential (ψ_w) brought about by increasing the solute (e.g. salt, sugar) concentration. This is referred to as **hyperosmotic shock** (or 'upshock'). Conversely, when such a medium is replaced with one of higher osmotic potential (due to reducing the solute concentration), cells experience a **hypoosmotic shock** (or 'downshock') (Jennings, 1995).

What happens to yeast cells when the osmotic potential of their growth medium is reduced or increased? With reference firstly to cell viability, although yeasts are generally able to survive short-term osmotic shock (Rose, 1975), Marechal *et al.* (1995) have found that in *S. cerevisiae* the intensity and the rate of change of water stress are very important in determining cell survival. Concerning growth, Barnett *et al.* (1990) reported that the majority of yeast strains displayed a moderate growth tolerance to conditions of low water potential, while relatively few had the capacity to grow at very low water potentials. Yeasts able to adapt to such environments are called osmophiles (or xerophiles) and are found in natural solute-rich habitats (e.g. honey, tree exudates).

Tables 4.17 and 4.23 list some osmotolerant yeasts. Note that *S. cerevisiae* is regarded as a non-osmotolerant yeast. For example, few cells of *S. cerevisiae* will survive when transferred from salt-free to saline media – a phenomenon described as 'water stress hypersensitivity'. The term 'osmoduric' is sometimes used to describe yeasts that can withstand low osmotic potential conditions but do not require those conditions for growth (e.g. halotolerant *Debaryomyces hansenii*). Those yeasts unable to adapt to growth in high water potential environments are termed obligately osmophilic and include the yeast-like fungus *Eremascus albus* and certain marine species of *Metschnikowia* which require, respectively, high levels of sugar and salt in order to grow well (Phaff *et al.*, 1978). Figure 4.33 schematically depicts the relationship between yeast growth rate and media water activity (a_w) in strains with varying osmotolerance.

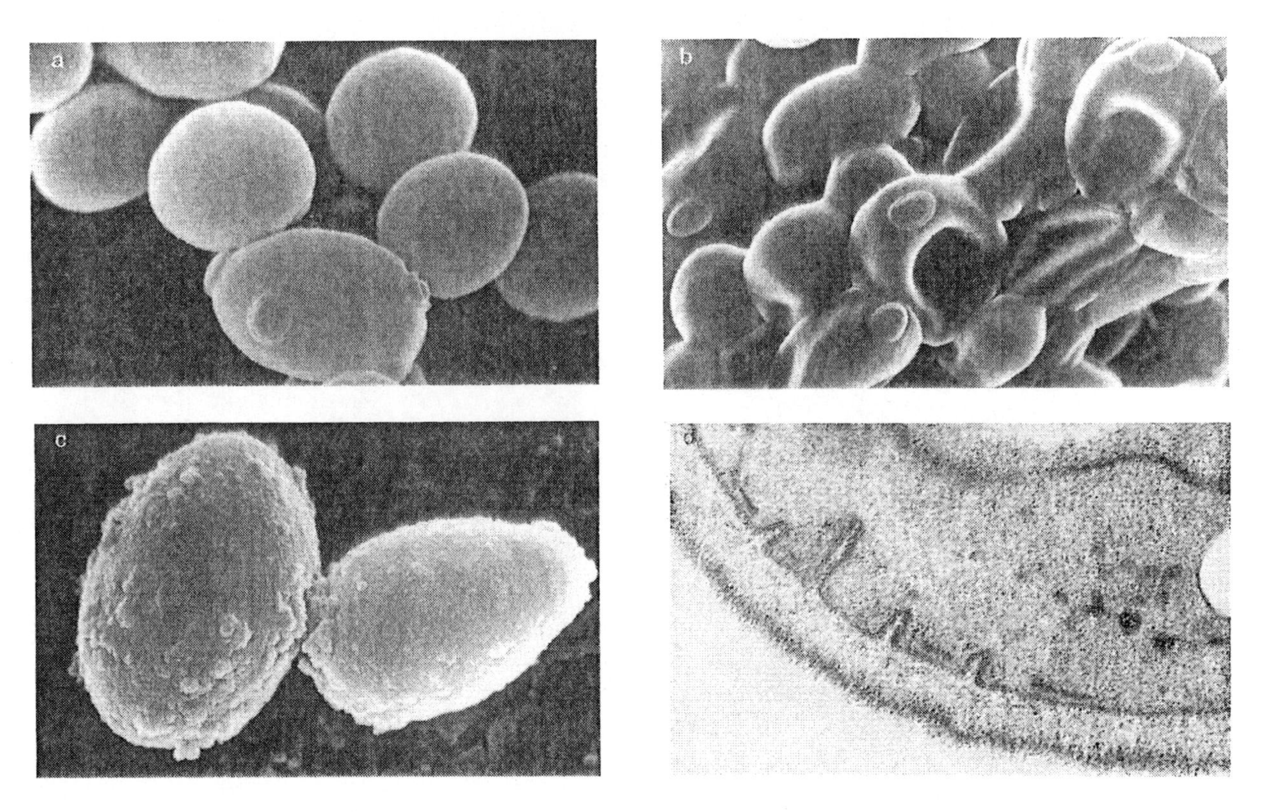

Figure 4.31. Morphological changes in *S. cerevisiae* following dehydration. (a) Scanning EM of intact *S. cerevisiae* cells (× 7000); (b) Scanning EM of *S. cerevisiae* cells after dehydration (× 5000); (c) Scanning EM of rehydrated *S. cerevisiae* cells (× 16 000); (d) Transmission EM showing cytoplasmic invaginations from the plasma membrane of a dehydrated cell of *S. cerevisiae* (× 170 000). Reproduced by kind permission of Professor A. Rapoport (University of Latvia, Riga, Latvia).

With reference to the gross morphological changes and physical cellular damage of water-stressed yeast cells, overall cell volume is observed to change when the osmotic potential of the medium changes. For example, cytoplasmic volume in *S. cerevisiae* alters dramatically depending on the external osmotic potential due to the fact that the cell walls of this yeast are relatively elastic and therefore weakly buffered against water loss (Rose, 1975; Meikle *et al.*, 1988). Hyperosmotic shock in *S. cerevisiae* thus results in a loss of cell turgor pressure and a rapid and substantial decrease in cytoplasmic water content and cell volume (Marechal *et al.*, 1995), the rate of recovery of which depends on temperature (Niedermeyer *et al.*, 1977) and the permeability of the osmoticum (Jennings, 1995). More osmotolerant yeasts like *Zygosaccharomyces rouxii* have a greater resistance to water loss and are better able to retain their volume in media of low water potential. Exposure of *S. cerevisiae* cells to a hypertonic solution also results in an irreversible decrease in vacuolar volume, but it is not clear if this reduction is greater than that of the osmotic volume of the cytoplasm (Morris *et al.*, 1986; Niedermeyer *et al.*, 1977). Cellular shrinkage of osmotically shocked *S. cerevisiae* cells does not result in plasmolysis due to the fact that the cell wall and the protoplast shrink together. That is, the yeast plasma mem-

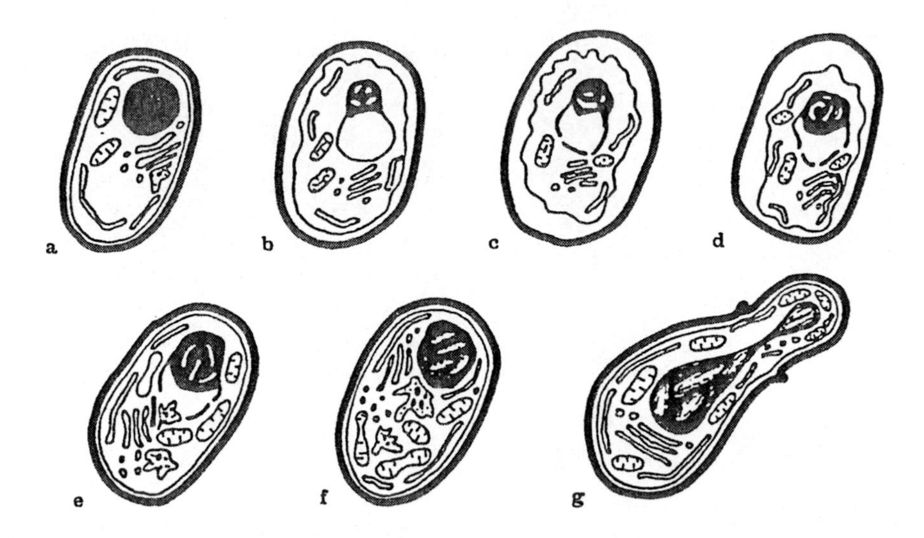

Figure 4.32. A scheme of several structural transformations in a yeast cell during dehydration and reactivation. (a) Initial cell. (b–d) Formation of cytoplasmic membrane invaginations during dehydration, plasmolysis, chromatin condensation and separation of chromosome-containing part of the nucleus. (e,f) Moistening of cells during reactivation, a further degradation of part of the nucleus to be separated and its elimination by phagosomes. (g) Division of a reactivated cell. Reproduced with permission from Beker and Rapoport (1987) and Springer-Verlag GmbH and Co.

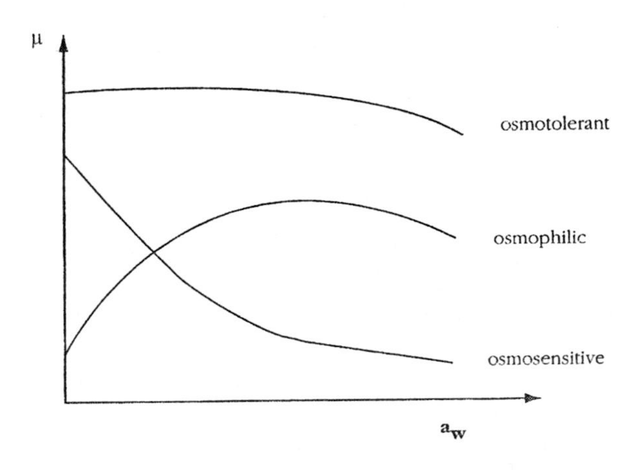

Figure 4.33. Schematic diagram relating growth responses of different yeast types to osmostress.

brane appears to remain in close contact with the cell wall when cells shrink.

Considering the converse situation (i.e. downshock), when osmotically dehydrated cells are resuspended in high water potential media, they quickly increase in volume due to the high water permeability of the plasma membrane. Hypoosmotically shocked cells may also increase their internal osmotic potential by reducing intracellular levels of K^+ or glycerol, the latter being effluxed by a specific glycerol facilitator in *S. cerevisiae* (see below). Figure 4.34 summarizes some of the osmoregulatory membrane transport processes which occur in yeast.

Yeasts respond to the effects of solute concentration in their growth medium by physiologically adapting in several ways. This is referred to as the **osmostress response** (Mager and Varela, 1993). Since yeasts have no means for actively transporting water, they transport, synthesize or prevent the loss of **compatible solutes** in order to maintain low cytosolic water activity when the external solute concentration is high. Compatible solutes are osmolytes which can effectively replace cellular water, restore cell volume and enable enzyme activity to continue. They alleviate the detrimental effects of high ionic strength by

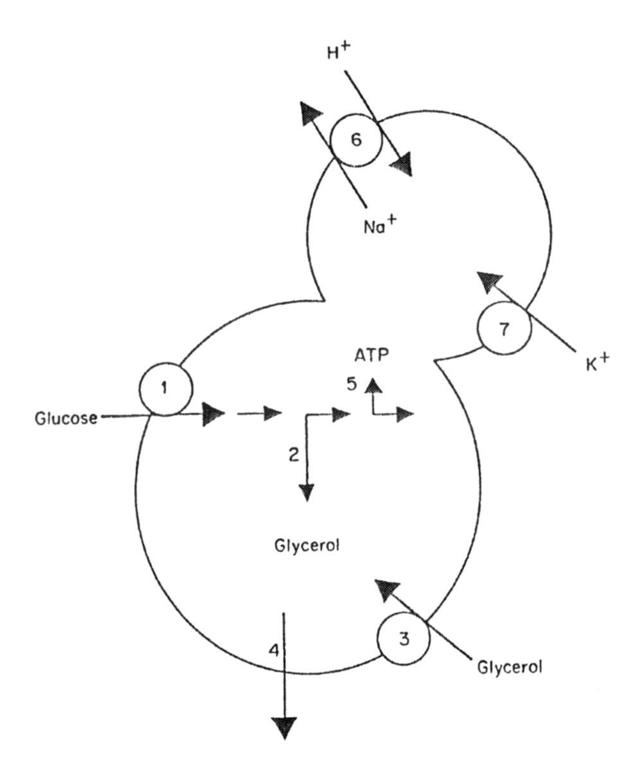

Figure 4.34. Schematic representation of cellular functions involved in osmoregulation in yeast. **1**, Indicates glucose uptake; **2**, glycerol production; **3**, glycerol uptake; **4**, glycerol efflux (Fps1p glycerol facilitator); **5**, energy supply; **6**, sodium efflux (P-type ATPase, encoded by the *ENA1* gene in *S. cerevisiae*); **7**, potassium influx. Reproduced with permission from Blomberg and Adler (1992) and Academic Press.

stabilizing proteins within the cell. They also stabilize membrane proteins and maintain the liquid crystal state of membrane phospholipids during dehydration. The most effective osmoregulatory compatible solutes in yeasts are polyols, particularly glycerol (Table 4.23). In bacteria, the compatible solutes are more commonly amino acids like betaine (*N*-trimethylglycine), glutamic acid and proline. In the presence of reduced water potential many yeasts have the ability preferentially to retain and/or induce the synthesis of glycerol in order to control intracellular solute potential relative to that of the medium, and thus counteract the deleterious effects of dehydration on cells. For example, when *Debaryomyces hansenii, Saccharomyces cerevisiae* or *Zygosaccharomyces rouxii* are grown in the presence of salt with glucose as the carbon source, net glycerol synthesis can effectively counterbalance the external osmotic pressure due to salt concentration (Reed *et al.*, 1987). Note, however, that in *S. cerevisiae*, net accumulation of glycerol following salt-induced water stress only takes place after several hours adaptation during which K^+, Na^+ and trehalose content increases (Singh and Norton, 1991). Sunder *et al.* (1996) have proposed a regulatory mechanism whereby yeast cells maintain their turgor under osmotic stress by coordinating both intracellular glycerol and cationic (K^+, Na^+) levels. Glycerol homeostasis must be regulated in some way since the polyol accumulates under hyperosmotic stress and is released under hypoosmotic stress. Luyten *et al.* (1995) have shown that the cytosolic concentration of glycerol in *S. cerevisiae* is controlled by a plasma membrane channel (encoded by the *FPS1* gene) which opens and closes depending on the absence and presence, respectively, of hyperosmotic stress. In *Z. rouxii*, an active glycerol transport mechanism is implicated in the osmotic stress response (van Zyl *et al.*, 1990) and one difference between osmotolerant yeasts like *Z. rouxii* and moderately tolerant yeasts like *S. cerevisiae* is the ability of the former yeast preferentially to retain glycerol. In addition to compatible solute accumulation, Batiza *et al.* (1996) have shown that *S. cerevisiae* cells respond to hypotonic shock by transiently elevating intracellular Ca^{2+} levels. This was mediated by stretch-activated channels whose gating promoted release of Ca^{2+} from intracellular stores. These results indicate an important role for Ca^{2+} ions in the response of yeast cells to hypotonicity.

In recent years, much progress has been made concerning our understanding of the molecular biology of osmotic and ionic stress responses in yeast cells (Varela and Mager, 1996). Changes in external osmotic pressure have been shown to induce corresponding changes in the expression of certain genes in yeast which are involved in controlling the levels of compatible solutes like glycerol. In *S. cerevisiae*, transcriptional control elements, termed the stress response elements (STREs), have been identified (core consensus sequences are AGGGG or CCCCT) which are activated by a wide variety of stress conditions, including osmotic shock (Figure 4.35). It is now known that a signalling system called the *HOG* (*h*igh *o*smolarity *g*lycerol) pathway specifically signals increases in external osmotic pressure to STREs (Schüller *et al.*, 1994). The *HOG* pathway, whose function appears limited to high osmolarity stress, rather than general stress signalling, involves various protein kinase cascades (Ruis and Schüller, 1995). The *HOG1* gene product encodes a mitogen-activated protein kinase (MAPK) and the pathway starts with the activation of plasma membrane bound receptor proteins which act as **osmosensors** to enable yeast to sense the external osmolarity. The genes encoding osmosensors (e.g. *SLN1*), which are upstream regulators of the *HOG* pathway, have been identified and the gene products characterized (Maeda *et al.*, 1995). The MAPK cascades transduce signals in yeast (and other eukaryotic cells) which are triggered not only by osmotic stress but also by other environmental stresses (heat shock, UV irradiation, etc.). These cascades

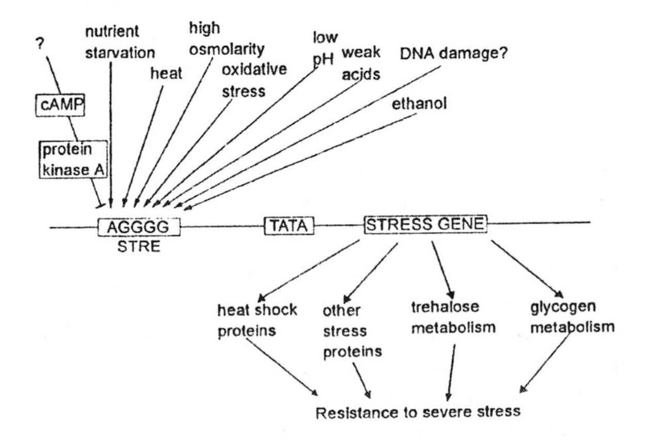

Figure 4.35. Factors controlling stress response elements (STREs) and effects triggered by STRE activation in yeast. Reproduced with permission from Ruis and Schüller (1995) and ICSU Press.

are also thought to be responsive to growth-related stimuli such as nutrients, growth factors or mating pheromones (see section 4.2.3). The relationship between stress tolerance and growth control in yeast makes sense since the main function of a stress response is to enable cells to continue growing under potentially lethal conditions. Current understanding of hyperosmotic stress signalling and response regulation has been reviewed by Varela and Mager (1996).

Other compounds, including other polyols (arabinitol, sorbitol, mannitol and erythritol), ions (e.g. K^+) and the disaccharide trehalose may accumulate in yeasts in response to water stress caused by specific solutes (Table 4.23). However, yeasts do not adjust their intracellular solute potential by accumulating salt, as is the case in certain other cell types (Serrano, 1996).

Concerning trehalose, this disaccharide has been shown to be one of the most effective saccharides in the stabilization of biological membranes against osmotic shock (Crowe *et al.*, 1984). Levels of trehalose are known to increase in stationary phase cultures of *S. cerevisiae* grown under low water potential and such cells are much more osmotolerant than actively dividing cells (Gadd *et al.*, 1987). MacKenzie *et al.* (1988) have shown that trehalose accumulation in stationary phase cells can protect cells from osmotic shock when they are transferred to low water-activity agar plates. With regard to exponentially growing yeasts, Singh and Norton (1991) have shown that when active *S. cerevisiae* cells are transferred to a medium with 8% NaCl, cellular trehalose accumulated immediately and rose 10-fold a few hours after transfer. van Dijck *et al.* (1995) have noted that any correlation between cellular trehalose and yeast stress resistance appears to hold only for non-fermenting cells. Although some authors have questioned the evidence implicating trehalose in the osmotic protection of yeast cells (Serrano, 1996), trehalose is now widely recognized as a general 'stress metabolite' in yeast. In fact, Wiemken (1990) has proposed that the primary role of trehalose in yeast is as a stress-protectant molecule. This is because trehalose has been shown to act in yeast not only

Table 4.23. Responses of yeasts to changes in media osmotic potential.

Yeast	Treatment[a]	Intracellular solute
Saccharomyces cerevisiae	Upshock with NaCl	K^+, Na^+ glycerol and trehalose increased
S. cerevisiae	Upshock with sorbitol	Glycerol initially, then trehalose increased
Zygosaccharomyces rouxii	Upshock with NaCl	Glycerol, arabinitol increased
Z. rouxii	Upshock with glucose	Arabinitol initially, then trehalose increased
Debaryomyces hansenii	Upshock with various solutes	Glycerol increased
D. hansenii	Downshock by reducing NaCl concentrations	Glycerol decreased

[a] Upshock refers to the lowering of osmotic potential in the medium, whereas downshock refers to the raising of osmotic potential due to decreasing solute concentrations. Adapted from Jennings (1995).

as an osmoprotectant, but also as an antidesiccant (Gadd *et al*., 1987), cryoprotectant (Hino *et al*., 1990), thermoprotectant (De Virgillo *et al*., 1994) and chemical detoxicant (Attfield, 1987).

There are several aspects relating to trehalose in osmotically stressed yeast cells which are of direct relevance in yeast biotechnology, particularly with regard to the baking and brewing industries. Concerning the former, two technological developments which have occurred in recent years involve the production of 'Instant Active Dry Yeast' and frozen dough. Baker's yeast with elevated trehalose levels (>10% of dry weight) is resistant to the drying process and does not require rehydration before mixing with flour (Trivedi and Jacobson, 1986). In general, a correlation exists between the cellular trehalose content and leavening capabilities of active dry baker's yeast (van Laere, 1989). In frozen dough preparations, yeast with a high trehalose content maintains its viability and is more freeze-resistant (Meric *et al*., 1995; van Dijck *et al*., 1995). Special production measures are therefore undertaken (e.g. dough–yeast mixing at low temperatures) to minimize trehalose loss through premature fermentative activity (Rose and Vijayalakshmi, 1993). Contrary to the stress metabolite role of trehalose in baker's yeast, Lewis *et al*. (1997) have argued that there is a minimal contribution of trehalose to stress tolerance in baking strains of *S. cerevisiae*. They found that high cellular trehalose contents failed to correlate with the ability of cells to tolerate stresses imposed by chemical (e.g. ethanol, hydrogen peroxide) and physical (e.g. heat shock, rapid freezing, osmostress) insults. It may be possible to reconcile these findings with the work of others if one assumes that lower levels of cellular trehalose are sufficient to protect yeasts against stress; higher levels above a certain threshold may not therefore confer any *extra* protection.

In relation to brewing, yeasts are subjected to osmotic stress when fermenting high-gravity wort (e.g. around 18°P) and this may reduce cell growth and viability which will have a deleterious effect on the progress of fermentation (Casey and Ingledew, 1983; D'Amore *et al*., 1988). Majara, O'Connor-Cox and Axcell (1996) have shown that yeast trehalose content was proportional to the original gravity of the wort. Trehalose levels were observed to rise higher than those of glycerol early in fermentation and it was proposed that trehalose was an important osmoprotectant and stress indicator in brewing yeasts during fermentation of high- and very high-gravity worts (Majara *et al*., 1996). In addition to trehalose and glycerol, Thomas *et al*. (1994) have found that other compounds such as amino acids, as well as particulate material, may also act as osmoprotectants during very high-gravity fermentations by *S. cerevisiae*.

4.4.2.3 Other Physical Stresses

In addition to temperature and water stresses, a variety of other physical stresses may be experienced by yeasts when growing in natural habitats, the laboratory or in industrial fermenters. These include, pressure, radiation, centrifugal force and mechanical shear stress. Other external physical treatments, such as high-voltage electrical pulses and high-frequency ultrasonic waves, have applications in yeast biotechnology.

High-pressure stress

With regard to pressure, yeast cells grown in large-capacity bioreactors may face quite severe stress due to both hydrostatic and gaseous pressure. The former would be encountered, for example, in tall cylindroconical fermentation vessels employed in the brewing industry, while the latter would result from endogenous CO_2 produced by yeast during fermentation in pressurized vessels. Although yeasts are in general relatively resilient to the effects of mild hydrostatic pressure, cells of *S. cerevisiae* failed to grow above 10 MPa (Perrier-Cornet *et al*., 1995) and could not be described as 'barotolerant'. Studies by Kobori *et al*. (1995) have revealed that the cytoskeleton and mitotic apparatus of *S. cerevisiae* were particularly susceptible to high hydrostatic pressure. For example, spindle microtubules and spindle pole bodies were severely damaged above 150 MPa, preventing mitosis and cell division. Hamada *et al*. (1992) have further shown that in industrial strains of *S. cerevisiae*, hydrostatic pressure treatment leads to the induction of polyploidy at high frequency. Pressures above 100 MPa in *Sch. pombe* have been shown by Hamada *et al*. (1996) to induce stable diploids, presumably by damaging the nuclear division apparatus in a similar manner to that proposed in *S. cerevisiae*. In *Candida tropicalis*, nuclear membranes and mitochondrial cristae were disrupted by hydrostatic pressures of around 200 MPa, with hyphal forms of this dimorphic yeast being slightly more resistant to pressure stress compared with yeast cell forms (Sato *et al*., 1995). Interestingly, Iwashashi *et al*. (1991) have shown that heat shock treatment may induce a degree of barotolerance in yeast, indicating a physiological linkage between the two stresses. Furthermore, Iwahashi *et al*. (1993) have compared physiological changes in yeast subjected to hydrostatic pressure, high temperature and oxidative stress and concluded that cellular damage induced by such insults was essentially the same in each case. Fujii *et al*. (1996) have studied a barotolerant mutant of *S. cerevisiae* which had elevated levels of cellular trehalose, but did not synthesize major heat-shock proteins. It was suggested that a combination of membrane fluidity and trehalose accumulation was essential for barotolerance in *S. cerevisiae*.

The influence of hydrostatic pressure stress in yeast cells is directly pertinent to certain industrial processes. For example, food industries are interested in the possible use of high hydrostatic pressure as an alternative method of sterilization. In this regard, Hashizume *et al*. (1995) have shown that pressurization at 190 MPa and −20°C inactivated *S. cerevisiae* cells to the same extent as treatments of 320 MPa at room temperature. It was implied that mild pressure–low temperature food preservation may be feasible due to the retention of the foods organoleptic qualities following such treatments.

With regard to CO_2-induced pressure effects, Sigler and Höfer (1991) have noted that CO_2 partial pressures in excess of 50 kPa ($\sim$0.5 atm) have an inhibitory effect on a variety of enzymes in *S. cerevisiae*. In the brewing industry, pressure due to the elevation of CO_2 levels during fermentation can stress yeast cells, particularly in combination with ethanol (Jones and Greenfield, 1982). Brewer's yeast cell membrane integrity, cell division cycle progression and beer flavour are compromised under CO_2 pressures which simulate those which develop in brewing fermenters (Arcay-Ledezma and Slaughter, 1984; Lumsden *et al.*, 1987; Slaughter *et al.*, 1987).

Radiation stress

Both ultraviolet and gamma ray irradiation are known to cause DNA damage in yeast cells. UV causes dimerization, nicks and lesions while penetrating γ-rays induce double-strand breaks, nicks and other damaging effects on yeast DNA (Friedberg *et al.*, 1991). Both forms of radiation induce intrachromosomal recombination in *S. cerevisiae* in a cell cycle-dependent manner (Galli and Shiestl, 1995). Stewart and Enoch (1996) have reviewed the mechanisms that control progression through radiation-induced DNA-damage checkpoints in the cell cycles of both *S. cerevisiae* and *Schizosaccharomyces pombe*. X-rays similarly damage yeast DNA causing specific cell cycle arrest and this response is under the control of the *RAD9* gene (Weinert and Hartwell, 1993). The main effects of ionizing radiation on yeast cells are indirect effects mediated by reactive oxygen species (hydroxyl radicals generated by water radiolysis) which damages DNA. Siede (1995) has discussed the value of yeast as a model for studying the fundamental mechanisms of radiation-induced cell cycle arrest in higher eukaryotic cells. Ramotar and Masson (1996) have discussed the central importance of *S. cerevisiae* in fundamental studies of DNA repair following radiation damage. Work with fission yeast has revealed the importance of a checkpoint gene, *CHK1*, in monitoring radiation-induced damage of DNA. *CHK1* encodes a protein kinase which prevents G2–M progression in the cell cycle if DNA is damaged. Concerning yeast biotechnology, Clacharkar et *al.* (1996) have shown that low doses of γ-radiation enhanced alcohol production by around 25% in sugar cane juice fermentations by *S. cerevisiae*. The mechanism for such a stimulation of yeast fermentative metabolism by γ-rays is unclear.

The influence of near-infrared high-power laser light (at 1064 nm) on yeasts has been studied by Frucht-Pery *et al.* (1993) and Ward *et al.* (1996). *C. albicans* appears particularly susceptible to laser light and this may have applications in dentistry for treating yeast-associated caries.

With regard to visible light irradiation, diffuse daylight has been known for many years to inhibit budding in *S. cerevisiae* (Guilliermond, 1920). Work by Edmunds and co-workers (e.g. Edmunds *et al.*, 1979) has shown that light can affect the progress of the cell division cycle in *S. cerevisiae* and can also inhibit certain membrane transport functions, presumably by affecting membrane lipid phase transitions and by interacting with photosensitive chromogenic molecules within the cell. Treatment of *Kluyveromyces marxianus* with photosensitive dyes which generate reactive oxygen species in the presence of light (i.e. photodynamic treatment) results in plasma membrane damage, inhibition of protein synthesis and a reduction in cellular ATP levels (Paardekooper *et al.*, 1995). Yeasts may therefore be useful in model studies of photodynamically induced cellular damage of certain tumours.

Mechanical and gravitational stresses

Due to their thick cell walls, yeasts are very resilient to shear stresses caused by agitators in stirred-tank bioreactors. In fact, yeast cells require quite severe mechanical stress in order to rupture cell walls to extract cell components. For example, glass bead homogenization (using a vortex mixer, Bead Beater or Braun Homogenizer) or high-pressure extrusion of frozen cells (using an Eaton Press, French Press or X-Press) is normally required (see Middelberg, 1995). Ultrasound frequencies, which would normally be capable of rupturing bacterial cells, are generally ineffectual in breaking yeasts.

Centrifugal force is not normally regarded as posing a particular stress on yeast cells and is routinely carried out in the laboratory and industrial plants to harvest yeast biomass. Nevertheless, Walker *et al.* (1980) found that continuous centrifugation of *Schizosaccharomyces pombe* cells resulted in metabolic perturbations, as exemplified by alterations in the activity of glutamine synthetase. Furthermore, Komatsu *et al.* (1996) have shown that viability sharply decreased in ultra-centrifuged *Sch. pombe* cells (at 250 000 *g*) in which chromatin regions were dislocated to one end due to the gravitational forces. Bruschi and Esposito (1995) and Walther *et al.* (1996) have studied the influence of sub-gravitational forces on the growth and metabolism of *S. cerevisiae* during a Spacelab mission. Results, particularly those from a miniature chemostat culture showing altered yeast budding patterns, are relevant to space biotechnologists studying growth of single cells in microgravity bioreactors.

Electrical stress

Treatment of yeast cells with electrical fields is thought primarily to affect the permeability properties of the cell membrane. This is exploited when attempting to introduce foreign genes into yeast using electroporation or electrofusion. The electrosensitivity of yeast cells depends on the amplitude, frequency and duration of exposure of electrical pulses. Gaskova *et al.* (1996) have established the minimum high-voltage pulse conditions (amplitude, duration, etc.) required to kill *S. cerevisiae* cells. Such knowledge is considered very important in the improvement of protocols for electroporation experiments in yeast recombinant DNA technology. Electrostimulatory effects on yeast cell physiology have also been observed by several workers. For example, Fiedler *et al.* (1995) and Simpson *et al.* (1995) have shown electrostimulation of yeast cell proliferation and ethanol production, respectively. Of particular relevance to yeast fermentation biotechnology are the findings of McHale and co-workers (e.g. Simpson *et al.*, 1995). They have shown that in the lactose and cellobiose-fermenting yeast, *Kluyveromyces marxianus*, electropermeabilization of cells using single electric field pulses of 2.4 kV/cm resulted in a dramatic stimulation of the yeasts' fermentative activity. With regard to yeast growth, Crombie *et al.* (1990) have described **galvanitropic** behaviour in *Candida albicans*. This is when germ tubes or buds of this yeast display directional growth towards the cathode of an externally applied electric field. Remarkably, *C. albicans* cells retain a memory of electric field-induced polarized growth after switching off the field, indicating maintenance of field-induced asymmetry in cellular organization.

Ultrasonic stress

A few studies have investigated the influence of ultrasound on yeast physiology and biotechnology (Sinisterra, 1992). High-intensity ultrasonic waves may exert a deleterious effect on cells due to cell rupture (rarely in yeast) and enzyme denaturation. Low-intensity ultrasonic waves, however, may improve mass transfer of substrates into and metabolites out of yeast cell envelopes. For example, ultrasonic irradiation of intact *S. cerevisiae* cells (50 W for 90 min at 0–5°C) stimulated the enzymic cyclization of imidazo-fused heterocycles (Kamal *et al.*, 1990). Ultrasound may therefore have a beneficial role to play in improving the efficiency of certain yeast-mediated biotransformations.

4.4.3 Effects of Chemical Stresses on Yeast Growth

Yeast cells may be confronted by numerous chemical stresses which result from natural and man-made activities (Table 4.24). For example, during yeast growth and metabolism, chemical stress may arise from compounds pre-existing in the growth environment or from toxic metabolites produced by the yeasts themselves. Examples of the latter include ethanol and acetaldehyde. Chemical stress may also result from reduced availability of essential nutrients through limitation or starvation, and these are conditions frequently encountered by yeasts growing in natural environments. In addition, during industrial processes, several chemical treatments, such as acid-washing of yeast in breweries or addition of yeast preservatives (e.g. weak acids) in certain foods, will also impose chemical stress on yeast cells. In

Table 4.24. Environmental stresses experienced by yeasts.

Physical stress
Temperature shock
Osmotic shock
Desiccation/dehydration
High hydrostatic/gaseous pressures
G-force/shear stress
Radiation
Chemical stress
Ethanol and other metabolite toxicity
Nutrient limitation/starvation
Oxidative stress
pH shock
Metal ion stress (toxicity and limitation)
Chemical mutagenesis
Biological stress
Cellular ageing
Genotypic changes (e.g. chromosome loss)
Competition from other organisms

the natural environment, yeasts will also be confronted by a multitude of chemical stresses, many due to the activities of mankind as in the case of chemical pollutants. Several chemical mutagens are known to cause DNA damage in yeast and mutants resistant to nitrogen mustard, photoactivated psoralens, methylmethanesulphonate and *N*-methyl-N^1-nitro-*N*-nitrosoguanidine are useful in yeast genetics to study DNA repair processes (Cooper and Kelly, 1987). With regard to heavy-metal toxicity in yeast, various metal ions represent a severe chemical stress toward yeast cells due to their effects on enzyme inactivation and membrane damage. Gadd (1993) has discussed the mechanisms employed by yeast to modulate intracellular concentrations of potentially toxic metals (e.g. Cu, Cd) and the role of yeast metallothioneins in metal ion sequestration was discussed in Chapter 3.

With regard to nutrient availability as a chemical stress, deprivation of several essential nutrients is known to adversely affect yeast growth and the progress of the cell division cycle. Concerning the latter point, nutrient starvation, in general, arrests yeast cells in the G1 phase of the cell cycle.

Yeast cells starved for particular amino acids or treated with the protein synthesis inhibitor, cycloheximide, may exhibit a phenomenon referred to as the *stringent response*. In bacteria, this relates to the inhibition of both rRNA and tRNA synthesis and is triggered (in *Escherichia coli*) by binding of uncharged tRNA to ribosomes. In *S. cerevisiae*, the stringent response results in the rapid inhibition of rRNA synthesis (but not that of tRNA or mRNA) and, although the mechanism is unknown, is thought to be signalled when cells sense a rapid reduction in the overall rate of protein biosynthesis (Oliver and Warmington, 1987).

In industrial fermentations, nutrient-related stress in yeast has several very important practical implications. In brewing, for example, if yeast is nutrient-starved during extended periods of storage, certain cell surface properties such as flocculation capability are deleteriously affected (Smart *et al*., 1995). Furthermore, the availability of specific nutrients is very important for maintaining active yeast fermentations. This is exemplified when considering the levels of assimilable nitrogen sources. If available nitrogen drops below certain threshold levels (e.g. around 150 mg/l), then so-called 'stuck' fermentations will result when yeast growth and fermentative metabolism are prematurely halted. This is of particular significance in high gravity fermentations in brewing (O'Connor-Cox and Ingledew, 1989).

The following sections deal with the influences of, and physiological responses to, two particular chemical stresses in yeast which have been the focus of widespread attention in recent years; namely, ethanol stress and oxidative stress.

Ethanol stress

Ethyl alcohol is a major metabolic product of yeast fermentation and is quantitatively the premier product of biotechnology on a global scale. A dilemma confronting yeast cells, and yeast technologists alike, is that when ethanol accumulates during fermentation it acts as a potent chemical stress towards yeast cells. In fact, ethanol is quite an effective antizymotic agent. Increasing concentrations of ethanol will be initially inhibitory and latterly lethal to yeast. The mechanisms of ethanol toxicity have been extensively studied in yeast, mainly due to the fact that fermentation impairment by product inhibition is of distinct commercial significance for alcohol producers. Ethanol-induced toxicity and ethanol tolerance in yeast has been reviewed by Ingram and Bukkte (1984), Casey and Ingledew (1986), Oliver (1987), Jones (1987, 1990), D'Amore *et al.* (1990) and Mishra (1993).

With regard to growth, ethanol acts as a non-competitive inhibitor of yeast growth rate at relatively low concentrations. Glycolytic metabolism, on the other hand, is comparatively resistant to the inhibitory effects of ethanol. For example, in *S. cerevisiae* there is little effect on glycolytic enzyme denaturation below around 13% (w/v) ethanol. In fact, growth inhibition or 'replicative deactivation' (Jones, 1990) by ethanol is not due to enzyme inhibition since the higher intracellular concentrations required for the latter would not be achieved due to the passive diffusion of ethanol from yeast cells. Although intracellular ethanol increases as fermentation progresses (D'Amore *et al.*, 1990), it is generally accepted that because ethanol diffuses very rapidly across the cell membrane it does not 'accumulate' in yeast cells (Guijarra and Lagunas, 1984). Nevertheless, the fact that external ethanol is much less toxic than fermentatively-derived ethanol remains an anomaly (Jones, 1988).

Table 4.25 summarizes some of the principal inhibitory effects of ethanol on yeast cells. Although the influences are complex and we still have much to learn about the physiological basis of ethanol toxicity in yeast, membrane structure and function appears to be the predominant target of ethanol (Figure 4.36). Thus, the sites of

Table 4.25. Important effects of ethanol on yeast cell physiology.

Physiological function	Ethanol influence
Cell viability and growth	• General inhibition of growth, cell division and cell viability • Decrease in cell volume • Induction of morphological transitions (e.g. promotion of germ-tube formation in *Candida albicans*) • Enhancement of thermal death
Intermediary metabolism and macromolecular biosynthesis	• Denaturation of intracellular proteins and glycolytic enzymes • Lowered rate of RNA and protein accumulation • Reduction of V_{max} of main glycolytic enzymes • Enhancement of petite mutation • Induction of heat shock-like stress proteins • Increase in oxygen free radicals • Induced synthesis of cytochrome P450
Membrane structure and function	• Alteration of fatty acid and sterol composition • Induced lipolysis of cellular phospholipids • Increased ionic permeability • Inhibition of nutrient uptake • Inhibition of H^+-ATPase and dissipation of proton-motive force • Uncoupling of electrogenic processes by promoting passive re-entry of protons and consequential lowering of cytoplasmic pH • Hyperpolarization of plasma membrane

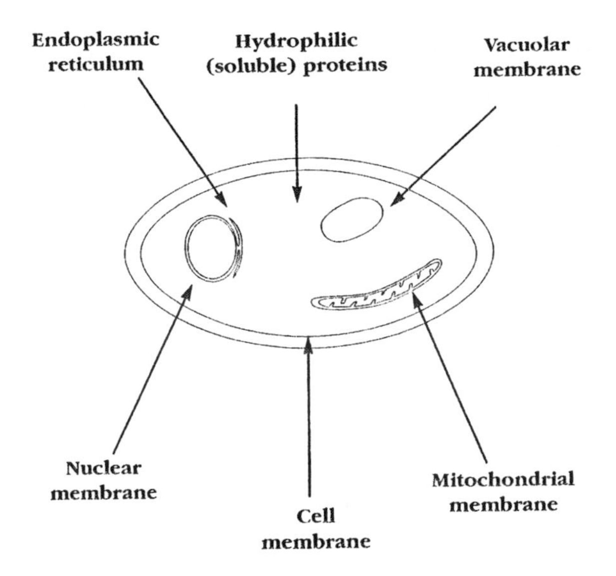

Figure 4.36. Possible cellular targets for ethanol in yeast. Adapted from D'Amore *et al.* (1990).

ethanol action in yeast cells are the plasma membrane, hydrophobic proteins of cell and mitochondrial membranes, nuclear membrane, vacuolar membrane, endoplasmic reticulum and hydrophilic proteins of the cytoplasm. Generally speaking, exposure to ethanol results in increased membrane fluidity and consequential decreased membrane structural integrity. Other alcohols are also toxic towards yeast and a correlation exists between the lipid solubility of the alcohols and their toxic effects (Leão and van Uden, 1982). In addition, the concentration of a particular alcohol needed to kill yeast decreases as its carbon chain length increases (Leão and van Uden, 1982). On a similar vein, chemical stresses in yeast due to aliphatic hydrocarbons are also observed and the toxicity of these compounds also correlates with their carbon chain length and hydrophobicity (Gill and Ratledge, 1972). Curran and Khalawan (1994) have additionally demonstrated that the concentration of alcohol required to induce the heat-shock response decreases as the carbon chain length, and thus hydrophobicity, increases. This, together with the fact that heat-shocked yeasts acquire thermotolerance (Watson and Cavicchioli, 1983), suggests that alcohol and heat both affect membrane lipids which play key roles in the stress physiology of yeast cells. Thus, heat shock and ethanol stress exhibit 'functional overlap' in yeast cells (Piper, 1995). The interrelationships between high temperature and ethanol in affecting yeast physiology have been discussed at length by Slapack *et al.* (1987).

The inhibitory effects of ethanol are enhanced synergistically not only with high temperatures, but also nutrient limitation (especially of Mg^{2+} ions) and other metabolic by-products such as other alcohols, aldehydes, esters, organic acids (especially octanoic and decanoic acids), fatty acids, carbonyl and phenolic compounds. Jones (1990) has proposed, rather controversially, that it is not ethanol *per se* which affects yeast cell replication, but rather acetaldehyde. Nevertheless, doubt has been cast on acetaldehyde as being the causative agent of ethanol toxicity by Stanley *et al.* (1993, 1997) who have shown that low acetaldehyde concentrations are actually stimulatory towards yeast cell growth rate. Furthermore, Stanley and Pamment (1993) have demonstrated that acetaldehyde can accumulate intracellularly in yeast at concentrations greater than those observed in yeast fermentation broth.

How do yeast cells respond to ethanol stress? Fermenting cells in particular must maintain viability and metabolic activity in the presence of increasing levels of ethanol. Various physiological adaptations occur in yeast which are thought to confer protection against ethanol and these are outlined in Table 4.26.

These adaptations range from alterations in membrane fluidity to synthesis of detoxification enzymes. For example, *S. cerevisiae* cells grown in the presence of ethanol exhibit an increase in monounsaturated fatty acids (especially oleic acid, 18 : 1) and a decrease in saturated fatty acid residues (palmitic acid, 16:0). In this regard, Mizoguchi and Hara (1997) have shown that palmitic acid-enriched cells exhibit enhanced ethanol tolerance. A common response of yeasts exposed to ethanol is an increase in fatty acyl chain length and an increase in the proportion of

Table 4.26. Adaptive physiological responses of yeasts to ethanol.

- Decrease in membrane saturated fatty acids (e.g. palmitic acid)
- Increase in membrane unsaturated fatty acids (e.g. oleic acid)
- Acceleration of squalene and ergosterol biosynthesis
- Increase in the phospholipid: protein ratio (e.g. enhanced phosphatidylinositol biosynthesis)
- Enhanced mitochondrial superoxide dismutase activity
- Elevated levels of cellular trehalose
- Stimulation of stress protein biosynthesis and acquisition of thermotolerance
- Increased synthesis of cytochrome P450 and increased ethanol metabolism

unsaturated fatty acids and sterols in the plasma membrane. Other responses to ethanol in yeast relate to the biosynthesis of heat shock-like proteins, or 'ethanol stress proteins'. Common heat shock and ethanol stress proteins with relative molecular masses of 70, 38, 26 and 23 kDa have been observed in brewing strains of *Saccharomyces* (Odumeru *et al.*, 1992). In addition, Shankar *et al.* (1996) have identified a 90 kDa protein in a distiller's strain of *S. cerevisiae* which was elicited both by ethanol and acetaldehyde stress. Both heat and ethanol stress in yeast leads to the acquisition of thermotolerance and ethanol tolerance, although the precise involvement of stress proteins in conferring environmental stress protection is unclear. Perhaps such proteins act as a damage limitation exercise or in a damage repair role (perhaps through 'chaperoned' protein folding), rather than a directly protective role. Additional metabolic responses to ethanol stress have been observed in *S. cerevisiae* cells including elevated levels of cellular trehalose (Kim *et al.*, 1996). Mansure *et al.* (1994) have demonstrated a positive correlation between the intracellular concentration of trehalose and the tolerance of *S. cerevisiae* to the membrane-damaging effects of ethanol. Increased activity of mitochondrial superoxide dismutase has also been observed in this yeast in response to ethanol (Costa *et al.*, 1993). The latter observation indicates that ethanol-induced oxygen free radical synthesis may be associated with ethanol toxicity and that antioxidant enzymes play a defensive role in protecting cells from the damaging effects of ethanol.

The various physiological changes observed in yeast in response to ethanol are key to our understanding of the phenomenon of ethanol tolerance. Although there is no unified definition of ethanol tolerance in yeasts, it may generally be described as the ability of a particular strain of yeast to withstand higher levels of ethanol without any deleterious effects on its viability and metabolic activities. There is also no generally accepted manner in which to measure the phenomenon (D'Amore *et al.*, 1990). Nevertheless, numerous techniques have been employed to assess ethanol tolerance including: suppression of growth, changes in fermentation rate and rate of CO_2 evolution, and extracellular acidification in the presence of exogenous ethanol. However, such assessments fail to take account of the fact that endogenously produced ethanol is more toxic than exogenous ethanol. A more accurate reflection of ethanol tolerance during fermentation is actually to measure the final amounts of ethanol produced by fermenting yeast cells. Although the effect of ethanol on growth and viability is not considered by this approach, it is quite simple and is widely used by yeast biotechnologists to assess ethanol tolerance in industrial yeasts. Such yeasts are known to differ markedly with respect to their tolerance to ethanol such that saké yeasts are generally regarded as being very tolerant while pentose-fermenting yeasts are ethanol-intolerant. Intrinsic factors are therefore undoubtedly important in influencing yeast ethanol tolerance which is known to be under polygenic control. This, together with the complexities of ethanol inhibitory effects, has severely hampered the isolation of ethanol-tolerant mutants of yeast. Nevertheless, Brown and Oliver (1982) were able successfully to isolate ethanol-tolerant variants of *S. uvarum* in continuous culture in which the ethanol feed was controlled by the rate of CO_2 evolution. This continuous selection with feed-back enabled viable mutants to be isolated which showed higher fermentation rates in the presence of ethanol (12%, w/v) compared with wild-type cells.

Environmental factors, as well as intrinsic factors, also play important roles in dictating how yeast cells tolerate ethanol. For example, osmotic pressure, media composition (especially metal ions, fatty acids and assimilable nitrogen constituents), modes of substrate feeding and by-product formation are all involved (Casey and Ingledew, 1986; D'Amore *et al.*, 1990). In addition, prior exposure of yeast cells to ethanol will confer a degree of resistance to subsequent exposure which would otherwise be toxic. Such physiological pre-conditioning of yeast is similarly observed by sub-lethal exposure to heat and oxidants and is collectively referred to as the **adaptive stress response**. As previously mentioned, the presence of unsaturated fatty acids and sterols in the yeast cell membrane is very important in determining tolerance to ethanol and supplementation of fermentation broths with membrane lipids will enhance cells tolerance to ethanol (Table 4.27). Metal ions have also been implicated and Table 4.28 provides evidence for a key involvement of Mg^{2+} ions in protecting yeasts from ethanol stress and in conferment of ethanol tolerance. Birch and Walker (1996) have found that Mg^{2+}-supplemented yeast cultures retain high viabilities in the face of otherwise lethal heat shocks or ethanol treatments. Elevated Mg^{2+} was also shown to prevent stress protein biosynthesis, indicating that the need for damage repair mechanisms had been precluded by Mg^{2+}. Such protective effects of Mg^{2+} ions against environmental stress were assumed to be elicited at the level of plasma membrane stabilization.

Oxidative stress

Another major chemical stress confronted by yeast cells, particularly during aerobic growth, is that due to reactive oxygen species such as the superoxide anion ($O_2^{\cdot -}$), hydrogen peroxide (H_2O_2) and the hydroxyl radical ($OH^{\cdot}$). Such species may be generated by mitochondrial respiration or by increases in dissolved oxygen tension of growth media or by exposure to ionizing radiation. They cause oxidative damage to proteins, lipids and DNA.

Yeast cells require to maintain a reduced intracellular state when faced with oxidative stress and they possess various antioxidant chemicals and enzymes to detoxify active oxygen (reviewed by Mager and Moradas-Ferreira, 1993; Jamieson, 1995; Moradas-Ferreira *et al.*, 1996). These antioxidant defences are summarized in Table 4.29.

Table 4.27. Membrane lipids as modulators of ethanol tolerance in *S. cerevisiae*.

Lipids supplemented	Product tolerance
Protein–phospholipid complex	Increased ethanol productivity
Phospatidylcholine, palmitic acid and cholesterol	Increased growth
Phosphatidylserine	Increased fermentative activity
Ergosterol or campesterol and linoleic acid	Increased viability and nutrient uptake
Oleic, linoleic or linolenic acid	Increased fermentative activity
Linseed/cotton seed or soyabean oil or their fatty acid extracts	Increased fermentation rates
Yeast hulls (mixture of sterols and unsaturated fatty acids) in cell envelope preparations	Enhanced growth rate and fermentative activity

Information from Singh and Mishra (1995).

Table 4.28.Role of Mg^{2+} ions in ethanol stress protection in yeasts.

Experimental observations	References
Mg^{2+} partially prevents the increase in proton and anion permeability caused by ethanol	Petrov and Okorokov (1990)
Mg^{2+} supplementations reduce the decline in yeast fermentative activity	Dombek and Ingram (1986)
Mg^{2+} increases ethanol production during fermentation of high-sugar media	D'Amore *et al.* (1988) Walker *et al.* (1996)
Mg^{2+} maintains high cell viability and ethanol production in rapid fermentations	Dasari *et al.* (1990)
Mg^{2+} is responsible for the difference in toxicity between produced and added ethanol	Dasari *et al.* (1990)
Mg^{2+} protects cells from ethanol toxicity and prevents ethanol-stress protein synthesis	Birch and Walker (1996) Ciesarova *et al.* (1996)

Glutathione (γ-L-glutamyl-L-cystinylglycine) is a thiol compound which plays a key role in protecting cells against oxidants, free radicals and alkylating agents and which represents a major non-enzymic oxidant defence system in yeast. Glutathione scavenges oxygen radicals and acts to maintain the redox state of yeast cells (Jamieson, 1995; Powis *et al.*, 1995; Grant *et al.*, 1996; Stephen and Jamieson, 1996). Glutathione-deficient mutants of *S. cerevisiae* were shown to be hypersensitive to hydrogen peroxide (Izawa *et al.*, 1995) and superoxide anions (Stephen and Jamieson, 1996). Grant *et al.* (1996) have shown that expression of the gene encoding glutathione reductase, *GLR1*, which catalyses reduction of the oxidized form of glutathione, is a key requirement for the protection of *S. cerevisiae* cells against oxidative stress. Other antioxidant molecules are the metallothioneins. For example, as well as acting to detoxify cellular copper ions,

Table 4.29. Main antioxidant defences of yeasts.

Defence enzyme or chemical	Function
Enzymes	
Cu/Zn superoxide dismutase	Dismutation of superoxide anion (cytoplasm)
Mn superoxide dismutase	Dismutation of superoxide anion (mitochondria)
Catalase A	Decomposition of hydrogen peroxide (peroxisome)
Catalase T	Decomposition of hydrogen peroxide (cytoplasm)
Cytochrome C peroxidase	Reduction of hydrogen peroxide
Glutathione reductase	Reduction of oxidized glutathione
Chemicals	
Glutathione	Scavenging of oxygen free radicals
Metallothionein	Cu^{2+}-binding, scavenging of superoxide and hydroxyl radicals
Thioredoxin	Reduction of protein disulphides
Polyamines	Protection of lipids from oxidation

Modified from Moradas-Ferreira *et al.* (1996).

copper–metallothionein in yeast (encoded by the *CUP1* gene) can also protect cells against the damaging effects of oxidants. In the red-pigmented yeast, *Phaffia rhodozyma*, the carotenoid astaxanthin acts to protect the cells against toxic oxygen metabolites generated by intracellular oxidative metabolism (Johnson and Schroeder, 1996).

Yeasts also possess enzymic mechanisms to neutralize the effects of active oxygen radicals: peroxisomal and cytosolic catalases (encoded by *CTA1* and *CTT* genes, respectively); mitochondrial and cytoplasmic superoxide dismutases (Mn Sod and Cu/Zn Sod, respectively, encoded by *SOD2* and *SOD1* genes) and peroxidases (e.g. cytochrome C peroxidase). With regard to the superoxide dismutases, it appears that the physiological role of Mn Sod is to protect mitochondria from oxidative stress during aerobic growth, whilst the role of the Cu/Zn Sod is to remove superoxide anions from the yeast cytoplasm.

When the concentration of reactive oxygen species exceed the cells' antioxidant capability, oxidative damage can occur. Nevertheless, when cells of *S. cerevisiae* (Jamieson, 1992) or *Schizosaccharomyces pombe* (Lee *et al.*, 1995; Mutoh, *et al.*, 1995) are pretreated with sublethal levels of oxidants, increased oxidative stress protection is encountered. This is referred to as the adaptive oxidant stress response. The molecular mechanisms of this response are presently unknown but a link has been suggested between the regulation of metal ion (e.g. Cu^{2+}, Mn^{2+}) homeostasis and oxidative stress. A very practical way of preventing oxidative stress in stored dried baker's yeast has been noted by Moradas-Ferreira *et al.* (1996). Thus, if cells are stored in inert gas (nitrogen) they survive for years, whereas they rapidly lose their viability if stored in air.

Davidson *et al.* (1996) have provided evidence of the involvement of oxidative stress in heat-induced death of *S. cerevisiae* cells. Mutants of this yeast which were deleted for antioxidant genes (catalase, superoxide dismutase and cytochrome C peroxidase) were more sensitive to lethal heat compared with wild-type cells. Similarly, cells in which catalase and superoxide dismutase genes were overexpressed were more thermotolerant. In addition, anaerobically grown cells maintained viability at 50°C but this thermotolerance was immediately abolished upon oxygen exposure. Jamieson (1992) has conversely shown that heat-shocked *S. cerevisiae* cells were protected against oxidative stress. These observations, together with the reported involvement of the Mn superoxide dismutase in the acquisition of ethanol tolerance (Costa *et al.*, 1993), indicates that a close interrelationship exists between various physiological stress responses in yeast cells (Mager and De Kruijff, 1995; Piper, 1995).

4.5 BIOTIC FACTORS INFLUENCING YEAST GROWTH

Interactions between yeasts and other life forms are studied intensively by yeast ecologists and clinical mycologists and a full coverage of the area is beyond the scope of this book. Nevertheless, many of the interactions are relevant to yeast growth and are also of practical significance in environmental, medical and industrial biotechnology. Yeasts interact with different plants, animals and other microorganisms in both beneficial and detrimental ways. Interactive relationships may be variously described as saprophytism, parasitism, commensalism, competition, mutualism, antagonism, etc. Yeast growth may be stimulated or inhibited by other organisms and, likewise, host organisms may benefit or be adversely affected by their association with yeasts. Several aspects of the relationships between yeasts and other organisms which pertain to yeast cell physiology and biotechnology are discussed in the following sections.

4.5.1 Yeast–Plant Interactions

Vascular plants are often the preferred natural habitat for yeasts where they are usually found growing at the interface between soluble nutrients

Table 4.30. Some plant pathogenic yeasts.

Yeast species	Plant disease	References
Nematospora coryli and *Ashbya gossypii*	Several diseases of tropical and subtropical plants, including: cotton boll disease, yeast spot of legumes, coffee bean diseases, citrus fruit 'inspissosis'	Phaff *et al.* (1978)
Ophiostoma ulmi	Dutch elm disease	Brasier (1991); Sticklen and Sherald (1993)
Saccharomyces paradoxus, *Kluyveromyces* spp., *Candida* spp. etc.	Black knot disease of trees (yeasts not the causative agents but may be involved in the development of tree galls)	Lachance (1990); Naumov *et al.* (1996)

and the septic world (e.g. rotting fruit, tree exudates). Most yeasts live a saprophytic life in association with dead and decaying plant material, but a few plant parasitic (endophytic) yeasts are known (see Table 4.30). Both saprophytic and parasitic yeasts living on vegetation depend on the host to supply essential nutrients for yeast growth. However, yeast growth can also be detrimentally affected by plants. For example, many plants contain selective chemical compounds which determine the make-up of the yeast community in plants (Lachance, 1990). Examples are given in Table 4.31. Several yeasts, most notably *Pichia minuta*, *Candida ernobii* and *Cryptococcus skinneri* appear to be truly wood-inhabiting (Hutchison and Miratsuka, 1994), whereas other yeast-wood associations involve bark beetles (Leufven and Nehls, 1986), *Drosophila* (Lachance *et al.*, 1995), tree exudates (Lachance *et al.*, 1993) or decomposing wood (Gonzalez *et al.*, 1989).

An interesting linkage between yeast and plant physiology has recently been shown to have practical implications in environmental biotechnology. Plant defence mechanisms against fungal pathogens involve the synthesis of specific low-molecular weight secondary metabolites known as phytoalexins. Cell wall material from *S. cerevisiae* has been demonstrated to elicit this phytoalexin response in certain plants. For example, work at the Scottish Crop Research Institute has

Table 4.31. Some plant chemicals inhibitory to yeasts.

Chemical compounds	Plant source	Influence on yeast growth
Oligoterpenes	Pine tree sap	Inhibitory to most yeasts, except some species of *Hansenula* and *Pichia*
Monoterpenes	Spruce trees	Inhibitory to most yeasts except *Candida diddensii*
Fatty acid esters	Cactus	Toxic to most yeasts
Tannins and phenolics	Oak, cherry trees	Toxic to most yeasts
Thionin peptides	Barley	Toxic to yeasts

Information from: Lachance (1990) and Holz and Stahl (1995).

revealed that elicitor-active yeast extracts may act as crop protectants in the control of powdery mildew infection (*Erysiphe graminis* f.sp. *hordei*) in barley (Reglinski *et al.*, 1994) and *Botrytis cinerea* and *Rhizoctonia solani* infection in lettuce (Reglinski *et al.*, 1995). Phytopathogenic elicitor proteins have now been identified (e.g. β-cryptogein) and successfully cloned and expressed in the methylotrophic yeast, *Pichia pastoris* (O'Donohue *et al.*, 1996).

4.5.2 Yeast–Animal Interactions

Yeasts may interact in numerous beneficial and detrimental ways with insects, crustaceans, protozoa and warm-blooded animals. Research by Phaff and his colleagues over many years has shown that insects are very important vectors in the distribution of yeasts in nature (Phaff *et al.*, 1978). Yeasts are also of significant nutritional importance in the life cycles of many insect species such as fruit flies, bark beetles and fig wasps. Generally, the association here is one which is detrimental to yeast growth since fungivorous insects are direct yeast predators which have intestinal enzymes capable of hydrolysing yeast cell walls. The digestive juices of the snail, *Helix pomatia*, have long been used by yeast biologists to digest cell walls in the preparation of yeast protoplasts. 'Helicase', the commercial preparation of snail juice, is a cocktail of lytic enzymes (glucanases, proteases, etc.) but nowadays purer preparations of yeast cell wall lytic enzymes are available from microbial sources (e.g. 'Zymolyase', 'Novozyme').

Regarding the interactions between yeasts and warm-blooded animals, the gastrointestinal tract is often a favoured niche for yeasts. Several yeast species (e.g. *Candida slooffii*, *Cyniclomyces guttulatus*, *Candida pintolopesii*) are known to be obligately associated with the intestinal tracts of horses, cattle, pigs, rabbits and rodents (Phaff *et al.*, 1978). Some yeasts form part of the normal commensal flora of animal intestines, whereas others are responsible for certain pathological conditions. Facultative intestinal yeast parasites are known, including *Candida krusei*, *C. tropicalis*, *C. parapsilosis* and *Trichosporon cutaneum*. Several yeasts are associated with animal skin and human scalp (e.g. *Pityrosporum* spp.). Certain yeasts, but most notably *Saccharomyces boulardii*, have been shown to act as growth stimulants and rumen metabolism stabilizers in domestic animals (e.g. Fiems *et al.*, 1995). In humans, *S. boulardii* has been shown to exert a probiotic effect by inhibiting intestinal bacteria and neutralizing microbial toxins and this may prevent antibiotic-associated diarrhoea (Surawitz *et al.*, 1989) and Crohn's disease-associated chronic diarrhoea (Plein and Hortz, 1991). McFarland and Elmer (1995) have reviewed the use of yeasts as biotherapeutic agents in the treatment and prevention of a range of human diseases.

Detrimental effects of yeasts on their animal hosts are apparent when considering human and animal pathogenic yeasts in the field of medical and veterinary mycology (Vanden Bossche *et al.*, 1993). Opportunistically infective yeasts in humans include *Candida albicans* and other *Candida* spp. (the causative agents of candidosis) and *Cryptococcus neoformans* (cryptococcosis). Medical fungi which assume a yeast phase in the body are *Histoplasma capsulatum* (histoplasmosis), *Blastomyces dermatitidis* (blastomycosis) and *Coccidioides immitis* (coccidioidomycosis). Considering *Candida* spp., *C. albicans* is by far the most common cause of yeast infections in humans and is a very frequent nosocomial (hospital-derived) pathogen (Mahayni *et al.*, 1995; Odds, 1996a). The role of *C. albicans* as an opportunistic human pathogen, particularly in immunocompromised patients, has been extensively reviewed (e.g. Odds, 1988, 1994; Meunier, 1989; Dupont, 1995; McCullough *et al.*, 1996). Other *Candida* species are known to cause human mycoses, including *C. krusei*, *C. parapsilosis* and *C. lusitaniae* (Nguyen *et al.*, 1996). Hazen (1995) has reviewed new and emerging yeast pathogens with species of *Malassezia*, *Rhodotorula*, *Hansenula* and *Trichosporon* representing the most frequent 'emerging' infective yeast isolates (see Chapter 1, Table 1.6). Even *S. cerevisiae* can colonize human mucosal surfaces and this yeast has been reported to cause several mycotic infections in immunocompromised individuals (e.g. Eng *et al.*, 1984; Aucott *et al.*, 1990; McCusker *et al.*, 1994). A principal distinguishing feature of clinical isolates of *S. cerevisiae* is their ability to grow at 42°C (McCusker *et al.*, 1994). It appears that disruption of the gastrointestinal tract appears to be one of the most important predisposing factors which leads to *S. cerevisiae* infections in humans. In this regard, it is interesting to note that patients with Crohn's disease (inflammatory bowel disease) appear to be hypersensitive to dietary *S. cerevisiae* antigens (Main *et al.*, 1988).

4.5.3 Yeast–Microbe Interactions

A realm of yeast ecology is concerned with the interactions between yeasts and fungi, protozoa and bacteria as well as with other yeasts. Several of the interactions are relevant to yeast biotechnology.

Microorganisms can influence the growth of yeasts in a number of ways. These range from nutrient competition to direct killing of cells. With regard to yeast–bacterial interactions, several bacteria are known which secrete antizymotic agents and antizymosis may be a naturally significant phenomenon. *Streptomyces* spp., in particular, produce compounds such as cyloheximide, chloramphenicol, lomofungin, and tunicamycin and various polyene macrolides (e.g. nystatin, amphotericin B) which are inhibitory to yeast growth. Some of these agents, notably the polyenes, have found use as pharmaceuticals in the treatment of candidosis and other yeast infections (see Table 4.35).

Yeast–bacterial interactions are also relevant in food and fermentation industries. Apart from detrimental activities of bacteria in yeast fermentations, beneficial interactions between yeasts and lactic acid bacteria are to be found in the production of traditional fermented foods and beverages; for example, soy sauce, rye bread, sour dough, cocoa, Belgian beer (lambic and gueuze), German beer (weisse biere) and Scotch malt whisky. Lievens *et al.* (1994) have studied interactions between *S. cerevisiae* and *Pediococcus damnosus* and have characterized a yeast proteinaceous factor which stimulated bacterial flocculation. Such a finding may explain the stable association between yeasts and lactic acid bacteria in the production of traditional beverages by spontaneous fermentation. In other fermentation processes, bacterial products may directly curtail the growth of yeasts. In vinegar production, for example, the synthesis of acetic acid by *Acetobacter* spp. may kill yeasts already present from the preceding (i.e. alcoholic fermentation) stage in the process. Occasionally, acetic acid bacteria and yeasts interact in a symbiotic relationship in, for example, oriental kombucha (or Russian teakvass) tea fermentations. Yeasts can conversely be detrimental to bacterial growth. For example, Polonelli and Morace (1986) have shown that certain killer yeasts secrete compounds which are inhibitory to the growth of a wide range of Gram-positive and -negative bacteria.

Considering yeast–fungal interactions, yeast growth may benefit or be adversely affected by the presence of fungi. An example of yeast–fungal mutualistic growth has been reported between the yeasts, *Pichia pini, Hansenula capsulata* and *H. holstii* and blue-stain fungi, *Ceratocystis montia* and *Europhium clavigerum* when both inhabit North American pinewood infected by mountain pine beetles (Whitney, 1971). Another beneficial aspect relates to nutritional cooperation from fungi towards yeast. Fungi in natural environments may make otherwise unassimilable nutrients available to yeasts. For example, many yeasts can grow readily on the fungal hydrolysis products of starch, cellulose, xylan, etc., but are unable to utilize the undigested polymers. Nevertheless, several associations between fungi and yeast are detrimental to the growth and survival of the latter. At the most extreme, yeasts can be directly parasitized by certain fungi. For example, Hutchison and Baron (1996) have shown that many wood-decay Basidiomycota are able to attack microcolonies and

Table 4.32. Some killer yeasts and their genetic basis.

Killer species	Genetic basis of killer character	References
Saccharomyces cerevisiae	Two double-stranded RNA plasmids	Wickner (1996)
Kluyveromyces lactis	Linear DNA plasmids	Stark *et al.* (1990)
Pichia acaciae	Linear DNA plasmids	McCracken *et al.* (1994)
Williopsis (*Hansenula*) *mrakii*	Chromosomal	Kimura *et al.* (1993)
W. saturnus	Chromosomal	Kimura *et al.* (1993)
Pichia kluyveri	Chromosomal	Radler *et al.* (1985)
Hanseniaspora uvarum	Chromosomal	Radler *et al.* (1985)
Sporidiobolus pararoseus	Chromosomal	Golubev *et al.* (1988)
Cryptococcus humicola	Chromosomal	Golubev and Shabalin (1994)
Bullera sinensis	Chromosomal	Golubev and Nakase (1996)
Metschnikowia spp.	Unknown	Vadkertiova and Slavikova (1995)
Rhodotorula rubra and R. glutinis	Unknown	Middelbeek *et al.* (1980)
Cryptococcus laurentii	Unknown	Young and Yagui (1978)
Torulopsis, Debaryomyces and	Unknown	Young and Yagui (1978)
Candida spp.	Unknown	Young and Yagui (1978)

single cells of *Candida*, *Cryptococcus*, *Pichia*, *Rhodotorula* and *Sporidiobolus* species. The process is referred to as necrotrophic mycoparasitism and is of ecological significance in that yeasts provide vitamins and nitrogenous nutrients for wood-inhabiting fungi. Mycotoxins (e.g. fusariotoxin) produced by certain fungal contaminants have been shown to be inhibitory to yeasts (Sukroongreung *et al.*, 1984). For example, Whitehead and Flannigan (1989) have shown that the trichothecene mycotoxin, deoxynivalenol (DON), which is produced by *Fusarium* spp. in contaminated barley and malt, is inhibitory to brewing yeast growth at 50 μg/ml. Fermentation, however, was little affected by DON at concentrations found in contaminated grain. A converse relationship may exist between certain yeast toxins and fungal growth. For example, Walker *et al.* (1995) have shown that certain killer yeasts are inhibitory to several wood-decay and plant pathogenic fungi. This has raised the possibility of using killer yeasts or their toxins as novel biological control agents against fungi of environmental and agronomical significance. Several yeast species have already been shown to be effective biological control agents in protecting plants against various fungal diseases. For example, *Debaryomyces hansenii* and *Candida guilliermondii* protect grapefruits against *Penicillium digitatum* (Droby *et al.*, 1989; McGuire, 1994), *Pichia guilliermondii* protect grapes against *Botrytis cinerea* (Wisniewski *et al.*, 1991), *Kloeckera apiculata* and *Candida guilliermondii* protect grapes and other fruits against *Botryotina fuckeliana* and *Rhizopus stolonifer* (McLaughlin *et al.*, 1992), *Candida oleophila* protects apples against *Botrytis cinerea* (Mercier and Wilson, 1994) and *Metschnikowia pulcherrima* protects peaches and apples against *Monilinia laxa* and *Botryotina fuckeliana* (De Curtis *et al.*, 1996). The antagonistic impact of certain yeasts against plant fungal pathogens is assumed to be due to the yeasts out-competing the fungi for nutrients and space (Filonow *et al.*, 1996).

Yeast–yeast interactions will now be considered with particular emphasis on their biotechnological significance. Several important yeast–yeast interactions have already been mentioned in this book in relation to: yeasts competing with each other for nutrients (Chapter 3, section 3.3.1); yeast–yeast flocculation (Chapter 2, section 2.4.1.3) and sexual conjugation in yeasts of opposite mating type (Chapter 4, section 4.2.3). Another well-known yeast–yeast interaction is the **killer phenomenon** which will now be considered in relation to aspects of yeast cell physiology and biotechnology. Genetic and molecular genetic aspects of killer yeasts and their toxins have been reviewed by Young (1987); Bussey (1991); Wickner (1992, 1996) and Vondrejs, Janderova and Valesek (1996).

Killer yeast species, depicted in Table 4.32, secrete proteinaceous toxins which are lethal to sensitive strains but to which the killers themselves are immune. A classification of killer yeasts (K_1–K_{11} and K_{28}) exists which is based on cross-reactivity between different species. Killer yeasts are widespread in laboratory culture collections (Philliskirk and Young, 1975) and in natural habitats (Stumm *et al.*, 1975; Starmer *et al.*, 1987). Starmer *et al.* (1987) have indicated that the existence of killer yeasts may play a role in determining the distribution of different yeast species in plants. The best studied killer species is *S. cerevisiae* which secretes three toxin types; K_1, K_2 and K_{28}. The K_1 killer toxin acts by firstly binding by the β-subunit to cell wall receptors (1→6-β-D-glucan moieties) on sensitive cell walls followed by the creation of channels in the cell membrane by the α-subunit. The latter quickly kills sensitive cells by affecting membrane permeability for protons, K^+ ions, ATP and amino acids (Skipper and Bussey, 1973; de la Pena *et al.*, 1980; Martinac *et al.*, 1990). The K_2 toxin acts similarly to K_1 as an ionophore, but the K_{28} toxin binds to cell wall mannoproteins then rapidly inhibits DNA synthesis in sensitive yeast cells rather than disrupting membrane function (Schmitt *et al.*, 1996). All killer toxins of *S. cerevisiae* are encoded on linear dsRNA plasmids. The K_1 toxin is encoded on killer viruses: MI and L-A. MI encodes the killer toxin and resistance determinant, and L-A encodes the coat protein which encapsulates both M1 and L-A dsRNA molecules into virus-like particles. The M1 dsRNA expresses a preprotoxin peptide which is proteolytically cleaved to yield a secreted αβ peptide dimer, the active toxin, and a glycosylated γ peptide which is thought to act as the immunity determinant which binds to receptors on killer cell membranes (Zhu *et al.*, 1993). The γ glycoprotein thus presumably prevents killer yeasts from committing suicide.

The genetic basis of killer character in non-*Saccharomyces* yeasts is quite different from the K_1 system in *S. cerevisiae* (Table 4.32). For example, *Kluyveromyces lactis* killer toxin is encoded by linear DNA plasmids (Stark *et al.*, 1990), while *Williopsis mrakii* toxin is chromosomally inherited (Kimura *et al.*, 1993).

Killer yeasts and their toxins have numerous potential applications in medical, environmental and industrial biotechnology, as summarized in Table 4.33.

Another antagonistic interaction between different yeast cells is that of **predation.** This hitherto unreported phenomenon has recently been described by Lachance and Pang (1997) and resembles necrophytic mycoparasitism observed in some filamentous fungi. Several yeasts isolated from natural environments have been shown to possess predatory characteristics towards other yeasts. These include *Arthroascus javanensis*, *Botryoascus synaedendrus*, *Guilliermondella selenospora*, *Saccharomycopsis fibuligera* and certain *Candida* species. A predacious *Candida* spp. (95-67.4) isolated from a beetle associated with the Australian *Hibiscus* tree, was fortuitously found by Lachance and Pang (1997) to produce feeding appendages called *haustoria* which penetrate and kill a variety of ascomycetous and basidiomycetous yeasts (see Figure 4.37). Of the target yeast strains assessed, only *Schizosaccharomyces pombe* was not vulnerable to predation by *Candida* isolate 95-697.4. A common feature of all predacious yeasts identified by Lachance and Pang (1997) appeared to be their requirements for organic sulphur, which was deemed an important factor in yeast prey–predator interrelationships.

Table 4.33. Biotechnological potential of killer yeasts and their toxins.

Biotechnological field	Application	References
Eukaryotic biological research	Fundamental studies of the biosynthesis, processing and secretion of eukaryotic proteins (e.g. hormones, neuropeptides)	Sossin *et al.* (1989); Wickner (1996)
Yeast taxonomy	Classification of heterobasidiomycetous yeasts based on killer toxin sensitivity	Golubev and Boekhout (1995)
Yeast genetics	Selection of intergeneric hybrids obtained by protoplast fusion	Palkova and Vondrejs (1996)
Recombinant DNA technology	Cloning vectors for directing the glycosylation and secretion of foreign proteins	Meinhardt *et al.* (1990); Dignard *et al.* (1991); Wickner (1992); Fukuhara (1995)
Medical bacteriology	Biotyping pathogenic bacteria	Morace *et al.* (1989)
Medical mycology	Biotyping pathogenic yeasts	Morace *et al.* (1989)
	Novel zymocide against human and animal yeast pathogens	Polonelli *et al.* (1986); Pfeiffer *et al.* (1988); Sawant *et al.* (1989); Hodgson *et al.* (1995); Séguy *et al.* (1996)
	Anti-idiotypic antibodies of killer toxins as therapeutic antifungal agents	Magliani *et al.*, (1997)
Beverage fermentations	Conferment of anti-wild yeast killer character in yeasts producing:	
	beer	Young (1982)
	wine	Bortol *et al.* (1986)
	saké	Rosini (1989)
	Fingerprinting of wine yeasts	Boone *et al.* (1990); Ouchi *et al.* (1979); Vaughan-Martini *et al.* (1996)
Food technology	Natural food preservatives	Holz and Stahl (1995)
Biological control in agriculture	Antifungal activity against wood-decay and plant pathogenic fungi	Koltin *et al.* (1993); Walker *et al.* (1995)

4.6 YEAST CELL DEATH

The death of yeast cells is of theoretical and practical interest for yeast cell physiologists and biotechnologists. From the fundamental biological viewpoint, yeasts represent very useful model systems in which to study phenomena such as cellular ageing and apoptosis. From a more applied perspective, yeast cell death is an important consideration for microbiologists in the food, fermentation and health-care industries. For example, yeast cell death needs to be minimized during industrial fermentations to maintain culture viabilities at high levels; while on the other hand, yeast cell death needs to be maximized to eradicate undesired yeasts in foods and beverages. Yeast death through autolysis is also an important consideration in the manufacture of yeast extracts for the food industry. Finally, effective measures for killing yeasts is a concern of medical mycologists treating human zymotic infections. Assessments of yeast cell

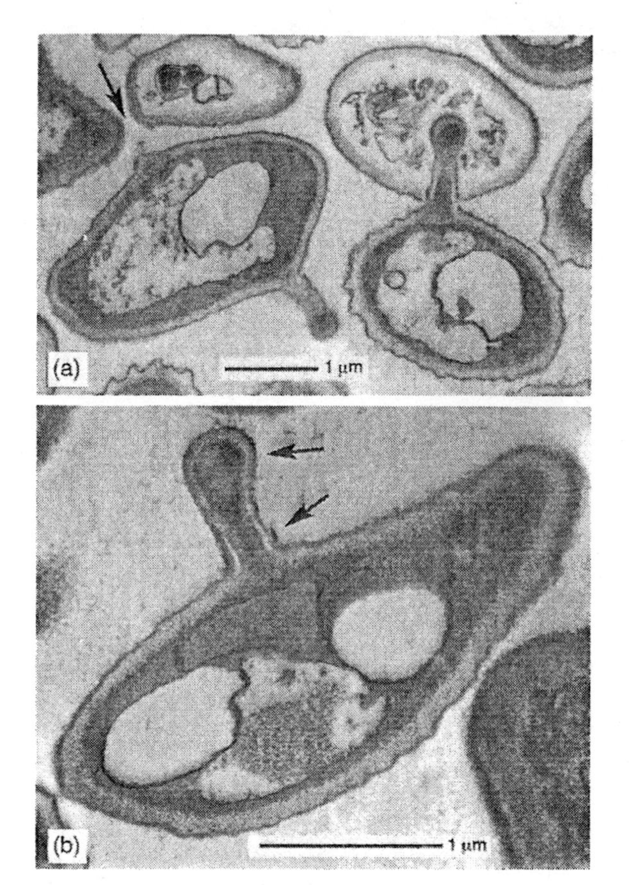

Figure 4.37. Yeast predation. The electron micrographs show predation in a mixture of *Candida* spp 95-697.4 (predator) and *Metschnikowia* spp. M 95-747.4 (prey). (a) The predator–prey pair shows disruption of prey cytoplasm. The other predator cell had two haustoria, one of which (arrow) was sectioned off. (b) A predator cell sectioned longitudinally across the centre of the haustorium. The arrows show a coiled membranous structure in the distal portion of the haustorium and parts of an electron-dense ring at the base, both of which were characteristic features. Reproduced with permission from Lachance and Pang (1997) © J. Wiley and Sons Ltd.

death have been discussed previously in this chapter (section 4.3.3). Some of the important physical, chemical and biological factors which influence yeast cell death will now be described and discussed with regard to their practical implications.

4.6.1 Physical Parameters and Yeast Cell Death

The effects of certain physical stresses on yeast cell growth and physiology have been discussed in section 4.5. However, it is apparent that yeasts will die if confronted with excessive heat, extreme cold, high-voltage electricity, ionizing radiation, and high hydrostatic and osmotic pressures. Death results when yeast cells' physiological protection responses (enzymic and non-enzymic) are insufficient to counteract the cellular damage caused by severe physical stress. Several physical treatments can be used to eradicate contaminant yeasts in industry. For example, yeasts exposed to supramaximal growth temperatures will lead to thermal death of the cells and this is exploited in the pasteurization of foods and beverages containing undesired yeasts. Beer is commonly pasteurized at 70°C for 20 seconds to eliminate yeasts and bacteria in the finished product. Some food and beverage spoilage yeasts are listed in Table 4.34. A much more comprehensive coverage of food and beverage spoilage yeasts is provided by Tudor and Board (1993) and Thomas (1993).

The thermal death rate of yeast cells is exponential and a semilogarithmic plot of viability versus time will be linear (van Uden, 1984). Depending on the temperature, this linear relationship may be preceded by a 'shoulder' which indicates that either several cellular target sites for heat exist in yeast or that cellular repair mechanisms operate. Semilogarithmic survival plots may also exhibit a 'tail' when thermal death rate slows and ceases to be exponential. This may be due to the presence of a subpopulation of yeasts with greater heat resistance, or the protection of living yeasts with material from dead cells. Exponential death of yeasts may be expressed as:

$$N_t = N_0 e^{-kdt}$$

where N_t and N_0 are the numbers of viable cells after time t and time zero, respectively and kd is the specific death rate constant (van Uden, 1984). Note that kd depends on the composition, pH

Table 4.34. Some food and beverage spoilage yeasts.

Yeast species	Food spoilage
Cryptococcus laurentii and *Candida zeylanoides*	Frozen poultry carcasses
Zygosaccharomyces bailii and *Z. rouxii*	Fruits, fruit juices and vegetables
Kluyveromyces spp., *Rhodotorula* spp. and *Candida* spp.	Dairy products (milk, yogurt, cheese)
Saccharomyces cerevisiae var. *turbidans, S. diastaticus, Torulopsis* spp., *Pichia* spp., *Candida* spp., *Hansenula* spp.	'Wild' yeast contaminants of beer
Numerous yeasts spoil wine with *Zygosaccharomyces bailii* being of prime importance.	Wine and other acidified beverages with ethanol $> 10\%$ v/v

Information from Fleet (1993); Kalathenos *et al.* (1995); Jakobsen and Narvhus (1996); Priest and Campbell (1996); Roosita and Fleet (1996).

and dissolved oxygen tension of yeast growth media, the presence of death-enhancing chemicals and the physiological state of the yeast cells. On this latter point, it is known that stationary phase cells are more heat-resistant than exponentially dividing cells. For a given yeast strain at a given temperature, *kd* is thus a constant only under defined and controlled conditions. The use of a chemostat is therefore particularly advantageous in designing experiments to investigate the thermal death kinetics of yeasts.

Besides heat, the effectiveness of several other physical treatments in killing yeasts in industry has been evaluated. For example, high pressure–low temperature treatments inactivate yeast cells (Hashizume *et al.*, 1995) and this has potential applications as an alternative to heat in the sterilization of foods (see Section 4.4.2.3).

4.6.2 Chemical Factors and Yeast Cell Death

There are numerous external chemicals and internal chemical/biochemical reactions which occur outwith and within yeast cells which may result in their death. External chemical factors lethal to yeasts include: toxic organic compounds, abnormal pH, nutrient starvation, oxygen free radicals, heavy metals, etc. Internal chemical factors lethal to yeast include: endogenously produced ethanol and other toxic metabolites (e.g. acetaldehyde), excessive intracellular acidity or alkalinity, inability to protect against oxidative damage or sequester toxic metals etc. The influences of nutrient starvation, ethanol toxicity and oxidative stress on yeast cell physiology were discussed in section 4.5. As with physical stresses, if yeast cells are unable to counteract detrimental effects of chemicals, they will die. The effects of certain natural and synthetic chemicals, or 'zymocides' which are pertinent in food and fermentation industries and in medical mycology will now be considered

Chemical preservatives are commonly employed to control undesired yeasts which spoil foods and beverages. Abnormally low pH values are often lethal to yeast cells and weak acid preservatives such as sorbic, benzoic and acetic acids have wide uses as anti-yeast agents in foods and beverages. Weak acids, which are transported into yeasts in their undissociated form (see section 3.3), act by dissipating plasma membrane proton gradients and depressing cell pH when they dissociate into ions in the yeast cytoplasm. However, they are generally zymostatic, rather than zymocidal in their mode of action. On the other hand, medium-chain fatty acids like decanoic acid have been shown to cause the

Table 4.35. Some antizymotic agents of clinical significance and potential.

Antizymotic agent	Cellular target	Mode of action
Polyenes: Amphotericin B, Nystatin	Plasma membrane	Complexation with ergosterol causing increased membrane permeability. Polyenes may also cause oxidative damage
Morpholines	Plasma membrane	Inhibition of ergosterol synthesis and alteration of membrane properties
Azoles: Miconazole, Itraconazole, Ketoconazole, Fluconazole	Endoplasmic reticulum and plasma membrane	Inhibition of ergosterol synthesis, resulting in membrane disruption making it vulnerable to further damage
Omeprazole	Plasma membrane	Inhibition of proton-pumping ATPase
Allylamines and thiocarbamates	Plasma membrane	Inhibition of squalene epoxidase, resulting in ergosterol depletion and perturbation of membrane integrity
Tunicamycin	Cell wall	Blocks mannoprotein synthetase
Polyoxin and nikkomycin	Cell wall	Blocks chitin synthase
Allosamidin	Cell wall	Blocks chitinase
Papulacaudrin, cilofungin and echinocandin	Cell wall	Block β-(1,3)-D-glucan synthase
Flucytosine	Nucleus	Mis-coding of RNA and inhibition of DNA biosynthesis

rapid cell death of yeast, presumably by severe disruption of cell membrane integrity (Stratford and Anslow, 1996). Sulphur dioxide has a long history of usage as a yeast (and bacterial) preservative in the manufacture of alcoholic beverages, especially wine. Sulphur dioxide (SO_2) dissociates within the yeast cell to SO_3^{2-} and HSO_3^- and the resulting decline in intracellular pH forms the basis of inhibitory action of sulphiting agents against yeast. Some yeasts are resistant to sulphite due perhaps to complexation of sulphite (e.g. by acetaldehyde) or an enhanced pH-buffering capacity within the cells.

Yeast plasma membrane structure and function is a common target for many anti-yeast agents used clinically in the control of human mycoses such as candidosis (Vanden Bossche, 1991). Table 4.35 summarizes the mode of action of some of these drugs. For example, polyene macrolide antibiotics obtained from *Streptomyces* spp. (e.g. nystatin and amphotericin B), together with synthetic azoles (e.g. miconazole and ketoconazole) act by disrupting membrane permeability of pathogenic yeasts like *Candida albicans*. Increased permeability in such yeast treated with these antizymotic agents causes cessation of growth and ultimately cell death. Ergosterol biosynthesis and accumulation in the yeast cell membrane is often the site of polyene and azole action. Allylamines such as naftifine (Ryder, 1984) and the thiocarbamate, tolnaftate are similarly known adversely to affect ergosterol biosynthesis in *C. albicans* (Georgopapadakou and Walsh, 1996). Many other yeast cell envelope targeting drugs are being research and developed, particularly those demonstrating lethality in *C. albicans*. Specific surface sites targeted in yeast include: synthesis of cell wall component subunits (e.g. glucan, mannoprotein and chitin); plasma membrane structure (e.g. ergosterol and phospholipid synthesis) and function (e.g. proton pumping). With regard to this latter target, Monk *et al.* (1995) have shown that inhibition of the plasma membrane H^+-ATPase by omeprazole is lethal to yeast.

Note that the effectiveness of certain antizymotic agents currently in clinical use (e.g. azoles) is being compromised by the emergence of resistant yeast strains (see Joseph-Horne and Holloman, 1997). Odds (1996b) has identified the resistance to azoles in *C. albicans* in HIV-positive individuals as an increasingly important problem.

4.6.3 Biological Factors Influencing Yeast Cell Death

Several external (allochthonous) and endogenous (autochthonous) factors can result in the death of yeast cells. Considering the former, yeasts in the proximity of other macro- and microorganisms may be killed by predation, parasitism and intoxication. Examples of lethal biotic interactions are: direct ingestion of yeasts as food by insects and protozoa; engulfment and lysis by mycoparasitizing fungi; haustoria-mediated predation by other yeasts and cell membrane disruption by yeast killer toxins or bacterial antimycotics. These external biotic factors were discussed in section 4.5.

With regard to endogenous factors, numerous physiological, morphological, genetic and biochemical events take place in yeast cells which may lead to their 'self-inflicted' death. With regard to biochemical events, yeast autolysis and autophagy may be regarded as cellular suicide when cells are deprived of nutrients or when anabolism ceases but catabolism continues. Autophagic death occurs when vacuolar hydrolytic proteases cause dissolution of the protoplasm and autolysis occurs when carbohydrases cause lysis of the cell envelope. Yeast cell wall lytic enzymes include glucanases, chitinases and mannanases (Fleet, 1984) and are sometimes referred to as 'yeast autolysins' (Nombela and Santamaria, 1984). In the commercial production yeast extracts, autolytic enzyme activity is encouraged by using relatively high temperatures (e.g. 45°C), salt (as a plasmolysing agent) or lipid solvents (e.g. ethyl acetate). Additional hydrolytic enzymes (e.g. papain) may sometimes be employed to aid in the dissolution of the cell wall in the manufacture of yeast extracts. Yeast autolysis is also important in winemaking where diffusion of autolytic products gradually modify wine characteristics (Fleet and Heard, 1993).

Genetic factors are important in influencing yeast cell death. Although a large number of *S. cerevisiae* genes may be deleted or disrupted without apparent loss of replicative ability on rich media (Oliver, 1996), many mutations do prove lethal in yeast. For example, some cell division cycle mutations (e.g. Russell and Nurse, 1987; Thomspon-Jaeger *et al.*, 1991; Motizuki *et al.*, 1995), mutations affecting the actin cytoskeleton (Adams *et al.*, 1993), respiratory metabolism (e.g. van Dijken and Scheffers, 1986) and mitochondrial RNA processing (Rinaldi *et al.*, 1995) are lethal. Certain antibiotic-sensitive/temperature-sensitive mutants of *S. cerevisiae* are even referred to as 'kamikaze' mutants (Spencer and Spencer, 1996). In addition, Lydall and Weinert (1996) have described how yeast cells may commit suicide when their DNA is damaged, presumably to avoid the risk of producing genetically altered progeny. Cell death in DNA damaged cells is thought to be due to the unbalanced activity of certain key cell cycle checkpoint control genes (e.g. *RAD* genes).

The final biological aspect pertaining to yeast cell death to be considered is the phenomenon of **cellular ageing**. Yeasts, especially *S. cerevisiae* cells, are now being widely studied at the molecular genetic level in an effort to understand the ageing process in mammalian cells (Bemis *et al.*, 1995; Kennedy *et al.*, 1995; Austriaco, 1996; Kennedy and Guarente, 1996). The beauty of *S. cerevisiae* in this respect is that aged and senescent populations of this yeast can be easily isolated and mutants displaying age-related phenotypes are now available (Kennedy and Guarente, 1996). Austriaco (1996) has discussed mutations in *S. cerevisiae* which modulate longevity. For example *UTH* (yo*u*t*h*) genes have been

Table 4.36. Some phenotypes of aged *S. cerevisiae* cells.

Phenotype	References
Mean cell volume of both mothers and daughters increased	Mortimer and Johnston (1959); Kennedy *et al.* (1994); Woldringh *et al.* (1995)
Accumulation of bud scars	Johnston (1966); Jazwinski (1990)
Arrest of old cells in G1 phase	Johnston (1966)
Granular appearance of cells and eventual cell lysis	Mortimer and Johnston (1959)
Accumulation of intracellular 'blebs'	Kennedy *et al.* (1994)
Cell surface wrinkling	Müller (1971); Barker and Smart (1996)
Increased resistance to mechanical separation of daughter buds from mother cells	Barker and Smart (1996)
Altered resistance to UV stress/chemical mutagenesis	Kale and Jazwinski (1996)
Cell RNA and protein content increase with age, but protein synthetic *rate* decreases linearly with age	Motizuki and Tsurugi (1992)
Pseudohyphal growth (after 25 generations)	Woldringh *et al.* (1995)
Mating sterility (loss of α-factor responsiveness)	Smeal *et al.* (1996)
Extended generation times (in both haploid and polyploid cells)	Mortimer and Johnston (1959); Egilmez *et al.* (1990); Barker and Smart (1996)

identified which increase both stress resistance and longevity in this yeast. Perhaps such genes encode a senescence factor which accumulates in old mother cells and is inherited by daughter cells (Kennedy *et al.*, 1994). Yeast ageing also has practical significance. For example, Barker and Smart (1996) and Soares and Mota (1996) have studied physiological changes which occur in brewing strains of *S. cerevisiae* during ageing and senescence and have discussed the relevance of changes in the flocculation characteristics of aged yeasts for brewing fermentations.

The age, or more strictly the 'lifespan', of a yeast cell is defined as the number of times it undergoes division. The maximum age for a particular yeast strain therefore refers to the highest number of cell division it can undergo. The Hayflick limit (Hayflick, 1965) refers to this maximum division capacity. The finite replicative capacity of yeast is governed by genetic and environmental factors and mean lifespans varying between 13 and 30 divisions have been reported in *S. cerevisiae* (Kennedy *et al.*, 1995). Nevertheless, a Hayflick limit of 25 in *S. cerevisiae* may be regarded as typical for this yeast. Beyond such a limit, cells can generate no further progeny and enter a senescent physiological state leading to death. The mortality of mother cells of *S. cerevisiae* increases in an exponential fashion with the number of cell cycles completed (Jazwinski *et al.*, 1989). Ageing in *S. cerevisiae* is also associated with extended generation times, as depicted in Figure 4.38.

The very fact that ageing in *S. cerevisiae* is dependent on the number of cell division cycles, rather than chronological time, has made this organism an attractive model for those interested in apoptotic cell death. Another reason is that aged *S. cerevisiae* cells possess characteristic morphological phenotypes (Table 4.36). Many of the

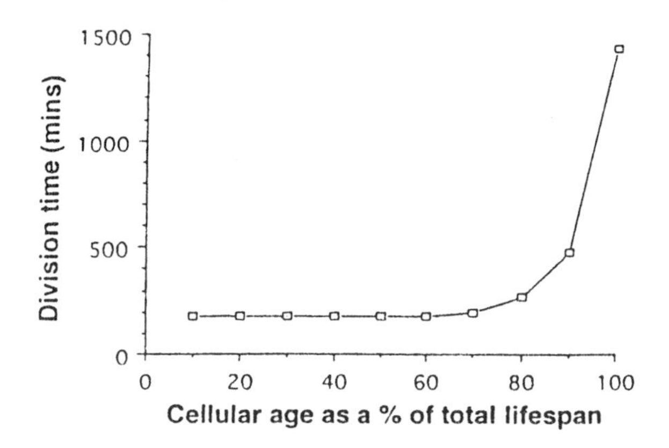

Figure 4.38. Relationship between cell age and generation time in *S. cerevisiae.* The figure shows the relationship between generation time and replicative age of individual cells of *S. cerevisiae.* The time taken to complete one round of cell division was recorded from 40 cells throughout their lifespans and generation time plotted as a function of age. Reproduced with permission from Barker and Smart (1996) and the American Society of Brewing Chemists.

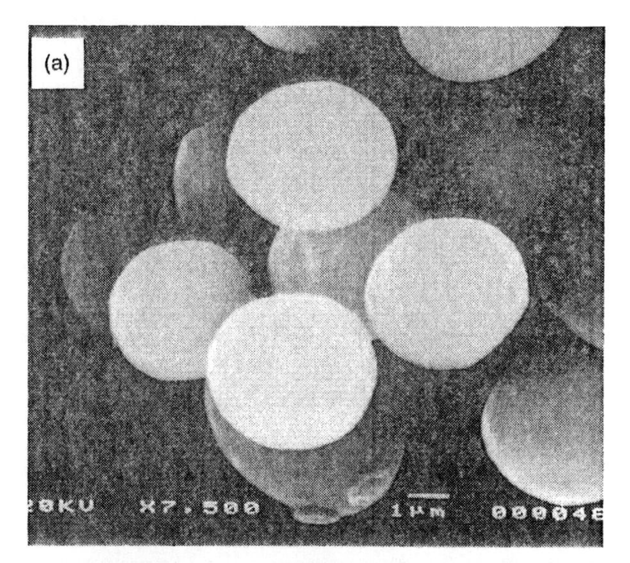

Figure 4.39. Cell surface topology of virgin and aged cells of *S. cerevisiae* The figure shows scanning electron micrographs of populations of *S. cerevisiae* cells taken from a set of sucrose gradients used to fractionate an eight-division-old cell preparation. (a) Cells from the uppermost (virgin) band. (b) Cells from the eight-division-old mother band. Reproduced with permission from Barker and Smart (1996) and the American Society of Brewing Chemists.

morphological changes associated with ageing in budding yeast occur in the cell wall. One of the most striking cell surface features in ageing *S. cerevisiae* is the accumulation of chitin-containing bud scars. These can be visualized in the scanning electron microscope (Figure 4.39) or by staining with Calcofluor White dye which, with the aid of a fluorescence microscope (Hasek and Streiblova, 1996) enable the age of cells to be determined (by a bud scar count).

4.7 SUMMARY

Yeast cells exhibit considerable diversity in their growth behaviour. Budding is very common and in *S. cerevisiae*, individual bud growth is known to depend critically on the dynamism of the cytoskeletal actin fibres and microtubules. Bud site selection in this yeast is governed by several genes which predispose cells to a non-random budding pattern. Fission yeasts, as typified by *Sch. pombe*, differ markedly from budding yeasts in that they divide symmetrically (rather than asymmetrically) and in so doing resemble the cytokinetic cleavage of higher eukaryotic cells. The medial fission of *Sch. pombe* cylinders and

Table 4.37. Physiological responses of yeasts to environmental stress.

General response	Examples
Thermoprotection	• Heat-shock protein biosynthesis • Decreased membrane lipid unsaturation • Altered cell pH • Polyamine biosynthesis
Cryoprotection	• Enhanced trehalose accumulation • Increased membrane lipid polyunsaturation
Osmoprotection	• Accumulation of compatible solutes (glycerol, trehalose) • Increased K^+ uptake/Na^+ efflux
Antidesiccation	• Trehalose accumulation
Antioxidation	• Enzymic: Superoxide dismutase, catalase, cytochrome peroxidase • Non-enzymic: Glutathione, thioredoxin, metallothionein, polyamines, carotenoids
Detoxification	• Ethanol: Stress proteins, altered membrane transport, mitochondrial superoxide dismutase • Xenobiotics: Glutathione • Heavy metals: Stress proteins, metallothioneins

their highly polarized growth makes this yeast particularly attractive to cell biologists studying eukaryotic cellular morphogenesis. Other yeasts, for example *C. albicans*, are typically dimorphic in that they can grow with both budding yeast and filamentous hyphal morphologies. Even *S. cerevisiae* can be induced by the prevailing growth conditions to develop in a pseudohyphal manner. It is apparent that yeast growth responds variably to environmental signals (including sex pheromones) and, therefore, yeast cells should be described as being truly polymorphic.

The underlying control processes in the yeast cell division cycle are gradually being unravelled at the molecular genetic level. However, several questions remain unanswered with respect to yeast cell cycle physiology. For example, how do yeasts monitor their cell size in response to nutrients? Nevertheless, intriguing insights are now being gleaned into cell size-related controls operating in both budding and fission yeasts at the G1–S and G2–M cell cycle checkpoints. DNA replication during S phase and chromosome segregation during M phase are key events of the eukaryotic cell cycle and work with *S. cerevisiae* and *Sch. pombe* has revealed specific cyclin-dependent protein kinases to be key players in their regulation. Despite many differences in fine detail, aspects of cell cycle control in budding and fission yeasts are basically quite similar (Huberman, 1996). Molecular genetic and cell physiological analyses of the cell cycles of *S. cerevisiae* and *Sch. pombe* will doubtless continue to provide detailed information of eukaryotic cell division mechanisms. Knowledge accrued to date indicates that several key elements of cell cycle regulation in yeast cells appear to operate in all eukaryotic cells and this serves to emphasize the importance of yeasts in the study of fundamental life processes.

The yeast cell physiologist can manipulate the growth behaviour of cells in both solid and liquid media. For example, by controlling the supply of nutrients and the physical growth environment, yeasts can be induced to grow randomly or non-randomly with respect to the cell cycle, and continuously or non-continuously with respect to growth rate. Yeast growth can also be restricted to a pseudostationary phase when immobilized on and in suitable inert matrices. Such manipulations of growth rates and growth phases in yeast can be extended to large-scale bioreactors to facilitate optimum production of various yeast-derived commodities.

During industrial yeast fementations, individual cells may be subject to several physical and chemical stresses which may impair normal growth and metabolism. For yeast biotechnology, it is vitally important to understand cellular stress physiology in order to alleviate as far as possible the detrimental influences of changes in the growth environment. Table 4.37 summarizes the physiological responses of yeasts to the commonly encountered physical and chemical stresses.

Several molecular-level responses overlap (e.g. trehalose accumulation) and there may be fundamental similarities in the way by which yeasts adapt to survive. In addition to those biotechnologically relevant phenomena, studies of the stationary phase and stress damage and stress protection in yeast will provide a better understanding of physiological processes involved in human diseases such as ageing, cancer and neurological degeneracy.

Numerous interactions between yeasts and other organisms have been described in this chapter and those with relevance in medical, environmental and industrial biotechnology further discussed. For example, the killer yeast phenomenon is now well described in terms of molecular biology and cell physiology and there are several areas of yeast biotechnology where killer toxins may be effectively exploited. The death of yeast cells, whether by predation, zymocidal toxins, chemical or physical agents, or through endogenous biological processes, is of interest to both cell physiologists and biotechnologists. For example, knowledge of cellular ageing in *S. cerevisiae* can provide insight into senescence and apoptosis in higher eukaryotes, but it is also of direct practical relevance for the continued success of industrial yeast fermentations.

4.8 REFERENCES

Aarnio, T. H., Suihko, M.L. and Kauppinen, V.S. (1991) Isolation of acetic acid-tolerant baker's yeast variants in a turbidostat. *Applied Biochemistry and Biotechnology*, **27**, 55–63.

Abel, C., Linz, F., Scheper, T. and Schügerl, K. (1994) Transient behaviour of continuously cultivated baker's yeast during enforced variations of dissolved oxygen and glucose concentrations. *Journal of Biotechnology*, **33**, 183–193.

Adami, A., Cavazzoni, V., Trezzi, M. and Craveri, R. (1988) Cellobiose hydrolysis by *Trichosporon pullulans* cells immobilized in calcium alginate. *Biotechnology and Bioengineering*, **32**, 391–396.

Adams, A.E., Cooper, J.A. and Drubin, D.G. (1993) Unexpected combinations of null mutations in genes encoding the actin cytoskeleton are lethal in yeast. *Molecular Biology of the Cell*, **4**, 459–468.

Agar, D.W. and Bailey, J.E. (1982) Measurements and models of synchronous growth of fission yeast induced by temperature oscillations. *Biotechnology and Bioengineering*, **24**, 217–236.

Alfa, C., Fantes, P., Hyams, J., McLeod, M. and Warbrick, E. (1993) *Experiments with Fission Yeast*. Cold Spring Harbor Press, Cold Spring Harbor, New York.

American Society of Brewing Chemists (1980) Report of sub-committee in microbiology. *Journal of the American Society of Brewing Chemists*, **38**, 109–110.

Amon, A. (1996) Mother and daughter are doing fine – asymmetric cell division in yeast. *Cell*, **84**, 651–654.

Anraku, Y., Ohya, O. and Iida, H. (1991) Cell cycle control by calcium and calmodulin in *Saccharomyces cerevisiae*. *Biochimica et Biophysica Acta*, **1093**, 169–177.

Arcay-Ledezma, G.J. and Slaughter, J.C. (1984) The response of *Saccharomyces cerevisiae* to fermentation under carbon dioxide pressure. *Journal of the Institute of Brewing*, **90**, 81–84.

Argüelles, J.C. (1997) Thermotolerance and trehalose accumulation induced by heat shock in yeast cells of *Candida albicans*. *FEMS Microbiology Letters*, **146**, 65–71.

Attfield, P.V. (1987) Trehalose accumulates in *Saccharomyces cerevisiae* during exposure to agents that induce heat shock response. *FEBS Letters*, **225**, 259–263.

Aucott, J.N., Fayen, J., Grossnicklas, H., Morrissey, A., Lederman, M.M. and Salata, R.A. (1990) Invasive infection with *Saccharomyces cerevisiae*: report of three cases and review. *Reviews of Infectious Diseases*, **12**, 406–411.

Austriaco, N.R. (1996) To bud until death – the genetics of aging in the yeast, *Saccharomyces*. *Yeast*, **12**, 623–630.

Bailey, D.A., Feldmann, P.J.F., Bovey, M., Gow, N.A.R. and Brown, A.J.P. (1996) The *Candida albicans HYR1* gene, which is activated in response to hyphal development, belongs to a gene family encoding yeast cell wall proteins. *Journal of Bacteriology*, **178**, 5353–5360.

Bajpai, P. and Margaritis, M. (1986) Optimization studies for production of high fructose syrup from Jerusalem artichoke using calcium alginate immobilized cells of *Kluyveromyces marxianus*. *Process Biochemistry*, **21**, 16–18.

Bakke, O. and Pettersen, E.O. (1976) A fast and accurate method for calculating Engelberg's synchronization index. *Cell and Tissue Kinetics*, **9**, 389–393.

Balasundaram, D., Tabor, C.W. and Tabor, H. (1996) Sensitivity of polyamine-deficient *Saccharomyces cerevisiae* to elevated temperatures. *Journal of Bacteriology*, **178**, 2721–2724.

Banat, I.M., Nigam, P. and Marchant, R. (1992) Isolation of a thermotolerant yeast growing at 52°C and producing ethanol at 45°C and 50°C. *World Journal of Microbiology and Biotechnology*, **8**, 259–263.

Barker, M.G. and Smart, K.S. (1996) Morphological changes associated with the cellular aging of a brewing yeast strain. *Journal of the American Society of Brewing Chemists*, **54**, 121–126.

Barnett, J.A., Payne, R.W. and Yarrow, D. (1990) *Yeasts: Characteristics and Identification*. 2nd edn. Cambridge University Press, Cambridge.

Baroni, M.D., Monti, P. and Alberghina, L. (1994) Repression of growth-related G1 cyclin by cyclic AMP in budding yeast. *Nature*, **371**, 339–342.

Barth,G. and Gaillardin, C. (1997) Physiology and genetics of the dimorphic fungus *Yarrowia lipolytica*. *FEMS Microbiology Reviews*, **19**, 219–237.

Bartholomew, J.W. and Mittwer, T. (1953) Demonstration of yeast bud scars with the electron microscope. *Journal of Bacteriology*, **65**, 272–275.

Batiza, A.F., Schulz, T. and Masson, P.J. (1996) Yeasts respond to hypotonic shock with a calcium pulse. *Journal of Biological Chemistry*, **271**, 23357–23362.

Beaudette, K.N., Lew, J. and Wang, J.H. (1993) Substrate specificity characterization of a cdc2-like protein kinase purified from porcine brain. *Journal of Biological Chemistry*, **268**, 20825–20830.

Beker, M. J. and Rapaport, A.I. (1987) Conservation of yeasts by dehydration. *Advances in Biochemical Engineering/Biotechnology*, **35**, 127–171.

Bellgardt, K.-H. (1994) Analysis of synchronous growth of baker's yeast. Part II: Comparison of model prediction and experimental data. *Journal of Biotechnology*, **35**, 35–49.

Bemis, L.T., Geske, F.J. and Strange, R. (1995) Use of the yeast two-hybrid system for identifying the cascade of protein interactions resulting in apoptotic cell death. In *Methods in Cell Biology. Vol. 46: Cell Death* (eds L.M. Schwartz and B.A. Osborne), pp. 139–151. Academic Press, San Diego.

Beran, K. (1968) Budding of yeast cells, their scars and ageing. *Advances in Microbial Physiology*, **2**, 143–171.

Beyeler, W., Einsele, A. and Fiechter, A. (1981) On-line measurements of culture fluorescence: method and application. *European Journal of Applied Microbiology and Biotechnology*, **13**, 10–14.

Birch, R. and Walker, G.M. (1996) The role of magnesium in yeast stress responses. Abstracts of the 9th International Symposium on Yeasts (Sydney). pp. 46–47.

Birol, G. and Özergen-Ulgen, K. (1995) A study of the effect of inoculum on yeast cell growth and ethanol production. *Turkish Journal of Chemistry*, **19**, 250–257.

Black, G.M., Webb, C., Mathews, T. and Atkinson, B. (1984) Practical reactor systems for yeast cell immobilization using biomass support particles. *Biotechnology and Bioengineering*, **26**, 134–141.

Blacketer, M.J., Madaule, P. and Myers, A.M. (1995) Mutational analysis of morphologic differentiation in *Saccharomyces cerevisiae*. *Genetics*, **140**, 1259–1275.

Blomberg, A. and Adler, L. (1992) Physiology of osmotolerance in fungi. *Advances in Microbial Physiology*, **33**, 145–212.

Blondel, M. and Mann, C. (1996) G2 cyclins are required for the degradation of G1 cyclin in yeast. *Nature*, **384**, 279–282.

Boiteaux, A. (1995) Metabolic studies on synchronized yeast cultures in continuous culture. *Folia Microbiologica*, **39**, 509–511.

Boone, C., Sdicu, A.-M., Wagner, J., Degre, R., Sanchez, C. and Bussey, H. (1990) Integration of the yeast K1 killer toxin gene into the genome of marked wine yeasts and its effect on vinification. *American Journal of Enology and Viticulture*, **41**, 37–42.

Bortol, A., Nudel, C., Fraile, E., de Torres, R., Giuletti, A., Spencer, J.F.T. and Spencer, D. (1986) Isolation of the yeast with killer activity and its breeding with an industrial brewing strain by protoplast fusion. *Applied Microbiology and Biotechnology*, **24**, 414–416.

Boulton, C.A., Maryan, P.S., Loveridge, D. and Kell, D.B. (1989) The application of a novel biomass sensor to the control of yeast pitching rate. European Brewery Convention. Proceedings of the 22nd Congress, Zurich, pp. 653–661. IRL Press at Oxford University Press, Oxford, UK.

Brasier, C.M. (1991) *Ophiostoma novo-ulmi sp. nov.*,

causative agent of current Dutch elm disease pandemics. *Mycopathologia*, **115**, 151–161.

Breeden, L. (1996) Start-specific transcription in yeast. *Current Topics in Microbiology and Immunology*, **208**, 95–127.

Breeuwer, P., Drocourt, J.L., Rombouts, F.M. and Abee, T. (1994) Energy-dependent, carrier-mediated extrusion of carboxyfluorescein from *Saccharomyces cerevisiae* allows rapid assessment of cell viability by flow cytometry. *Applied and Environmental Microbiology*, **60**, 1467–1472.

Broek, D., Bartlett, R., Crawford, K. and Nurse, P. (1991) Involvement of $p34^{cdc2}$ in establishing the dependency of S phase on mitosis. *Nature*, **349**, 388–393.

Brown, S.W. and Oliver, S.G. (1982) Isolation of ethanol-tolerant mutants of yeast by continuous selection. *European Journal of Applied Microbiology and Biotechnology*, **16**, 119–122.

Bruschi, C.V. and Esposito, M.S. (1995) Cell division, mitotic recombination and onset of meiosis by diploid yeast cells during space flight. In *Biorack on Spacelab IML-1. ESA SP-1162*. pp. 83–93. ESA Noordwijk, The Netherlands.

Bulawa, C.E. (1993) Genetics and molecular biology of chitin synthesis in fungi. *Annual Review of Microbiology*, **47**, 505–534.

Bussey, H. (1991) K1 killer toxin, a pore-forming protein from yeast. *Molecular Microbiology*, **5**, 2339–2343.

Cahill, G., Walsh, P.K. and Ryan, T.P. (1990) Studies on the production of β-glucanase by free and immobilized recombinant yeast cells. In *Physiology of Immobilized Cells* (eds J.A.M. de Bont, J. Visser, B. Mattiasson and J. Tramper), pp. 405–410. Elsevier Science Publishers, Amsterdam.

Calleja, G.B., Zucker, M. and Johnson, B.F (1980) Analyses of fission scars as permanent records of cell division in *Schizosaccharomyces pombe*. *Journal of Theoretical Biology*, **84**, 523–544.

Casey, G.P. and Ingledew, W.M. (1983) High-gravity brewing: influence of pitching rate and wort gravity on early yeast viability. *Journal of the American Society of Brewing Chemists*, **41**, 148–152.

Casey, G.P. and Ingledew, W.M. (1986) Ethanol tolerance in yeast. *CRC Critical Reviews in Microbiology*, **13**, 219–280.

Cashin, M.-M. (1996) Comparative studies of five porous supports for yeast immobilization by adsorption/attachment. *Journal of the Institute of Brewing*, **102**, 5–10.

Champluvier, B., Kamp, B. and Rouxhet, P.G. (1988) Immobilization of β-galactosidase retained in yeast: adhesion of the cells on a support. *Applied Microbiology and Biotechnology*, **27**, 464–469.

Chant, J. (1995) Control of cell polarity by internal programs and external signals in yeast. *Seminars in Developmental Biology*, **6**, 13–23.

Chant, J. (1996a) Septin scaffolds and cleavage planes in *Saccharomyces*. *Cell*, **84**, 187–190.

Chant, J. (1996b) Generation of cell polarity in yeast. *Current Opinion in Cell Biology*, **8**, 557–565.

Chant, J. and Herskowitz, I. (1991) Genetic control of bud site selection in yeast by a set of gene products that constitute a morphogenetic pathway. *Cell*, **65**, 1203–1212.

Chant, J. and Pringle, J.R. (1995) Patterns of bud-site selection in the yeast *Saccharomyces cerevisiae*. *Journal of Cell Biology*, **129**, 751–765

Chant, J. and Stowers, L. (1995) GTPase cascades choreographing cellular behaviour: movement, morphogenesis and more. *Cell*, **81**, 1–4.

Characklis, W.G. and Marshall, K.C. (1990) *Biofilms*. J. Wiley and Sons, New York.

Cheng, C. and Yang, S.T. (1996) Dynamics and modelling of temperature-regulated gene product expression in recombinant yeast fermentation. *Biotechnology and Bioengineering*, **50**, 663–674.

Ciesarova, Z., Smorrovicova, D. and Domeny, Z. (1996) Enhancement of yeast ethanol tolerance by calcium and magnesium. *Folia Microbiologica*, **41**, 485–488.

Clacharkar, M.P., Tak, B.B. and Bhati, J. (1996) Alteration of yeast activity by gamma-radiation. *Journal of Radioanalytical Nuclear Chemistry, Letters*, **213**, 79–86.

Cook, A.H. (1963) Yeast and the work of the yeast group. European Brewery Convention. Proceedings of the 9th Congress, Brussels, pp. 477–486. Elsevier Science Pub. Co., Amsterdam.

Cooper, A.J. and Kelly, S.L. (1987) DNA repair and mutagenesis in *Saccharomyces cerevisiae*. In *Enzyme Induction, Mutagen Activation and Carcinogen Testing in Yeast* (ed. A. Wiseman), pp. 73–114. Ellis Horwood, Chichester, UK.

Coote, P.J., Cole, M.B. and Jones, M.V. (1991) Induction of increased thermotolerance in *Saccharomyces cerevisiae* may be triggered by a mechanism involving intracellular pH. *Journal of General Microbiology*, **137**, 1701–1708.

Costa, V., Reis, E., Quintanilha, A. and Moradas-Ferreira, P. (1993) Acquisition of ethanol tolerance in *Saccharomyces cerevisiae*: the key role of mitochondrial superoxide dismutase. *Archives of Biochemistry and Biophysics*, **300**, 608–614.

Coutinho, C., Bernardes, E., Felix, D. and Panek, A.D. (1988) Trehalose as cryoprotectant for preservation of yeast strains. *Journal of Biotechnology*, **7**, 23–32.

Creanor, J. and Toyne, J. (1993) Preparation of synchronous cultures of the yeasts *Saccharomyces cerevisiae* and *Schizosaccharomyces pombe*. In *The*

Cell Cycle. A Practical Approach (eds P. Fantes and R. Brooks), pp. 25–44. IRL Press, Oxford.

Crombie, T., Gow, N.A.R. and Gooday, G.W. (1990) Influence of applied electric fields on yeast and hyphal growth of *Candida albicans*. *Journal of General Microbiology*, **136**, 311–317.

Crowe, J.H., Crowe, L.M. and Chapman, D. (1984) Preservation of membranes in anhydrobiotic organisms. The role of trehalose. *Science*, **223**, 701–703.

Curran, B.P.G. and Khalawan, S.A. (1994) Alcohols lower the threshold temperature for the maximal activation of a heat shock expression vector in the yeast *Saccharomyces cerevisiae*. *Microbiology* (*UK*), **140**, 2225–2228.

D'Amore, T., Panchal, C.J. and Stewart, G.G. (1988) Intracellular ethanol accumulation in *Saccharomyces cerevisiae* during fermentation. *Applied and Environmental Microbiology*, **54**, 110–114.

D'Amore, T., Panchal, C.J., Russell, I. and Stewart, G.G. (1990) A study of ethanol tolerance in yeast. *Critical Reviews in Biotechnology*, **9**, 287–304.

D'Amore, T. Crumplen, R. and Stewart, G.G. (1991) The involvement of trehalose in yeast stress tolerance. *Journal of Industrial Microbiology*, **7**, 191–196.

Dasari, G., Worth, M.A., Connor, M.A. and Pamment, N.B. (1990) Reasons for the apparent difference in the effects of produced and added ethanol in culture viability during rapid fermentations by *Saccharomyces cerevisiae*. *Biotechnology and Bioengineering*, **35**, 109–122.

Davey, H.M., Davey, C.L., Woodward, A.M. Edmonds, A.N., Lee, A.W. and Kell, D.B. (1996) Oscillatory, stoichastic and chaotic growth rate fluctuations in permittistatically controlled yeast cultures. *BioSystems*, **39**, 43–61.

Davidson, J.F., Whyte, B., Bissinger, P.H. and Schiestl, R.H. (1996) Oxidative stress is involved in heat-induced cell death in *Saccharomyces cerevisiae*. *Proceedings of the National Academy of Sciences* (*USA*), **93**, 5116–5121.

Dawson, P.S.S. (1972) Continuously synchronized growth. *Journal of Applied Chemistry and Biotechnology*, **22**, 79–103.

Dawson, P.S.S. (1985) Continuous cultivation of microorganisms. *Critical Reviews in Biotechnology*, **2**, 315–372.

De Curtis, F., Torriani, S., Rossi, E. and De Cicco, V. (1996) Selection and use of *Metschnikowia pulcherrima* as a biological control agent for postharvest rots of peaches and table grapes. *Annali di Microbiologia ed Enzimologia*, **46**, 45–55

De la Pena, P., Barros, F., Gascon, S., Ramos, S. and Lazo, P.S. (1980) Primary effects of yeast killer toxin. *Biochemical and Biophysical Research Communications*, **96**, 544–550.

Demuyakor, B. and Ota, Y. (1992) Promotive action of ceramics on yeast ethanol production and its relationship to pH, glycerol and alcohol dehydrogenase activity. *Applied Microbiology and Biotechnology*, **36**, 717–721.

De Virgillo, C., Hottiger, T., Dominguez, J., Boller, T. and Wiemken, A. (1994) The role of trehalose synthesis for the acquisition of thermotolerance in yeast. I. Genetic evidence that trehalose is a thermoprotectant. *European Journal of Biochemistry*, **219**, 179–186.

Dickinson, J.R. (1991) Metabolism and biosynthesis. In *Saccharomyces* (eds M.F. Tuite and S.G. Oliver), pp. 59–100. Plenum Press, New York and London.

Diffley, J.F.X. (1995) The initiation of DNA duplication in the budding yeast cell division cycle. *Yeast*, **11**, 1651–1670.

Dignard, D., Whiteway, M., Germain, D., Tessier, D. and Thomas, D.Y. (1991) Expression in yeast of a cDNA copy of the K2 killer toxin gene. *Molecular and General Genetics*, **227**, 127–136.

Dinsdale, M.B., Lloyd, D. and Jarvis, B. (1995) Yeast vitality during cider fermentation: two approaches to the measurement of membrane potential. *Journal of the Institute of Brewing*, **101**, 453–458.

Divies, C., Cachon, R., Cavin, J.-F. and Prevost, H. (1994) Theme 4: Immobilized cell technology in wine production. *Critical Reviews in Biotechnology*, **14**, 135–153.

Dombek, K.M. and Ingram, L.O. (1986) Nutrient limitation as a basis for the apparent toxicity of low levels of ethanol during fermentation. *Journal of Industrial Microbiology*, **1**, 219–225.

Doran, P.M. and Bailey, J.E. (1986) Effects of immobilization on growth, fermentation properties and macromolecular composition of *Saccharomyces cerevisiae*. *Biotechnology and Bioengineering*, **28**, 73–87.

Douglas, L.J. (1995) Adhesin-receptor interactions in the attachment of *Candida albicans* to host epithelial cells. *Canadian Journal of Botany*, **73**, S1147–S1153.

Driedonks, R.A., Toschka, H.Y., van Almkerk, J.W., Schaffers, I.M. and Verbakel, J.M.A. (1995) Expression and secretion of antifreeze peptides in the yeast *Saccharomyces cerevisiae*. *Yeast*, **11**, 849–864.

Droby, S., Chalutz, E., Wilson, C.L. and Wisniewoski, M.E. (1989) Characterization of the biological control activity of *Debaryomyces hansenii* in the control of *Penicillium digitatum* on grapefruit. *Canadian Journal of Microbiology*, **35**, 794–800.

Duboc, P., Marison, I. and von Stockar, V. (1996) Physiology of *Saccharomyces cerevisiae* during cell cycle oscillations. *Journal of Biotechnology*, **51**, 57–72.

Duffus, J.H. and Patterson, L.J. (1974) Control of cell division in yeast using the ionophore A23187 with calcium and magnesium. *Nature*, **251**, 626–627.

Dupont, P.F. (1995) *Candida albicans*, the opportunist – a cellular and molecular perspective. *Journal of the American Podiatric Medical Association*, **85**, 104–115.

Edmunds, L.N., Apter, R.I., Rosenthal, P.J., Shen, W.K. and Woodward, J.R. (1979) Light effects in yeast: persisting oscillations in cell division activity and amino acid transport in cultures of *Saccharomyces cerevisiae* entrained by light–dark cycles. *Photochemistry and Photobiology*, **30**, 595–601.

Edwards, C., Porter, J., and West, M. (1996) Fluorescent probes for measuring physiological fitness of yeast. *Ferment*, **9**, 288–293.

Egilmez, N.K., Chen, J.B. and Jazwinski, S.M. (1990) Preparation and partial characterization of old yeast cells. *Journal of Gerontology*, **45**, B9–B17.

Elliot, B. and Futcher, B. (1993) Stress resistance of yeast cells is largely independent of cell cycle phase. *Yeast*, **9**, 33–42.

Elliot, B., Haltiwanger, R.S. and Futcher, B. (1996) Synergy between trehalose and Hsp104 for thermotolerance in *Saccharomyces cerevisiae*. *Genetics*, **144**, 923–933.

Eng, R.H.K., Drehmel, R., Smith, S.M. and Goldstein, E.J.C. (1984) *Saccharomyces cerevisiae* infections in man. *Sabouraudia: Journal of Medical and Veterinary Mycology*, **22**, 403–407.

Enoch, T., Peter, M., Nurse, P. and Nigg, E.A. (1991) p34^{cdc2} acts as a lamin kinase in fission yeast. *Journal of Cell Biology*, **112**, 797–807.

Evans, H.A.V. and Cleary, P. (1985) Direct measurements of yeast and bacterial viability. *Journal of the Institute of Brewing*, **91**, 73.

Fankhauser, C. and Simanis, V. (1994) Cold fission: splitting the *pombe* cell at room temperature. *Trends in Cell Biology*, **4**, 96–101.

Fantes, P.A. (1984) Temporal control of the *Schizosaccharomyces pombe* cell cycle. In *Cell Cycle Clocks* (ed. L.N. Edmunds), pp. 233–252. Marcel Dekker Inc., New York and Basel.

Fantes, P.A. and Nurse, P. (1977) Control of cell size at division in fission yeast by a growth modulated size control over nuclear division. *Experimental Cell Research*, **107**, 377–386.

Fargher, J. and Smith, N.A. (1995) Evidence of cold shock sensitivity in brewing yeast strains. European Brewery Convention. Proceedings of the 25th Congress, Brussels, pp. 345–352. IRL Press at Oxford University Press, Oxford, UK.

Feilotter, H., Lingner, C., Rowley, R. and Young, P.G. (1992) Regulation of the G2–mitosis transition. *Biochemical and Cellular Biology*, **70**, 954–971.

Fiechter, A., Käppeli, D. and Meussdoerffer, F. (1987) Batch and continuous culture. In *The Yeasts. Vol. 2*. 2nd edn (eds A.H. Rose and J.S. Harrison), pp. 99–129. Academic Press, London.

Fiedler, U., Gröbner, U. and Berg, H. (1995) Electrostimulation of yeast proliferation. *Bioelectrochemistry and Bioenergetics*, **38**, 423–425.

Fiems, L.O., Cottyn, B.G. and Boucque, C. V. (1995) Effect of yeast supplementation on health, performance and rumen fermentation in beef bulls. *Archives of Animal Nutrition*, **47**, 295–300.

Filonow, A.B., Vishniac, H.S., Anderson, J.A. and Janisiewicz, W.J. (1996) Biological control of *Botrytis cinerea* in apple by yeasts from various habitats and their putative mechanisms of antagonism. *Biological Control*, **7**, 212–220.

Fleet, G.H. (1984) The occurrence and function of endogenous wall-degrading enzymes in yeast. In *Microbial Cell Wall Synthesis and Autolysis* (ed C. Nombela), pp. 227–238. Elsevier Science Publishers, Amsterdam.

Fleet, G.H. (1993) (ed.) *Wine Microbiology and Biotechnology*. Harwood Academic Publishers, Chur, Switzerland.

Fleet, G.H. and Heard, G.M. (1993) Yeasts – growth during fermentation. In *Wine Microbiology and Biotechnology* (ed. G.H. Fleet), pp. 27–54. Harwood Academic Publishers, Chur, Switzerland.

Forsburg, S.L. and Nurse, P. (1991) Cell cycle regulation in the yeasts *Saccharomyces cerevisiae* and *Schizosaccharomyces pombe*. *Annual Review of Cell Biology*, **7**, 227–256

Friedberg, E.C., Siede, W.A. and Cooper, A.J. (1991) Cellular responses to DNA damage in yeast. In *The Molecular and Cellular Biology of the Yeast* Saccharomyces (eds J.R. Broach, J.R. Pringle and E.W. Jones), pp. 147–191. Cold Spring Harbor Laboratory Press, Cold Spring Harbor, New York.

Frucht-Pery, J., Mor, M., Evron, R., Lewis, A. and Zauberman, H. (1993) The effect of the ArF excimer laser on *C. albicans, in vitro*. *Graefes Archives of Clinical Opthalmology*, **231**, 413–415.

Fuge, E.K., Braun, E.L. and Werner-Washburne, M. (1994) Protein synthesis in long-term stationary-phase cultures of *Saccharomyces cerevisiae*. *Journal of Bacteriology*, **176**, 5802–5813.

Fujii, S., Iwahashi, H., Obuchi, K., Fujii, T. and Komatsu, Y. (1996) Characterization of a barotolerant mutant of the yeast *Saccharomyces cerevisiae*: importance of trehalose content and membrane fluidity. *FEMS Microbiology Letters*, **141**, 97–101.

Fukuhara, H. (1995) Linear DNA plasmid of yeasts. *FEMS Microbiology Letters*, **131**, 1–9.

Futcher, B. (1995) Analysis of the cell cycle in *Saccharomyces cerevisiae*. In *The Cell Cycle. A Practical Approach* (eds P. Fantes and R. Brooks), pp. 69–92. IRL Press, Oxford.

Futcher, B. (1996) Cyclins and the wiring of the yeast cell cycle. *Yeast*, **12**, 1635–1646.

Gadd, G.M. (1993) Interactions of fungi with toxic metals. *New Phytologist*, **124**, 25–60.

Gadd, G.M., Chalmers, K. and Reed, R.H. (1987) The role of trehalose in dehydration resistance of *Saccharomyces cerevisiae. FEMS Microbiology Letters*, **48**, 249–254.

Galli, A. and Shiestl, R.H. (1995) On the mechanism of UV and γ-ray-induced intrachromosomal recombination in yeast cells synchronized in different stages of the cell cycle. *Molecular and General Genetics*, **248**, 301–310.

Gaskova, D., Sigler, K., Janderova, B. and Plasek, J. (1996) Effect of high voltage electric pulses on yeast cells – factors influencing the killing efficiency. *Bioelectrochemistry and Bioenergetics*, **39**, 195–202.

Georgopapadakou, N.H. and Walsh, T.J. (1996) Antifungal agents: chemotherapeutic targets and immunological strategies. *Antimicrobial Agents and Chemotherapy*, **40**, 279–291.

Gervais, P., Molin, P., Marechal, P.A. and Herail-Foussereau, C. (1996) Thermodynamics of yeast cell osmoregulation: passive mechanisms. *Journal of Biological Physics*, **22**, 73–86.

Gill, C.O. and Ratledge, C. (1972) Toxicity of *n*-alkanes, *n*-alk-l-enes, *n*-alkan-1-ols and *n*-alkyl-l-bromides towards yeasts. *Journal of General Microbiology*, **72**, 165–172.

Gimeno, C.J., Ljungdahl, P.O., Styles, C.A. and Fink, G.R. (1992) Unipolar cell divisions in the yeast *S. cerevisiae* lead to filamentous growth: regulation by starvation and RAS. *Cell*, **68**, 1077–1090.

Gimeno, C.J., Ljungdahl, P.O., Styles, C.A. and Fink, G.R. (1993) Characterization of *Saccharomyces cerevisiae* pseudohyphal growth. In *Dimorphic Fungi in Biology and Medicine* (eds H. Vanden Bossche, F.C. Odds and D. Kerridge), pp. 83–103. Plenum Press, New York.

Godia, F., Casas, C., Castellano, B. and Sola, C. (1987) Immobilized cells: behaviour of carrageenan entrapped yeast during continuous ethanol fermentation. *Applied Microbiology and Biotechnology*, **26**, 342–346.

Godia, F. Casas, C. and Sola, C. (1991) Application of immobilized yeast cells to sparkling wine fermentation. *Biotechnology Progress*, **7**, 468–470.

Golubev, W.I. and Boekhout, T. (1995) Sensitivity to killer toxins as a taxonomic tool among heterobasidiomycetous yeasts. *Studies in Mycology*, **38**, 47–58.

Golubev, W.I. and Nakase, T. (1996) Taxonomic specificity of sensitivity to mycocin produced by *Bullera sinensis* L1. *Microbiological Culture Collections*, **12**, 45.

Golubev, W.I. and Shabalin, Y. (1994) Microcin produced by the yeast *Cryptococcus humicola. FEMS Microbiology Letters*, **119**, 105–110.

Golubev, W.I., Tsiomenko, A.B. and Tikbomirova, L.P. (1988) Plasmid-free killer strains of the yeast *Sporidiobolus pararoseus. Microbiologiya*, **57**, 805–809.

Gonzalez, A.E., Martinez, A.T., Almendros, G. and Grinbergs, J. (1989) A study of yeasts during the delignification and fungal transformation of wood into cattle feed in Chilean rain forest. *Antonie van Leeuwenhoek*, **55**, 229–236.

Gould, K.L. and Nurse, P. (1989) Tyrosine phosphorylation of the fission yeast cdc2$^+$ protein kinase regulates entry into mitosis. *Nature*, **342**, 39–45.

Gow, N.A.R. (1994) Growth and guidance of the fungal hypha. *Microbiology (UK)*, **140**, 3193–3205.

Gow, N.A.R. (1996) *Candida albicans*: morphogenesis and pathogenesis. *Japanese Journal of Medical Mycology*, **37**, 49–58.

Gow, N.A.R., Perera, T.H.S., Sherwood-Higham, J., Gooday, G.W., Gregory, D.W. and Marshall, D. (1994) Investigation of touch-sensitive responses by hyphae of the human pathogenic fungus *Candida albicans. Scanning Microscopy*, **8**, 705–710.

Gow, N.A.R., Hube, B., Bailey, D.A., Schofield, D.A., Munro, C., Swoboda, R.K., Bertram, G., Westwater, C., Broadbent, I., Smith, R.J., Gooday, G.W. and Brown, A.J. (1995) Genes associated with dimorphism and virulence of *Candida albicans. Canadian Journal of Botany*, **73**, 5335–5342.

Granot, D. and Snyder, M. (1993) Carbon source induces growth of stationary phase yeast cells, independent of carbon source metabolism. *Yeast*, **9**, 465–479.

Grant, C.M., MacIver, F.H. and Dawes, I.W. (1996) Glutathione is an essential metabolite required for resistance to oxidative stress in the yeast *Saccharomyces cerevisiae. Current Genetics*, **29**, 511–515.

Grant, C.M., Collinson, L.P., Roe, J.-H. and Dawes, I.W. (1996) Yeast glutathione reductase is required for protection against oxidative stress and is a target gene for yAP-1 transcriptional regulation. *Molecular Microbiology*, **21**, 171–179.

Gray, B.F. and Kirwan, N.A. (1974) Growth rates of yeast colonies on solid media. *Biophysical Chemistry*, **1**, 204–213.

Gregory, M.E., Bulmer, M., Bogle I.D.L. and Titchener-Hooker, N. (1996) Optimizing enzyme production by bakers yeast in continuous culture: physiological knowledge useful for process design and control. *Bioprocess Engineering*, **15**, 239–245.

Groboillot, A., Boadi, D.K., Poncelet, D. and Neufeld, R.J. (1994) Immobilization of cells for application in the food industry. *Critical Reviews in Biotechnology*, **14**, 75–107.

Gross, C. and Watson, K. (1996) Heat shock protein

synthesis and trehalose accumulation are not required for induced thermotolerance in derepressed *Saccharomyces cerevisiae*. *Biochemical and Biophysical Research Communications*, **220**, 766–772.

Guijarra, J.M. and Lagunas, R. (1984) *Saccharomyces cerevisiae* does not accumulate ethanol against a concentration gradient. *Journal of Bacteriology*, **160**, 874–878.

Guillermond, A. (1920) *The Yeasts* (translation by F.W. Tanner). J. Wiley, New York.

Guillot, J. Gueho, E. and Prevost, M.C. (1995) Ultrastructural features of the dimorphic yeast *Malassezia furfur*. *Journal of Mycological Medicine*, **5**, 86–91.

Hagan, I.M. and Hyams, J.S. (1996) Forces acting on the fission yeast anaphase spindle. *Cell Motility and the Cytoskeleton*, **34**, 69–73.

Hamada, K., Nakatomi, Y. and Shimada, S. (1992) Direct induction of tetraploids or homozygous diploids in the industrial yeast *Saccharomyces cerevisiae* by hydrostatic pressure. *Current Genetics*, **22**, 371–376.

Hamada, K., Nakatomi, Y., Osumi, M. and Shimada, S. (1996) Direct induction of homozygous diploidization in the fission yeast *Schizosaccharomyces pombe* by pressure stress. *FEMS Microbiology Letters*, **136**, 257–262.

Hamada, T., Ishiyama, T. and Motai, H. (1989) Continuous fermentation of soy sauce by immobilized cells of *Zygosaccharomyces rouxii* in an airlift reactor. Applied *Microbiology and Biotechnology*, **31**, 346–352.

Hansen, E.C. (1886) Recherches sur la physiologie et al morphologie des ferments alcoholiques V. Méthodes pour obtenir des cultures pures de *Saccharomyces* et de microorganismes analogues. *Comptes Rendus des Travaux du Laboratoire Carlsberg, Série Physiologie*, **2**, 92–105.

Harold, F.M. (1995) From morphogenes to morphogenesis. *Microbiology* (*UK*), **141**, 2765–2778.

Hartwell, L.H. and Unger, M.W. (1977) Unequal division *in Saccharomyces cerevisiae* and its implications for the control of cell division. *Journal of Cell Biology*, **75**, 422–435.

Hartwell, L.H., Culotti, J., Pringle, J.R. and Reid, B.J. (1974) Genetic control of the cell division cycle in yeast. *Science*, **183**, 46–51.

Hasal, P., Cejkova, A. and Vojtisek, V. (1992) Continuous sucrose hydrolysis by an immobilized whole cell biocatalyst. *Enzyme and Microbial Technology*, **14**, 1007–1012

Hasek, J. and Streiblová, E. (1996) Fluorescence Microscopy Methods. In *Methods in Molecular Biology, Vol. 53: Yeast Protocols* (ed. I. Evans), pp. 391–405. Humana Press Inc., Totowa, N.J., USA.

Hashida, M., Sakai, K. and Kogame, M. (1995) Analytical methods of yeast morphological character and evaluation of yeast activity by image processing. European Brewery Convention. Proceedings of the 25th Congress, Brussels, 1995, pp.353–360. IRL Press at Oxford University Press, Oxford.

Hashizume, C., Kimura, K. and Hayashi, R (1995) Kinetic analysis of yeast inactivation by high pressure treatment at low temperatures. *Bioscience Biotechnology and Biochemistry*, **59**, 1455–1458.

Hawser, S.P. and Douglas, L.J. (1994) Biofilm formation by *Candida* species on the surface of catheter materials *in vitro*. *Infection and Immunity*, **52**, 915–921.

Hawser, S.P. and Douglas, L.J. (1995) Resistance of *Candida albicans* biofilms to antifungal agents *in vitro*. *Antimicrobial Agents and Chemotherapy*, **39**, 2128–2131.

Hayflick, L. (1965) The limited *in vitro* lifetime of human diploid cell strains. *Experimental Cell Research*, **37**, 614–636.

Hazen, K.C. (1995) New and emerging yeast pathogens. *Clinical Microbiology*, **8**, 462–478.

Herskowitz, I., Park, H.-O, Sanders, S., Valtz, N. and Peter, M. (1995) Programming of cell polarity in budding yeast by endogenous and exogenous signals. *Cold Spring Harbor Symposia on Quantitative Biology*, **60**, 717–727.

Hill, G.A. and Robinson, C.W. (1988) Morphological behaviour of *Saccharomyces cerevisiae* during continuous fermentation. *Biotechnology Letters*, **10**, 815–820.

Hino, A., Mihara, K., Nakashima, K. and Takano, H. (1990) Trehalose levels and survival ratio of freeze-tolerant versus freeze-sensitive yeasts. *Applied and Environmental Microbiology*, **56**, 1386–1391.

Hirano, T. and Yanagida, M. (1989) Controlling elements in the cell division cycle of *Schizosaccharomyces pombe*. In *Molecular and Cell Biology of Yeasts* (eds E.F. Walton and G.T. Yarranton), pp. 223–245. Blackie, Glasgow and London.

Hodgson, V.J., Button, D. and Walker, G.M. (1995) Anti-*Candida* activity of a novel killer toxin from the yeast *Williopsis mrakii*. *Microbiology* (*UK*), **141**, 2003–2012.

Holz, C.M. and Stahl, U. (1995) Ribosomally synthesized antimicrobial peptides in prokaryotic and eukaryotic organisms. *Food Biotechnology*, **9**, 85–117.

Hough, J.S. (1985) *The Biotechnology of Malting and Brewing*. Cambridge University Press, Cambridge.

Hough, J.S., Briggs, D.E., Stevens, R. and Young, T.W. (1982) *Malting and Brewing Science. Vol. II. Hopped Wort and Beer*. Chapman & Hall, London.

Hsiao, H.-Y., Chiang, L.-C., Yang, C.-M., Chen, L.-F. and Tsao, G.T. (1983) Preparation and performance of immobilized yeast cells in columns containing no inert carrier. *Biotechnology and Bioengineering*, **25**, 363–375.

Huberman, J.A. (1996) Cell cycle control of S phase: a comparison of 2 yeasts. *Chromosoma*, **105**, 197–203.

Huisman, G.W. and Kolter, R. (1994) Sensing starvation – a homoserine lactone-dependent signalling pathway. *Science*, **265**, 537–539.

Hunter-Cevera, J.C. and Belt, A. (1996) *Maintaining Cultures for Biotechnology and Industry*. Academic Press, San Diego.

Hutchison, L.J. and Baron, G.L. (1996) Parasitism of yeasts by lignicolous basidiomycota and other fungi. *Canadian Journal of Botany*, **74**, 735–742.

Hutchison, L.J. and Miratsuka, Y. (1994) Some wood-inhabiting yeasts of trembling aspen (*Populus tremuloides*) from Alberta and north eastern British Colombia. *Mycologia*, **86**, 386–391.

Hysert, D.W. and Morrison, N.M. (1977) Studies on ATP, ADP and AMP concentrations in yeast and beer. *Journal of the American Society of Brewing Chemists*, **35**, 160–167.

Igarashi, M., Nagata, A., Jinno, S., Suto, K. and Okayama, H. (1991) *Wee1*$^{+}$ -like gene in human cells. *Nature*, **353**, 80–83.

Imai, T. and Ohno, T. (1995) The relationship between viability and intracellular pH in the yeast *Saccharomyces cerevisiae*. *Applied and Environmental Microbiology*, **61**, 3604–3608.

Imai, T., Nakajima, I. and Ohno, T. (1994) Development of a new method for evaluation of yeast vitality by measuring intracellular pH. *Journal of the American Society of Brewing Chemists*, **52**, 5–8.

Ingram, L.A. and Bukkte, T.M. (1984) Effects of alcohols on microorganisms. *Advances in Microbial Physiology*, **25**, 253–303.

Iwahashi, H., Kaul, S.C., Obuchi, K. and Komatsu, Y. (1991) Induction of barotolerance by heat shock treatment in yeast. *FEMS Microbiology Letters*, **80**, 325–328.

Iwashashi, H., Fujii, S., Obuchi, K., Kaul, S.C., Sato, A. and Komatsu, Y. (1993) Hydrostatic pressure is like high temperature and oxidative stress in the damage it causes to yeast. *FEMS Microbiology Letters*, **108**, 53–58.

Iwahashi, H., Obuchi, K., Fujii, S. and Komatsu, Y. (1995) The correlative evidence suggesting that trehalose stabilizes membrane structures in the yeast *Saccharomyces cerevisiae*. *Cellular and Molecular Biology*, **41**, 763–769.

Izawa, S., Inoue, Y. and Kimura, A. (1995) Oxidative stress-response in yeast – effect of glutathione on adaptation to hydrogen peroxide stress in *Saccharomyces cerevisiae*. *FEBS Letters*, **368**, 73–76.

Jakobsen, M. and Narvhus, J. (1996) Yeasts and their possible beneficial and negative effects on the quality of dairy products. *International Dairy Journal*, **6**, 755–768.

Jackson, C.L. and Hartwell, L.H. (1990) Courtship in *S. cerevisiae*: both cell types choose mating partners by responding to the strongest pheromone signal. *Cell*, **63**, 1039–1051.

Jamieson, D.J. (1992) *Saccharomyces cerevisiae* has distinct adaptive responses to both hydrogen peroxide and menadione. *Journal of Bacteriology*, **174**, 6678–6681.

Jamieson, D.J. (1995) The effects of oxidative stress on *Saccharomyces cerevisiae*. *Redox Report*, **1**, 89–95.

Jazwinski, S.M. (1990) Ageing and senescence of the budding yeast *Saccharomyces cerevisiae*. *Molecular Microbiology*, **4**, 377–344.

Jazwinski, S.M., Egilmez, N.K. and Chen, J.B. (1989) Replication control and cellular life span. *Experimental Gerontology*, **24**, 423–436.

Jennings, D.H. (1995) *The Physiology of Fungal Nutrition*. Cambridge University Press, Cambridge.

Jirku, V. and Turkova, J. (1987) Cell immobilization by covalent linkage. *Methods in Enzymology*, **135**, 341–357.

Johnson, A.D. (1995) Molecular mechanisms of cell-type determination in budding yeast. *Current Opinion in Genetics and Development*, **5**, 552–558.

Johnson, B.F. and McDonald, I.J. (1983) Cell division: a separable cellular sub-cycle in the fission yeast *Schizosaccharomyces pombe*. *Journal of General Microbiology*, **129**, 3411–3419.

Johnson, B.F., Calleja, G.B., Yoo, B.Y., Zuker, M. and McDonald, I.J. (1982) Cell division: key to cellular morphogenesis in the fission yeast *Schizosaccharomyces pombe*. *International Review of Cytology*, **75**, 167–208

Johnson, B.F., Miyata, M. and Miyata, H. (1989) Morphogenesis of fission yeasts. In *Molecular Biology of the Fission Yeast* (eds A. Nassim, P. Young and B.F. Johnson), pp. 331–366. Academic Press, San Diego.

Johnson, B.F., Yoo, B.Y. and Calleja, G.B. (1995) Smashed fission yeast cell walls. Structural discontinuities related to wall growth. *Cell Biophysics*, **26**, 57–75.

Johnson, E.A. and Schroeder, W.A. (1996) Biotechnology of astaxanthin production in *Phaffia rhodozyma*. *American Chemical Society Symposia Series*, **637**, 39–50.

Johnston, J.R. (1966) Reproductive capacity and mode of death of yeast cells. *Antonie van Leeuwenhoek*, **32**, 94–98.

Jones, J.C. and Gray, B.F. (1978) Surface colony growth in a controlled nutrient environment 2. The effect of ethanol. *Microbios*, **22**, 195–210.

Jones, R.P. (1987) Factors affecting deactivation of

yeast cells exposed to ethanol. *Journal of Applied Bacteriology*, **63**, 153–164.

Jones, R.P. (1988) Intracellular ethanol-accumulation and exit from yeast and other cells. *FEMS Microbiology Reviews*, **54**, 239–258.

Jones, R.P. (1990) Roles for replicative inactivation in yeast ethanol fermentations. *Critical Reviews in Biotechnology*, **10**, 205–222.

Jones, R.P. and Greenfield, P.F. (1982) Effect of carbon dioxide on yeast growth and fermentation. *Enzyme and Microbial Technology*, **4**, 210–223.

Joseph-Horne, T. and Holloman, D.W. (1997) Molecular mechanisms of azole resistance in fungi. *FEMS Microbiology Letters*, **149**, 141–149.

Jozwichk, Z. and Leyko, W. (1992) Role of membrane components in thermal injury of cells and development of thermotolerance. *International Journal of Radiation Biology*, **62**, 743–756.

Kalathenos, P., Sutherland, J.P. and Roberts, T.A. (1995) Resistance of some wine spoilage yeasts to combinations of ethanol and acids in wine. *Journal of Applied Bacteriology*, **78**, 245–250.

Kale, S.P. and Jazwinski, S.M. (1996) Differential response to UV stress and DNA damage during the yeast replicative life span. *Developmental Genetics*, **18**, 154–160.

Kamal, A., Rao, M.V. and Rao, A.B. (1990) Enzymatic cyclizations mediated by ultrasonically stimulated baker's yeast: synthesis of imidazo-fused heterocycles. *Journal of the Chemical Society Perkins Transactions*, **1**, 2755 2757.

Kamath, R.S. and Bungay, H.R. (1988) Growth of yeast colonies on solid media. *Journal of General Microbiology*, **134**, 3061–3069.

Käppeli, O., Arreguin, M. and Rieger, M. (1985) The respirative breakdown of glucose by *Saccharomyces cerevisiae*: an assessment of a physiological state. *Journal of General Microbiology*, **131**, 1411–1416.

Kara, B.V., Dauod, I. and Searle, B. (1987) Assessment of yeast quality. European Brewery Convention. Proceedings of the 21st Congress, Madrid, 1987, pp. 409–416. IRL Press at Oxford University Press, Oxford, UK.

Kaul, S.C., Obuchi, K. and Komatsu, Y. (1992) Cold shock response of yeast cells: induction of a 33 kDa protein and protection against freezing injury. *Cellular and Molecular Biology*, **38**, 553–559.

Kennedy, B.K. and Guarente, L. (1996) Genetic analysis of aging in *Saccharomyces cerevisiae*. *Trends in Genetics*, **12**, 355–359.

Kennedy, B.K., Austriaco, N.R. and Guarente, L. (1994) Daughter cells of *Saccharomyces cerevisiae* from old mothers display reduced life span. *Journal of Cell Biology*, **127**, 1985–1993.

Kennedy, B.K. Austriaco, N.R., Zhang, J. and Guarente, L. (1995) Mutation in the silencing gene *SIR4* can delay aging in *S. cerevisiae*. *Cell*, **80**, 485–496.

Kim, J., Alizadeh, P., Harding, T., Hefner-Gravink, A. and Klionsky, D.J. (1996) Disruption of the yeast *ATH1* gene confers better survival after dehydration, freezing and ethanol shock – potential commercial applications. *Applied and Environmental Microbiology*, **62**, 1563–1569.

Kimura, T., Kitamoto, N., Matsuoka, K., Nakamura, K., Imura, Y. and Kito, Y. (1993) Isolation and nucleotide sequences of the genes encoding killer toxins from *Hansenula mrakii* and *H. saturnus*. *Gene*, **137**, 265–270.

Kirsop, B. and Henry, J. (1984) Development of a miniaturized cryopreservation method for the maintenance of a wide range of yeasts. *Cryo Letters*, **5**, 191–200.

Kobori, H. Sato, M., Tameike, A., Hamada, K. Shimada, S. and Osumi, M. (1995) Ultrastructural effects of pressure stress to the nucleus in *Saccharomyces cerevisiae*: a study by immunoelectron microscopy using frozen thin sections. *FEMS Microbiology Letters*, **132**, 253–258.

Kocková-Kratochvilová, A. (1990) *Yeasts and Yeast-Like Organisms*. VCH Publishers, New York.

Koltin, Y., Ginzberg, I. and Finkler, A. (1993) Fungal 'killer' toxins as potential agents for biocontrol. In *Biotechnology in Plant Disease Control* (ed. I. Chet), pp. 257–274. Wiley-Liss Inc., New York.

Komatsu, Y., Kaul, S.C., Iwashashi, H., and Obuchi, K. (1990) Do heat shock proteins provide protection against freezing? *FEMS Microbiology Letters*, **72**, 159–162.

Komatsu, Y., Kodama, O. and Fujita, K. (1996) Heat shock treatment reduces *in situ* temperature in yeast at sublethal high temperature. *Cellular and Molecular Biology*, **42**, 839–845.

Kramhøft, B., Hamburger, K., Nissen, S.B. and Zeuthen, E. (1978) The cell cycle and glycolytic activity of *Schizosaccharomyces pombe* synchronized in defined medium. *Carlsberg Research Communications*, **43**, 227–239.

Kron, S.J. and Gow, N.A.R. (1995) Budding yeast morphogenesis: signalling, cytoskeleton and cell cycle. *Current Opinion in Cell Biology*, **7**, 845–855.

Kron, S.J., Styles, C.A. and Fink, G.R. (1994) Symmetric cell division in pseudohyphae of the yeast *Saccharomyces cerevisiae*. *Molecular Biology of the Cell*, **5**, 1003–1022.

Kruuv, J., Lepock, J.R. and Keith, A.D. (1978) The effect of fluidity of membrane lipids on freeze–thaw survival of yeast. *Cryobiology*, **15**, 73–79.

Kunduru, M.R. and Pometto, A.L. (1996) Continuous ethanol production by *Zymomonas mobilis* and *Saccharomyces cerevisiae* in biofilm reactors. *Journal of Industrial Microbiology*, **16**, 249–256.

Küntzel, H., Schultz, A. and Ehbrecht, I-M. (1996) Cell cycle control and initiation of DNA replication in *Saccharomyces cerevisiae. Biological Chemistry*, **377**, 481–487.

Kuriyama, H. and Slaughter, J.C. (1995) Control of cell morphology of the yeast *Saccharomyces cerevisiae* by nutrient limitation in continuous culture. *Letters in Applied Microbiology*, **20**, 37–40.

Kurtzman, C.P. (1987) Molecular taxonomy of industrial yeasts. In *Biological Research on Industrial Yeasts. Vol. 1* (eds G.G. Stewart, I. Russell, R.D. Klein and R.R. Hiebsch), pp. 27–45. CRC Press, Boca Raton, Florida.

Kyung, K.H. and Gerhardt, P. (1984) Continuous production of ethanol by yeast 'immobilized' in a membrane-contained fermentor. *Biotechnology and Bioengineering*, **26**, 252–256.

Labib, K. and Moreno, S. (1996) *rum 1*: a CDK inhibitor regulating G1 progression in fission yeast. *Trends in Cell Biology*, **6**, 62–66.

Lachance, M.-A. (1990) Yeast selection in nature. In *Yeast Strain Selection* (ed. C. Panchal), pp. 21–41. Marcel Dekker Inc., New York.

Lachance, M.-A. and Pang, W.-M. (1997) Predacious yeasts. *Yeast*, **13**, 225–232.

Lachance, M.-A., Phaff, H.J. and Starmer, W.T. (1993) *Kluyveromyces bacillisporus* sp. *nov*., a yeast from Enony oak exudate. *International Journal of Systematic Bacteriology*, **43**, 115–119.

Lachance, M.-A., Gilbert, D.G. and Starmer, W.T. (1995) Yeast communities associated with *Drosophila* species and related flies in an Eastern oak-pine forest – a comparison with Western communities. *Journal of Industrial Microbiology*, **14**, 484–494.

Lagunas, R. (1981) Is *Saccharomyces cerevisiae* a typical facultative anaerobe? *Trends in Biochemical Sciences*, **8**, 201–202.

Lagunas, R. (1986) Misconceptions about the energy metabolism of *Saccharomyces cerevisiae. Yeast*, **2**, 221–228.

Laszlo, A. (1988) Evidence for two states of thermotolerance in mammalian cells. *International Journal of Hyperthermia*, **4**, 513–526.

Leão, C. and van Uden, N. (1982) Effect of ethanol and other alcohols on the kinetics and the activation parameters of thermal deaths in *Saccharomyces cerevisiae. Biotechnology and Bioengineering*, **24**, 1581–1590.

Leberer, E., Thomas, D.Y. and Whiteway, M. (1997) Pheromone signalling and polarized morphogenesis in yeast. *Current Opinion in Genetics and Development*, **7**, 59–66.

Lee, J., Dawes, I.W. and Roe, J.-H. (1995) Adaptive response of *Schizosaccharomyces pombe* to hydrogen peroxide and menadione. *Microbiology* (*UK*), **141**, 3127–3132.

Lee, M.G. and Nurse, P. (1987) Complementation used to clone a human homolog of the fission yeast cell cycle control gene *cdc2*$^+$. *Nature*, **327**, 31–35.

Leufven, A. and Nehls, L. (1986) Quantification of different yeasts associated with the bark beetle, *Ips typographus*, during its attack on a spruce tree. *Microbial Ecology*, **12**, 237–243.

Lew, D.J. and Reed, S.I. (1995) Cell cycle control of morphogenesis in budding yeast. *Current Opinion in Genetics and Development*, **5**, 17–23.

Lewis, J.G., Learmonth, R.P., Attfield, P.V. and Watson, K. (1997) Stress co-tolerance and trehalose content in baking strains of *Saccharomyces cerevisiae. Journal of Industrial Microbiology and Biotechnology*, **18**, 30–36.

Lewis, M.J. and Young, T.W. (1995) *Brewing*. Chapman & Hall, London

Lievens, K., Devogel, D., Iserentant, D. and Verachtert, H. (1994) Evidence for a factor produced by *Saccharomyces cerevisiae* which causes flocculation of *Pediococcus damnosus* 12A7 cells. *Colloids and Surfaces B: Biointerfaces*, **2**, 189–198.

Lillie, S.H. and Pringle, J.R. (1980) Reserve carbohydrate metabolism in *Saccharomyces cerevisiae*: responses to nutrient limitation. *Journal of Bacteriology*, **143**, 1384–1394.

Liu, H., Styles, C.A. and Fink, G.R. (1993) Elements of the yeast pheromone response pathway required for filamentous growth of diploids. *Science*, **262**, 1741–1744.

Lindquist, S. and Kim, G. (1996) Heat shock protein 104 expression is sufficient for thermotolerance in yeast. *Proceedings of the National Academy of Sciences* (*USA*), **93**, 5301–5306.

Longo, V.D., Gralla, E.B. and Valentine, J.S. (1996) Superoxide dismutase activity is essential for stationary phase survival in *Saccharomyces cerevisiae. Journal of Biological Chemistry*, **271**, 2275–2280.

Lumsden, W.B., Duffus, J.H. and Slaughter, J.C. (1987) Effects of CO_2 on budding and fission yeasts. *Journal of General Microbiology*, **133**, 877–881.

Luyten, K., Albertyn, J., Skibbe, W.F., Prior, B.A., Ramos, J., Thevelein, J.M. and Hohmann, S. (1995) *Fps1*, a yeast member of the *MIP* family of channel proteins is a facilitator of glycerol uptake and efflux and is inactive under osmotic stress. *EMBO Journal*, **14**, 1360–1371.

Lydall, D. and Weinert, T. (1996) From DNA damage to cell cycle arrest and suicide – a budding yeast perspective. *Current Opinion in Genetics*, **6**, 4–11.

MacKenzie, K.F., Singh, K.K. and Brown, A.D. (1988) Water stress plating hypersensitivity of yeasts: protective role of trehalose in *Saccharomyces cerevisiae*. *Journal of General Microbiology*, **134**, 1661–1666.

Maeda, T., Takekawa, M. and Saito, H. (1995) Activation of yeast PBS2 MAPKK by MAPKKKs or by binding of an SH3-containing osmosensor. *Science*, **269**, 554–558.

Mager, W.H. and De Kruijff, A.J.J. (1995) Stress-induced transcriptional activation. *Microbiological Reviews*, **59**, 506–531.

Mager, W.H. and Moradas-Ferreira, P. (1993) Stress response of yeast. *Biochemical Journal*, **290**, 1–13.

Mager, W.H. and Varela, J.C.S. (1993) Osmostress response of the yeast *Saccharomyces cerevisiae*. *Molecular Microbiology*, **10**, 253–258.

Magliani, W., Conti, S., DeBernardis, F., Gerloni, M., Bertolotti, D., Mozzoni, P., Cassone, A. and Polonelli, L. (1997) Therapeutic potential of anti-idiotypic single chain antibodies with yeast killer toxin activity. *Nature Biotechnology*, **15**, 155–158.

Magyar, I. and Panyik, I. (1989) Biological deacidification of wine with *Schizosaccharomyces pombe* entrapped in Ca-alginate gel. *American Journal of Enology and Viticulture*, **40**, 233–240.

Mahayni, R., Vazquez, J.A. and Zervos, M.J. (1995) Nosocomial candidiasis: epidemiology and drug resistance. *Infectious Agents and Disease*, **4**, 248–253.

Main, J., McKenzie, H., Yeaman, G.R., Kerr, M.A., Robson, D., Pennington, C.R. and Parratt, D. (1988). Antibody to *Saccharomyces cerevisiae* (baker's yeast) in Crohn's Disease. *British Medical Journal*, **297**, 1105–1106.

Majara, M., O'Connor-Cox, E.S.C. and Axcell, B.C. (1996) Trehalose – an osmoprotectant and stress indicator compound in high and very high gravity brewing. *Journal of the American Society of Brewing Chemists*, **54**, 149–154.

Manson, D.H. and Slaughter, J.C. (1986) Methods for predicting yeast fermentation activity. In *Proceedings of the Second Aviemore Conference on Malting, Brewing and Distilling* (eds I. Campbell and F.G. Priest), pp. 295–297. Institute of Brewing, London.

Mansure, J.J.C., Panek, A.D., Crowe, L.M. and Crowe, J.H. (1994) Trehalose inhibits ethanol effects on intact yeast cells and liposomes. *Biochimica et Biophysica Acta*, **1191**, 309–316.

Marechal, P.A., Martinez de Marañon, I., Molin, P. and Gervais, P. (1995) Yeast cell responses to water potential variations. *International Journal of Food Microbiology*, **28**, 277–287.

Maresca, B., and Kobayashi, G.S. (1989) Dimorphism in *Histoplasma capsulata*: a model for the study of a cell differentiation in pathogenic fungi. *Microbiological Reviews*, **53**, 186–209.

Martegani, E., Porro, D., Ranzi, M.R. and Alberghina, L. (1990) Involvement of a cell size control mechanism in the induction and maintenance of oscillations in continuous cultures of budding yeast. *Biotechnology and Bioengineering*, **36**, 453–459.

Martinac, B., Zhu, H., Kukalski, A., Zhou, X., Culbertson, M. Bussey, H. and Kung, C. (1990) Yeast K1 killer toxin forms ion channels in sensitive yeast spheroplasts and in artificial liposomes. *Proceedings of the National Academy of Sciences* (*USA*), **87**, 6228–6232.

Martinez de Marañon, I., Marechal, P.A. and Gervais, P. (1996) Passive response of *Saccharomyces cerevisiae* to osmotic shifts: cell volume variations depending on the physiological state. *Biochemical and Biophysical Research Communications*, **227**, 519–523.

Masschelein, C.A., Ryder, D.S. and Simon, J.-P. (1994) Immobilized cell technology in beer production. *Critical Reviews in Biotechnology*, **14**, 155–177.

McCaig, R. (1990) Evaluation of the fluorescent dye 1-anilino-8-naphthalene sulfonic acid for yeast viability determination. *Journal of the American Society of Brewing Chemists*, **48**, 22–25.

McCracken, D.A., Martin, V.J., Stark, M.J.R. and Bolen, P.L. (1994) The linear plasmid-encoded toxin produced by the yeast *Pichia acaciae*: characterization and comparison with the toxin of *Kluyveromyces lactis*. *Microbiology* (*UK*), **140**, 425–531.

McCullough, M.J., Ross, B.C. and Reade, P.C. (1996) *Candida albicans* – review of its history, taxomomy, epidemiology, virulence attributes and methods of strain differentiation. *International Journal of Oral and Maxillofacial Surgery*, **25**, 136–144.

McCusker, J.H., Clemons, K.V., Stevens, D.A. and Davis, R.W. (1994) Genetic characterization of pathogenic *Saccharomyces cerevisiae* isolates. *Genetics*, **136**, 1261–1269.

McFarland, L.V. and Elmer, G.W. (1995) Biotherapeutic agents: past, present and future. *Microecology and Therapy*, **23**, 46–73.

McGuire, R.G. (1994) Application of *Candida guilliermondii* in commercial citrus coatings for biological control of *Penicillium digitatum* on grapefruits. *Biological Control*, **4**, 1–7.

McLaughlin, R.J., Wilson, C.L., Droby, S., Ben-Arie, R. and Chalutz, E. (1992) Biological control of postharvest diseases of grape, peach, and apple with yeasts *Kloeckera apiculata* and *Candida guilliermondii*. *Plant Disease*, **76**, 470–473.

Meikle, A.J., Reed, R.J. and Gadd, G.M. (1988) Osmotic adjustment and the accumulation of organic solutes in whole cells and protoplasts of *Saccharomyces cerevisiae*. *Journal of General Microbiology*, **134**, 3049–3060.

Meinhardt, F., Kempen, F., Kaemper, J. and Esser, K. (1990) Linear plasmids among eukaryotes: fundamentals and application. *Current Genetics*, **17**, 89–95.

Mercier, J. and Wilson, C.L. (1994) Colonization of apple wounds by naturally occurring microflora and introduced *Candida oleophila* and their effect on infection by *Botrytis cinerea* during storage. *Biological Control*, **4**, 138–144.

Meric, L., Lambert-Guilois, S., Neyreneuf, O. and Richard-Molard, D. (1995) Cryoresistance of baker's yeast *Saccharomyces cerevisiae* in frozen dough: contribution of cellular trehalose. *Cereal Chemistry*, **72**, 23–32.

Meunier, F. (1989) Candidiasis. *European Journal of Clinical Microbiology and Infectious Diseases*, **8**, 438–447.

Meyer, C. and Beyeler, W. (1984) Control strategies for continuous bioprocesses based on biological activities. *Biotechnology and Bioengineering*, **26**, 916–925.

Meyer, H.P., Käppeli, O. and Fiechter, A (1985) Growth control in microbial cultures. *Annual Review of Microbiology*, **39**, 299–319.

Middelbeek, E.J., Peters, J.W.G.H., Stumm, C. and Vogels, G.D. (1980) Properties of a *Cryptococcus laurentii* killer toxin and conditional killing effect of the toxin on *Cryptococcus albidus*. *FEMS Microbiology Letters*, **9**, 81–84.

Middelberg, A.P.J. (1995) Process-scale disruption of microorganisms. *Biotechnology Advances*, **13**, 491–551.

Mishra, P. (1993) Tolerance of fungi to ethanol In *Stress Tolerance of Fungi* (ed. D.H. Jennings), pp. 189–208. Marcel Dekker, New York.

Mitchison, J.M. (1957) The growth of single cells. I. *Schizosaccharomyces pombe*. *Experimental Cell Research*, **13**, 244–262.

Mitchison, J.M. (1970) Physiological and cytological methods for *Schizosaccharomyces pombe*. *Methods in Cell Physiology*, **4**, 131–165.

Mitchison, J.M. (1971) *The Biology of the Cell Cycle*. Cambridge University Press, London.

Mitchison, J.M. (1984) Dissociation of cell cycle events. In *Cell Cycle Clocks* (ed. L.N. Edmunds), pp. 163–172. Marcel Dekker, New York and Basel.

Mitchison, J.M. (1988) Synchronous cultures and age fractionation. In *Yeast: A Practical Approach* (eds I. Campbell and J.H. Duffus), pp. 51–63. IRL /Press, Oxford and Washington.

Mizoguchi, H. and Hara, S. (1997) Ethanol-induced alterations in lipid composition of *Saccharomyces cerevisiae* in the presence of exogenous fatty acid. *Journal of Fermentation and Bioengineering*, **83**, 12–16.

Mochaba, F.M., O'Connor-Cox, E.S.C. and Axcell, B.C. (1997) A novel and practical yeast viability method based on magnesium ion release. *Journal of the Institute of Brewing*, **103**, 99–102.

Moradras-Ferreira, P., Costa, V., Piper, P. and Mager, W. (1996) The molecular defenses against reactive oxygen species in yeast. *Molecular Microbiology*, **19**, 651–658.

Molzahn, S.W. (1992) Fermentation control – the key to quality and efficiency. Proceedings of the 22nd Convention of the Institute of Brewing, Australia and New Zealand Section (Melbourne, 1992), pp. 84–88. Institute of Brewing, Adelaide, Australia.

Monk, B.C., Mason, A.B., Lardos, T.B. and Perlin, D.S. (1995) Targeting the fungal plasma membrane proton pump. *Acta Biochimica Polonica*, **42**, 481–496.

Morace, G., Manzara, S., Getteri, G., Fanti, F., Conti, S., Campana, L., Polonelli, L. and Chezzi, C. (1989) Biotyping of bacterial isolates using the killer system. *European Journal of Epidemiology*, **5**, 303–310.

Morris, E.O. and Hough, J.S. (1956) Some aspects of the giant colony characteristics of *Saccharomyces cerevisiae*. *Journal of the Institute of Brewing*, **62**, 466–469.

Morris, G.J., Winters, L., Coulson, G.E. and Clarke, K.J. (1986) Effect of osmotic stress on the ultrastructure and viability of the yeast *Saccharomyces cerevisiae*. *Journal of General Microbiology*, **132**, 2023–2034.

Mortimer, R.K. and Johnston, J.R. (1959) Life-span of individual yeast cells. *Nature*, **183**, 1751–1752.

Motizuki, M. and Tsurugi, K. (1992) The effect of ageing on protein synthesis in the yeast *Saccharomyces cerevisiae*. *Mechanisms of Ageing and Development*, **64**, 235–245.

Motizuki, M., Yokota, S. and Tsurugi, K. (1995) Autophagic death after cell cycle arrest at the restrictive temperature in temperature-sensitive cell division cycle and secretory mutants of the yeast *Saccharomyces cerevisiae*. *European Journal of Cell Biology*, **68**, 275–287.

Mulholland, J., Preuss, D., Moon, A., Wong, A., Drubin, D. and Botstein, D. (1994) Ultrastructure of the yeast actin cytoskeleton and its association with the plasma membrane. *Journal of Cell Biology*, **125**, 381–391.

Müller, I. (1971) Experiments on ageing in single cells of *Saccharomyces cerevisiae*. *Archiv für Mikrobiologie*, **77**, 20–25.

Münch, T., Sonnleitner, B. and Fiechter, A. (1992a) New insights into the synchronization mechanisms with forced synchronous cultures of *Saccharomyces cerevisiae*. *Journal of Biotechnology*, **24**, 299–314.

Münch, T., Sonnleitner, B. and Fiechter, A. (1992b) The decisive role of the *Saccharomyces cerevisiae* cell cycle behaviour for dynamic growth characterization. *Journal of Biotechnology*, **22**, 329–352.

Murray, A.W. (1995) The genetics of cell cycle checkpoints. *Current Opinion in Genetics and Development*, **5**, 5–11.

Murray, A. and Hunt, T. (1993) *The Cell Cycle. An Introduction*. Oxford University Press, New York and Oxford.

Mutoh, N., Nakagawa, C.W. and Hayashi, Y. (1995) Adaptive response of *Schizosaccharomyces pombe* to hydrogen peroxide. *FEMS Microbiology Letters*, **132**, 67–72.

Nagashima, M., Azuma, M., Nogushi, S., Inuzuka, K. and Samejima, H. (1987) Large scale preparation of calcium alginate-immobilized yeast cells and its application to industrial ethanol production. *Methods in Enzymology*, **135**, 394–405.

Nasmyth, K. (1993) Control of the yeast cell cycle by the Cdc28 protein kinase. *Current Opinion in Cell Biology*, **5**, 166–179.

Nasmyth, K. (1996) At the heart of the budding yeast cell cycle. *Trends in Genetics*, **12**, 405–412.

Naumov, G.I., Naumov, E.S. and Sancho, E.D. (1996) Genetic re-identification of *Saccharomyces* strains associated with black knot disease of trees in Ontario and *Drosophila* species in California. *Canadian Journal of Microbiology*, **42**, 335–339.

Neufeld, R.J., Peleg, Y., Roken, J.S., Pines, O. and Goldberg, I. (1991) L-Malic acid formation by immobilized *Saccharomyces cerevisiae* amplified for fumarase. *Enzyme and Microbial Technology*, **13**, 991–996.

Neves, M.-J. and Francois, J. (1992) On the mechanism by which a heat shock induces trehalose accumulation in *Saccharomyces cerevisiae*. *Biochemical Journal*, **288**, 859–2864.

Nguyen, M.H., Morris, A.J., Dobson, M.E., Snydman, D.R., Peacock, J.E., Rinaldi, M.C. and Yu, V.L. (1996) *Candida lusitaniae* – an important emerging cause of candidemia. *Infectious Disease in Clinical Practice*, **5**, 273–278.

Niedermeyer, W., Parish, G.R. and Moor, H. (1977) Reactions of yeast cells to glycerol treatment. Alterations to membrane structure and glycerol uptake. *Protoplasma*, **92**, 177–193.

Nishimoto, F and Yamashoji, S. (1994) Rapid assay of cell activity of yeast cells. *Journal of Fermentation and Bioengineering*, **77**, 107–108.

Nombela, C. and Santamaria, C. (1984) Genetics of yeast cell wall autolysis. In *Microbial Cell Wall Synthesis and Autolysis* (ed. C. Nombela), pp. 249–259. Elsevier Science Publishers, Amsterdam.

Norton, S. and D'Amore, T. (1994) Physiological effects of yeast cell immobilization: applications for brewing. *Enzyme and Microbial Technology*, **16**, 365–375.

Norton, S. and Vuillemand, J.-C. (1994) Food bioconversions and metabolite production using immobilized cell technology. *Critical Reviews in Biotechnology*, **14**, 193–224.

Nurse, P. (1975) Genetic control of cell size at division in yeast. *Nature*, **256**, 547–551.

Nurse, P. (1990) Universal control mechanism regulating onset of M-phase. *Nature*, **344**, 503–508.

O'Connor-Cox, E.S.C. and Ingledew, W.M. (1989) Wort nitrogenous sources – their use by brewing yeasts: A review. *Journal of the American Society of Brewing Chemists*, **47**, 102–108.

Odds, F.C. (1985) Morphogenesis in *Candida albicans*. *CRC Critical Reviews in Microbiology*, **12**, 45–93.

Odds, F.C. (1988) Candida *and Candidosis*. 2nd edn. Baillière Tindall, London.

Odds, F.C. (1994) *Candida* species and virulence. *American Society of Microbiology News*, **60**, 313–318.

Odds, F.C. (1996a) Epidemiological shifts in opportunistic and nosocomial *Candida* infections: mycological aspects. *International Journal of Antimicrobial Agents*, **6**, 141–144.

Odds, F.C. (1996b) Resistance of clinically important yeasts to antifungal agents. *International Journal of Antimicrobial Agents*, **6**, 145–147.

O'Donohue, M.J., Boissy, G., Huck, J.C. Nespoulous, C., Brunie, S. and Pernollet, J.C. (1996) Overexpression in *Pichia pastoris* and crystallization of an elicitor protein secreted by the phytopathogenic fungus, *Phytophthora cryptogea*. *Protein Expression and Purification*, **8**, 254–261.

Odumeru, J.A., D'Amore, T., Russell, I. and Stewart, G.G. (1992) Changes in protein composition of *Saccharomyces* brewing strains in response to heat shock and ethanol stress. *Journal of Industrial Microbiology*, **9**, 229–234.

O'Kennedy, R., Houghton, C.J. and Patching, J.W. (1995) Effect of growth environment on recombinant plasmid stability in *Saccharomyces cerevisiae* grown in continuous culture. *Applied Microbiology and Biotechnology*, **44**, 126–132.

Oliver, S.G. (1987) Physiology and genetics of ethanol tolerance in yeast. In *Biological Research on Industrial Yeasts. Vol. I* (eds G.G. Stewart, I. Russell, R.D. Klein and R.R. Hiebsch), pp. 81–98. CRC Press, Boca Raton.

Oliver, S.G. (1996) From DNA sequence to biological function. *Nature*, **379**, 597–600.

Oliver, S.G. and Warmington, J.R. (1987) Transcription. In *The Yeasts. Vol. 3* (eds A.H. Rose and J.S. Harrison), pp. 117–160. Academic Press, London.

Omori, T., Ogawa, K., Umemoto, Y., Yuki, Y., Kajihara, Y., Shimoda, M. and Wada, H. (1996) Enhancement of glycerol production by brewing yeast (*Saccharomyces cerevisiae*) with heat-shock treatment. *Journal of Fermentation and Bioengineering*, **82**, 187–190.

Opekarova, M. and Sigler, K. (1982) Acidification power: indicator of metabolic activity and autolytic changes in *Saccharomyces cerevisiae*. *Folia Microbiologica*, **27**, 395–403.

O'Shea, D.G. and Walsh, P.K. (1996) Morphological characterization of dimorphic yeast *Kluyveromyces marxianus* var. *marxianus* NRRLy 2415 by semi-automated image analysis. *Biotechnology and Bioengineering*, **51**, 679–690.

Ouchi, K., Wickner, R.B., Tohe, A. and Akiyama, H. (1979) Breeding of killer yeasts for saké brewing by cytoduction. *Journal of Fermentation Technology*, **57**, 483–487.

Paardekooper, M., Van Grompel, A.E., van Steveninck, J. and van den Broek, P.J.A. (1995) The effect of photodynamic treatment of yeast with sensitizer chloroaluminium phthalocyanine on various cellular parameters. *Photochemistry and Photobiology*, **62**, 561–567.

Palmieri, M.C., Greenhalf, W. and Laluce, C (1996) Efficient flotation of yeast cells grown in batch culture. *Biotechnology and Bioengineering*, **50**, 248–256.

Palkova, Z. and Vondrejs, V. (1996) Killer plaque technique for selecting hybrids and cybrids obtained by induced protoplast fusion. In *Methods in Molecular Biology. Vol. 53: Yeast Protocols* (ed. I.H. Evans), pp. 339–343. Humana Press, Totowa, N.J., USA.

Parascandola, P. and de Alteriis, E. (1996) Pattern of growth and respiratory activity of *Saccharomyces cerevisiae* (baker's yeast) cells growing entrapped in an insolubilized gelatin gel. *Biotechnology and Applied Biochemistry*, **23**, 7–12.

Parascandola, P., de Alteriis, E. and Scardi, V. (1990) Immobilization of viable yeast cells within polyaldehyde-hardened gelatin gel. *Biotechnology Techniques*, **4**, 237–242.

Parascandola, P., de Alteriis, E., Farris, G.A., Budroni, M. and Scardi, V. (1992) Behaviour of grape must ferment *Saccharomyces cerevisiae* immobilized within insolubilized gelatin. *Journal of Fermentation and Biotechnology*, **74**, 123–125.

Parsell, D.A. and Lindquist, S. (1994) Heat shock proteins and stress tolerance. In *The Biology of Heat Shock Proteins and Molecular Chaperones*. pp. 457–489. Cold Spring Harbor Laboratory Press, Cold Spring Harbor, New York.

Parulekar, S.J., Semones, G.B., Rolf, M.J., Lievense, J.C. and Lim, H.C. (1986) Induction and elimination of oscillations in continuous cultures of *Saccharomyces cerevisiae*. *Biotechnology and Bioengineering*, **28**, 700–710.

Perrier-Cornet, J.-M., Marechal, P.-A. and Gervais, P. (1995) A new design intended to relate high pressure treatment to yeast cell mass transfer. *Journal of Biotechnology*, **41**, 49–58.

Petrov, V.V. and Okorokov, L.A. (1990) Increase of the anion and proton permeability of *Saccharomyces carlsbergensis* plasmalemma by n-alcohols as a possible cause for de-energization. *Yeast*, **6**, 311–318.

Pfaller, M.A. and Barry, A.L. (1994) Evaluation of a novel colorimetric broth microdilution method for antifungal susceptibility testing of yeast isolates. *Journal of Clinical Microbiology*, **32**, 1992–1996.

Pfeiffer, P., Radler, F., Caspritz, G. and Hänel, H. (1988) Effect of killer toxin of yeast on eukaryotic systems. *Applied and Environmental Microbiology*, **54P**, 1068–1069.

Phaff, H.J., Miller, M.W. and Mrak, E.M. (1978) *The Life of Yeasts*. 2nd edn. Harvard University Press, Cambridge (USA) and London (UK).

Philliskirk, G. and Young, T.W. (1975) The occurrence of killer character in yeasts of various genera. *Antonie van Leeuwenhoek*, **41**, 147–151.

Pichova, A., Kohlwein, S.D. and Yamamoto, M. (1995) New arrays of cytoplasmic microtubules in the fission yeast *Schizosaccharomyces pombe*. *Protoplasma*, **188**, 252–257.

Pierce, J.S. (1970) Institute of Brewing: Analysis Committee measurement of yeast viability. *Journal of the Institute of Brewing*, **76**, 442–443.

Piper, P.W. (1993) Molecular events associated with acquisition of heat tolerance by the yeast *Saccharomyces cerevisiae*. *FEMS Microbiology Reviews*, **11**, 339–356.

Piper, P.W. (1995) The heat shock and ethanol stress responses of yeast exhibit extensive similarity and functional overlap. *FEMS Microbiology Letters*, **134**, 121–127.

Piper, P.W. and Kirk, N. (1991) Inducing heterologous gene expression in yeast as fermentations approach maximal biomass. In *Genetically Engineered Proteins and Enzymes from Yeast: Production Control* (ed. A. Wiseman), pp. 147–184. Ellis Horwood, Chichester, UK.

Plein, K. and Hortz, J. (1991) Therapeutic effects of *Saccharomyces boulardii* on mild residual symptoms in a stable phase Crohn's disease with special respect to chronic diarrhea – a pilot study. *Zeitschrift für Gastroenterologie*, **31**, 129–134.

Podgornik, A., Koloini, T. and Raspor, P. (1997) Online measurement and analysis of yeast flocculation. *Biotechnology and Bioengineering*, **53**, 179–184.

Polonelli, L. and Morace, G. (1986) Re-evaluation of the killer phenomenon. *Journal of Clinical Microbiology*, **24**, 866–869.

Polonelli, L., Lorenzini, R., de Bernadis, F. and Morace, G. (1986) Potential therapeutic effect of yeast killer toxin. *Mycopathologia*, **96**, 103–107.

Polonelli, L., Conti, S., Campani, L., Morace, G. and Fanti, F. (1989) Yeast killer toxins and dimorphism. *Journal of Clinical Microbiology*, **27**, 1423–1425.

Pons, M.N., Litzen, A. Kresbach, G.M., Ehrat, M. and Vivier, H. (1997) Study of *Saccharomyces cerevisiae* yeast cells by field-flow fractionation and image analysis. *Separation Science and Technology*, **32**, 1477–1492.

Porro, D. and Alberghina, L. (1996) Use of protein distribution to analyze budding yeast population structure and cell cycle progression population structure and cell cycle progression. In *Flow Cytometry Applications in Cell Culture* (eds M. Al-Rubeai and A.N. Emery), pp. 225–240. Marcel Dekker Inc., Basel.

Porro, D., Ranzi, B.M., Smeraldi, C., Martegani, E. and Alberghina, L. (1995) A double flow cytometric tag allows tracking of the dynamics of cell cycle progression of newborn *Saccharomyces cerevisiae* cells during balanced exponential growth. *Yeast*, **11**, 1157–1169.

Powis, G., Briehl, M. and Oblong, J. (1995) Redox signalling and the control of cell growth and death. *Pharmaceutical Therapy*, **68**, 149–173.

Priest, F.G. and Campbell, I. (1996) *Brewing Microbiology*. 2nd edn. Chapman & Hall, London.

Pringle, J.R., Bi, E., Harkins, H.A., Zahner, J.E., DeVirgillo, C., Chant, J., Corrado, K. and Fares, H. (1995) Establishment of cell polarity in yeast. *Cold Spring Harbor Symposia on Quantitative Biology*, **60**, 729–744.

Quain, D.E. (1988) Studies in yeast physiology – impact on fermentation performance and product quality. *Journal of the Institute of Brewing*, **95**, 315–323.

Radler, F., Pfeiffer, P. and Dennert, M. (1985) Killer toxins in new isolates of the yeasts *Hanseniaspora uvarum* and *Pichia kluyveri*. *FEMS Microbiology Letters*, **29**, 269–272.

Ramotar, D. and Masson, J.-Y. (1996) *Saccharomyces cerevisiae* DNA repair processes: an update. *Molecular and Cellular Biochemistry*, **158**, 65–78.

Rapoport, A.I., Khrustaleva, G.M., Chammanis, G.Y. and Beker, M.E. (1995) Yeast anhydrobiosis: permeability of the plasma membrane. *Microbiology (Mikrobiologiya)*, **64**, 229–232.

Rasmussen, L., Christensen, S.T., Schousboe, P. and Wheatley, D.N. (1996) Cell survival and multiplication. The overriding need for signals: from unicellular to multicellular systems. *FEMS Microbiology Letters*, **137**, 123–128.

Reed, R.H., Chuydek, J.A., Foster, R. and Gadd, G.M. (1987) Osmotic significance of glycerol accumulation in exponentially growing yeasts. *Applied and Environmental Microbiology*, **53**, 2119–2123.

Reglinski, T., Lyon, G.D. and Newton, A.C. (1994) Induction of resistance mechanisms in barley by yeast-derived elicitors. *Annals of Applied Biology*, **124**, 509–517.

Reglinski, T., Lyon, G.D. and Newton, A.C. (1995) The control of *Botrytis cinerea* and *Rhizoctonia solani* in lettuce using elicitors extracted from yeast cell walls. *Z. Pflanzenk*, **102**, 257–266.

Rinaldi, T., Bolotin-Fukuhara, M. and Frontali, L. (1995) A *Saccharomyces cerevisiae* gene essential for viability has been conserved in evolution. *Gene*, **160**, 135–136.

Robinow, C.F. and Hyams, J.S. (1989) General cytology of fission yeasts. In *Molecular Biology of the Fission Yeast* (eds A. Nassim, P. Young and B.F. Johnson), pp. 273–330. Academic Press, San Diego.

Rodriguez, C. and Dominguez, A. (1984) The growth characteristics of *Saccharomycopsis lipolytica*: morphology and induction of mycelium formation. *Canadian Journal of Microbiology*, **30**, 605–612.

Rodriguez de Sousa, H., Madeira-Lopes, A. and Spencer-Martins, I. (1995) The significance of active fructose transport and maximum temperature for growth in the taxonomy of *Saccharomyces sensu stricto*. *Systematic and Applied Microbiology*, **18**, 44–51.

Roeder, G.S. (1995) Sex and the single cell: meiosis in yeast. *Proceedings of the National Academy of Sciences (USA)*, **92**, 10450–10456.

Roemer, T., Vallier, L.G. and Snyder, M. (1996) Selection of polarized growth sites in yeast. *Trends in Cell Biology*, **6**, 434–441.

Rojas, A.P. and Vondrejs, V. (1995) Tumors on colonies of rad 6 mutants in *Saccharomyces cerevisiae*. *Folia Microbiologica*, **40**, 560–561.

Roosita, R. and Fleet, G.H. (1996) Growth of yeasts in milk and associated changes to milk composition. *International Journal of Food Microbiology*, **31**, 205–219.

Rose, A.H. and Vijayalakshmi, G. (1993) Baker's yeast. In *The Yeasts. 2nd edn. Vol. 5: Yeast Technology* (eds A.H. Rose and J.S. Harrison), pp. 357–398. Academic Press, London.

Rose, D. (1975) Physical responses of yeast cells to osmotic shock. *Journal of Applied Bacteriology*, **38**, 169–175.

Rosini, G. (1989) Killer yeasts: notes on properties and technical use of the character. In *Biotechnological Applications in Beverage Production* (eds C. Cantarelli and G. Lanzarini), pp. 41–48. Elsevier, Amsterdam.

Rothen, S.A., Saner, M., Meenakshisundaram, S., Sonnleitner, B. and Fiechter, A. (1996) Glucose uptake kinetics of *Saccharomyces cerevisiae* monitored with a newly developed FIA. *Journal of Biotechnology*, **50**, 1–12.

Roukas, T. (1996) Ethanol production from non-sterilized beet molasses by free and immobilized *Saccharomyces cerevisiae* cells using fed-batch culture. *Journal of Food Engineering*, **27**, 87–96.

Ruggeri, B., Sassi, G., Specchia, V., Bosco, F. and Marzona, M. (1991) Alginate beads coated with polyacrylamide resin: potential as a biocatalyst. *Process Biochemistry*, **26**, 331–335.

Ruis, H. and Schüller, C. (1995) Stress signalling in yeast. *BioEssays*, **17**, 959–965.

Russell, P. and Nurse, P. (1987) Negative regulation of mitosis by *wee1*$^{+}$, a gene encoding a protein kinase homolog. *Cell*, **49**, 559–567.

Ryder, N.S. (1984) Selective inhibition of squalene epoxidation by allylamine antimycotic agents. In *Microbial Cell Wall Synthesis and Autolysis* (ed. C. Nombela), pp. 313–321. Elsevier Science Publishers, Amsterdam.

Salter, G.J., Kell, D.B., Ash, L.A., Adams, J.M., Brown, A.J. and James, R. (1990) Hydrodynamic deposition: a novel method of cell immobilization. *Enzyme and Microbial Technology*, **12**, 419–430.

Sato, M., Kobori, H., Shimada, S. and Osumi, M. (1995) Pressure–stress effects on the ultrastructure of cells of the dimorphic yeast, *Candida albicans*. *FEMS Microbiology Letters*, **131**, 11–15.

Sawant, A.D., Abdelal, A.T. and Ahearn, D.G. (1989). Purification and characterization of the anti-*Candida* toxin of *Pichia anomala* WC65. *Antimicrobial Agents and Chemotherapy*, **33**, 48–52.

Scheffers, W.A. (1987) Alcoholic fermentation. *Studies in Mycology*, **30**, 321–332.

Schmitt, M.J., Klavehn, P., Wang, J., Schönig, I. and Tipper, D.J. (1996) Cell cycle studies on the mode of action of yeast K28 killer toxin. *Microbiology (UK)*, **142**, 2655–2662.

Schu, P. and Reith, M. (1995) Evaluation of different preparation parameters for the production and cryopreservation of seed cultures with recombinant *Saccharomyces cerevisiae*. *Cryobiology*, **32**, 379–388.

Schüller, C., Brewster, J.L., Alexander, M.R., Gustin, M.C. and Ruis, H.(1994) The HOG pathway controls osmotic regulation of transcription via the stress response element (STRE) of *Saccharomyces cerevisiae CTT1* gene. *EMBO Journal*, **13**, 4382–4389.

Schultz, J., Ferguson, B. and Sprague, G.F. (1995) Signal transduction and growth control in yeast. *Current Opinion in Genetics Development*, **5**, 31–37.

Schwob, E., Bohm, T., Mendenhall, M. and Nasmyth, K. (1994) The B-type cyclin kinase inhibitor p40^{sic1} controls the G1 to S transition in *S. cerevisiae*. *Cell*, **79**, 233–244.

Scott, J.A. and O'Reilly, A.M. (1995) Use of a flexible sponge material to immobilize yeast for beer fermentation. *Journal of the American Society of Brewing Chemists*, **53**, 67–71.

Sedlak, B.J. (1997) Biotechnology companies aim at cell cycle for novel drug design targets. *Genetic Engineering News*, **17**, 11, 36, 43.

Séguy, N., Cailliez, J.-C., Polonelli, L., Dei-Cas, E. and Camus, D. (1996) Inhibitory effect of a *Pichia anomala* killer toxin on *Pneumocystis carinii* infectivity to the SCID mouse. *Parasitology Research*, **82**, 114–116.

Serrano, R. (1996) Salt tolerance in plants and microorganisms: toxicity targets and defense responses. *International Review of Cytology*, **165**, 1–52.

Shankar, C.S., Aneez Ahamad, P.Y., Ramakrishnan, M.S. and Umesh-Kumar, S. (1996) Mitochondrial NADH dehydrogenase activity and ability to tolerate acetaldehyde determine faster ethanol production in *Saccharomyces cerevisiae*. *Biochemistry and Molecular Biology International*, **40**, 145–150.

Sheldrick, K.S. and Carr, A.M. (1993) Feedback controls and G2 checkpoints: fission yeast as a model system. *BioEssays*, **15**, 775–782.

Sherwood, J., Gow, N.A.R., Gooday, G.W., Gregor, D.W. and Marshall, D. (1992) Contact sensing in *Candida albicans*: a possible aid to epithelial penetration. *Journal of Medical and Veterinary Mycology*, **30**, 461–49.

Sherwood-Higham, J., Zhu, W.-Y., Devine, C.A., Gooday, G.W., Gow, N.A.R. and Gregory, G.W. (1994) Helical hyphae of *Candida albicans*. *Journal of Medical and Veterinary Mycology*, **32**, 437–445.

Shimizu, H., Miura, K., Shioya, S. and Suga, K. (1996) Online recognition of physiological state in a yeast fed batch culture. *Journal of Process Control*, **6**, 373–378.

Shuler, M.L. (1987) Immobilization of cells by entrapment in membrane reactors. *Methods in Enzymology*, **135**, 372–387.

Siede, W. (1995) Cell cycle arrest in response to DNA damage-lessons from yeast. *Mutation Research DNA Research*, **337**, 73–84.

Sigler, K. and Höfer, M. (1991) Mechanism of acid extrusion in yeast. *Biochimica et Biophysica Acta*, **1071**, 375–391.

Simon, M.N., De Virgillo, C., Souza, B., Pringle, J.R., Abo, A. and Reed, S.I. (1995) Role for the Rho-family GTPase Cdc42 in yeast mating-pheromone signal pathway. *Nature*, **376**, 702–705.

Simpson, J. Brady, D., Rollan, A., Barron, N., McHale, L. and McHale, A.P. (1995) Increased ethanol production during growth of electric-field stimulated *Kluyveromyces marxianus* IMB3 during growth on lactose-containing media at 45°C. *Biotechnology Letters*, **17**, 757–760.

Singh, A. and Mishra, P. (1995) *Microbial Pentose Utilization. Progress in Industrial Microbiology. Vol. 33*. Elsevier, Amsterdam.

Singh, K.S. and Norton, R.S. (1991) Metabolic changes induced during adaptation of *Sacchar-*

omyces cerevisiae to water stress. *Archives of Microbiology*, **156**, 38–42.

Sinisterra, J.V. (1992) Application of ultrasound to biotechnology: an overview. *Ultrasonics*, **30**, 180–185.

Sipiczki, M. (1995) Phylogenesis of fission yeasts. Contradictions surrounding the origin of a century old genus. *Antonie van Leeuwenhoek*, **68**, 119–149.

Skipper, N. and Bussey, H. (1973) Mode of action of yeast toxins: energy requirement for *Saccharomyces cerevisiae* killer toxin. *Journal of Bacteriology*, **129**, 668–677.

Slapack, G.E., Russell, I. and Stewart, G.G. (1987) Thermophilic microbes in ethanol production. In *Thermotolerant Yeasts*. Chapter 6. CRC Press, Boca Raton, USA.

Slaughter, J.C. and Nomura, T. (1992) Intracellular glycogen and trehalose contents as predictors for yeast vitality. *Enzyme and Microbial Technology*, **14**, 64–67.

Slaughter, J.C., Flint, P.W.N. and Kular, K.S. (1987) The effect of CO_2 on the absorption of amino acids from a malt extract medium by *Saccharomyces cerevisiae*. *FEMS Microbiology Letters*, **40**, 239–243.

Smart, K.A., Boulton, C.A., Hinchliffe, E., and Molzahn, S. (1995) Effect of physiological stress on the surface properties of brewing yeasts. *Journal of American Society of Brewing Chemists*, **53**, 33–38.

Smeal, T., Claus, J., Kennedy, B., Cole, F. and Guarente, L. (1996) Loss of transcriptional silencing causes sterility in old mother cells of *S. cerevisiae*. *Cell*, **84**, 633–642.

Soares, E.V. and Mota, M. (1996) Flocculation onset, growth phase and genealogical age in *Saccharomyces cerevisiae*. *Canadian Journal of Microbiology*, **42**, 539–547.

Sossin, W.S., Fisher, J.M. and Scheller, R.H. (1989) Cellular and molecular biology of neuropeptide processing and packaging. *Neuron*, **2**, 1407–1417.

Speers, R.A. and Ritcey, L.L. (1995) Towards an ideal flocculation assay. *Journal of the American Society of Brewing Chemists*, **53**, 174–177.

Spencer, J.F.T. and Spencer, D.M. (1996) Mutagenesis in yeast. In *Methods in Molecular Biology. Vol. 53: Yeast Protocols* (ed. I.H. Evans), pp. 17–38. Humana Press Inc., Totowa, N.J., USA.

Sprague, G.F. (1995) Mating and mating-type interconversion in *Saccharomyces cerevisiae* and *Schizosaccharomyces pombe*. In *The Yeasts. 2nd edn. Vol. 6* (eds A.H. Rose, J.S. Harrison and A.H. Wheals), pp. 411–459. Academic Press, London.

Stanley, G.A. and Pamment, N.B. (1993) Acetaldehyde transport and intracellular accumulation in *Saccharomyces cerevisiae*. *Biotechnology and Bioengineering*, **42**, 24–29.

Stanley, G.A., Douglas, N.G., Every, E.J., Tzanatos, T., and Pamment, N.B. (1993) Inhibition and stimulation of yeast growth by acetaldehyde. *Biotechnology Letters*, **15**, 1199–1204.

Stanley, G.A., Hobley, T.J. and Pamment, N.B. (1997) Effect of acetaldehyde on *Saccharomyces cerevisiae* and *Zymomonas mobilis* subjected to environmental shocks. *Biotechnology and Bioengineering*, **53**, 71–78.

Stark, M.J.R., Boyd, A., Mileham, A.J. and Romanos, M.A. (1990) The plasmid-encoded killer system of *Kluyveromyces lactis*: A review. *Yeast*, **6**, 1–29.

Starmer, W.T., Ganter, P.F., Aberdeen, V., Lachance, M.-A. and Phaff, H.J. (1987) The ecological role of killer yeasts in natural communities of yeasts. *Canadian Journal of Microbiology*, **33**, 783–796.

Stephen, D.W.S. and Jamieson, D.J. (1996) Glutathione is an important antioxidant molecule in the yeast *Saccharomyces cerevisiae*. *FEMS Microbiology Letters*, **141**, 207–212.

Stern, B. and Nurse, P. (1996) A quantitative model for the *cdc2* control of S phase and mitosis in fission yeast. *Trends in Genetics*, **12**, 345–350.

Stewart, E. and Enoch, T. (1996) S-phase and DNA-damage checkpoints: a tale of two yeasts. *Current Opinion in Cell Biology*, **8**, 781–787.

Stewart, G.G. (1996) Yeast performance and management. *The Brewer*, **82**, 211–215.

Sticklen, M.B. and Sherald, J.L. (1993) (eds) *Dutch Elm Disease. Cellular and Molecular Approaches*. Springer-Verlag, New York.

Stoldt, V., Rademacher, F., Kehren, V., Ernst, J.F. and Sherman, F. (1996) Review: The Cct eukaryotic chaperonin subunits of *Saccharomyces cerevisiae* and other yeasts. *Yeast*, **12**, 523–529.

Stratford, M. and Anslow, P.A. (1996) Comparison of the inhibitory action on *Saccharomyces cerevisiae* of weak-acid preservatives, uncouplers and medium-chain fatty acids. *FEMS Microbiology Letters*, **142**, 53–58.

Streiblová, E. (1981) Fission. In *Yeast Cell Envelopes: Biochemistry, Biophysics and Ultrastructure* (ed. W.N. Arnold), pp. 79–92. CRC Press, Boca Raton, Florida, USA

Streiblová, E. and Beran, K. (1963) Types of multiplication scars in yeasts, demonstrated by fluorescence microscopy. *Folia Microbiologica*, **8**, 221–227.

Streiblová, E. and Bonaly, R (1995) Yeast motor proteins. *Folia Microbiologica*, **40**, 571–582.

Stumm, C., Hermans, J.M.H., Middelbeek, E.J., Croes, A.F. and de Vries, G.J.M.L. (1975) Killer sensitive relationships in yeasts from natural habitats. *Antonie van Leeuwenhoek*, **43**, 125–128.

Sukroongreung, S., Schappert, K.T. and Khachatourians, G.G. (1984) Survey of sensitivity of twelve yeast genera toward T–2 toxin. *Applied and Environmental Microbiology*, **48**, 416–419.

Sunder, S., Singh, A.J., Gill, S. and Singh, B. (1996) Regulation of intracellular level of Na^+, K^+ glycerol in *Saccharomyces cerevisiae* under osmotic stress. *Molecular and Cellular Biochemistry*, **158**, 121–124.

Surawitz, C.M., Ehmer, G.W., Speelman, P., McFarland, L.V., Chinn, J. and Belle, G. (1989) Prevention of antibiotic-associated diarrhea by *Saccharomyces boulardii*: a prospective study. *Gastroenterology*, **96**, 981–988.

Sveiczer, A., Novak, B. and Mitchison, J.M. (1996) The size control of fission yeast revisited. *Journal of Cell Science*, **109**, 2947–2957.

Svoboda, A. and Masa, J. (1970) The protoplasts of the dimorphic yeast *Endomycopsis fibuligera*. *Folia Mirobiologica*, **15**, 199.

Syamsu, K., Greenfield, P.F. and Mitchell, D.A. (1996) The use of dilution rate cycling to stabilize recombinant plasmids in continuous cultures of recombinant *Saccharomyces cerevisiae*. *Journal of Biotechnology*, **45**, 205–210.

Szajani, B., Buzas, Z., Dallmann, K., Gimesij, I., Kirsch, J. and Toth, M. (1996) Continuous production of ethanol using yeast cells immobilized in performed cellulose beads. *Applied Microbiology and Biotechnology*, **46**, 122–125.

Taipa, M.A., Cabral, J.M.S. and Santos, H. (1993) Comparison of glucose fermentation by suspended and gel-entrapped yeast cells: an *in vivo* nuclear magnetic resonance study. *Biotechnology and Bioengineering*, **41**, 647–653.

Tellier, R., Krajden, M., Grigoriew, G.A. and Campbell, I. (1992) Innovative end point determination system for antifungal susceptibility testing of yeasts. *Antimicrobial Agents and Chemotherapy*, **36**, 1619–1625.

Thevelein, J.M. (1992) The RAS-adenylate cyclase pathway and cell cycle control in *Saccharomyces cerevisiae*. *Antonie van Leeuwenhoek*, **62**, 109–130.

Thevelein, J.M. (1994) Signal transduction in yeast. *Yeast*, **10**, 1753–1790.

Thevelein, J.M. and Hohmann, S. (1995) Trehalose synthetase: guard to the gate of glycolysis in yeast? *Trends in Biochemical Sciences*, **20**, 3–10.

Thomas, D.S. (1993) Yeast as spoilage organisms in beverages. In *The Yeasts. 2nd edn. Vol. 5: Yeast Technology* (eds A.H. Rose and J.S. Harrison), pp. 517–561. Academic Press, London.

Thomas, K.C., Hynes, S.H. and Ingledew, W.M. (1994) Effects of particulate materials and osmoprotectants in very high gravity ethanolic fermentations by *Saccharomyces cerevisiae*. *Applied and Environmental Microbiology*, **60**, 1519–1524.

Thompson-Jaeger, S., Francois, J., Gaughram, J.P. and Tatchell, K. (1991) Deletion of *SNF1* affects the nutrient response of yeast and resembles mutations which activate the adenylate cyclase pathway. *Genetics*, **129**, 697–706.

Tokiwa, G., Tyers, M., Volpe, T. and Futcher, B. (1994) Inhibition of G1 cyclin activity by the Ras/cAMP pathway in yeast. *Nature*, **371**, 342–345.

Trevors, J.T., Merrick, R.L., Russell, I. and Stewart, G.G. (1983) A comparison of methods for assessing yeast viability. *Biotechnology Letters*, **5**, 131–134.

Trivedi, N.B. and Jacobson, G. (1986). Recent advances in baker's yeast. *Progress in Industrial Microbiology*, **23**, 45–71.

Tschopp, J.F., Hansen, W., Emr, S.D. and Schekman, R. (1987) Plasma membrane assembly in yeast. In *Biological Research on Industrial Yeasts. Vol. III* (eds G.G. Stewart, I. Russell, R.D. Klein and R.R. Hiebsch), pp. 87–104. CRC Press, Boca Raton, Florida.

Tsiomenko, A.B. and Tuymetova, G.P. (1995) Secretory heat shock proteins of yeasts: a novel family of stress proteins? *Biochemistry (Moscow)*, **60**, 625–628.

Tudor, E.A. and Board, R.G. (1993) Food-spoilage yeasts. In *The Yeasts. 2nd edn. Vol. 5: Yeast Technology* (eds A.H. Rose and J.S. Harrison), pp. 435–516. Academic Press, London.

Tyers, M., Tokiwa, G. and Futcher, B. (1993) Comparison of the *Saccharomyces cerevisiae* G1 cyclins: Cln3 may be an upstream activator of Cln1, Cln2 and other cyclins. *EMBO Journal*, **12**, 1955–1968.

Tzeng, J.W., Fan, L.S., Gan, Y.R. and Hu, T.T. (1991) Ethanol fermentation using immobilized cells in a multistage fluidized bed reactor. *Biotechnology and Bioengineering*, **38**, 1253–1258.

Vadkertiova, R. and Slavikova, E. (1995) Killer activity of yeasts isolated from the water environment. *Canadian Journal of Microbiology*, **41**, 759–766.

Vanden Bossche, H. (1991) Anti-*Candida* drugs – mechanism of action. In *Candida and Candidamycosis* (ed. E. Tumbay), pp. 83–95. Plenum Press, New York.

Vanden Bossche, H., Odds, F.C. and Kerridge, D. (1993) (eds) *Dimorphic Fungi in Biology and Medicine*. Plenum Press, New York.

Van Dijck, P., Colavizza, D., Smet, P. and Thevelein, J.M. (1995) Differential importance of trehalose in stress resistance in fermenting and non-fermenting *Saccharomyces cerevisiae* cells. *Applied and Environmental Microbiology*, **61**, 109–115.

Van Dijken, J.P. and Scheffers, W.A. (1986) Redox balances in the metabolism of sugars by yeasts. *FEMS Microbiology Reviews*, **32**, 199–224.

Van Laere, A. (1989) Trehalose: reserve and/or stress metabolite? *FEMS Microbiology Reviews*, **63**, 201–210.

Van Uden, N. (1984) Temperature profiles of yeasts. *Advances in Microbial Physiology*, **25**, 195–251.

Van Zyl, P.J., Kilian, S.G. and Prior, B.A. (1990) The role of an active transport mechanism in glycerol accumulation during osmoregulation in *Zygosaccharomyces rouxii*. *Applied Microbiology and Biotechnology*, **34**, 231–235.

Varela, J.C.S. and Mager, W.H. (1996) Response of *Saccharomyces cerevisiae* to changes in external osmolarity. *Microbiology (UK)*, **142**, 721–731.

Vaughan-Martini, A. and Martini, A. (1993) A taxonomic key for genus *Saccharomyces*. *Systematic and Applied Microbiology*, **16**, 113–119.

Vaughan-Martini, A., Cardinali, G. and Martini, A. (1996) Differential killer sensitivity as a tool for fingerprinting wine yeast strains of *Saccharomyces cerevisiae*. *Journal of Industrial Microbiology*, **17**, 124–127.

Verde, F., Mata, J. and Nurse, P. (1995) Fission yeast cell morphogenesis – identification of new genes and analysis of their role during the cell cycle. *Journal of Cell Biology*, **131**, 1529–1538.

Vondrejs, V., Janderova, B. and Valesek, L. (1996) Yeast killer toxin K 1 and its exploitation in genetic manipulations. *Folia Microbiologica*, **41**, 379–393.

Waddle, J.A., Carpova, T.S., Waterston, R. and Cooper, J.A. (1996) Movement of cortical actin patches in yeast. *Journal of Cell Biology*, **132**, 861–870.

Walker, G.M. (1986) Magnesium and cell cycle control: an update. *Magnesium*, **5**, 9–23.

Walker, G.M. and Duffus, J.H. (1980) Magnesium ions and control of the cell cycle in yeast. *Journal of Cell Science*, **42**, 324–356.

Walker, G.M. and Duffus, J.H. (1983) Magnesium as the fundamental regulator of the cell cycle. *Magnesium*, **2**, 1–16.

Walker, G.M. and Maynard, A.I. (1996) Magnesium-limited growth of *Saccharomyces cerevisiae*. *Enzyme and Microbial Technology*, **18**, 455–459.

Walker, G.M. and O'Neill, J.D. (1990) Morphological and metabolic changes in the yeast *Kluyveromyces marxianus* var. *marxianus* NRRLy 2415. *Journal of Chemical Technology and Biotechnology*, **49**, 75–89.

Walker, G.M., Thompson, J.C., Slaughter, J.C. and Duffus, J.H. (1980) Perturbation of enzyme activity in cells of the fission yeast, *Schizosaccharomyces pombe*, subjected to continuous-flow centrifugation. *Journal of General Microbiology*, **119**, 543–546.

Walker, G.M., Sullivan, P.M. and Shepherd, M.G. (1984) Magnesium and the regulation of germ-tube formation in *Candida albicans*. *Journal of General Microbiology*, **130**, 1941–1945.

Walker, G.M., MacLeod, A.M. and Hodgson, V.J. (1995) Interactions between killer yeasts and pathogenic fungi. FEMS *Microbiology Letters*, **127**, 213–222.

Walker, G.M., Birch, R.M., Chandrasena, G. and Maynard, A.I. (1996) Magnesium, calcium and fermentative metabolism in industrial yeasts. *Journal of American Society of Brewing Chemists*, **54**, 13–18.

Walsh, P.K. and Malone, D.M. (1995) Cell growth patterns in immobilization matrices. *Biotechnology Advances*, **13**, 13–43.

Walther, I., Bechler, B., Müller, O., Hunzinger, E. and Cogoli, A. (1996) Cultivation of *Saccharomyces cerevisiae* in a bioreactor in microgravity. *Journal of Biotechnology*, **47**, 113–127.

Ward, G.D., Watson, I.A., Stewart-Tull, D.E., Wardlaw, A.C. and Chatwin, C.R. (1996) Inactivation of bacteria and yeasts on agar surface with high power ND-YAG laser light. *Letters in Applied Microbiology*, **23**, 136–140.

Watson, C.D. and Berry, D.R. (1977) Fluctuations in cAMP levels during the cell cycle of *Saccharomyces cerevisiae*. *FEMS Microbiology Letters*, **1**, 175–178.

Watson, K. (1987) Temperature relations. In *The Yeasts. Vol. 2* (eds A.H. Rose and J.S. Harrison), pp. 41–72. Academic Press, London.

Watson, K. (1990) Microbial stress proteins. *Advances in Microbial Physiology*, **31**, 183–223.

Watson, K. and Cavicchioli, R. (1983) Acquisition of ethanol tolerance in yeast cells by heat shock. *Biotechnology Letters*, **4**, 683–688.

Webber, J.F. (1993) D factors and their potential for controlling Dutch elm disease. In: *Dutch Elm Disease Research. Cellular and Molecular Approaches* (eds M.B. Sticklen and J.L. Sherald), pp. 322–332. Springer-Verlag, New York.

Weinert, T.A. and Hartwell, L.H. (1993) Cell cycle arrest of cdc mutants and specificity of the RAD9 checkpoint. *Genetics*, **134**, 63–80.

Werner-Washburne, M., Braun, E.L., Johnston, G.C. and Singer, R.A. (1993) Stationary phase in the yeast *Saccharomyces cerevisiae*. *Microbiological Reviews*, **57**, 383–401.

Werner-Washburne, M., Braun, E.L., Crawford, M.E. and Peck, V.M. (1996) Stationary phase in *Saccharomyces cerevisiae*. *Molecular Microbiology*, **19**, 1159–1166.

Wheals, A.H. (1987) Biology of the cell cycle in yeasts. In *The Yeasts. Vol. 1* (eds A.H. Rose and J.S. Harrison), pp. 283–377. Academic Press, London.

Whitehead, M.P. and Flannigan, B. (1989) The *Fusarium* mycotoxin deoxynivalenol and yeast growth and fermentation. *Journal of Institute of Brewing*, **95**, 411–413.

Whitney, H.S. (1971) Association of *Dendroctonus ponderosae* (*Coleoptera: Scolytidae*) with blue stain fungi and yeasts during brood development in lodgepole pine. *The Canadian Entomologist*, **103**, 1495–1503.

Wickner, R.B. (1992) Double-stranded and single-stranded RNA viruses of *Saccharomyces cerevisiae*. *Annual Review of Microbiology*, **46**, 346–375.

Wickner, R.B. (1996) Double-stranded RNA viruses of *Saccharomyces cerevisiae*. *Microbiological Reviews*, **60**, 250–265.

Wiemken, A. (1990) Trehalose in yeast: stress protectant rather than reserve carbohydrate? *Antonie van Leeuwenhoek*, **58**, 209–217.

Winkler, M. (1991) Time-profiling and environmental design in computer-controlled fermentation and enzyme production. In *Genetically Engineered Proteins and Enzymes from Yeast: Process Control* (ed. A. Wiseman), pp. 96–146. Ellis Horwood, Chichester, UK.

Wisniekwski, M.E., Biles, C.L., Droby, S., McLaughlin, R.J., Wilson, C.L. and Chalutz, E. (1991) Mode of action of the post harvest biocontrol yeast *Pichia guilliermondii*. Characterization of attachment to *Botrytis cinerea*. *Physiological and Molecular Plant Pathology*, **39**, 245–258.

Wittenberg, C. and Reed, S.I. (1996) Plugging it in: signalling circuits and the yeast cell cycle. *Current Opinion in Cell Biology*, **8**, 223–230.

Woldringh, C.L., Fluiter, K. and Juls, P.G. (1995) Production of senescent cells of *Saccharomyces cerevisiae* by centrifugal elutriation. *Yeast*, **11**, 361–369.

Yamamoto, M. (1996) The molecular control mechanisms of meiosis in fission yeast. *Trends in Biochemical Sciences*, **21**, 18–22.

Yang, S., Ayscough, K.R. and Drubin, D.G. (1997) A role for the actin cytoskeleton of *Saccharomyces cerevisiae* in bipolar bud-site selection. *Journal of Cell Biology*, **136**, 111–123.

Yeh, E., Skibbens, R.V., Cheng, J.W., Salmon, E.D. and Bloom, K. (1995) Spindle dynamics and cell cycle regulation of dynein in the budding yeast, *Saccharomyces cerevisiae*. *Journal of Cell Biology*, **130**, 687–700.

Young, T.W. (1982) The properties and brewing performance of brewing yeasts possessing killer character. *Journal of the American Society of Brewing Chemists*, **42**, 1–4.

Young, T.W. (1987) Killer yeasts. In *The Yeasts. Vol. 2* (eds A.H. Rose and J.S. Harrison), pp. 131–164. Academic Press, London.

Young, T.W. and Yagui, M. (1978) A comparison of the killer character in different yeasts and its classification. *Antonie van Leeuwenhoek*, **44**, 59–77.

Zahner, J.E., Harkins, H.A. and Pringle, J.R. (1996) Genetic analysis of the bipolar pattern of bud site selection in the yeast *Saccharomyces cerevisiae*. *Molecular and Cellular Biology*, **16**, 1857–1870.

Zhang, Z., Moo-Young, M. and Chisti, Y. (1996) Plasmid stability in recombinant *Saccharomyces cerevisiae*. *Biotechnology Advances*, **14**, 401–435.

Zhu, Y.-S., Kane, J., Zhang, X.-Y., Zhang, M. and Tipper, D.J. (1993) Role of the γ component of preprotoxin in expression of the yeast K1 killer phenotype. *Yeast*, **9**, 251–266.

5

YEAST METABOLISM

5.1 INTRODUCTION
5.2 CARBON AND ENERGY METABOLISM
5.2.1 Sugar Catabolism and its Regulation
5.2.1.1 Sugar Catabolism
5.2.1.2 Regulation of Sugar Catabolism
5.2.1.3 Respiration versus Fermentation
5.2.1.4 Oscillatory Metabolism in Yeasts
5.2.2 Gluconeogenesis and Carbohydrate Biosynthesis
5.2.2.1 Structural Polysaccharide Synthesis
5.2.2.2 Storage Carbohydrate Synthesis
5.2.3 Metabolism of Non-Hexose Carbon Sources
5.2.3.1 Biopolymer Metabolism
5.2.3.2 Pentose Sugar Metabolism
5.2.3.3 Lower Aliphatic Alcohol Metabolism
5.2.3.4 Sugar Alcohol Metabolism
5.2.3.5 Hydrocarbon Metabolism
5.2.3.6 Fatty Acid and Lipid Metabolism
5.2.3.7 Sterol Biosynthesis
5.2.3.8 Organic Acid Metabolism
5.3 NITROGEN METABOLISM
5.3.1 Nitrogen Assimilation by Yeasts
5.3.2 Amino Acid Metabolism
5.3.3 Protein Metabolism
5.4 PHOSPHORUS AND SULPHUR METABOLISM
5.4.1 Phosphorus Metabolism
5.4.2 Sulphur Metabolism
5.5 SPECIALIZED METABOLISM
5.6 SUMMARY
5.7 REFERENCES

5.1 INTRODUCTION

Yeast metabolism refers to the biochemical assimilation and dissimilation of nutrients by yeast cells. The subject therefore encompasses all enzymic reactions within the yeast cell and how such reactions are regulated. Assimilatory, anabolic pathways are energy-consuming, reductive processes which lead to the biosynthesis of new cellular material. Dissimilatory, catabolic pathways are oxidative processes which remove electrons from intermediates and use these to generate energy. The reductive and oxidative processes of anabolism and catabolism are mediated by dehydrogenase enzymes which predominantly use NADP and NAD, respectively, as redox cofactors. Such biosynthetic and degradative pathways, however, do not operate in isolation and should be regarded as components of the integrated processes which are associated with the growth and survival of the yeast cell. Figure 5.1 summarizes the linkage of anabolic and catabolic processes which occur in yeasts.

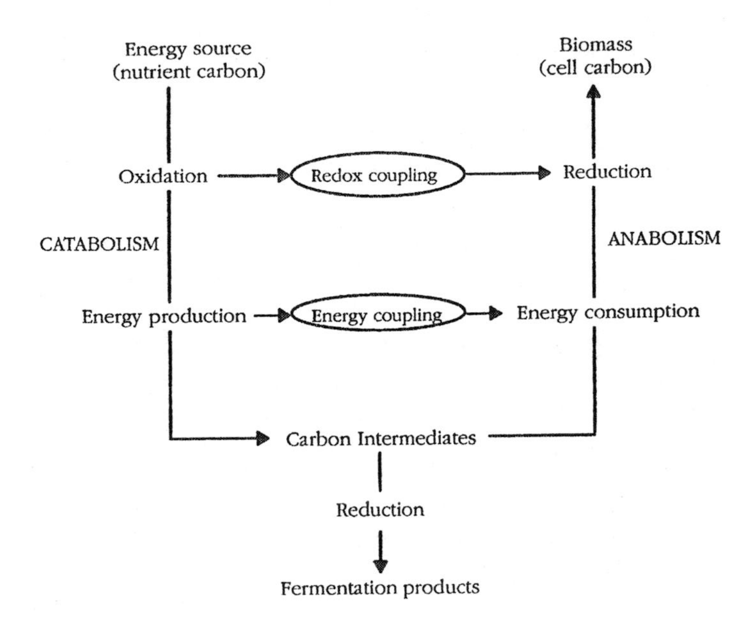

Figure 5.1. Simplistic overview of yeast carbon metabolism.

Yeasts are chemoorganotrophic microorganisms which derive their chemical energy, in the form of ATP, from the breakdown of organic compounds. However, there is metabolic diversity in the way in which yeasts generate and consume energy from carbon substrates. Section 5.2 of this chapter describes different pathways involved in yeast carbon and energy metabolism and how these pathways are regulated. Knowledge of such regulation is crucial in yeast biotechnologies which exploit carbon metabolism in the production of industrial commodities. Further discussion is therefore given of the relevance of control of yeast sugar metabolism for optimizing yeast biomass and fermentation metabolite production. Section 5.2 also covers the metabolism of non-conventional carbon sources by *Saccharomyces cerevisiae* and non-*Saccharomyces* yeasts since these are important processes pertinent to modern yeast biotechnology.

Nitrogen metabolism in yeasts is covered in section 5.3, which describes the assimilation of simple nitrogenous sources and how these are utilized to biosynthesize amino acids and proteins. Metabolic regulation by protease enzymes has received considerable attention in recent years and this will be discussed in relation to specific intracellular proteolytic reactions in yeasts. The regulation of protein biosynthesis and the extracellular secretion of certain proteins is a very important consideration in yeast recombinant DNA technology and the roles of environmental growth conditions in such regulation are discussed in section 5.3.

Aspects of the utilization of other inorganic molecules by yeasts are discussed in section 5.4 with particular reference to the assimilation and metabolism of phosphorus and sulphur sources. The final section (5.5) in this chapter deals with specialized metabolic pathways in yeasts and these will be discussed in the context of novel processes in yeast biotechnology.

Rather than providing an exhaustive coverage of metabolic pathways and enzyme mechanisms, the intention of this chapter is to describe the metabolism of principal nutrients by yeasts and to discuss physiological and biotechnological aspects of such metabolism.

5.2 CARBON AND ENERGY METABOLISM

5.2.1 Sugar Catabolism and its Regulation

Most yeasts employ sugars as their preferred carbon and energy sources. The ways in which sugars are translocated by yeasts from their growth environment and into the cell were discussed in Chapter 3. The following section considers the principal metabolic fates of sugars in yeasts, namely: the dissimilatory pathways of fermentation and respiration, and the assimilatory pathways of gluconeogenesis and carbohydrate synthesis. The regulation of sugar catabolism in yeasts will also be discussed in the context of industrial fermentations and yeast biomass propagations.

5.2.1.1 Sugar Catabolism

With regard to the catabolism of sugars, Figure 5.2 represents a simplified scheme of energy generation in yeast from glucose.

The sequence of enzyme-catalysed reactions that oxidatively convert glucose to pyruvic acid in the yeast cytoplasm is known as **glycolysis**, which is outlined in Figure 5.3. The study of yeast glycolysis by pioneering scientists at the turn of this century laid the foundations for the birth of biochemistry as a new scientific discipline. Glycolysis provides yeasts with energy, together with precursor molecules and reducing power for biosynthetic pathways. Products derived from yeast glycolytic intermediates have additionally provided mankind with valuable industrial commodities.

The key regulatory enzymes in glycolysis are the irreversible phosphofructokinase and pyruvate kinase whose activity is influenced by numerous effectors, including ATP (Dawes, 1986).

With regard to fermentation, several biotechnologically important yeasts are fermentative and as such are defined as organisms which use organic substrates anaerobically as electron donor, electron acceptor and as carbon source.

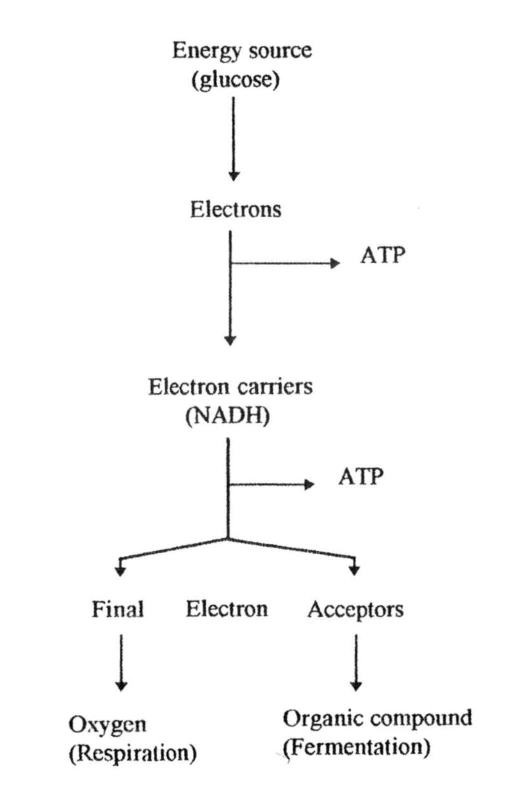

Figure 5.2. Summary of electron transfer and energy generation from glucose by yeast cells.

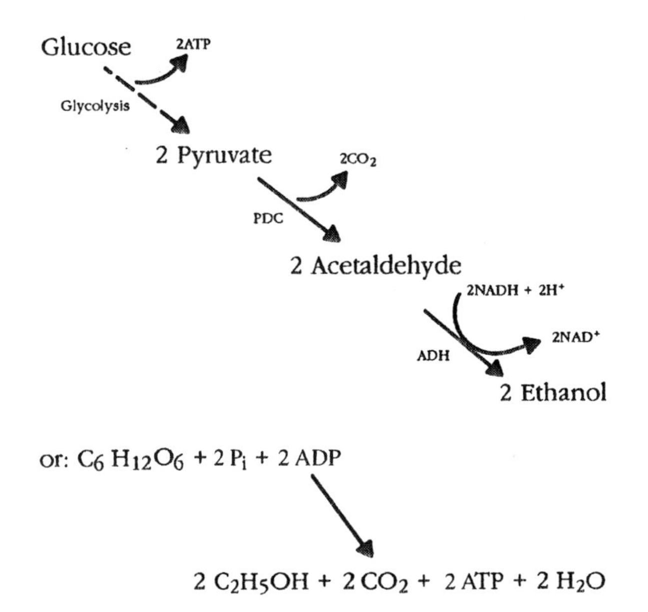

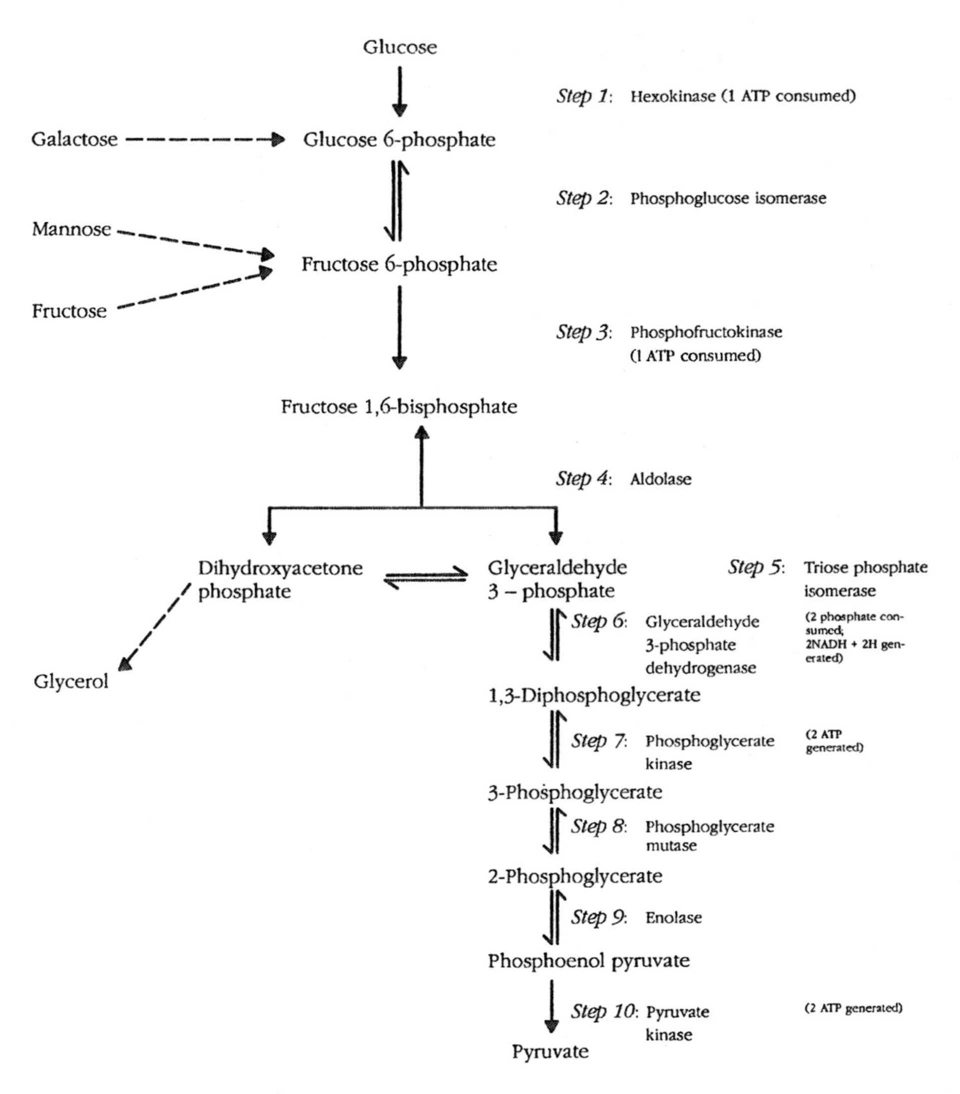

Figure 5.3. Glycolysis in yeast. In summary, glucose is phosphorylated in two stages (steps 1–3) using 2ATP to produce fructose 1,6-bisphosphate which is then split by aldolase to form two 3-carbon triose phosphates (steps 4 and 5). Inorganic phosphate is assimilated to form two triose diphosphates from which four H atoms are accepted by two molecules of oxidised NAD (step 6). In the latter stages (steps 7–10), 4ATPs are formed by transfer of phosphate from the triose diphosphates to ADP resulting in the formation of two molecules of pyruvic acid.
Overall summary: Glucose → 2 Pyruvate + 2ATP + 2NADH + H^+

During alcoholic fermentation of sugars, yeasts re-oxidize NADH to NAD in terminal step reactions from pyruvate. In the first of these reactions, catalysed by pyruvate decarboxylase (PDC), pyruvate is decarboxylated before a final reduction, catalysed by alcohol dehydrogenase (ADH) to ethanol (see scheme on page 205).

Note that the regeneration of NAD is necessary to maintain the redox balance and prevent the stalling of glycolysis (at Step 6 – see Figure

5.3). An alternative means of replenishing NAD is the pathway:

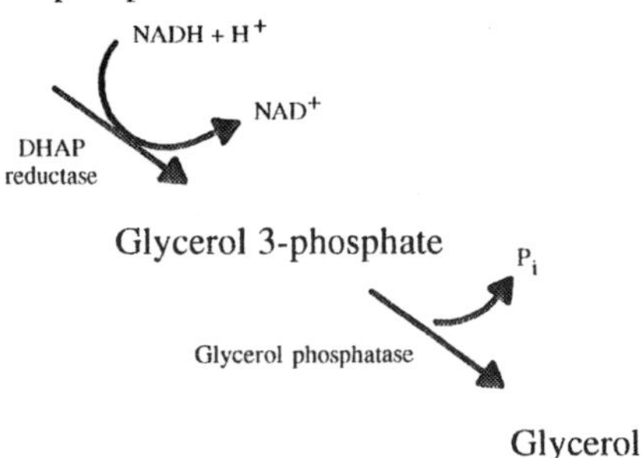

Glycerol is quantitatively one of the most important products of yeast alcoholic fermentation, besides ethanol and carbon dioxide. This production can be enhanced by trapping acetaldehyde with sodium sulphite which prevents ethanol synthesis via alcohol dehydrogenase:

$$\text{glucose} + HSO_3^- \rightarrow \text{glycerol} + \text{acetaldehyde-}HSO_3^- + CO_2$$

Dihydroxyacetone phosphate then becomes the preferred hydrogen acceptor and glycerol is eventually formed as a minor product.

Glycerol production thus maintains the redox balance of the cell but does so at the expense of carbohydrate conversion to ethanol. Such 'steered' fermentations were pioneered by the German chemist, Neuberg and led to the industrial production of glycerol by yeast fermentation around the time of World War I (glycerol being required to make nitroglycerine explosives).

Other minor fermentation metabolites are produced by yeast during the course of alcoholic fermentation. These products, which vary depending on the yeast strain and culture conditions, are very important in the development of flavour in alcoholic beverages such as beer, wine and whisky. Examples include fusel alcohols (e.g. isoamyl alcohol), esters, (e.g. ethyl acetate) organic acids (e.g. citrate, succinate, acetate) and aldehydes (e.g. acetaldehyde) and a summarized outline of their production is shown in Figure 5.4.

Succinic acid, together with glycerol, represent the major secondary fermentation products in *S. cerevisiae*. Succinic acid is thought to be synthesized and secreted by yeasts either following limited operation of the citric acid cycle or by

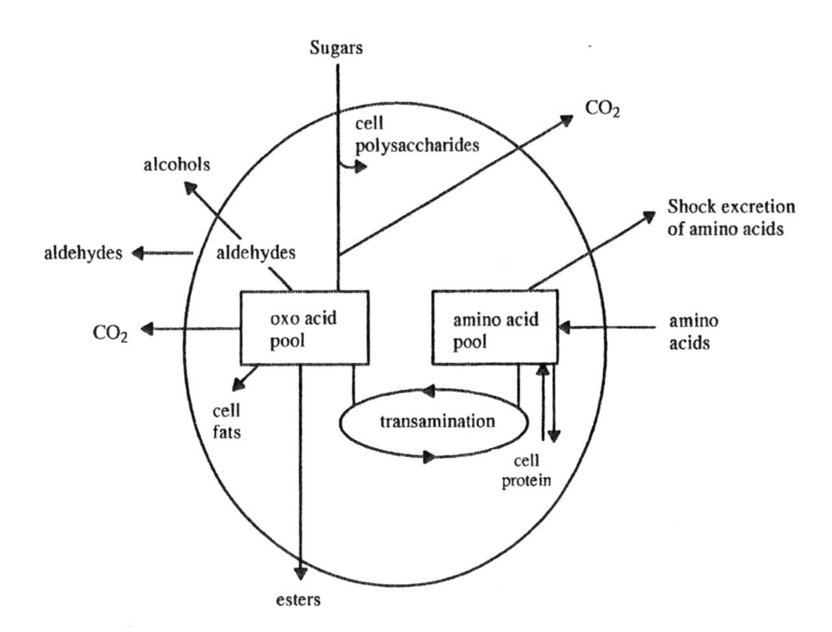

Figure 5.4. Production of minor fermentation products by *S. cerevisiae*. Adapted from Hough (1985).

reductive linear pathways involving some citric acid cycle enzymes (Hough *et al.*, 1982).

During anaerobic glucose dissimilation by yeasts, a pathway which accompanies glycolysis is known as the **hexose monophosphate pathway** (or Warburg–Dickens pathway). This pathway, depicted in Figure 5.5, exists primarily to generate cytosolic NADPH for biosynthetic reactions leading to the production of fatty acids, amino acids and sugar alcohols. The first step in the hexose monophosphate pathway, and the one which mainly regulates its progression, is the dehydrogenation of glucose 6-phosphate (to 6-phosphogluconolactone) using glucose 6-phosphate dehydrogenase and NADP as hydrogen acceptor. The reducing power generated by the hexose monophosphate pathway is not linked to terminal respiration. In *S. cerevisiae*, depending on whether cells are actively growing or not, 0–20% of total glucose may be degraded via the hexose monophosphate pathway. Although the enzymes of this pathway appear to be of limited importance for furnishing *S. cerevisiae* with its pentose sugar requirements, *Candida utilis* produces all its pentose sugars by the hexose monophosphate pathway (Horecker *et al.*, 1968).

Besides generating NADPH, the other major function of the hexose monophosphate pathway is the formation of ribose sugars which are important in the synthesis of nucleotide precursors for nucleic acids, RNA and DNA and for nucleotide coenzymes, NAD, NADP, FAD and FMN. The hexose monophosphate pathway is summarized in Figure 5.5 which also shows how two pentose catabolic sequences, the xylitol and phosphoketolase pathways, feed in to the hexose monophosphate pathway.

In some yeasts, a cyclic mechanism, the **oxida-**

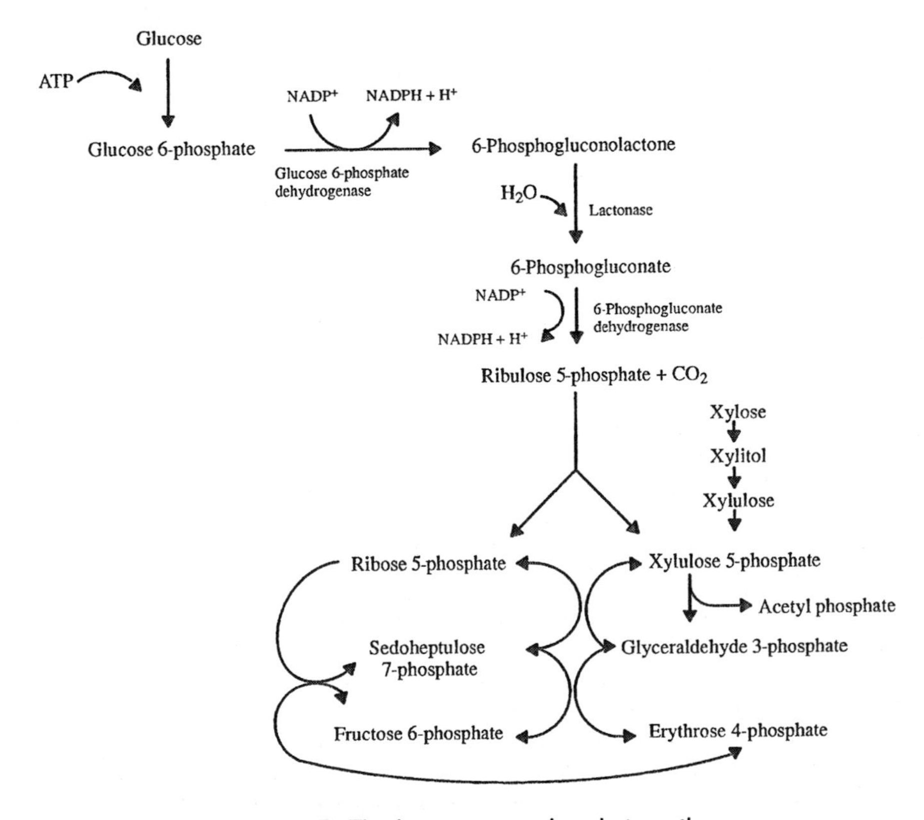

Figure 5.5. The hexose monophosphate pathway.

tive pentose phosphate cycle accounts for the complete oxidation of carbohydrates. In one turn of this cycle, glucose 6-phosphate is converted to ribulose 5-phosphate and CO_2:

$$\text{Glucose 6-phosphate} + 2NADP^+ \rightarrow \text{Ribulose 5-phosphate} + CO_2 + NADPH + 2H^+$$

and repetitive action of the cycle accounts for complete oxidation of glucose 6-phosphate:

$$\text{Glucose 6-phosphate} + 12NADP^+ \rightarrow 6CO_2 + 12NADPH + 12H^+ + Pi$$

The xylitol pathway is important in pentose-utilizing yeasts (e.g. *Candida shehatae* and *Pichia stipitis*) in that it provides more efficient conversion of pentose sugars to 2-carbon units for respiration and fermentation (Evans and Ratledge, 1984).

With regard to yeast respiration, glycolysis may be regarded as the prelude to the citric acid cycle, the electron transport chain and oxidative phosphorylation, which collectively harvest most of the energy (in the form of ATP) from glucose. Not only glucose can be respired by yeasts; in fact, a greater array of carbon sources can be respired than fermented. Those substrates which are respired by yeast cells include: pentoses (e.g. xylose), sugar alcohols (e.g. glycerol), organic acids (e.g. acetic acid), aliphatic alcohols (e.g. methanol, ethanol), hydrocarbons (e.g. *n*-alkanes) and aromatic compounds (e.g. phenol). The metabolism of some of these 'non-conventional' carbon sources by yeast is discussed further below.

Considering glucose respiration, if yeast growth conditions are conducive (i.e. presence of oxygen, absence of repression), pyruvate enters the mitochondrial matrix where it is oxidatively decarboxylated to acetyl CoA. This reaction is catabolysed by the pyruvate dehydrogenase multienzyme complex which acts as the link between glycolysis and the citric acid cycle. The activated acetyl unit is then completely oxidized to two molecules of CO_2 by the **citric acid cycle** (the Krebs cycle) which is the final common pathway for the oxidation of sugars and other carbon sources in yeast. The initial, and principal regulatory sequence, of the citric acid cycle (catalysed by citrate synthase) requires oxaloacetate which is regenerated by a cyclic series of enzyme catalysed reactions, as summarized in Figure 5.6.

The citric acid cycle is referred to as *amphibolic* since the pathway performs both catabolic and anabolic functions. The latter result from the provision of intermediates as precursors for the biosynthesis of amino acids and nucleotides. The removal of intermediates (e.g. oxaloacetate and α-ketoglutarate) necessitates their replenishment by *anaplerotic* reactions to ensure continued operation of the cycle. Examples of such reactions are the **glyoxylate cycle** and the fixation of carbon dioxide. The latter can be achieved by the actions of the enzymes pyruvate carboxylase:

$$\text{pyruvate} + CO_2 + ATP + H_2O \rightarrow \text{oxaloacetate} + ADP + Pi$$

and phosophoenolpyruvate carboxykinase:

$$\text{phosphoenolpyruvate} + CO_2 + H_2O \rightarrow \text{oxaloacetate} + H_3PO_4$$

The replenishment of oxaloacetate in this manner allows the continued flow of carbon through the citric acid cycle. Concerning the glyoxylate cycle, this anaplerotic sequence is important when yeasts (e.g. *S. cerevisiae*) are grown on 2-carbon substrates such as acetate or ethanol. The cycle can be regarded as a short-cut across the citric acid cycle which by-passes two decarboxylation steps and links isocitrate, succinate and malate (see Figure 5.7). The two additional enzymes required are isocitrate lyase:

$$\text{isocitrate} \rightarrow \text{succinate} + \text{glyoxylate}$$

and malate synthase:

$$\text{aetyl CoA} + \text{glyoxylate} + H_2O \rightarrow \text{malate} + CoA$$

When yeasts oxidize acetate, the operation of the glyoxylate cycle results in the net formation of C_4 dicarboxylic acids. The regulation of the glyoxylate cycle is principally through the activity

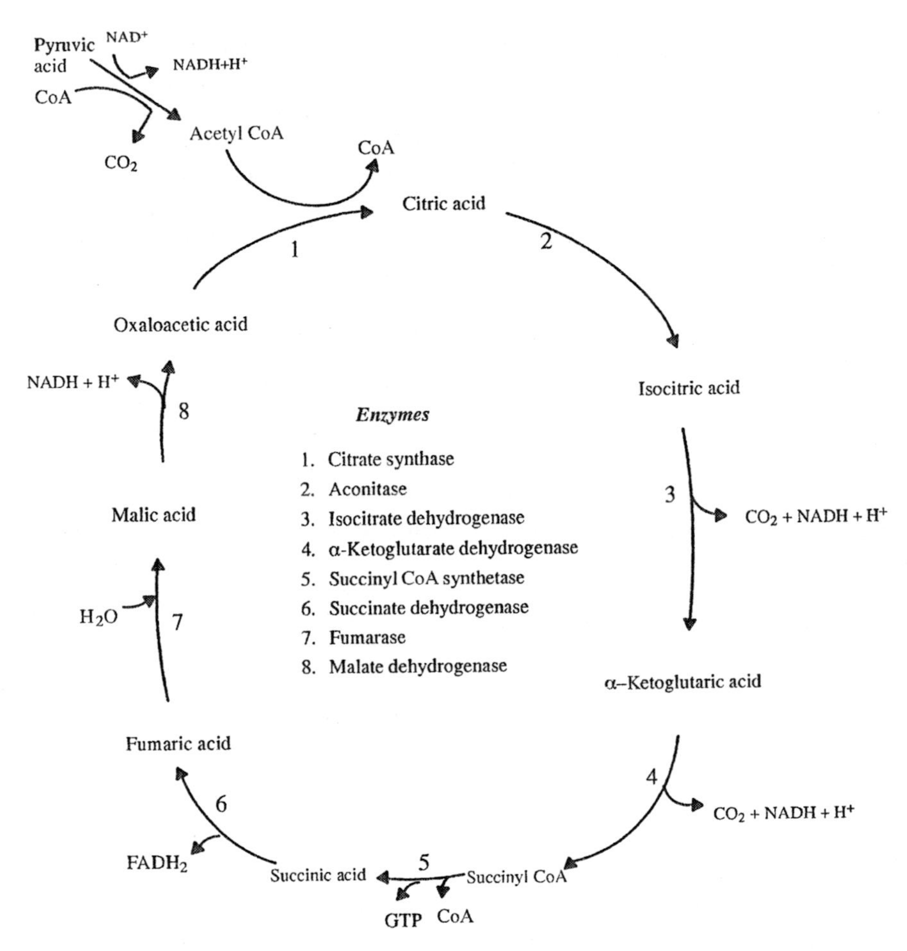

Figure 5.6. The citric acid cycle in yeasts. The net effect of the citric acid cycle is to produce: $2CO_2$, 3 NADH, 1 FADH2, 4 H^+ and 1 GTP from one pyruvate molecule.

of isocitrate lyase, which is in turn controlled by the availability of C_3 and C_4 compounds (Dawes, 1986). Galons *et al.* (1990) have estimated that in *S. cerevisiae*, 58% of acetate is metabolized through the glyoxylate cycle, whereas other yeasts, notably *Schizosaccharomyces pombe*, lack glyoxylate cycle enzymes and are unable to utilize acetate for cell growth (Tsai *et al.*, 1989). In alkane-grown yeasts the enzymes of the glyoxylate cycle are located in the peroxisomes (see Chapter 2) and these will be repressed in the presence of glucose.

During the citric acid cycle hydrogens are transferred, in processes mediated by dehydrogenase enzymes, to the redox carriers NAD and FAD, which become reduced. The reduced carriers are reoxidized and oxygen reduced to water via the **electron transport chain** on the inner mitochondrial membrane. The energy released by the transfer of electrons is used to synthesize ATP by a process called **oxidative phosphorylation** and chemiosmotic principles describe the linkage between oxidation and phosphorylation (see below). Glycolysis and the citric acid cycle in yeasts therefore encompasses oxidation reactions which produce a small amount of ATP by substrate-level phosphorylation and NADH. The latter is then processed by electron transport to

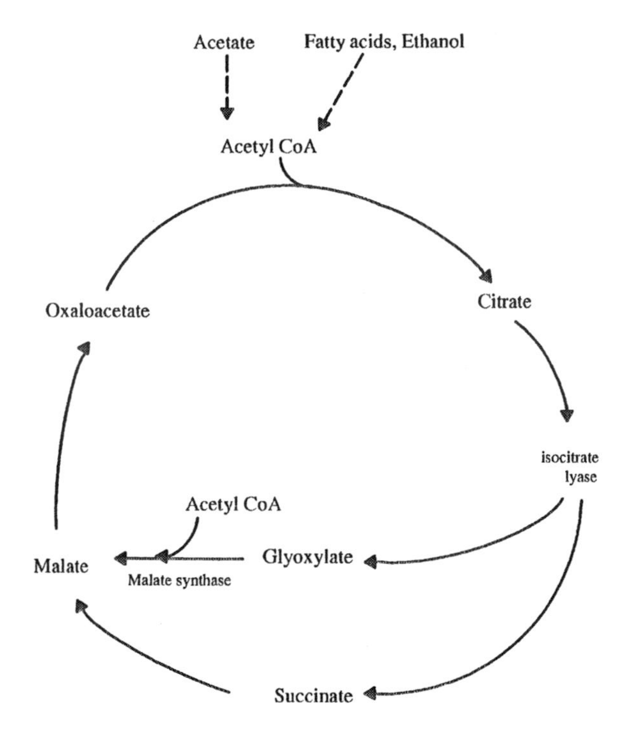

Figure 5.7. The glyoxylate cycle.

oxygen as terminal electron acceptor to yield much more ATP.

The electron transport chain (as depicted in Figure 5.8) pumps protons out of the mitochondrial matrix across the inner mitochondrial membrane to create a transmembrane proton gradient (ΔpH) and a membrane potential difference ($\Delta\Psi$), which together comprise the *proton-motive force* ($\Delta\tilde{\mu}H^+$). This is the driving force used to synthesize ATP (by H^+-transporting ATP synthase). The total yield of ATP for complete oxidation of glucose by yeast cells is around 30. This is because each pair of electrons in NADH generates about 2.5 ATP while succinate oxidation yields about 1.5 ATP (Hinkle *et al*., 1991; Alberts *et al*., 1994; Fell, 1997). The proportion of energy not retained for use by the yeast cell as ATP is largely dissipated as metabolic heat. This is one reason why industrial fermenters require cooling to maintain optimum conditions for yeast metabolism.

NADH produced in glycolysis cannot be oxidized directly within yeast mitochondria due to the impermeability of these organelles to NADH. Cytoplasmic 'shuttle' processes enable reduced cofactors to enter mitochondria. For example, the *glycerophosphate shuttle* utilizes NADH to reduce dihydroxyacetone phosphate to glycerol 3-phosphate and the *malate shuttle* utilizes NADH to reduce oxaloacetate to malate (Gancedo *et al*., 1968). These molecules then enter the mitochondrion where enzymes oxidize them to yield reduced cofactors which in turn are oxidized by the electron transport chain.

The manner by which yeast cells utilize molecular oxygen as terminal electron acceptor in aerobic respiratory metabolism appears to be species-dependent. For example, some yeasts, including *S. cerevisiae*, exhibit **alternative respiration** distinct from the cytochrome path which is insensitive to cyanide but sensitive to azide. Table 5.1 summarizes three types of respiration in yeasts and fungi.

The azide-sensitive pathway lacks proton transport capability and accepts electrons from NADH but not from succinate. The SHAM-sensitive pathway transports electrons to oxygen also without proton transport, and therefore does not phosphorylate ADP.

Mitochondria also play vital roles in the anabolic metabolism of *anaerobically* growing yeast cells (Visser *et al*., 1995). During anaerobic yeast fermentations, for example in brewing, mitochondria are not irrelevant to *S. cerevisiae* physiology even though no respiration is taking place. O'Connor-Cox *et al*. (1996) have discussed the roles of brewing yeast mitochondria and have highlighted several essential cellular functions which are not linked with aerobic respiration, but which are critical to fermentation performance and end product quality (see Table 2.12, Chapter 2).

5.2.1.2 Regulation of Sugar Catabolism

Control of carbohydrate metabolism in *S. cerevisiae* and other yeasts is of both fundamental and practical significance and has been the subject of

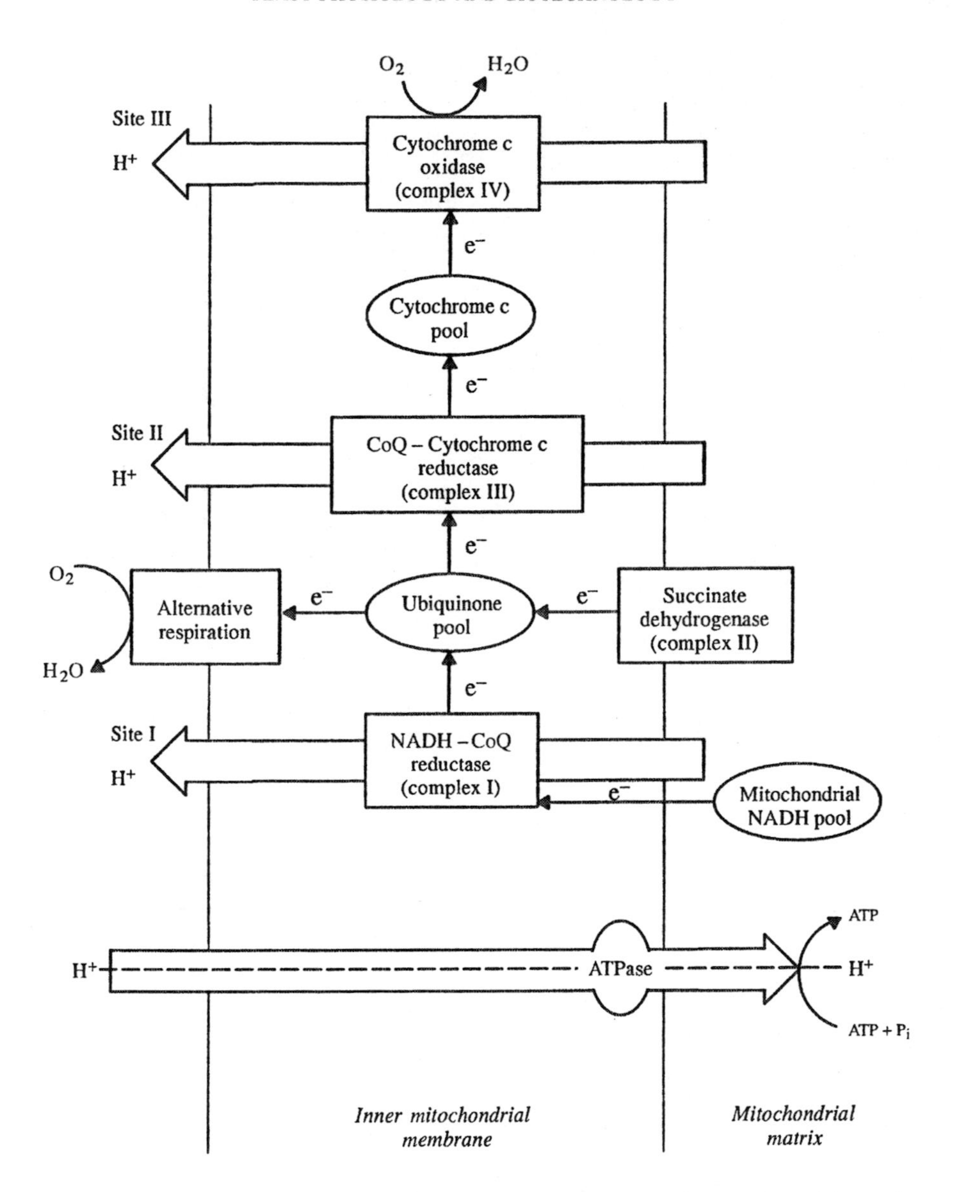

Figure 5.8. Generalized scheme of yeast mitochondrial electron transport. Adapted from Alexander and Jeffries (1990).

several reviews over the past decade or so (e.g. Käppeli and Sonnleitner, 1986; van Dijken and Scheffers, 1986; Gadd, 1988; Gancedo and Serrano, 1989; Alexander and Jeffries, 1990; Wills 1990; Entian and Barnett, 1992; Fiechter and Seghezzi, 1992; Gancedo, 1992; Johnston and Carlson, 1992; Chambers *et al.*, 1995; Pronk *et al.*, 1996). These reviews have discussed physiological, molecular and genetic aspects of the regulation of glycolysis, gluconeogenesis, hexose monophosphate pathway, pyruvate metabolism, glyoxylate cycle, citric acid cycle and oxidative phosphorylation. Sugar uptake systems and sugar influx into glycolysis are also known to

Table 5.1. Respiratory chain characteristics of yeasts and fungi.

Type	Typical species	Sensitive to	Insensitive to
Normal respiration	All aerobic fungi	Cyanide and low azide	SHAM[a]
Classic alternative	*Yarrowia lipolytica* (and in stationary phase cultures of several yeast species)	SHAM	Cyanide, high azide
New alternative	*Schizosaccharomyces pombe* *Saccharomyces cerevisiae* *Kluyveromyces lactis* *Williopsis saturnus*	High azide	Cyanide, low azide, SHAM

[a] SHAM = salicyl hydroxamate. Information from Goffeau and Crosby (1978); Alexander and Jeffries (1990).

exert regulatory influences over sugar catabolism (and *vice versa*) and these have been discussed by Lagunas (1993), Thevelein (1994), van Dam (1996) and in Chapter 3. The following section focuses on the control of respiration and fermentation in yeasts and the significance of such control for yeast biotechnologies.

5.2.1.3 Respiration versus Fermentation

Biochemical pathways in yeasts may be regulated at various levels. These include: enzyme synthesis (e.g. induction, repression and derepression of gene expression), enzyme activity (e.g. allosteric activation, inhibition or interconversion of isoenzymes) and cellular compartmentalization (e.g. mitochondrial localization of respiratory enzymes). The following discussion will focus primarily on external factors which influence respiratory and fermentative metabolism in yeasts of biotechnological significance, particularly *S. cerevisiae*.

Yeasts exhibit diversity in their modes of energy generation and Table 5.2 categorizes groups of yeasts with respect to their utilization of respiration and fermentation in ATP production. Figure 5.9 summarizes the principal reaction sequences in these two major sugar catabolic pathways.

Of the environmental factors that regulate respiration and fermentation in yeast cells, the availability of glucose and oxygen are the best documented. These factors are linked to the expression of several regulatory phenomena (Table 5.4). Note, however, that yeasts adapt to varying growth environments and even within a single yeast species, the presence of a particular regulatory phenomenon (e.g. Pasteur or Crabtree effect) will very much depend on the prevailing growth conditions. For example, *S. cerevisiae* utilizes glucose in several different ways which are dependent on the availability of oxygen and carbon source (Table 5.3).

The Pasteur effect

The Pasteur effect relates oxygen with the kinetics of yeast sugar catabolism and states that under anaerobic conditions, glycolysis proceeds faster than it does under aerobic conditions. Alternatively, it may be defined as a suppression of fermentation by oxygen. However, this phenomenon is only observable when glucose concentrations are low (e.g. below around 5 mM in *S. cerevisiae*) or under certain nutrient-limited conditions (Lagunas, 1979). It is associated with a decreased affinity for sugar uptake under aerobic conditions (Fiechter *et al.*, 1981; Lagunas, 1981). Changing resting or nutrient-starved cells from aerobic to anaerobic conditions results in an energetic discrepancy (i.e. less ATP) and so cells respond by

Table 5.2. Principal modes of sugar catabolism in yeasts.

Group Name	Examples	Respiration	Fermentation	Anaerobic growth
Obligate Respirers[a]	*Rhodotorula* spp. and *Cryptococcus* spp.	Yes	No	No
Aerobic respirers	*Candida* spp., *Kluyveromyces* spp., *Pichia stipitis and Pachysolen tannophilus*	Yes	Anaerobic in pre-grown cells	No
Aerobic fermenters	*Schizosaccharomyces pombe*	Limited	Aerobic and anaerobic	No
Facultative aerobic fermenters[b]	*Saccharomyces cerevisiae*	Limited	Aerobic and anaerobic	Facultative, but no growth in absence of sterols and fatty acids
Obligate fermenters[c]	*Torulopsis* (*Candida*) *pintolopesii*	No	Anaerobic	Yes

[a] About one third of the yeast species (over 400) studied by Barnett *et al.* (1979) were categorized as non-fermentative, although such distinctions may be dubious due to the choice of fermentation test (Scheffers, 1987).
[b] Lagunas (1986) has discussed the evidence which shows that *S. cerevisiae* is not a model facultative anaerobe.
[c] *T. pintolopesii* lacks a respiratory ability (altogether) and depends exclusively on the energy derived from fermentation. This a yeast is found in the intestinal tract of rodents (Phaff, Miller and Mrak, 1978).

increasing glycolytic flux. Nissen *et al.* (1997) have recently carried out a detailed quantification of metabolic fluxes in anaerobic glucose-limited chemostat cultures of *S. cerevisiae*. With regard to nitrogen-limitation, Busturia and Lagunas (1986) have shown that starving *S. cerevisiae* cells of nitrogen leads to an inactivation of sugar transport systems which in turn reduces the fermentation rate, but does not alter the rate of respiration. The reasons for the drop in fermentation rate with reduced sugar uptake rate may be explained by Holzer's *enzyme competition theory* (Holzer, 1961). Thus, pyruvate dehydrogenase (which channels carbon through the respiratory route) has a much higher affinity for pyruvate (the concentration of which is lowered following curtailed sugar uptake) than that of pyruvate decarboxylase of the fermentative route. A key role for Mg^{2+} ions in control of yeast metabolic flux at the level of pyruvate metabolism has been proposed by Walker (1994). Other extracellular and intracellular factors have been implicated in control of glycolytic flux in yeast cells and in expression of the Pasteur effect. These include: inorganic phosphate (Lagunas and Gancedo, 1983); ammonium ions (Lloyd *et al.*, 1983); C_6 and C_3 glycolytic intermediates (Müller *et al.*, 1995; Boles *et al.*, 1996); AMP (Dombek and Ingram, 1988) and intracellular pH (Gillies *et al.*, 1982; Rowe *et al.*, 1994). A possible target for modulation by such effectors is the enzyme phosphofructokinase which has often been considered the rate-limiting step or *pacemaker* of glycolysis and therefore a key regulator of the Pasteur effect.

However, physiological control over glycolysis most likely involves simultaneous regulation through action on several enzymes (Zimmerman, 1992; Fell and Thomas, 1995) and not solely phosphofructokinase (Davies and Brindle, 1992). In fact, overexpression of *PFK1* genes in *S. cerevisiae* leaves glycolytic flux unchanged (Brindle, 1996).

The situation is very different in *growing* cultures of *S. cerevisiae* where respiration only accounts for 3–20% of the sugar catabolized, compared with 25–100% in resting cells (Lagunas *et al.*, 1982). Irrespective of oxygen availability, fermentation is the predominant route of sugar metabolism in actively growing cells of *S. cerevisiae*. Therefore, ATP yields are

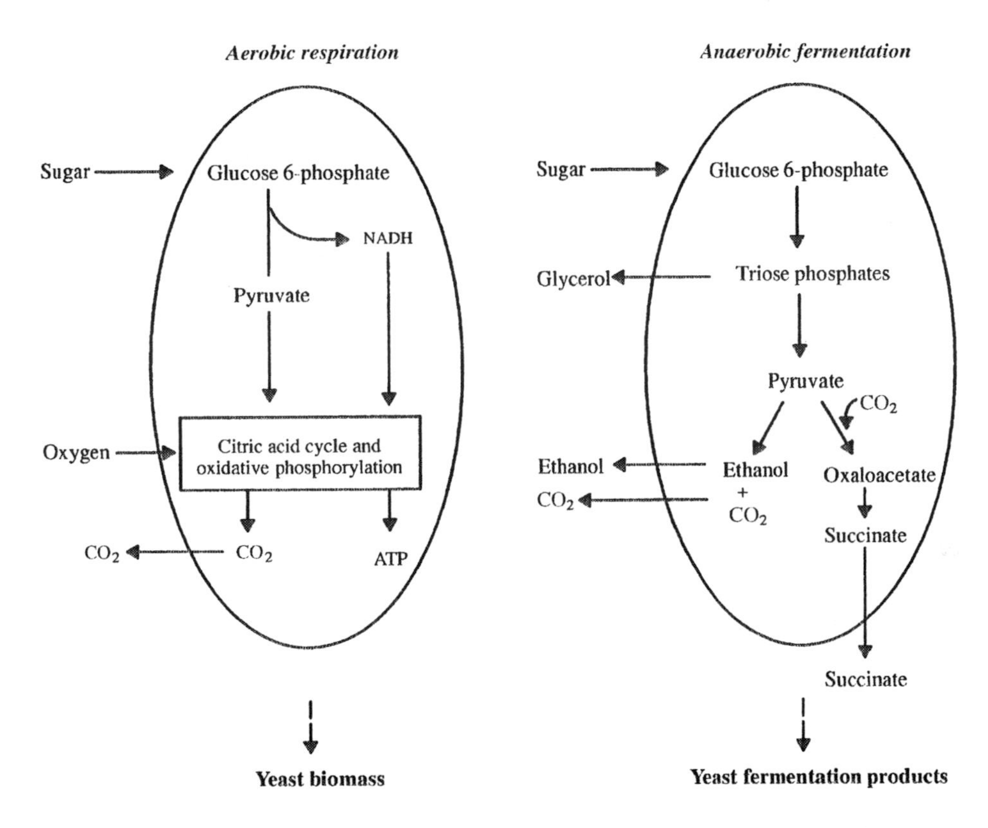

Figure 5.9. Summary of major sugar catabolic pathways in yeast cells.

Table 5.3. Types of glucose utilization in *S. cerevisiae*.

Conditions	Type of metabolism
Batch culture	
Aerobic growth Phase 1	
C-limited	Respirofermentative[a]
O-limited	Respirofermentative
Aerobic growth Phase 2	Assimilation of ethanol
Anaerobic growth	Fermentative
Continuous culture	
Aerobic growth at low dilution rates	
C-limited	Respirative
O-limited	Respirofermentative
Aerobic growth at high dilution rates	
C-limited	Respirofermentative
O-limited	Respirofermentative
Anaerobic growth	Fermentative

[a] See text for an explanation of respirofermentative metabolism. Adapted from Käppeli and Sonnleitner (1986).

Table 5.4. Regulatory phenomena in sugar metabolism of yeasts.

Phenomenon	Description	Comments	Examples
Pasteur effect	Activation of sugar consumption rate by anaerobiosis	Only observable in resting or nutrient-starved cells.	*S. cerevisiae*
Crabtree effect	Suppression of respiration by high glucose	Cells continue to ferment irrespective of oxygen availability due to glucose repressing/inactivating respiratory enzymes or due to the inherent limited respiratory capacity of cells.	*S. cerevisiae* *Sch. pombe*
Custers effect	Transient inhibition of fermentation by anaerobiosis	Oxygen stimulates ethanol production due to a lack of intracellular NAD^+ following secretion of acetate.	*Dekkera* and *Brettanomyces* spp.
Kluyver effect	Anaerobic fermentation of glucose, but not of other sugars, although they may be aerobically assimilated	This phenomenon may be linked to sugar transport limitations or altered activity of pyruvate decarboxylase.	*C. utilis*
Harden–Young effect	Addition of inorganic phosphate in yeast cell extracts stimulates the rate of fermentation	This results from phosphate incorporation into organic phosphate esters. ADP is then phosphorylated to ATP which allows phosphorylation of additional glucose.	Cell-free extracts of *S. cerevisiae.*

more or less the same in aerobically and anaerobically grown cultures. The main reasons for the absence of a Pasteur effect and the energetic irrelevance of aerobiosis in actively growing *S. cerevisiae* cells are the repressive concentrations of available glucose and the expression of the Crabtree effect.

The Crabtree effect

If the concentration of available glucose is high the Pasteur effect in *S. cerevisiae* is no longer operable. Then the Crabtree effect comes into play. This phenomenon (also referred to as the glucose effect or contre-effect Pasteur) relates glucose concentration with the particular catabolic route adopted by glucose-sensitive yeasts (like *S. cerevisiae*) in the presence of oxygen and states that, even under aerobic conditions, fermentation predominates over respiration. Thus, even though oxygen may be present, NADH generated during glycolysis is mainly oxidized by fermentation, rather than by respiration. A *short-term* Crabtree effect is observed as a sudden fermentative response when excess sugar is added to a non-fermenting yeast culture, whereas a *long-term* Crabtree effect is manifest by aerobic fermentation under full-adapted physiological steady-state conditions. The latter has been explained on the basis of the *limited respiratory capacity* of Crabtree-positive yeasts; and the former on the basis of saturation of respiration leading to overflow at pyruvate (see Pronk *et al.*, 1996). The Crabtree effect is not restricted to growth on glucose since *S. cerevisiae* will also aerobically ferment fructose. In the presence of mannose or galactose, aerobically grown *S. cerevisiae* will simultaneously respire and ferment these sugars (De Deken, 1966).

The Crabtree effect is not noticeable in glucose-insensitive yeasts (e.g. *Candida utilis*, *Kluyveromyces marxianus*, *Trichosporon cutaneum*) or in respiratory-deficient mutants (e.g. *S. cerevisiae* 'petites'). *C. utilis*, a Crabtree-negative yeast, may limit its glycolytic rate by accumulating intracellular reserve carbohydrates or the cells may exhibit altered regulation of sugar uptake (Postma *et al.*, 1988). In *S. cerevisiae*, glucose suppression of respiration in the Crabtree effect is thought to be due to glucose repressing respiratory enzyme synthesis and/or inactivating respiratory enzymes and sugar transport activity (Entian and Barnett, 1992; Trumbly, 1992; Lagunas, 1993; Wills, 1996).

Catabolite repression occurs when glucose, or an initial product of glucose metabolism, represses the *synthesis* of various respiratory and gluconeogenic enzymes (Fiechter *et al.*, 1981); whereas catabolite inactivation results in the rapid disappearance of such enzymes on addition of glucose. In catabolite repression, enzyme activity is lost by dilution with cell growth. That is, although enzymes are still present, they are no longer being synthesized due to gene repression by signals derived from glucose (see Figure 5.10) or other sugars (Gancedo, 1992). The nature of the signal(s) in question is at present unclear. In addition to glucose-repression mechanisms, normal yeast mitochondrial structures are disrupted when glucose levels are high. For example, inner membranes and cristae disappear, but re-appear once aerobic metabolism replaces alcoholic fermentation (see Chapter 2). Alteration of cellular compartmentalization in this manner clearly has profound metabolic consequences for yeast cells.

Glucose repression in yeasts like *S. cerevisiae* describes a long-term regulatory adaptation to degrade glucose exclusively to ethanol and CO_2. Therefore, when *S. cerevisiae* is grown aerobically on high concentrations of glucose, fermentation will account for the bulk of glucose consumption. In batch culture, however, when the levels of consumed glucose decline, cells will gradually become *derepressed*, resulting in induction of respiratory enzyme synthesis. This, in turn, results in oxidative consumption of accumulated ethanol when cells enter a second phase of growth known as **diauxie** (see Chapter 4).

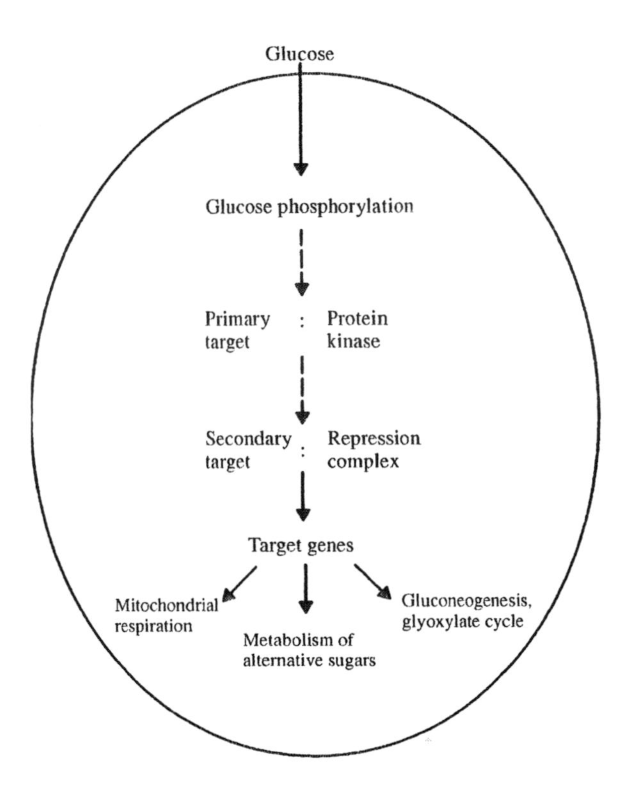

Figure 5.10. Summary of the glucose repression pathway in *S. cerevisiae*. Adapted from Thevelein (1994).

With regard to the involvement of glucose repression in control of sugar uptake, this has been described as part of a 'global regulatory system' in *S. cerevisiae* (Carlson, 1987). This is because levels of available glucose (generally above around 5 mM) affect not only respiratory enzymes but also the expression of several genes involved in sugar uptake and utilization (e.g.

SUC, *GAL* and *MAL* genes for sucrose, galactose and maltose utilization, respectively).

Catabolite inactivation, which is more rapid than repression, is thought to be due to glucose-induced deactivation of a limited number of key enzymes like fructose 1,6-bisphosphatase (Holzer, 1976). Enzyme inactivation occurs firstly (and rapidly) by enzyme phosphorylation, followed by a slower vacuolar proteolysis of the enzyme. François *et al.* (1984) have shown that reversible inactivation of fructose 1,6-bisphosphatase occurs in *S. cerevisiae* by phosphorylation mediated by a cyclic AMP-dependent protein kinase and triggered by glucose-stimulated increases in both cAMP and fructose 2,6-bisphosphate levels in the cell. Note that repression and inactivation are not necessarily mutually exclusive mechanisms. For example, both may be operating in the regulation of maltose uptake systems in *S. cerevisiae*.

The role of cyclic AMP as a second messenger in the regulation of catabolite repression and inactivation in *S. cerevisiae* has been discussed by Wills (1990) and Figure 5.11 summarizes the

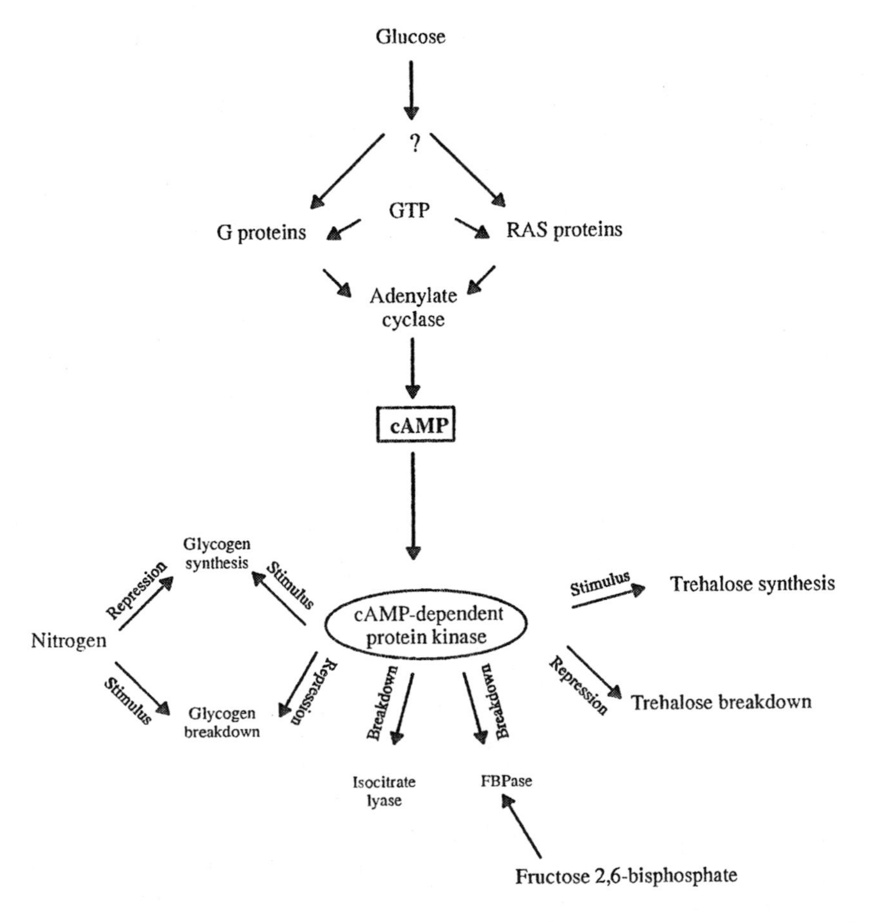

Figure 5.11. cAMP-mediated regulation of carbon metabolism in yeasts. The figure shows pathways leading to the regulation of the cAMP-dependent protein kinase, and the various cellular functions that it regulates in turn. Also shown are some of the other factors that are known to interact with this regulation, such as nitrogen and fructose 2,6-bisphosphate. Adapted from Wills (1990).

principal pathways of such regulation. Although Gancedo (1992) has questioned if cAMP plays a *critical* role in catabolite repression in *S. cerevisiae*, Hoffman and Winston (1991) have shown that the phenomenon in *Schizosaccharomyces pombe* is linked to the activation of a cAMP-dependent protein kinase.

The Crabtree effect may also be due to a saturation of the **limited respiratory capacity** of yeast cells (Käppeli and Sonnleitner, 1986). Thus, glucose-sensitive (Crabtree-positive) yeasts like *S. cerevisiae* may possess a limited oxidative capacity when grown on glucose which leads to an *overflow* reaction at pyruvate (see Figure 5.12). When the respiratory capacity is saturated, ethanol is formed.

The respiratory capacity of *S. cerevisiae* is not a fixed value and can be dynamically adapted to the needs of the cell. However, Sonnleitner and Hahnemann (1994) have shown that in this species there is always a minimal respiratory capacity which can be exploited if necessary. A more meaningful way of describing fermentative metabolism in the presence of oxygen would be to refer to it as **respirofermentative**, or oxidoreductive (Käppeli and Sonnleitner, 1986). Alexander and Jeffries (1990) have calculated the contributions to growth of ATP produced by fermentation and by respiration when these two pathways are operating simultaneously in different yeasts. For example, *S. cerevisiae* is respiration-limited and exhibits respirofermentation, but certain *Candida* spp. (e.g. *C. shehatae*) are not respiration-limited and do not exhibit respirofermentation. Fiechter and Seghezzi (1992) have discussed the mechanisms of such behaviour based on control of glucose transport, which in turn governs metabolic overflow reactions in the cell.

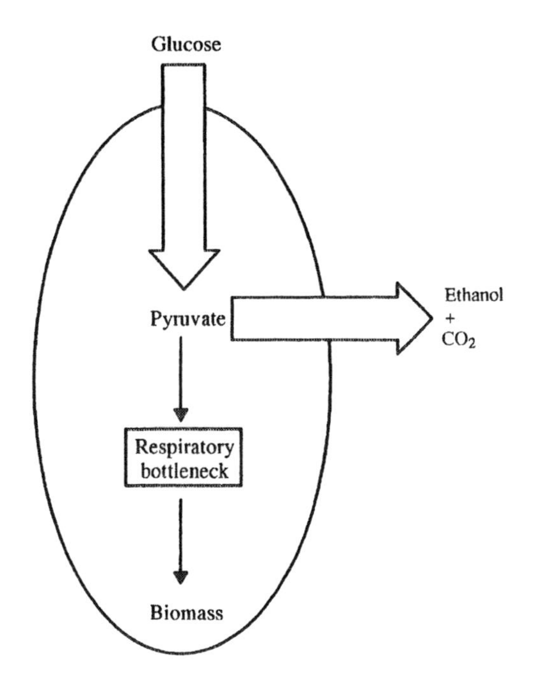

Figure 5.12. The respiratory bottleneck in *S. cerevisiae.* The scheme indicates overflow metabolism of glucose to ethanol when respiration of pyruvate is limited (due either to glucose overload – the Crabtree effect – or to anaerobic conditions – the Pasteur effect). Adapted from Ratledge (1991).

The regulation of respiration and fermentation is fundamental to the success of several industrial processes which exploit yeast metabolism. In *S. cerevisiae*, optimization of respiration is important in the production of yeast biomass (e.g. for the food industry), while optimization of fermentation is important in potable and industrial ethanol production. For example, in the propagation of baker's yeast, sugar levels (in molasses) must be controlled by an incremental feeding regime (i.e. fed-batch cultures – see Chapter 4) in order to avoid the Crabtree effect and circumvent suppression of respiration and the accumulation of ethanol.

The Custers effect

This may be defined as the transient inhibition of fermentation by anaerobiosis and is observed when small levels of oxygen or organic hydrogen-acceptors (e.g. acetoin) abolish this anaerobic inhibition of fermentation. In other words, oxygen actually stimulates ethanol production. This phenomenon, which occurs in *Brettanomyces* and *Dekkera* spp., is due to a lack of NAD^+ caused by secretion of acetate by cells which alters the NAD^+/NADH ratios and this results in an unfavourable redox balance in the cell:

$$C_6H_{12}O_6 + 4NAD^+ + 2H_2O \rightarrow 2CH_3COOH + 2CO_2 + 4NADH + 4H^+$$

NADH reoxidation can only occur very slowly under anaerobic conditions. Oxygen, as an exogenous hydrogen acceptor, can rectify the redox imbalance and regenerate NAD^+. The Custers effect is therefore explained on the basis of a disturbed redox balance in the absence of oxygen (Scheffers, 1966; Wijsman *et al.*, 1984; van Dijken and Scheffers, 1986).

The Kluyver effect

Several yeasts which can ferment glucose anaerobically are able aerobically to assimilate, but not ferment, other sugars such as galactose and certain disaccharides. For example, *Candida utilis*, *Kluyveromyces wickerhamii* and *Debaryomyces yamadae* exhibit a Kluyver effect for maltose, lactose and sucrose, respectively (Kaliterna *et al.*, 1995a; Castrillo *et al.*, 1996). In these yeasts, therefore, alcoholic fermentation in the presence of disaccharides does not occur. The inability to ferment glycosides other than glucose (the Kluyver effect) appears to be quite widespread in facultatively fermentative yeasts, having been reported in 97 species (Sims and Barnett, 1978). Although the precise mechanism responsible for the Kluyver effect is currently unresolved, the phenomenon does appear to be brought about by the interplay of several factors involving lowered rates of transport and metabolism of certain sugars (Barnett, 1992). For example, Sims and Barnett (1991) have identified a reduced activity of pyruvate decarboxylase in *Candida*, *Kluyveromyces* and *Debaryomyces* species exhibiting a Kluyver effect which results in a lowered flux of non-glucose sugars through glycolysis. Another indication of fermentative metabolism being important in the expression of the Kluyver effect comes from Weusthuis *et al.* (1994) who have shown that ethanol causes suppression of maltose utilization in *C. utilis* and *D. castellii*. Perturbed sugar transport may also account for the phenomenon and Kaliterna *et al.* (1995b) have suggested that regulation in *C. utilis* may occur at the level of synthesis of the maltose transporter.

The biotechnological relevance of the Kluyver effect lies in the production of yeast biomass or heterologous proteins on inexpensive disaccharide-based growth media such as lactose-rich cheese whey or sucrose-rich molasses. Thus, facultatively fermentative yeasts exhibiting this phenomenon need not necessarily be propagated using fed-batch nutrient delivery regimes which are designed to prevent alcohol formation. This is because expression of the Kluyver effect would result in the disaccharides not being *fermented*. There would also be no need to critically control dissolved oxygen tension during yeast propagation in an effort to avoid alcoholic fermentation (Castrillo *et al.*, 1996).

5.2.1.4 Oscillatory Metabolism in Yeasts

Several phenomena have been documented in yeasts which can collectively be described as *metabolic oscillations*. Such oscillatory behaviour can either be linked to cell cycle phases and have a time-scale periodicity of hours, or be independent of the cell cycle and last for seconds or minutes. The oscillations in question are those mainly associated with fluctuations in glycolytic metabolism in *Saccharomyces* spp.

With regard to cell cycle-related fluctuations in sugar metabolism, several workers have described oscillations during spontaneously synchronized growth of *S. cerevisiae* in continuous culture (e.g. Parulekar *et al.*, 1986; Porro *et al.*, 1988; Chen and McDonald, 1990; Filipini *et al.*, 1991; Münch *et al.*, 1992). Duboc *et al.* (1996) have shown that in synchronous populations of *S. cerevisiae* in a glucose-limited chemostat, glycogen content decreased but CO_2 evolution rate, O_2 consumption rate and metabolic heat production rate all increased during S phase. The onset of DNA synthesis in *S. cerevisiae* under these conditions appeared to be associated with a higher catabolic activity and greater reductive metabolism. This was most dramatically observed when cellular ethanol production was followed during synchronous cell divisions (Figure 5.13). Very rapid secretion of ethanol appeared at the onset of S phase and this was followed by its rapid consumption on entry into G2 phase. Similar cell cycle-related output of ethanol by *S. cerevisiae* has been observed previously (Heinritz *et al.*, 1985; Käppeli *et al.*, 1985; Parulekar *et al.*, 1986; Martegani *et al.*, 1990). It has been proposed that ethanol acts as an intercellular signal to trigger and sustain growth-related oscillations (Münch *et al.*, 1992).

Short-period autonomous oscillatory metabolism occurs in yeast cells on a much more rapid time-scale than cell cycle-related oscillations. These have been recognized for many years (Chance *et al.*, 1964; Hommes, 1964) and are usually linked to glycolytic metabolism. The phase and amplitude of the oscillations in question depend on the growth conditions (e.g. periodicity of glucose supply), cell density and whether or not intact cells or cell-free extracts are employed. Oscillations in the levels of several intracellular parameters (e.g. NADH, ATP, acetate, glycogen, pyruvate, intracellular pH) as well as metabolic changes (e.g. O_2 uptake rate, ethanol production rate, heat flux) have been recorded by several workers (e.g. Satroudinov *et al.*, 1992; Abel *et al.*, 1994; Bier *et al.*, 1996; Keulers *et al.*, 1996; Richard *et al.*, 1996b; Teusink *et al.*, 1996). Many of these oscillations are sinusoidal periodic fluctuations of the order of seconds or minutes.

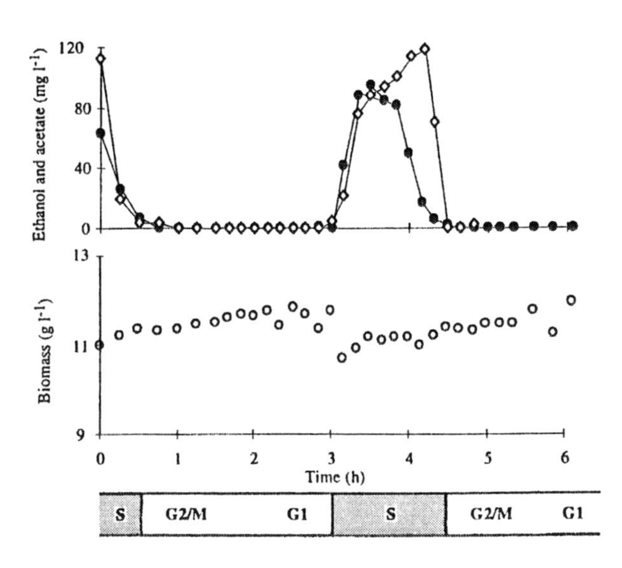

Figure 5.13. Oscillatory metabolism in a continuous-synchronous yeast culture. The graph shows the time course of ethanol (•), acetic acid (◇) and biomass (○) concentration during spontaneous oscillations in a synchronized *S. cerevisiae* population grown in a glucose-limited chemostat (dilution rate, $0.1\ h^{-1}$). Reprinted from Duboc *et al.* (1996) Physiology of *Saccharomyes cerevisiae* during cell cycle oscillations. *Journal of Biotechnology*, 51, 57–72, with kind permission of Elsevier Science-NL, Amsterdam, The Netherlands.

In contrast to transient damped glycolytic oscillations (e.g. Chance *et al.*, 1978) which are observable in batch culture situations (e.g. by glucose-pulsing), continued or sustained oscillations are observed in chemostat cultures at low dilution rates when the phase, period and amplitude of the oscillations remains stable for extended time intervals. Various factors have been proposed as agents which maintain the synchrony of sustained glycolytic oscillations in *S. cerevisiae* (see Table 5.5), although no conclusive evidence exists for any one intra – or intercellular signal.

5.2.2 Gluconeogenesis and Carbohydrate Biosynthesis

The growth of yeasts on non-carbohydrate substrates as sole carbon sources (e.g. ethanol, glycerol, succinate and acetate) necessitates the synthesis of sugars required for macromolecular biosynthesis (Figure 5.14). Several yeast species have the ability to synthesize complex cellular polysaccharides from simple short-chain carbon compounds. Central to this capability is the conversion of pyruvate to glucose which is referred to as gluconeogenesis. The process requires energy in the form of ATP and reducing power in the form of NADH.

Table 5.5. Agents implicated as synchronization affectors of glycolytic oscillations in *S. cerevisiae*.

Agent	References
Diffusible glycolytic intermediates	Aldridge and Pye (1976)
Ethanol	Aon *et al.* (1991)
CO_2	Keulers *et al.* (1996)
Acetaldehyde	Richard *et al.* (1996a)
ATP	Richard *et al.* (1996b)

Glycolysis is an amphibolic pathway (like the citric acid cycle) since it fulfils both catabolic and anabolic functions and the pathway of gluconeogenesis may be regarded as a reversal of glycolysis. Many yeasts, including *S. cerevisiae*, can alternate between a glycolytic mode of metabolism in the presence of glucose and a gluconeogenic mode in its absence. The following enzyme-catalysed steps are the gluconeogenic alternatives of the irreversible glycolytic steps catalysed by hexokinase, phosphofructokinase and pyruvate kinase:

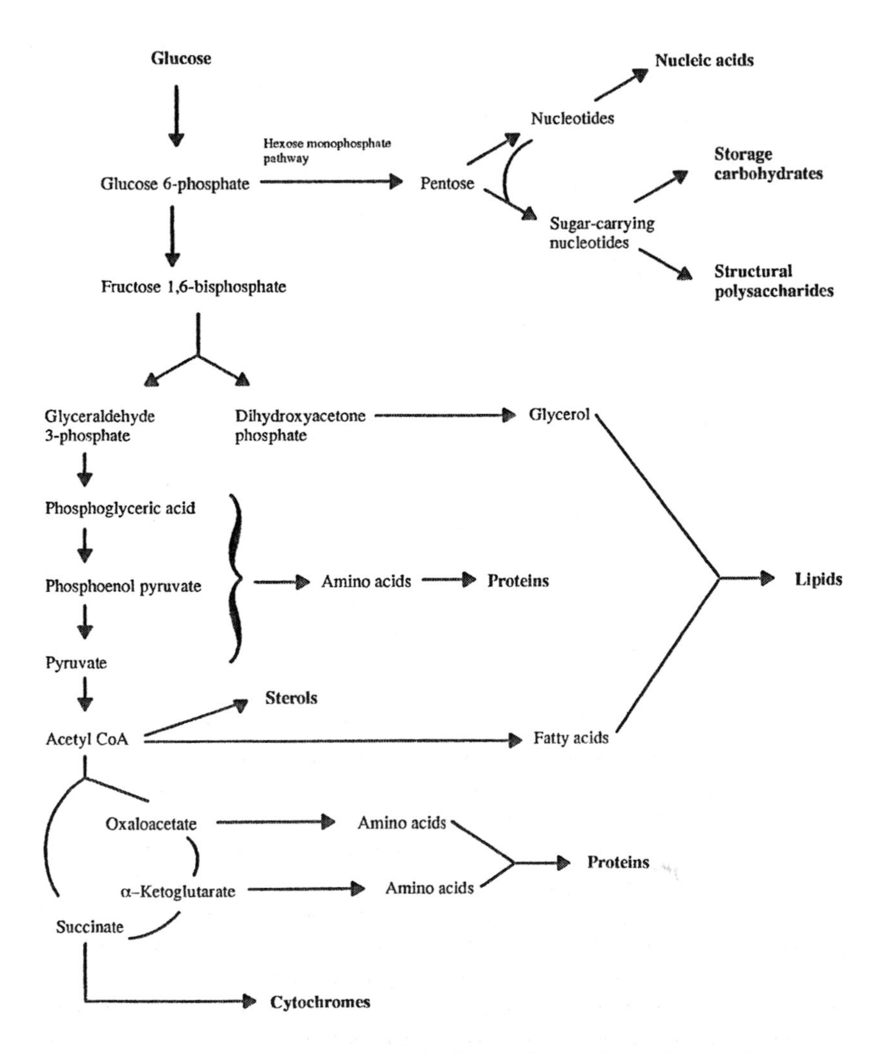

Figure 5.14. Outline of macromolecular biosyntheses from glucose in yeasts.

1. Oxaloacetate + GTP $\xrightarrow{\text{Phosphoenol pyruvate carboxykinase}}$ Phosphoenolpyruvate + CO_2 + GDP

2. Fructose bisphosphate + H_2O $\xrightarrow{\text{Fructose 1,6-bisphosphatase}}$ Fructose 6-phosphate + Pi

3. Glucose 6-phosphate + H_2O $\xrightarrow{\text{Glucose 6-phosphatase}}$ Glucose + Pi

Overall:

2Pyruvate + 4ATP + 2GTP + 2NADH + $2H^+$ + $4H_2O$

↓ Gluconeogenesis

Glucose + $2NAD^+$ + 4ADP + 2GDP + 6Pi

The pathway of gluconeogenesis from ethanol (being one of the most common gluconeogenic substrates for yeasts in nature) is shown in Figure 5.15. In *S. cerevisiae*, the first step in this pathway is catalysed by alcohol dehydrogenase II (ADH2) which is one of the four isoenzymes of ADH present in this yeast (Johnston and Carlson, 1992; see Table 5.6). Wills (1990) and Reid and Fewson (1994) have discussed the regulation of *ADH* genes in yeasts.

Gluconeogenic enzymes such as fructose 1,6-bisphosphatase are tightly regulated by yeast growth conditions. For example, in carbon-limited chemostat cultures of *S. cerevisiae*, DeJong-Gubbels *et al.* (1995) have shown that gluconeogenic enzymes are derepressed by ethanol and repressed by glucose. Glucose can also switch off gluconeogenesis by inactivation of key enzymes such as fructose 1,6-bisphosphatase as well as by repression of several genes encoding gluconeogenic enzymes. In this way, actively growing yeast cells benefit from continued glycolysis, whereas resting cells or cells with excess carbon source benefit by storing carbohydrates synthesized by gluconeogenesis. The availability of glucose clearly plays an integral role in governing gluconeogenesis and the switching between glycolysis and gluconeogenesis in yeasts.

Although some enzymes of gluconeogenesis and glycolysis are subject to induction and repression by glucose, several enzymes involved in both pathways (e.g. glyceraldehyde 3-phosphate dehydrogenase and enolase) are constitutively expressed (Chambers *et al.*, 1995).

5.2.2.1 Structural Polysaccharide Synthesis

The principal structural polysaccharides in yeasts are those associated with the cell wall and include: mannans, glucans and chitin. A typical chemical analysis of yeast cell walls would comprise around 30% each of glucans and mannans. Knowledge of the biosynthesis and assembly of cell wall components is fundamental to an understanding of yeast growth, cell division and morphogenic differentiation (see Chapters 2 and 4).

All sugar polymerization reactions employ *sugar nucleotides* as substrates which are formed by pyrophosphorylase enzymes. For example:

Glucose 1-phosphate + UTP → UDP-glucose + PPi

Mannose 1-phosphate + GTP → GDP-mannose + PPi

Polysaccharide biosynthesis then involves transglycosylase enzymes to transfer the glycosyl moieties to intermediate acceptor molecules, such as *dolichols*. Dolichols are monophosphates of isoprenoid alcohols (C_{14-18}) which act as lipid carriers in the initial stages of mannoprotein biosynthesis in yeasts.

Glucan synthesis in yeasts involves the activity of specific vesicular or plasma membrane-associated β-glucan synthetases which are responsible for the assembly of both β-1,3 linkages and β-1,6 branches of cell wall glucan. In *S. cerevisiae*, 1,3-β-D-glucan synthetase is involved in the biosynthesis of cell wall β-1,3 and β-1,6-glucan and consists of the following two subunits which are integral components of the plasma membrane:

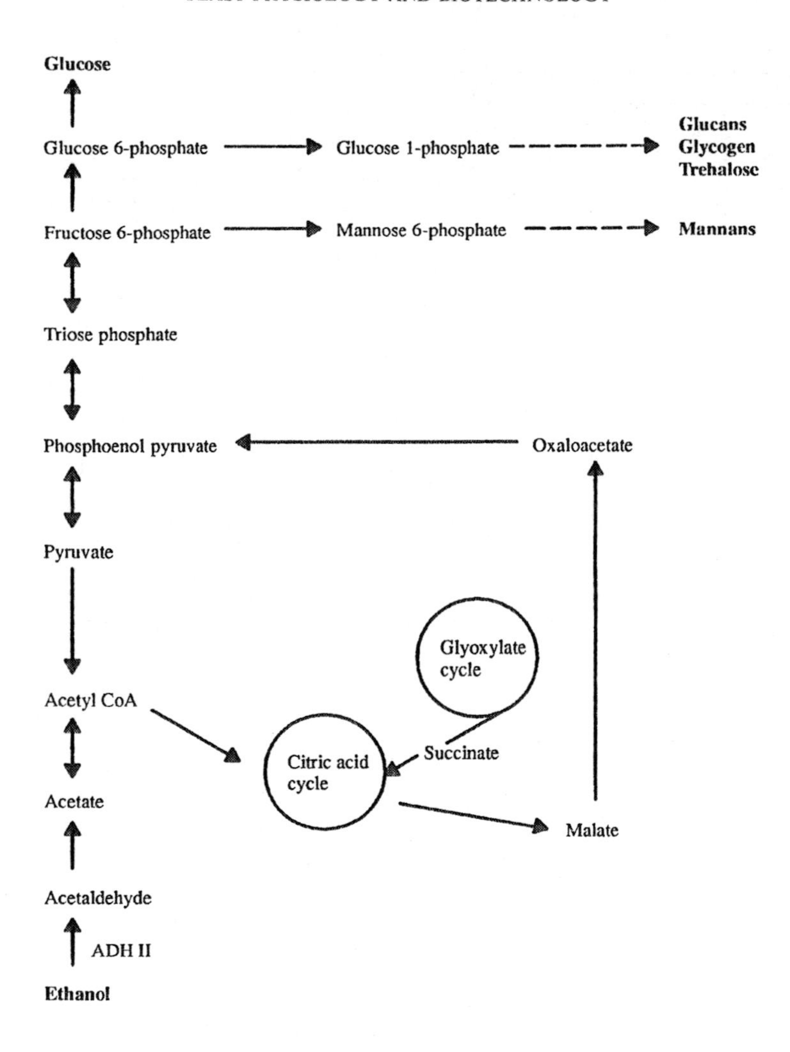

Figure 5.15. Gluconeogenesis from ethanol in yeasts.

- The GTP-binding *activating factor*, from the small G-protein family
- The UDP-glucose-binding *catalytic factor*.

Although several genes have been implicated in cell wall glucan biosynthesis in *S. cerevisiae* (e.g. *KNR4*, *HKR1*, *FKS1*, *ETG1*, *CWH51*, *CWH52*, *KRE* series, etc.), the structural gene for the catalytic subunit of glucan synthetase remains to be identified.

Chitin is a polymer of *N*-acetylglucosamine and its biosynthesis is important in the budding process and in dimorphism of yeasts. The activity and regulation of chitin synthetases have been reviewed by Cabib *et al.* (1990, 1996). Chitin synthetase catalyses the transfer of *N*-acetylglucosamine from UDP-*N*-acetylglucosamine to an elongating chitin chain. In *S. cerevisiae*, three distinct chitin synthetases with separate physiological functions have been identified (Table 5.7).

The direct precursor of chitin, UDP-*N*-acetylglucosamine, is synthesized by *N*-acetylglucosamine phosphate mutase, the gene for which in *S.*

Table 5.6. Isoenzymes of alcohol dehydrogenase in *S. cerevisiae.*

Enzyme	Gene	Location	Function
Alcohol dehydrogenase I	*ADH 1*	Cytoplasmic	Reduction of acetaldehyde to ethanol and regeneration of NAD. This constitutive enzyme catalyses the final step in fermentation.
Alcohol dehydrogenase II	*ADH 2*	Cytoplasmic	Formation of acetaldehyde by oxidation of ethanol. The enzyme, which is induced during respiratory growth on ethanol and repressed by glucose, catalyses the first step in gluconeogenesis from ethanol.
Alcohol dehydrogenase III	*ADH 3*	Mitochondrial	The activity of this mitochondrial enzyme increases during growth on glucose, but its physiological function is unclear.
Alcohol dehydrogenase IV	*ADH 4*		This is a cryptic enzyme of unknown function. It does reduce acetaldehyde to ethanol, but is undetectable in most laboratory strains of *S. cerevisiae.*

Table 5.7. Chitin synthetases of *S. cerevisiae.*

Enzyme	Gene	Function
Chitin synthetase 1	*CHS1*	Uncertain function, non-essential for growth.
Chitin synthetase 2	*CHS2*	Septum formation and budding, but not required for ascospore formation.
Chitin synthetase 3	*CHS3*	Synthesis of chitin at necks between mother and daughter bud, but not in septum.

Information from Cabib *et al.* (1990).

cerevisiae (*AGM1*) has been characterized by Hofmann *et al.* (1994).

5.2.2.2 Storage Carbohydrate Synthesis

Glycogen and trehalose are the main storage carbohydrates in yeast cells (Panek, 1991). The synthesis of both these compounds commences with the formation of UDP-glucose catalysed by UDP-glucose pyrophosphorylase:

UTP + glucose 1-phosphate → UDP-glucose + PPi

Glycogen is an α-1,4-glucan with α-1,6-branches similar to starch but with a higher degree of branching. The synthesis and degradation of glycogen are controlled by environmental factors (e.g. Rowen *et al.*, 1992). The biosynthesis of glycogen is effected by glycogen synthase which catalyses the sequential addition of glucose from UDP-glucose to a polysaccharide acceptor in a linear α-1,4 linkage. Branching enzymes are responsible for the formation of α-1,6 branches. Once the supply of exogenous sugars and other essential nutrients becomes limiting, yeasts can accumulate glycogen when their cell cycle is arrested. Utilization of stored glycogen under starvation conditions by slow endogenous fermentation contributes to the *maintenance metabolism* of yeast cells by furnishing ATP required for continued cell viability. However, glycogen degradation (to glucose 1-phosphate), which occurs rapidly following replenishment of nutrients, is mediated by glycogen phosphorylase. Cyclic AMP is known to be involved in the control of glycogen degradation in *S. cerevisiae* through phosphorylation of glycogen phosphorylase (Feng *et al.*, 1991). During the initial stages of fermentation of brewer's wort, *S. cerevisiae* rapidly degrades stored glycogen. Glycogen then reaccumulates in cells when fermentation is completed. This fine control of the activities of glycogen synthase and phosphorylase will doubtless be involved in the cyclic turnover of cellular glycogen during yeast fermentation.

Quain (1988) has discussed the importance of glycogen levels in brewing yeast (*S. cerevisiae*) for sterol biosynthesis. Glycogen breakdown is accompanied by sterol formation which is essential for yeast vitality and successful fermentation. Consequently, low levels of cellular glycogen may result in insufficient yeast sterols which may impair fermentation performance. The coordination of glycogen and lipid metabolism in *S. cerevisiae* has been studied by Anderson and Tatchell (1996) who found that cells which overproduced glycogen stored less esterified lipid, while glycogen-deficient strains stored more. The use of cellular glycogen as an indicator of vitality in brewing yeast has been discussed by Lodolo, O'Connor-Cox and Axcell (1995) who have employed the mammalian hormones, glucagon and insulin, to alter yeast glycogen levels and study subsequent effects on yeast fermentation.

Trehalose is a storage disaccharide (α, α-1,1-diglucose) present in particularly high concentrations in resting and in stressed yeast cells. Trehalose-phosphate is synthesized in yeast from glucose 6-phosphate and UDP-glucose by trehalose 6-phosphate synthetase and converted to trehalose by a phosphatase:

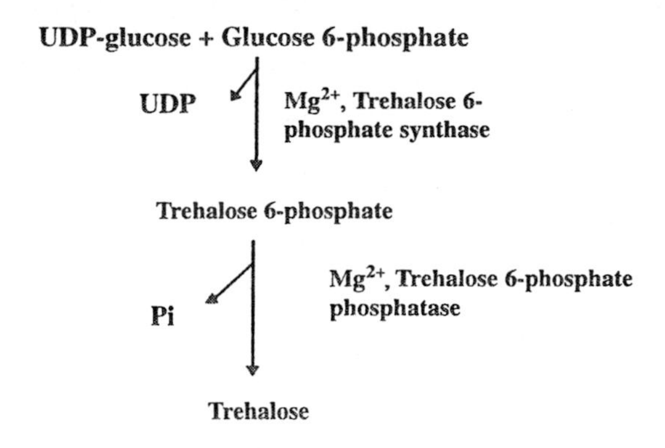

Thevelein (1996) has discussed the regulation of trehalose synthesis and breakdown (by trehalase) by cAMP-dependent phosphorylation mechanisms.

Jennings (1995) has discussed the involvement in yeast cells of a trehalose *futile cycle* in 'energy spillage' when cells move from nutrient limitation to energy limitation. This occurs when growth is no longer limited by a carbon source but by physical or chemical stress. Figure 5.16 shows a trehalose cycle in heat-shocked cells of *S. cerevisiae*. Under stressful growth conditions, *S. cerevisiae* channels carbon away from anabolic and into catabolic reactions. Although energy is still generated, it has to be dissipated due to the curtailment of cellular biosyntheses. A trehalose futile cycle may dissipate this energy.

Aspects of the role of trehalose as a stress metabolite in yeast cells were discussed in

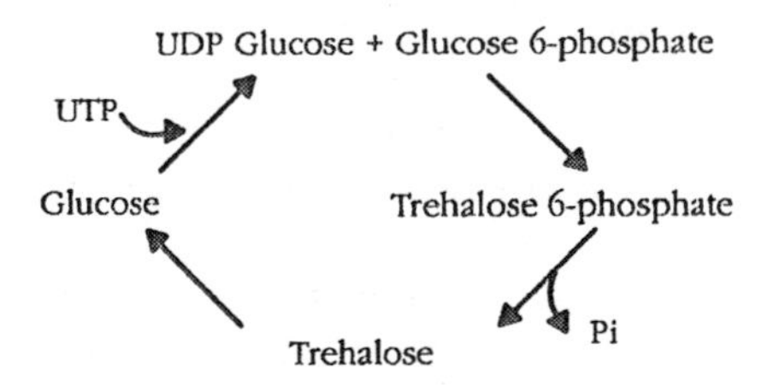

Figure 5.16. Treehalose futile cycle in *S. cerevisiae*.

Table 5.8. Roles of trehalose in yeast cell physiology.

Cellular function	Comments
Storage carbohydrate	Particularly during sporulation (e.g. trehalose is virtually the only sugar present in *S. cerevisiae* spores) and nutrient starvation. Mobilization of trehalose is also associated with growth resumption.
Growth control	Yeast growth rate and trehalose content are strongly correlated. Trehalase activity is associated with exit from the G0 phase of cell cycle. Activation of trehalase may be linked to nutrient-induced control of cell cycle progression.
Stress protectant	A positive correlation exists between cellular trehalose and resistance against dehydration, freezing, heating and osmo-stress in yeast. Toxic chemicals (ethanol, oxygen radicals, heavy metals) also induce trehalose accumulation. Slow mobilization of trehalose is important in the loss of activity of stored baker's yeast.
Glucose metabolism	Trehalose is implicated in control of glucose sensing, transport and initial stages of glucose metabolism in *S. cerevisiae*. Trehalose 6-phosphate synthase/phosphatase complex is also involved in control of glucose influx into glycolysis in other yeasts (e.g. *K. lactis* and *Sch. pombe*).

Chapter 4 and Thevelein (1996) has reviewed the involvement of trehalose in *S. cerevisiae* growth and metabolism. Table 5.8 summarizes some of the principal roles of trehalose in yeast cell physiology.

5.2.3 Metabolism of Non-Hexose Carbon Sources

The metabolism of several non-hexose (or 'non-conventional') carbon sources by yeasts will now be described and their biotechnological significance discussed. The compounds in question are biopolymers, pentoses, alcohols, polyols, hydrocarbons, fatty acids and organic acids. Chapter 3 covered membrane transport processes for many of these carbon sources and Table 3.2 represents a summary of carbon source utilization by various yeast species.

The preceding discussions in section 5.2.1 have focused primarily on the metabolism of hexose sugars, most notably glucose. Although carbon metabolism is considered primarily as glucose metabolism, it should be stressed that free glucose is quite scarce in many natural products including several yeast fermentation media derived from plants (e.g. malt wort, molasses, sulphite waste liquor, corn-steep liquor) and animals (e.g. cheese whey). Nevertheless, other saccharides can be catabolized by yeasts in the glycolytic pathway. For example, Figure 5.3 shows the entry of other hexoses (galactose, mannose and fructose) into the pathway. Several **disaccharides** can also flow into glycolysis following hydrolysis to their component monosaccharides; for example:

$$\text{Maltose} \xrightarrow[\textit{S. cerevisiae}]{\text{Transport, intracellular hydrolysis (maltase)}} \text{Glucose} + \text{Glucose} \rightarrow \text{Glycolysis}$$

$$\text{Sucrose} \xrightarrow[\textit{S. cerevisiae}]{\text{Extracellular hydrolysis (invertase), transport}} \text{Glucose} + \text{Fructose} \rightarrow \text{Glycolysis}$$

$$\text{Melibiose} \xrightarrow[\textit{S. carlsbergenisis}]{\text{Extracellular hydrolysis } (\alpha\text{-galactopyranosidase), transport}} \text{Glucose} + \text{Galactose} \rightarrow \text{Glycolysis}$$

$$\text{Lactose} \xrightarrow[\textit{Kluyveromyces marxianus}]{\text{Transport, intracellular hydrolysis } (\beta\text{-galactosidase})} \text{Glucose} + \text{Galactose} \rightarrow \text{Glycolysis}$$

$$\text{Cellobiose} \xrightarrow[\textit{Brettanomyces claussenii}]{\text{Extracellular hydrolysis } (\beta\text{-glucosidase})} \text{Glucose} + \text{Glucose} \rightarrow \text{Glycolysis}$$

5.2.3.1 Biopolymer Metabolism

Generally speaking, most yeasts require naturally occurring polymeric compounds to be extracellularly hydrolysed by non-yeast enzymes before their utilization as carbon and energy sources. For example:

$$\text{Starch} \xrightarrow{\text{Amylases, glucoamylase}} \text{Glucose}$$

$$\text{Inulin} \xrightarrow{\text{Inulinase}} \text{Fructose}$$

$$\text{Cellulose} \xrightarrow{\text{Cellulase}} \text{Glucose}$$

$$\text{Hemicellulose} \xrightarrow{\text{Hemicellulases, xylanase}} \text{Xylose, Glucose}$$

$$\text{Pectin} \xrightarrow[\text{Polygalacturonases, Pectin lyases}]{\text{Pectin esterases}} \text{Galacturonic acid}$$

$$\text{Proteins} \xrightarrow{\text{Proteinases}} \text{Amino acids}$$

$$\text{Lipids} \xrightarrow{\text{Lipases}} \text{Fatty acids}$$

Glucose-containing oligosaccharides (e.g. maltodextrins derived from starch) and polysaccharides (e.g. starch) generally require to be hydrolysed extracellularly by non-yeast enzymes before glycolysis, although some amylolytic yeasts are known. In fact, several specialized yeast species are known which have the ability to metabolize a variety of biopolymers directly, as outlined in Table 5.9. Many of these yeasts are of industrial significance in biomass-conversion biotechnology, either as a source of the hydrolytic enzymes themselves, or as a source of the appropriate genes for transfer to other organisms (including other yeasts) by recombinant DNA technology.

5.2.3.2 Pentose Sugar Metabolism

The hydrolysis of plant hemicellulose materials yields primarily D-xylose with L-arabinose as the second major pentose. The efficient utilization of these sugars by yeasts and other microorganisms is of fundamental importance in the bioconversion of renewable plant hemicelluloses to useful commodities such as single-cell protein, single-cell oil and ethanol (Olson and Hahn-Hägerdal, 1996).

The metabolism of pentose sugars by yeasts has been reviewed by Schneider (1988); Prior *et al.* (1989); Jeffries (1990); Webb and Lee (1990) and Hahn-Hägerdal *et al.* (1994). Very few yeasts are able to *ferment* pentoses to ethanol, although many yeasts can grow aerobically on pentose sugars (Toivola *et al.*, 1984). Those species not able to anaerobically ferment xylose exhibit the Kluyver effect (Jennings, 1995). For example, *Candida utilis* does not possess xylose isomerase and is unable to directly convert xylose to xylulose (Bruinberg *et al.*, 1983). Many yeasts which are unable to ferment xylose directly can, however, both respire and ferment its ketoisomer, D-xylulose in the presence of other sugars. This includes *S. cerevisiae* (Jeppsson *et al.*, 1996), although the rate of xylulose metabolism in this yeast is slow. In addition, many yeasts can convert xylose to xylitol, but are unable to grow on pentose sugars.

Table 5.10 lists several yeasts which have been identified as being able to ferment xylose to ethanol. Of these, *Candida shehatae*, *Pichia stipitis* and *Pachysolen tannophilus* have been extensively studied with regard to their biotechnological potential. These species also ferment L-arabinose, as depicted in Figure 5.18 (Singh and Mishra, 1995).

Nevertheless, several constraints exist in the industrial fermentation of pentoses to ethanol. For example, pentose-fermenting yeasts have a low ethanol tolerance. In addition, the presence

Table 5.9. Biopolymer degradation/utilization by yeasts.

Biopolymer	Examples of yeasts	Comments	References
Starch	*Candida* spp. (e.g. *C. antarctica, C. tsukubaensis*); *Schwanniomyces* spp. (e.g. *S. alluvius, S. castellii, S. occidentalis*); *Trichosporon* spp. (*e.g. T. pullulans*); *Lipomyces* spp. (e.g. *L. kononenkoae, L. starkeyi*); *Saccharomycopsis* spp. (e.g. *S. fibuligera*); *Saccharomyces cerevisiae var. diastaticus*; *Pichia* spp. (*e.g. P.anomala, P. holestii*); *Btrettanomyces naardenensis*	These yeasts possess extracellular α-amylase and occasionally, glucoamylase and cyclodextrinase activity.	Sills and Stewart (1982); Tubb (1986); De Mot (1990); Fogarty and Kelly (1990); Steyn and Pretorius (1995); Ray and Nanda (1996)
Inulin	*Kluyveromyces* spp. (e.g. *K. marxianus*); *Candida* spp. (e.g. *C. kefyr*)	These yeasts possess β-fructosidase activity to degrade inulin and levans to fermentable fructose.	Guiraud and Galzy (1990)
Hemicellulose	*Cryptococcus* spp. (e.g. *C. albidus*); *Aureobasidium* spp. (e.g. *A. pullulans*)/ *Trichosporon* spp. (e.g. *T. cutaneum*); *Pichia stipitis, Candida shehatae*	Direct conversion of hemicellulose is rare in yeasts, but a few species possess xylanase activity. Many other yeasts are able to utilize lignocellulosic hydrolysates containing pentoses and cellobiose.	Biely (1985); Jeffries (1990); Singh and Mishra (1995)
Pectin	*Candida* spp. (e.g. *C. kefyr, C. solani, C. pseudotropicalis*); *Kluyveromyces marxianus; Saccharomycopsis* spp. (e.g. *S. fibuligera, S. vini*); *Cryptococcus* spp. (e.g. *C. albidus*); *Aureobasidium pullulans*	These yeasts possess polygalacturonase activity. Some *S. cerevisiae* strains are able to utilize galacturonic acid.	Whitaker (1990); Biely and Slavikova (1994); Schwan and Rose (1994); Gainvors and Belarbi (1995).
Protein	*Candida* spp. (e.g. *C. albicans*); *Saccharomycopsis* spp. (e.g. *S. fibuligera*); *Kluyveromyces* spp.; *Rhodotorula* spp.; *Saccharomyces* spp. (e.g. *S. cerevisiae, S. carlsbergensis*)	*C. albicans* secretes acid proteases and some industrial *S. cerevisiae* strains possess very limited extracellular proteolytic activity.	Bilinski and Stewart (1990); Homma *et al.* (1990); White *et al.* (1993); Neklyudov *et al.* (1996)
Lipid	*Candida* spp. (e.g. *C. guilliermondii, C. cylindraceae*); *Pichia* spp. (e.g. *P. vini, P. miso*); *Yarrowia lipolytica; Torulopsis* spp. (e.g. *T. glabrata, T. ernobii*); *Schizosaccharomyces pombe*	These yeasts possess extracellular lipase activity.	Hirohara *et al.* (1985); Godfredsen (1990); Ratledge and Tan (1990)

Table 5.10. Xylose-fermenting yeasts.

Genus	Representative species
Candida	*C. shehatae, C. tenuis, C. blankii*
Pichia	*P. stipitis, P. segobiensis*
Pachysolen	*P. tannophilus*
Kluyveromyces	*K. marxianus, K. cellobiovorus*
Brettanomyces	*B. naardenensis*
Debaryomyces	*D. nepalensis, D. polymorpha*
Schizosaccharomyces	*Sch. pombe*

of glucose in hemicellulosic hydrolysates acts to repress xylose utilization genes. Sequential utilization of glucose then xylose increases the fermentation time (Webb and Lee, 1990).

The inability of *S. cerevisiae* to ferment xylose to ethanol can be circumvented by several approaches including the use of exogenous xylose isomerase or by recombinant DNA technology. With regard to the latter, transformation of *S. cerevisiae* has been achieved with genes encoding xylose reductase and xylitol dehydrogenase from pentose-fermenting yeasts such as *Pichia stipitis* (e.g. Kötter and Ciriacy, 1993), or with xylose isomerase genes from bacteria (Amore *et al.*, 1989). Nevertheless, such approaches have met with limited success in terms of the slow rate of xylose fermentation to ethanol. It appears that such rate limitation is due to an insufficiency of xylulose conversion in *S. cerevisiae* by the hexose monophosphate pathway (Kötter and Ciriacy, 1993).

D-Xylose metabolism in yeasts proceeds mostly via a two stage oxidative–reductive pathway (Figure 5.17). Xylose is firstly reduced to xylitol by an NADPH/NADH-linked xylose reductase, followed by xylitol oxidation to xylulose by an NAD-linked xylitol dehydrogenase. In xylose-fermenting yeasts an imbalance of the NAD$^+$:NADH redox system is avoided by reducing xylose to xylitol in the presence of NADH. NADH produced by oxidizing xylitol can then be used to reduce xylose to xylitol, thus creating a redox balance (see Bruinberg *et al.*, 1983):

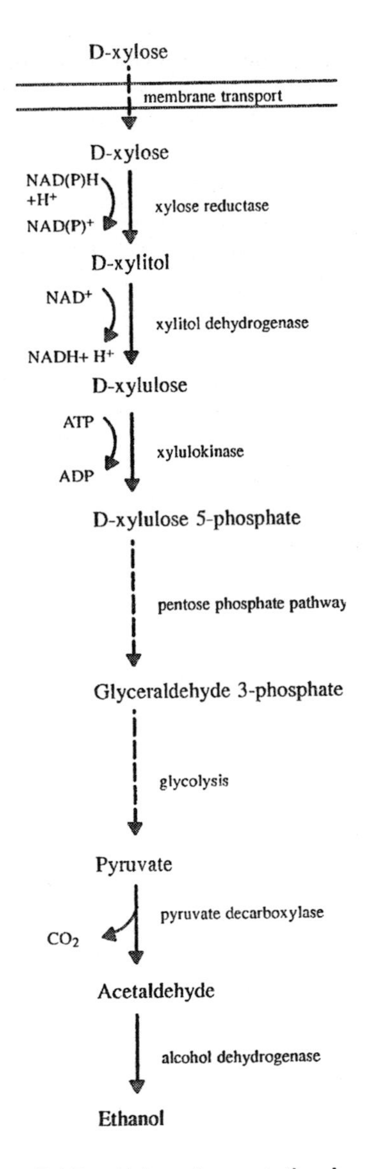

Figure 5.17. D-Xylose fermentation by yeasts.

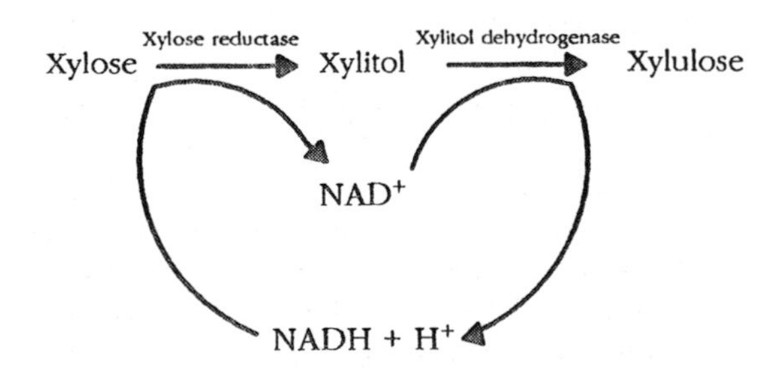

The redox cofactor requirement for xylose reductase varies in different genera of xylose-fermenting yeasts and with oxygen availability. In *Pachysolen tannophilus* and *Pichia stipitis*, there is a xylose reductase capable of functioning with both NADH and NADPH. In contrast, NADPH is the preferred cofactor for xylose reductase in anaerobic cultures of *Candida shehatae*. In most yeasts which do not or only slowly ferment xylose anaerobically, xylose reductase is NADPH-linked and they either lack or exhibit low NADH-linked reductase activity.

Very few yeasts have been reported to convert D-xylose to D-xylulose in a single step by xylose isomerase. *Candida utilis* and *Rhodotorula gracilis* are the exceptions but it is unclear how important xylose isomerase is to these yeasts in the overall catabolism of xylose (Singh and Mishra, 1995).

With regard to L-arabinose fermentation by yeasts, the pathway is quite similar to that of D-xylose fermentation and is outlined in Figure 5.18.

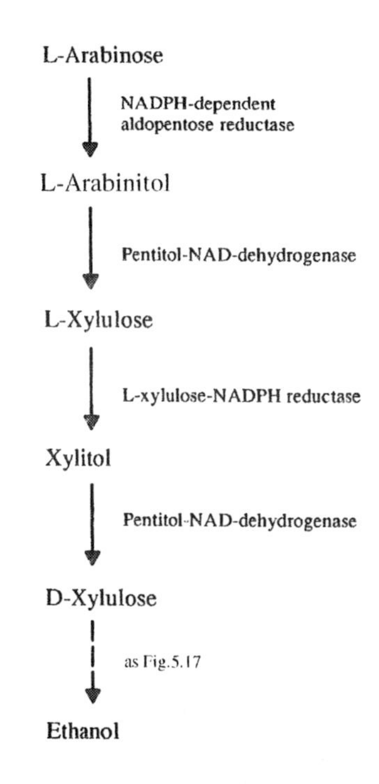

Figure 5.18. L-Arabinose fermentation by yeasts.

5.2.3.3 Lower Aliphatic Alcohol Metabolism

Many yeasts have the capability of metabolizing alcohols such as ethanol and methanol as their sources of carbon and energy. These yeasts are significant in biotechnology for biomass propagations (e.g. *Candida utilis*) and for high-level expression of heterologous genes (e.g. *Pichia pastoris*). The metabolism of **ethanol** by yeasts was considered in the discussion of gluconeogenesis in section 5.2.1. Being more reduced than glucose, ethanol is a very efficient aerobic growth substrate for yeasts. For example, Paalme *et al.* (1997) have calculated maximum growth yields (Y_{xs}) of 0.57 and 0.68 (C-mol/C-mol) for *S. cerevisiae* growing on glucose and ethanol, respectively. The practical relevance of ethanol utilization by yeasts (e.g. *C. utilis*) lies in the production of single-cell protein on ethanol-containing growth media derived from the petroleum refining industry. The following discussion will now focus on the utilization of **methanol** by yeasts.

The first report of a eukaryotic organism (namely, the yeast *Candida boidinii*) being able to utilize methanol as its sole carbon and energy source was reported by Ogata *et al.* (1969). It is now known that **methylotrophic** (i.e. methanol-utilizing) yeasts are found in at least four different genera (Table 5.11).

Methylotrophic yeasts are facultative in that they can grow on glucose as well as methanol but are generally unable to utilize other 1C compounds such as methylamine, formaldehyde and formate as sole C source (although some yeasts use methylamine as sole nitrogen source – see section 5.3.1). Some **methanotrophic** yeasts have been isolated which are capable of utilizing methane as energy source (Wolf and Hanson, 1979).

The metabolism of methanol-utilizing yeasts has been reviewed by Harder and Veenhuis (1989), Harder and Brooke (1990) and De Koning and Harder (1992). The pathways summarizing metabolism of methanol in the methylo-

Table 5.11. Methylotrophic yeasts.

Genus		Representative species
Ascomycetous genera	*Hansenula*	*H. polymorpha*
	Pichia	*P. pastoris, P. pinus*
Asporogenous genera	*Candida*	*C. utilis, C. maltosa, C. tropicalis, C. guilliermondii, C. boidinii,*
	Torulopsis	*T. sonorensis*

trophic yeast *Hansenula polymorpha*, are shown in Figure 5.19. It should be noted that methanol utilization serves to produce energy and assimilate carbon for generating yeast biomass. Methanol oxidation to CO_2 via formaldehyde provides the energy whereas fixation of formaldehyde via the **xylulose monophosphate pathway** provides glyceraldehyde phosphate from which cell constituents are built.

The first step in methanol utilization, the oxidation of methanol to hydrogen peroxide and formaldehyde, is catalysed by alcohol oxidase.

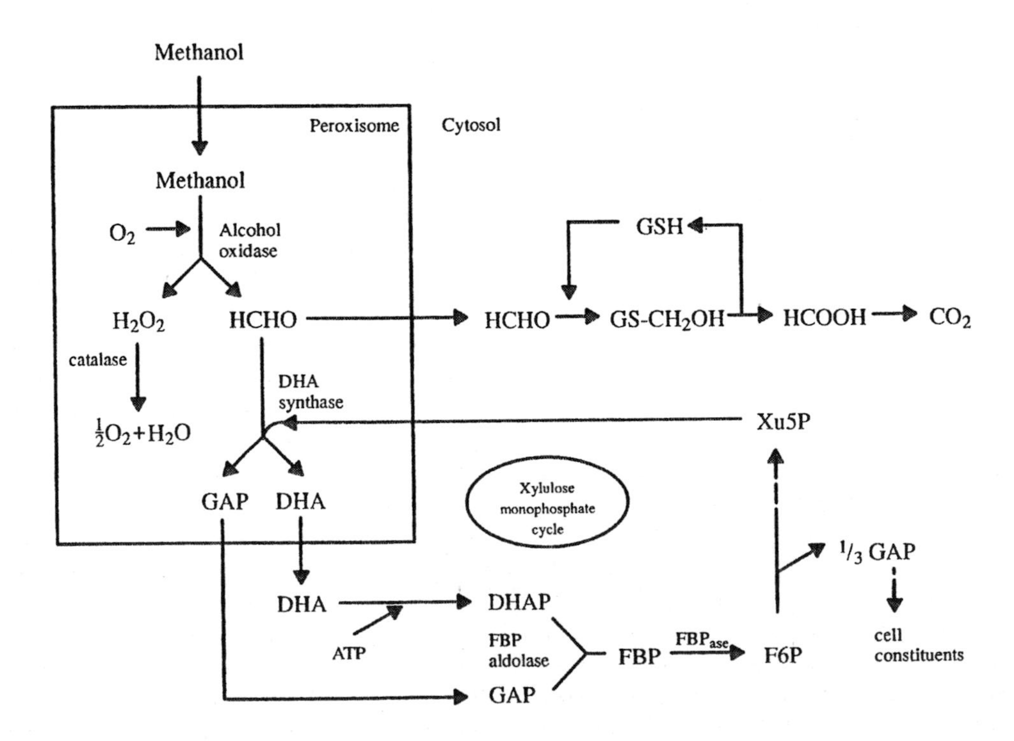

Figure 5.19. Methanol metabolism in *Hansenula polymorpha.* DHA, dihydroxyacetone; DHAP, dihydroxyacetone phosphate; GAP, glyceraldehyde 3-phosphate; FBP, fructose 1,6-bisphosphate; F6P, fructose 6-phosphate; Xu5P, xylulose 5-phosphate; GSH, reduced glutathione; GS-CH_2OH, S-hydroxymethylglutathione. Adapted from Harder and Veenhuis (1989).

This enzyme is regulated by a combination of repression (by glucose or ethanol) and induction (by methanol). During growth on methanol, methylotrophic yeasts such as *Hansenula polymorpha* and *Pichia pastoris* synthesize significant amounts of alcohol oxidase, which may account for around 30% of total cell protein. Methanol metabolism is highly compartmentalized in these yeasts. Alcohol oxidase is sequestered in specialized organelles called *peroxisomes* which can occupy up to 80% of yeast cell volume (see Chapter 2). The strong induction of alcohol oxidase by methanol in methylotrophic yeasts is of interest in recombinant DNA technology due to methanol-regulated promotion of heterologous gene expression and the fact that such yeasts can be propagated to very high cell densities with methanol as carbon source (see Faber *et al.*, 1995; Sudbery, 1995; Gellissen *et al.*, 1995; Gellissen and Melber, 1996; and Chapter 6). Other biotechnological applications of methylotrophic yeasts are discussed in Chapter 6.

The formaldehyde generated from methanol by alcohol oxidase can either be dissimilated to CO_2 by cytosolic dehydrogenase or peroxisomal catalase activity or be assimilated by the xylulose monophosphate pathway. The key enzyme of this pathway, as shown in Figure 5.19, is dihydroxyacetone synthase which catalyses the transfer of a glyceraldehyde group in the peroxisome from xylulose 5-phosphate to formaldehyde and forms glyceraldehyde 3-phosphate and dihydroxyacetone.

5.2.3.4 Sugar Alcohol Metabolism

Osmotolerant yeasts capable of growing in high sugar or salt environments synthesize polyols (sugar alcohols) such as glycerol which function as compatible solutes in osmoregulation (see Chapter 4). With regard to polyol catabolism, over 50% of yeasts surveyed (Barnett, 1981) can aerobically use glycerol, D-glucitol (sorbitol) and D-mannitol while fewer than 10% of species utilized galactitol.

With regard to **glycerol** catabolism, this polyol represents a non-fermentable carbon source for many species of yeast, including *S. cerevisiae*. Some yeasts, notably *Candida utilis*, can grow on glycerol as readily as on glucose (Gandedo, Gancedo and Sols, 1968). After translocation of glycerol by free diffusion or by facilitated permeation through glycerol channels (in *S. cerevisiae*) or by active transport (in osmophilic yeasts), glycerol is phosphorylated to glycerol 3-phosphate by glycerol kinase and then oxidized to dihydroxyacetone phosphate, before gluconeogenic reactions:

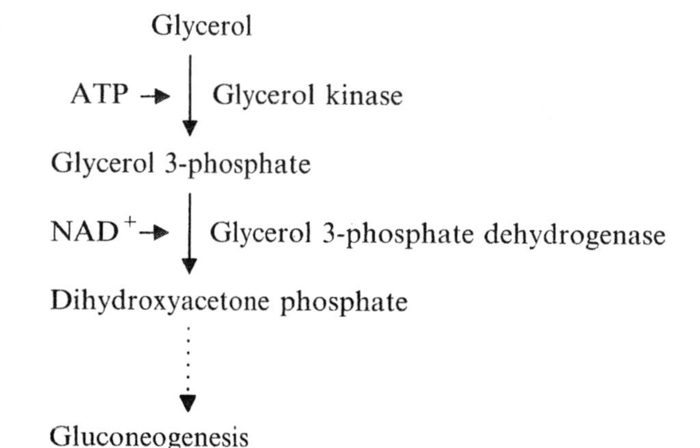

Glycerol utilization is important for the growth of *oleaginous* yeasts on lipids. Following enzymatic breakdown of lipids by extracellular lipases, acylglycerols and fatty acids accumulate. In lipolytic yeasts like *Yarrowia lipolytica*, diacylglycerols are preferentially hydrolysed over monoacylglycerols (Ratledge and Tan, 1990). Utilization of glycerol and fatty acids in oleaginous yeasts is concomitant, although glycerol metabolism may be inhibited by the yeasts' metabolic adjustment to fatty acid catabolism.

Metabolism of **mannitol** and other hexitols like D-glucitol (sorbitol) and D-galactitol involves specific hexitol phosphate dehydrogenase activity (Figure 5.20). Although *Candida utilis* has been shown to produce a hexitol dehydrogenase with broad specificity, Quain and Boulton (1987) reported a lack of a common dehydrogenase for both mannitol and sorbitol in *S. cerevisiae*.

Strains of *S. cerevisiae* differ in their ability to

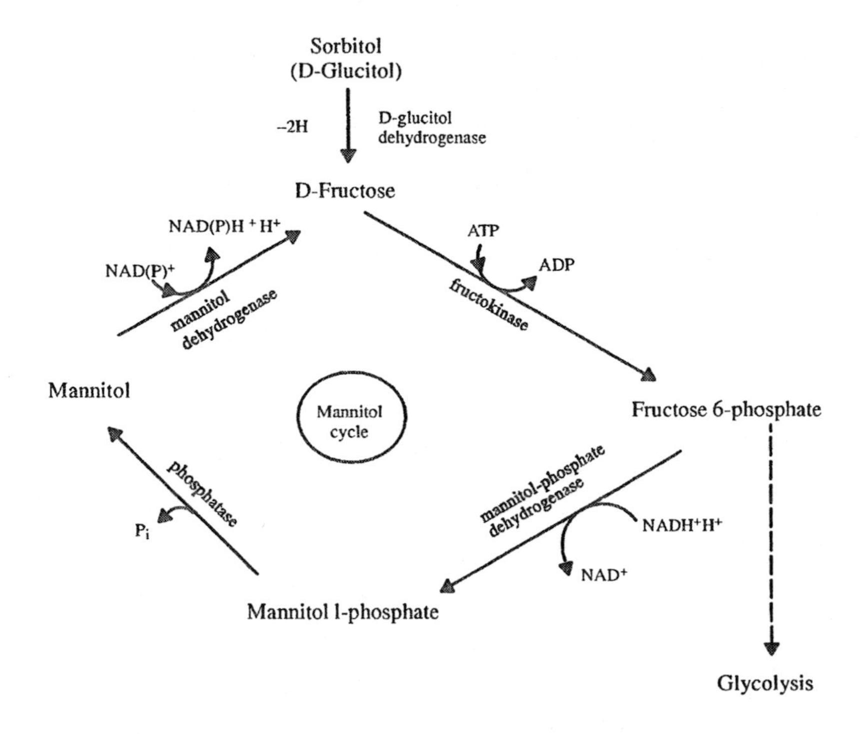

Figure 5.20. Pathways of mannitol and sorbitol metabolism in yeasts.

assimilate mannitol. Quain and Boulton (1987) have shown that several industrial strains of *S. cerevisiae* grow well with mannitol as sole carbon source. A few strains able to grow on mannitol were also able to grow on sorbitol. Yeasts unable to assimilate mannitol apparently have an impaired ability to transport this sugar alcohol. Aerobic respiration is absolutely essential for mannitol metabolism and in the presence of oxygen an NAD-dependent mannitol dehydrogenase was detected in *S. cerevisiae* by Quain and Boulton (1987). No activity was found when NADP replaced NAD.

With regard to **xylitol**, many yeasts convert xylose to xylitol, but xylitol is generally a poor carbon source for growth due to its limited permeability into yeast cells (Singh and Mishra, 1995). Xylitol can be re-oxidized to D-xylulose by most yeasts, including *S. cerevisiae*, by xylitol dehydrogenase. Pentitols like xylitol are major by-products of pentose fermentation and efficiency would be increased if xylitol could be converted to ethanol. *Pichia angophorae* appears to be one of the few xylitol-fermenting yeasts (Singh and Mishra, 1995).

5.2.3.5 *Hydrocarbon Metabolism*

The metabolism of hydrocarbons by yeasts and fungi is significant in several areas of environmental and industrial biotechnology. For example, in the biodeterioration of aircraft fuel, in the treatment of oil pollution, in oil-recovery systems and in the production of single-cell protein (Boulton and Ratledge, 1984).

The hydrocarbons most readily assimilable by yeasts are straight-chain alkanes in the C_{10}–C_{20} range (above C_{20}, alkanes are solid below 35°C). Several yeast genera have species which possess *n*-alkane-utilizing capabilities, most notably *Candida*, *Pichia*, *Kloeckera*, *Torulopsis*, *Rhodotor-*

ula and *Saccharomycopsis* (Shennan and Levi, 1974; Cartledge, 1987). The *Candida* species include *C. tropicalis*, *C. intermedia* and *C.* (*Yarrowia*) *lipolytica*.

The initial step in alkane utilization by yeasts is their incorporation into the cells. This has been studied in certain *Candida* spp. (e.g. *C. tropicalis*) and involves specialized adherence and emulsification mechanisms which reduce the interfacial tension of the hydrophobic liquid (see Chapter 3). Once the alkanes have been internalized, they are oxidized by yeasts to their corresponding fatty acids before further metabolism to acetyl CoA which can then enter the glyoxylate cycle and gluconeogenesis. The following scheme (Figure 5.21) shows the initial steps in *n*-alkane metabolism by yeasts and Figure 5.22 represents an overview of the oxidative pathways involved.

The initial oxidative stage is linked to an electron carrier system in which cytochrome P450 monooxygenase is the terminal oxidase which utilizes oxygen and electrons derived from either NADPH or NADH via a flavoprotein-cytochrome P450 reductase (Boulton and Ratledge, 1984). Rehm and Reiff (1981) and Käppeli (1986) have discussed the role of cytochromes P450 in the metabolism of hydrocarbons by yeasts.

As was the case with methanol utilization, the metabolism of *n*-alkanes by yeasts is a highly compartmentalized process. The initial oxidative steps take place in microsomes and peroxisomal enzymes carry out the remaining degradation to intermediates that are transported to mitochondria (see Figure 5.22).

5.2.3.6 *Fatty Acid and Lipid Metabolism*

Pathways of fatty acid catabolism and anabolism and their regulation have been studied extensively in yeasts (e.g. Ratledge and Evans, 1989; Paltauf *et al.*, 1992). Fatty acids available to yeasts for catabolism include those derived from microsomal alkane oxidation or extracellular lipolysis of fats or those supplied exogenously in the growth medium. The fatty acids are catabolized by β-oxidation in peroxisomes (Figure 5.23). Note that peroxisomal β-oxidation differs from the system in mitochondria in the involvement of catalase in re-oxidizing $FADH_2$ and in the mechanism of re-oxidizing NADH (Boulton and Ratledge, 1984).

With regard to fatty acids derived from fat breakdown, several lipolytic yeasts are known (e.g. *Candida rugosa*, *Yarrowia lipolytica*) which secrete inducible lipases to degrade triacylgycerol substrates to fatty acids and glycerol, as outlined in Figure 5.24.

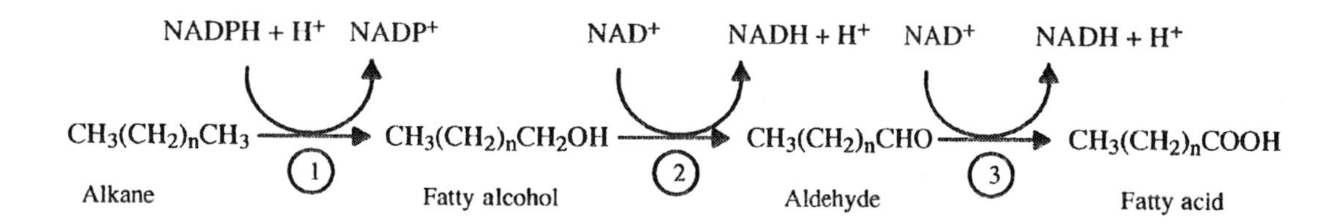

Figure 5.21. Initial steps in *n*-alkane metabolism by yeasts.

Step ① Hydroxylation of alkanes to their corresponding primary alcohols by a mixed-function monooxygenase containing cytochrome P450 and NADPH-cytochrome P450 reductase.

Step ② NAD-linked long-chain alcohol dehydrogenase.

Step ③ NAD-linked long-chain aldehyde dehydrogenase.

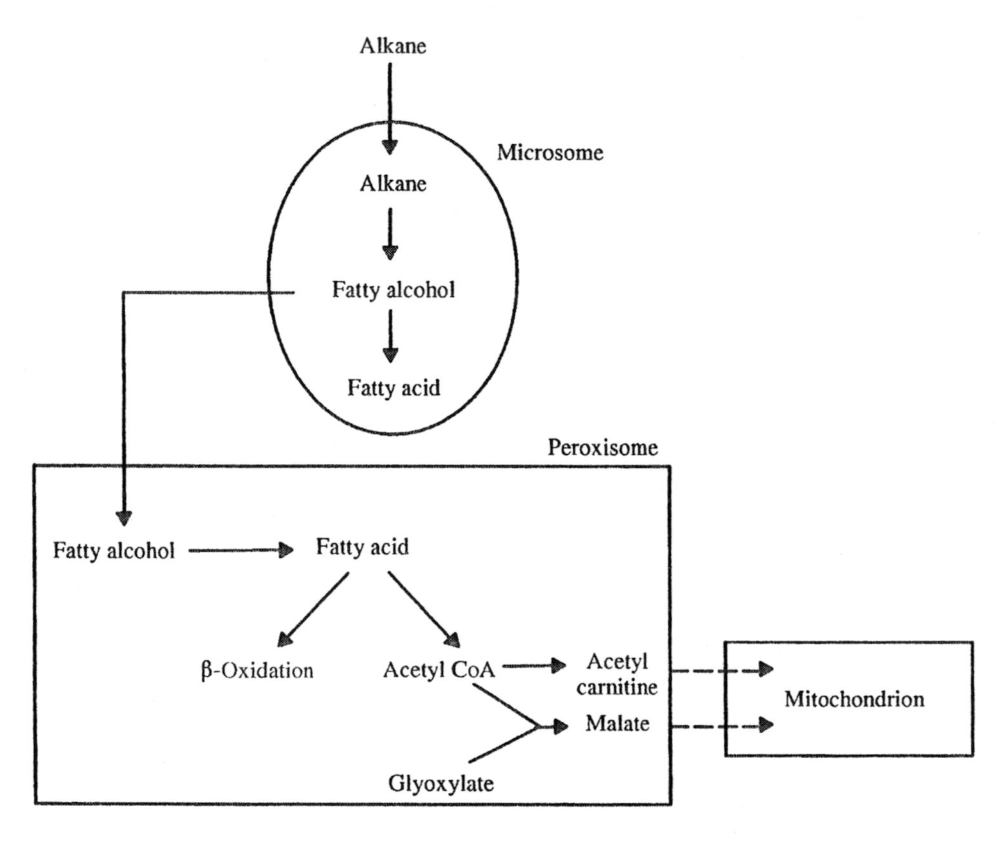

Figure 5.22. Compartmentalization of *n*-alkane metabolism in yeasts.

Fatty acid and lipid synthesis from acetyl CoA has been studied extensively in oleaginous yeasts which have the capability of growing on sugars and accumulating significant amounts of intracellular lipid. The lipid biosynthetic pathway in oleaginous yeasts is shown in Figure 5.25 and the biotechnological significance of such yeasts is discussed further below.

A key enzyme in oleaginous yeasts is ATP citrate lyase and it is the activity of this enzyme which determines whether a yeast is capable of lipid accumulation. The series of reactions leading to the synthesis of long-chain fatty acids takes place in a multienzyme complex known as *fatty acid synthase*, which exists as multiple high molecular mass subunits. The action of this complex in synthesizing yeast fatty acids has been described by Singh *et al.* (1985). The unsaturated fatty acids, which play important roles in yeast physiology (e.g. membrane integrity) and biotechnology (e.g. ethanol tolerance), include palmitoleic (16:1) and oleic (18:1) acids. Together, these compounds constitute the bulk (approx. 70%) of the fatty acids in *S. cerevisiae* cell membranes. The others consist mainly of the saturated fatty acids: primarily palmitic (16:0) with lesser amounts of stearic (18:0) and myristic (14:0) acids. The synthesis of long-chain monounsaturated fatty acids in yeasts involves an oxidative desaturation, by a fatty acid desaturase, to introduce double bonds into the saturated fatty acid. For example:

$$CH_3(CH_2)_7\text{–}\underset{H}{\overset{H}{C}}\text{–}\underset{H}{\overset{H}{C}}\text{–}(CH_2)_7\text{–CoA} \xrightarrow[\text{yeast desaturase}]{\frac{1}{2}O_2 \quad H_2O} CH_3(CH_2)_7\text{–}\overset{H}{C}\text{=}\overset{H}{C}\text{–}(CH_2)_7\text{–CoA}$$

Stearoyl CoA (18:0) (saturated fatty acid) → Oleoyl CoA (18:1) (unsaturated fatty acid)

The requirement for oxygen in the desaturase-mediated synthesis of essential unsaturated fatty acids means that *S. cerevisiae* cannot grow strictly anaerobically in the absence of these preformed compounds.

Oleaginous yeasts (Table 5.12) can accumulate appreciable quantities, occasionally as much as 70% of biomass weight, of intracellular lipid. In addition, *extracellular* lipids or glycolipids can be formed by certain yeasts in oxygenated culture conditions (Phaff, Miller and Mrak, 1978).

Lipid microdroplets which accumulate in oleaginous yeasts are composed of triacylglycerols with the following relative abundance of fatty acyl groups:

oleic (18 : 1) > palmitic (16 : 0) > linoleic (18 : 2) = stearic (18 : 0)

Limitation of nutrients (other than carbon) promotes the intracellular accumulation of lipid in oleaginous yeasts. Although yeasts do not produce any 'unusual' fatty acids such as the pharmacologically useful γ-linolenic acid (18 : 3), they nevertheless have biotechnological potential

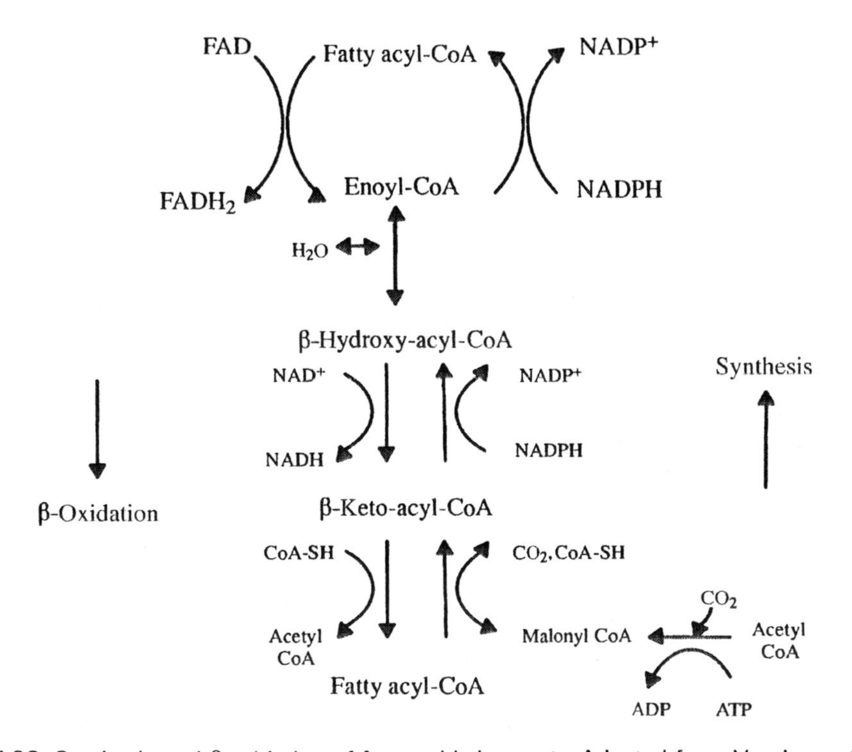

Figure 5.23. Synthesis and β-oxidation of fatty acids in yeasts. Adapted from Van Laere (1994).

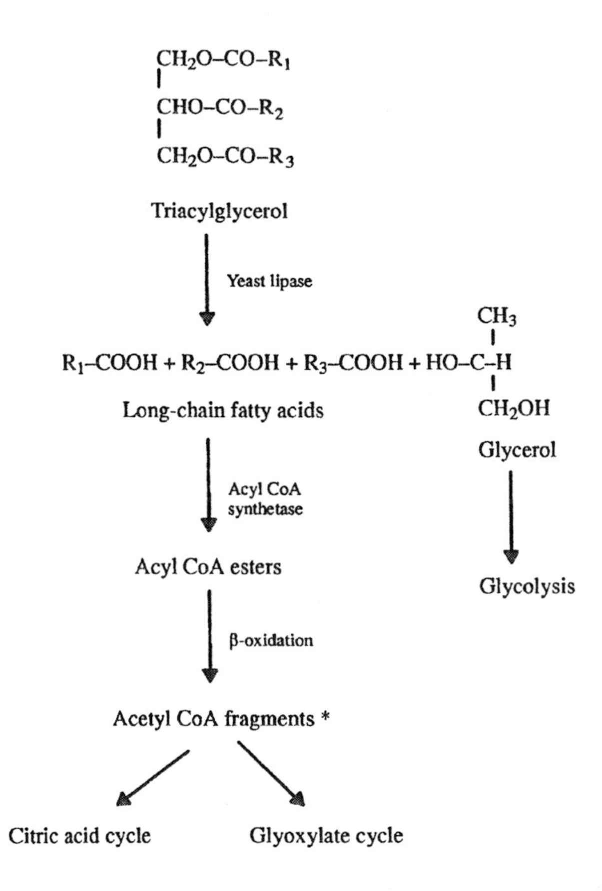

Figure 5.24. Metabolism of triacylglycerols by yeasts. *Propionyl CoA will also be formed if there was an odd number of carbon atoms in the fatty acid.

in the production of speciality/high value commodities such as cocoa butter (Hassan *et al.*, 1994) and biosurfactants (Cooper and Paddock, 1984; Johnson *et al.*, 1992). In addition, *Candida tropicalis* has been genetically modified to produce a range of long-chain dicarboxylic acids (Picataggio *et al.*, 1992). Single-cell oil production from yeasts is particularly attractive due to the fact that oleaginous yeasts grow on inexpensive and readily available fermentation feedstocks like cheese whey (Davies, 1988), molasses (Leman *et al.*, 1990) and crude oil (Olama *et al.*, 1990). Ratledge and Tan (1990) have discussed the metabolism of oleaginous yeasts and their biotechnological potential.

5.2.3.7 Sterol Biosynthesis

Sterols, particularly ergosterol, are integral structural components of yeast cell membranes and are essential in the maintenance of plasma membrane permeability. Other isoprenoid compounds produced by yeasts include ubiquinone and dolichols which are involved in respiratory metabolism and in protein modifications. Parks and Casey (1995) have reviewed the physiology of sterol metabolism in *S. cerevisiae.*

Mevalonic acid (3,5-dihydroxy-3-methylvaleric acid), the first intermediate in sterol biosynthesis, is formed by firstly condensing acetyl CoA with acetoacetyl CoA to form hydroxy-3-methylglutaryl CoA (HMG CoA) which is then reduced to mevalonic acid. The synthesis of HMG CoA by HMG CoA reductase is the first committed step in yeast isoprenoid biosynthesis. Mevalonic acid is converted to an isoprenoid structural form (dimethylallyl pyrophosphate) and isopentenyl pyrophosphate. Condensation reactions ultimately lead to squalene (a C_{30} compound) formation. Cyclization of squalene to lanosterol and further conversions lead to ergosterol (Figure 5.26). In *S. cerevisiae*, lanosterol is converted to zymosterol which is transported into the mitochondrion where it is methylated to give fecosterol. Fecosterol is then transported to the endoplasmic reticulum where it is finally converted to ergosterol (Gooday, 1994). Under anaerobic conditions, yeasts cannot synthesize ergosterol and are unable to grow well in media unsupplemented with this sterol. This is because oxygen is absolutely required for the cyclization of squalene and anaerobic cultures accumulate squalene.

Many antizymotic agents (e.g. imidazoles like miconazole and triazoles like fluconazole), which are used to combat human pathogenic yeasts (e.g. *Candida albicans*), are known to interfere specifically with sterol biosynthesis and are therefore able to disrupt yeast membrane function (Köller, 1992).

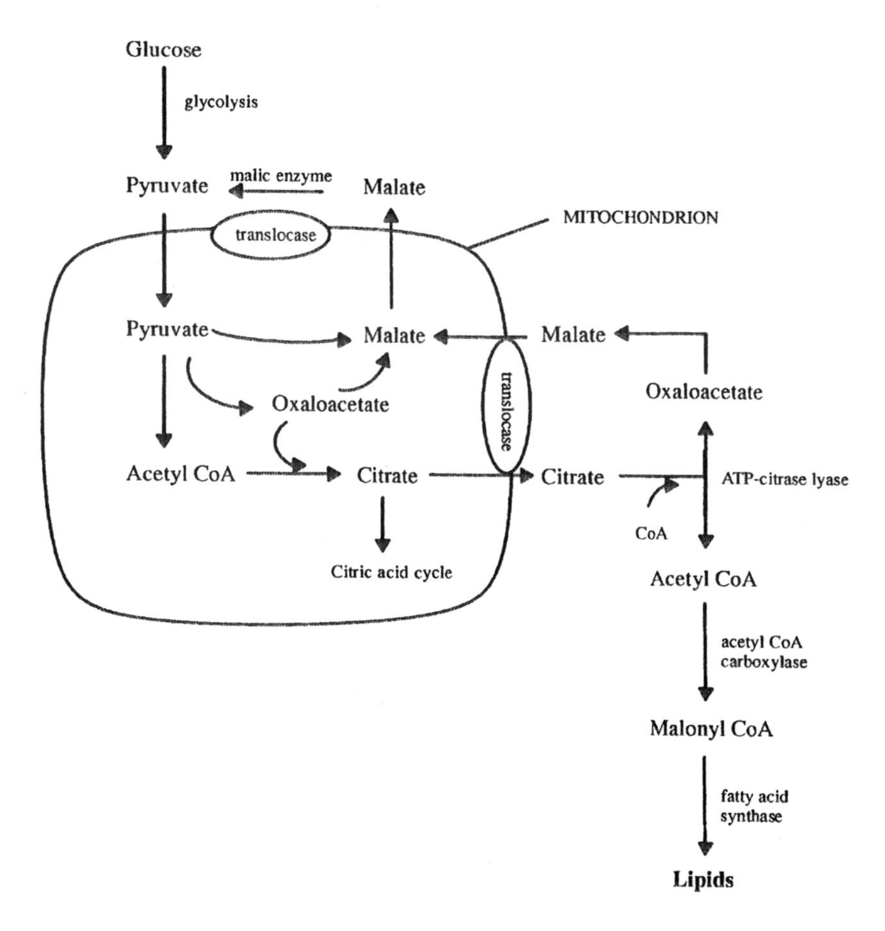

Figure 5.25. Summary of pathway of lipid biosynthesis in oleaginous yeasts.

Table 5.12. Some oleaginous yeasts.

Genus	Representative species
Candida	*C. curvata, C. diddensiae*
Cryptococcus	*C. albidus, C. laurentii*
Lipomyces	*L. lipofer, L. starkeyi*
Rhodotorula	*R. glutinis, R. graminis*
Trichosporon	*T. cutaneum*
Yarrowia	*Y. lipolytica*

Information from Ratledge and Tan (1990).

5.2.3.8 Organic Acid Metabolism

Numerous exogenously supplied organic acids (e.g. acetate, malate, lactate, pyruvate, citrate, succinate) may be utilized as growth substrates by yeasts. Even acids used as weak acid food preservatives (e.g. benzoic, propionic and sorbic acids) may be utilized by a few yeasts, including certain species of *Rhodotorula* and *Brettanomyces* (Phaff *et al.*, 1978). However, for yeasts to metabolize organic acids it is necessary for them to provide: NADPH for reductive biosynthesis and various intermediates (e.g. acetyl CoA, malate, pyruvate) for the synthesis of fatty acids, sterols and amino

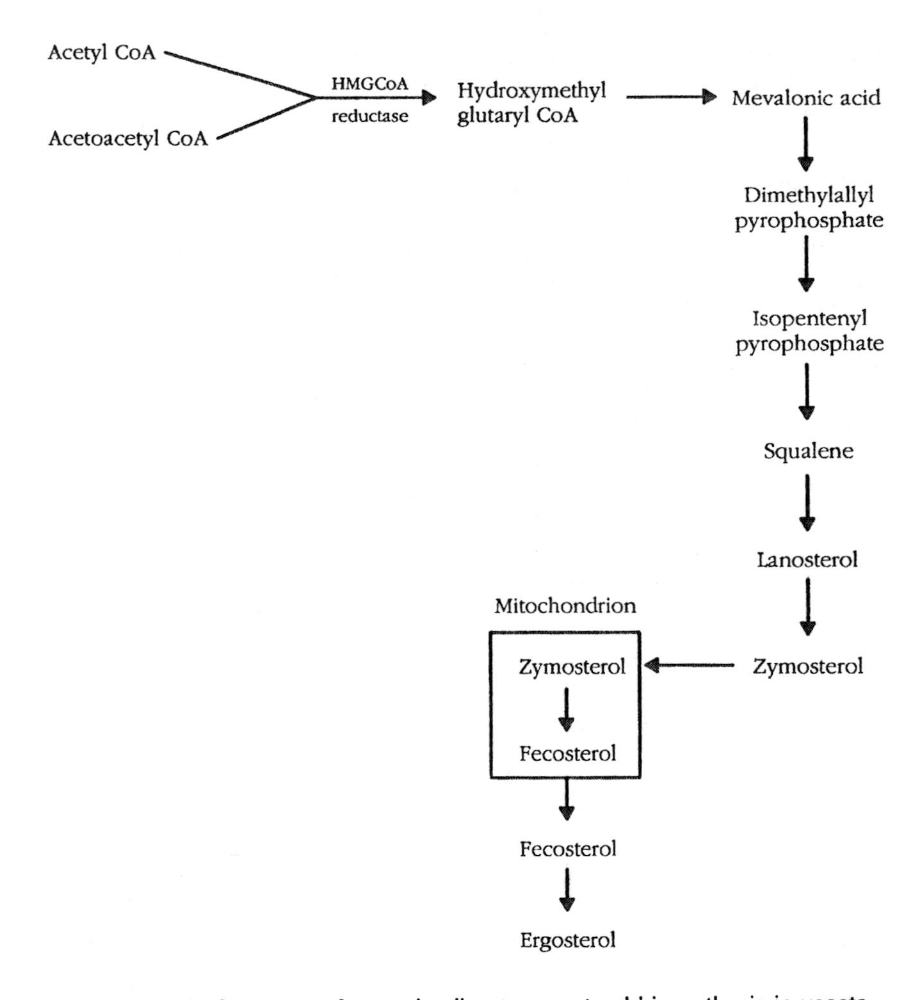

Figure 5.26. Summary of steps leading to ergosterol biosynthesis in yeasts.

acids. The glyoxylate cycle (see Figure 5.7) plays a key role in the generation of such intermediates during growth on organic acids. Often the limiting factor in organic acid metabolism by yeasts is their membrane transport (Ansanay *et al.*, 1996; Kovac *et al.*, 1996).

The utilization of malic acid by yeasts is relevant in wine fermentations since reduction in the levels of this acid is required to improve the organoleptic properties and microbiological stability of wines. Alternatives to bacterial *malolactic fermentations* (in which malate is converted to lactate and CO_2), which are occasionally difficult to control, may be achieved by using yeasts such as *Schizosaccharomyces pombe* which carry out *malo-ethanolic fermentations* (in which malate is converted to ethanol and CO_2). In anaerobically grown *Sch. pombe*, L-malate is translocated into the cell via malate permease and is then oxidatively decarboxylated to pyruvate by NAD-dependent malic enzyme (Subden and Osothsilp, 1987). Pyruvate is then decarboxylated and reduced to ethanol. Recombinant DNA technology has also been considered in the regulation of malate levels in wine fermentations. For example, wine strains of *S. cerevisiae* may be transformed with malic enzyme genes from *Sch. pombe* (Subden, 1990) or malolactic enzyme genes from *Lactococcus lactis* may be expressed in either *S. cerevisiae* or *Sch. pombe* (Ansanay *et al.*, 1996). The methylotrophic yeasts, *Hansenula polymorpha* and *Pichia pastoris*, have recently been transformed to overproduce pyruvic acid from L-lactate (Eisenberg *et al.*, 1997).

5.3 NITROGEN METABOLISM

5.3.1 Nitrogen Assimilation by Yeasts

Yeasts are capable of utilizing a range of different inorganic and organic sources of nitrogen for incorporation into the structural and functional nitrogenous components of the cell (as depicted in Figure 5.27). In industrial fermentation media, available nitrogen is usually in the form of complex mixtures of amino acids, rather than ammonium salts. Nevertheless, media are often supplemented with inexpensive inorganic nitrogen forms, such as ammonium sulphate.

Ammonium ions, either supplied in nutrient media or derived from the catabolism of other nitrogenous compounds, are actively transported and readily assimilated by all yeasts. Ammonium ions can be directly assimilated into a few amino acids, most notably glutamate and glutamine. These compounds then serve as precursors for the biosynthesis of other amino acids. Glutamate and glutamine are therefore primary products of ammonium assimilation and are key compounds in both nitrogen and carbon metabolism. Figure 5.28 shows the central role played by glutamine in this respect.

The major route of ammonium assimilation in yeasts is into glutamate through the action of NADP-dependent glutamate dehydrogenase (GDH):

$$\alpha\text{-Ketoglutarate} + NH_4^+ \xrightarrow{NADPH + H^+ \rightarrow NADP^+} \text{L-Glutamate} + H_2O$$

The NAD-dependent form of this enzyme carries out the catabolic deamination of glutamate in the reverse direction. However, if the ammonium ion availability is reduced (say, to below 1 mM), then other enzymes play more significant roles in ammonium assimilation. These are **glutamine synthetase**:

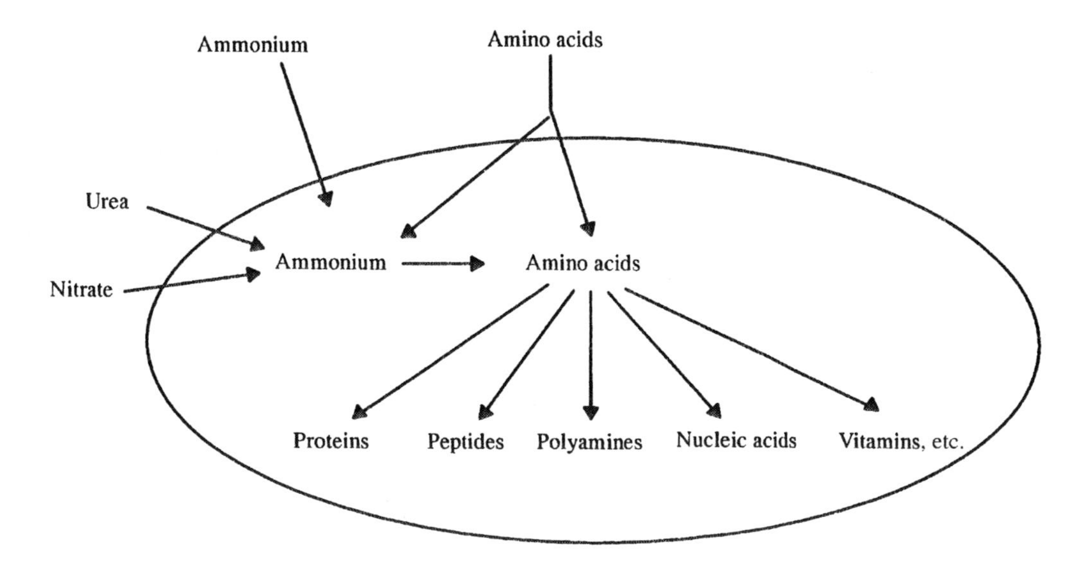

Figure 5.27. Overview of nitrogen assimilation in yeasts.

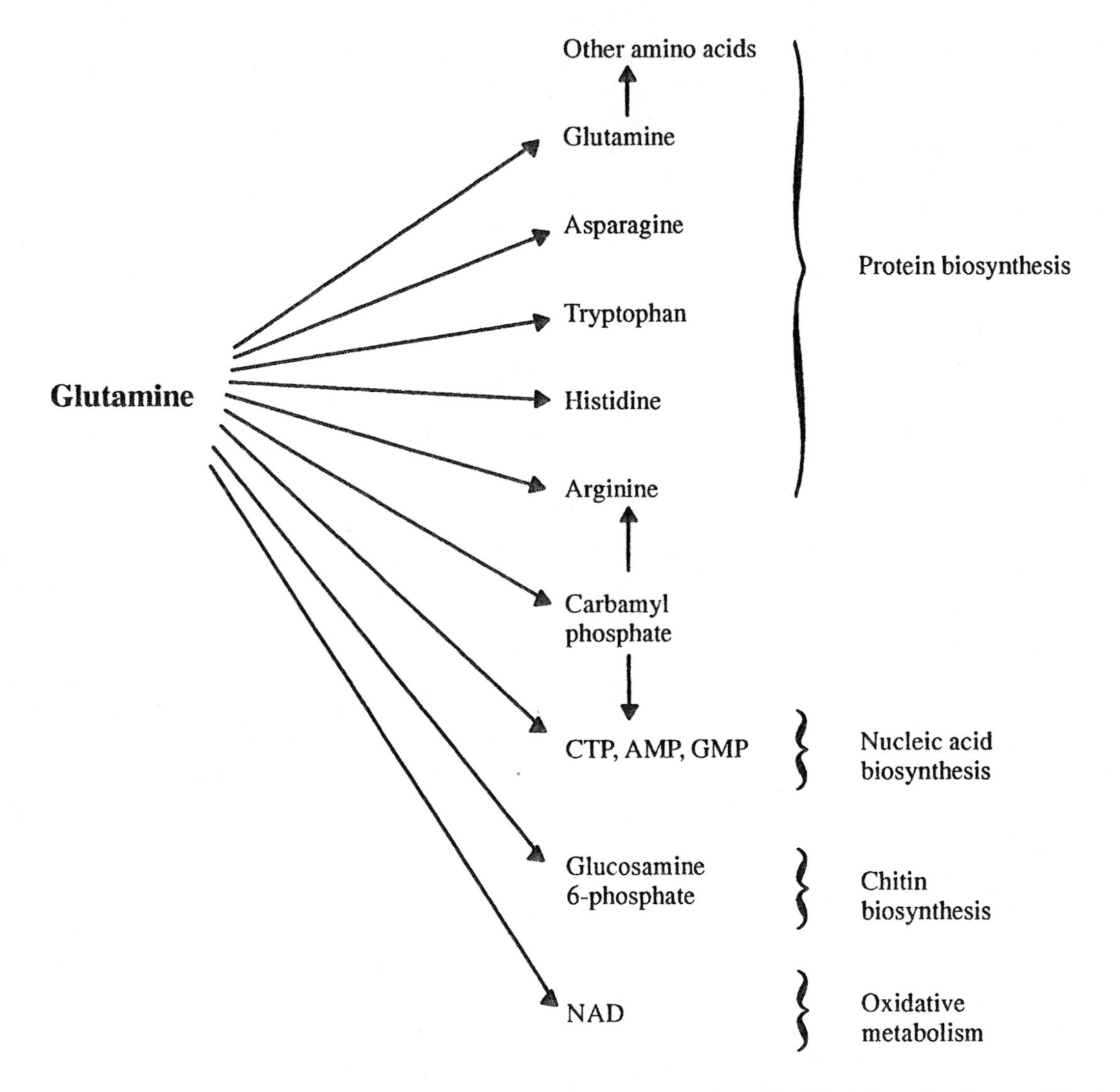

Figure 5.28. Central role of glutamine in nitrogen and carbon metabolism. Modified from Prusiner and Stadtman (1973).

$$\text{L-Glutamate} + NH_4^+ + \text{ATP} \rightarrow \text{L-Glutamine} + \text{ADP} + \text{Pi}$$

and **glutamate synthase**:

$$\alpha\text{-Ketoglutarate} + NH_4^+ + \text{ATP} \xrightarrow{\text{NADPH} + H^+ \;\rightarrow\; \text{NADP}^+} \text{L-Glutamate} + H_2O + \text{ADP} + P_i$$

Glutamine synthetase is of singular importance since it catalyses the first step in several pathways leading to the synthesis of many important cellular macromolecules (Figure 5.28). In *S. cerevisiae*, glutamine synthetase mediates an essential stage for growth in glucose and ammonia-containing media (Magasanik, 1992). Glutamate synthase (glutamine amide: 2-oxoglutarate aminotransferase, or GOGAT)-mediated assimilation of ammonia is clearly energetically less favourable than the GDH reaction because of ATP expenditure in the former process. The role of GOGAT in ammonium assimilation in *S. cerevisiae*, *Sch. pombe* and *C. albicans* has been discussed by Holmes *et al.* (1989).

When glutamine synthetase is coupled with glutamate synthase (GOGAT), this *glutamine pathway* not only provides an alternative route to glutamate compared with the glutamate dehydrogenase reaction (which possesses a high K_m for NH_4^+), but also represents a highly efficient process for yeasts to assimilate ammonia into alpha-amino nitrogen. In addition to this nitrogen-scavenging role, the glutamine pathway may play more important roles in yeast cell physiology such as maintenance of citric acid cycle intermediates, cell growth and morphology (Lacerda *et al.*, 1990).

In yeasts growing in the presence of ammonium salts, almost all the nitrogen is first assimilated into glutamate and glutamine. Since these amino acids derive their alpha-amino nitrogen directly from ammonia, they are synthesized at a rate sufficient to provide the alpha-amino nitrogen required for yeast cell growth. Other amino acids are formed by transamination reactions.

The GDH and glutamine synthetase–GOGAT pathways are highly regulated and the particular route(s) of ammonium assimilation adopted by yeasts will depend on various factors, not least the concentration of available ammonium ions and the intracellular amino acid pools. Slaughter (1988) has discussed the control of ammonium ion assimilation by yeasts through nitrogen catabolite repression.

With regard to **nitrate** ion assimilation, although *S. cerevisiae* is unable to utilize this form of nitrogen, several other yeasts (e.g. *Candida* and *Hansenula* spp.) can transport and assimilate nitrate as a sole source of nitrogen (Hipkin, 1989). Assimilation into organic nitrogen is through the activities of **nitrate reductase**:

$$NO_3^- \xrightarrow{NADPH + H^+ \;\rightarrow\; NADP^+} NO_2^-$$

and **nitrite reductase**:

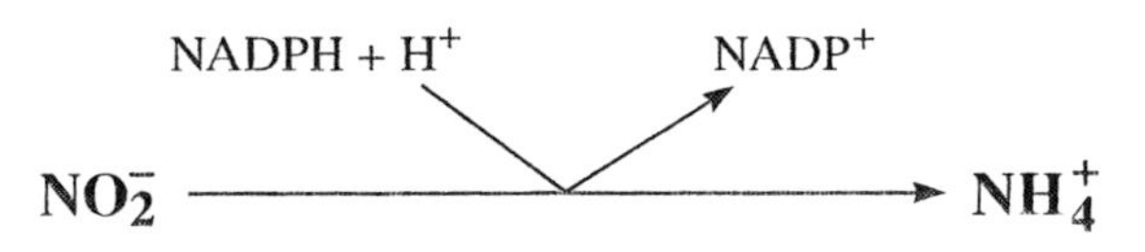

The ammonium ions formed from this latter reaction may then be assimilated as discussed above and so glutamate and glutamine can be envisaged as the end-products of nitrate assimilation by yeasts.

Note that exogenous nitrite is generally quite toxic towards yeasts, but in certain species (e.g. *Candida nitratophila*) it can be transported if present in low concentrations and reduced to ammonium (Hipkin, 1989).

Some yeasts have been shown to nitrify ammonium to nitrate and although the environmental role of yeasts in heterotrophic nitrification is thought to be of little environmental significance, Wainwright and Falih (1996) have shown that *Williopsis californica* may be important in nitrification of carbon-rich soils.

Urea is widely used as a nitrogen source by yeasts and, depending on its extracellular concentration, it may enter cells by active transport or by facilitated diffusion (see Chapter 3). In urease-negative *S. cerevisiae*, urea aminohydrolase (urea carboxylase plus allophanate hydrolase) hydrolyses urea to ammonium ions:

$$\underset{\text{Urea}}{NH_2CONH_2} + ATP + HCO_3^- \rightarrow \underset{\text{Allophanate}}{NH_2CONHCOO^-} \xrightarrow{3H_2O} 2NH_4^+ + 2HCO_3^-$$

Urea can be employed as an inexpensive nitrogen source in certain industrial fermentation feedstocks like molasses. However, urea would not be recommended as a nutritional supplement in fermentations for potable spirit beverage production due to the possible formation of carcinogenic ethylcarbamate which is formed as a reaction product between ethanol and residual urea during the distillation process. Ethylcarbamate may also arise as an intermediate from urea metabolism in wine yeast fermentations (Henschke and Jiranek, 1993).

Some yeasts (e.g. *Candida*, *Pichia* spp.) can also utilize **methylated amines** (methylamine) as sole nitrogen source (van Dijken and Bos, 1981).

Yeasts transport **L-amino acids** by both general and specific membrane processes (Chapter 3) into the cell where they can either be assimilated into proteins or dissimilated through various metabolic routes as discussed in section 5.3.2. In industrial growth media, the concentration and types of amino acids present plays an important role in dictating the progress of yeast fermentation and on the spectrum of metabolites produced. For example, Thomas and Ingledew (1990) showed that in wheat mash fermentation media, addition of mixtures of amino acids or glutamate on its own stimulated yeast growth and reduced fermentation times. In addition, Albers *et al.* (1996) showed that if amino acids were added to a chemically defined yeast growth medium, the yield of ethanol increased while the level of glycerol was reduced. Levels of higher alcohols are also influenced by the presence of amino acids available to yeast (see below).

Yeasts may synthesize **polyamines** (spermine, spermidine and putrescine) from ornithine and *S*-adenosyl methionine (Tabor and Tabor, 1985). Polyamines play several important roles in yeast physiology including: sporulation, dimorphism, stress responses, magnesium ion substitution, regulation of gene expression and metabolism (see Davis, 1996).

5.3.2 Amino Acid Metabolism

With regard to amino acid catabolism by yeasts, several dissimilatory pathways exist including decarboxylation, deamination (oxidative and non-oxidative), transamination and fermentation (where different amino acids act as hydrogen donors and acceptors in the *Stickland reaction*). Ultimately, the degradation of amino acids and other organic nitrogenous compounds by yeasts yields two end-products, ammonium and glutamate (Large, 1986).

An amino acid catabolic sequence which is pertinent in industrial yeast fermentations is the *Ehrlich pathway*. This is a deamination and decarboxylation process resulting in the formation of higher alcohols or *fusel oils* such as isobutanol and isopentanol. These alcohols, which are flavour constituents in fermented beverages, are formed via the Ehrlich pathway when amino acid availability in the medium is high or via the *biosynthesis pathway* when they are deficient (Figure 5.29). The latter route derives higher alcohols from biosynthetic pools of α-keto acid intermediates which arise due to a shortage of alpha-amino groups for transamination and also to an absence of feed-back inhibition of the biosynthetic routes (Oshita *et al.*, 1995).

With regard to amino acid biosynthesis in yeasts, nitrogen from simple nitrogenous compounds such as ammonium may be assimilated into amino acid *families*. The carbon skeletons of these amino acids originate from common precursors of intermediary carbon metabolism as depicted in Figure 5.30.

The pathways of amino acid biosynthesis are under general and specific control. The former involves coordinate repression and derepression of many enzymes in different pathways based on the availability of amino acids. Specific control mechanisms operate on particular amino acid pathways and involve end-products and intermediates of individual biosynthetic routes. The regulation of amino acid biosynthesis in *S. cerevisiae* at the transcriptional and translational level has been discussed by Hinnebusch (1990, 1992) and Sachs (1996). The biotechnological potential of yeasts in the production of amino acids by fermentation has been addressed by Niederberger (1989), particularly with regard to regulatory strategies for overproducing tryptophan in *S. cerevisiae*.

5.3.3 Protein Metabolism

Yeast cells are rich sources of protein. Generally, 40–60% of yeast biomass comprises protein and this fact, together with the relatively rapid growth of yeasts, has made them attractive

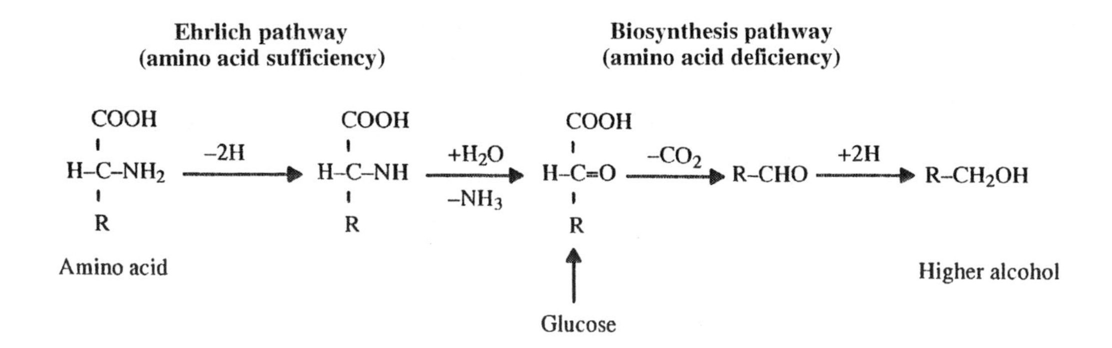

For example:

Figure 5.29. Origin of higher alcohols from yeast metabolism.

organisms in the production of *single-cell protein* (SCP) (see Chapter 6).

Gene expression in yeasts, from DNA replication to transcription and translation has been widely covered in the literature (e.g. Tuite, 1989; Hinnebusch and Liebman, 1991; Tyler and Holland, 1996). Only the salient features of protein biosynthesis will be described here. The first stage in the process is when DNA nucleotide sequences are translated into mRNA sequences in the nucleus. This is then followed by processes on clusters of ribosomes (polysomes) in which translation of mRNA nucleotide sequences into amino acid sequences in a polypeptide occurs. A summary of these latter stages of translation in *S. cerevisiae* is given in Figure 5.31.

Of interest in yeast recombinant DNA technology is the export of synthesized proteins out of the cell. Protein secretory pathways and their regulation were discussed in Chapter 2.

Protein degradation within yeast cells has been extensively studied, particularly in *S. cerevisiae*, due to its importance in many aspects of yeast physiology (e.g. Rendueles and Wolf, 1988; Jones, 1991; Van Den Hazel *et al.*, 1996). Yeast proteases, which are involved in several vital cellular functions, are either compartmentalized in organelles or are cytoplasmically located in *proteasomes*. Some proteases, especially proteinases A and B, carboxypeptidase Y and aminopeptidase in the vacuole, fulfil relatively non-specific roles as general protein hydrolases, whereas others act more specifically and catalytically to cleave proteins in order to render them biologically active. Mitochondrial proteases are essential for the biogenesis of fully functional yeast mitochondria (see Rep and Grivell, 1996). Table 5.13 summarizes some of the roles of proteases in yeast cell physiology.

The *ubiquitination* system is responsible for

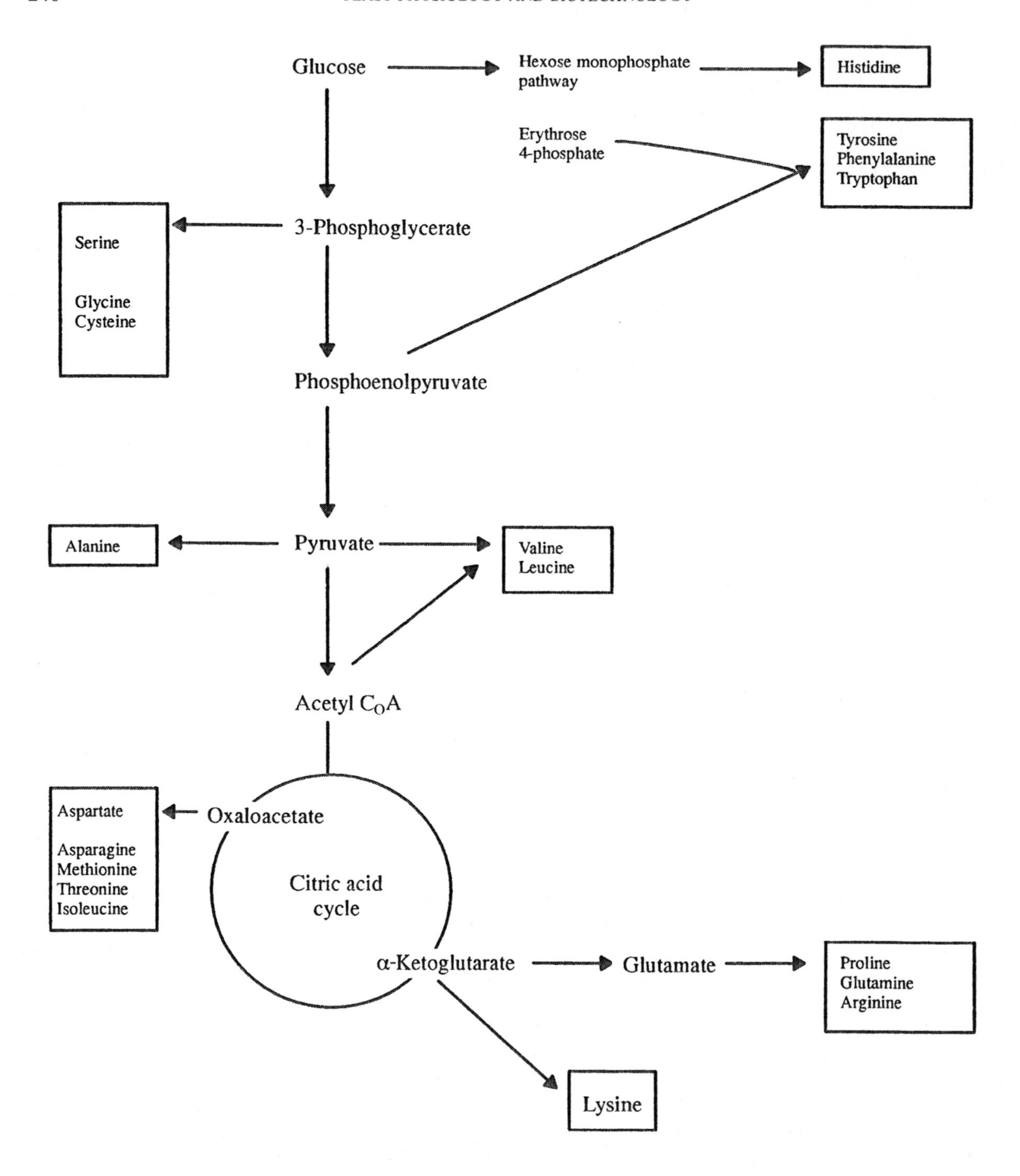

Figure 5.30. Summary of amino acid biosynthetic pathways.

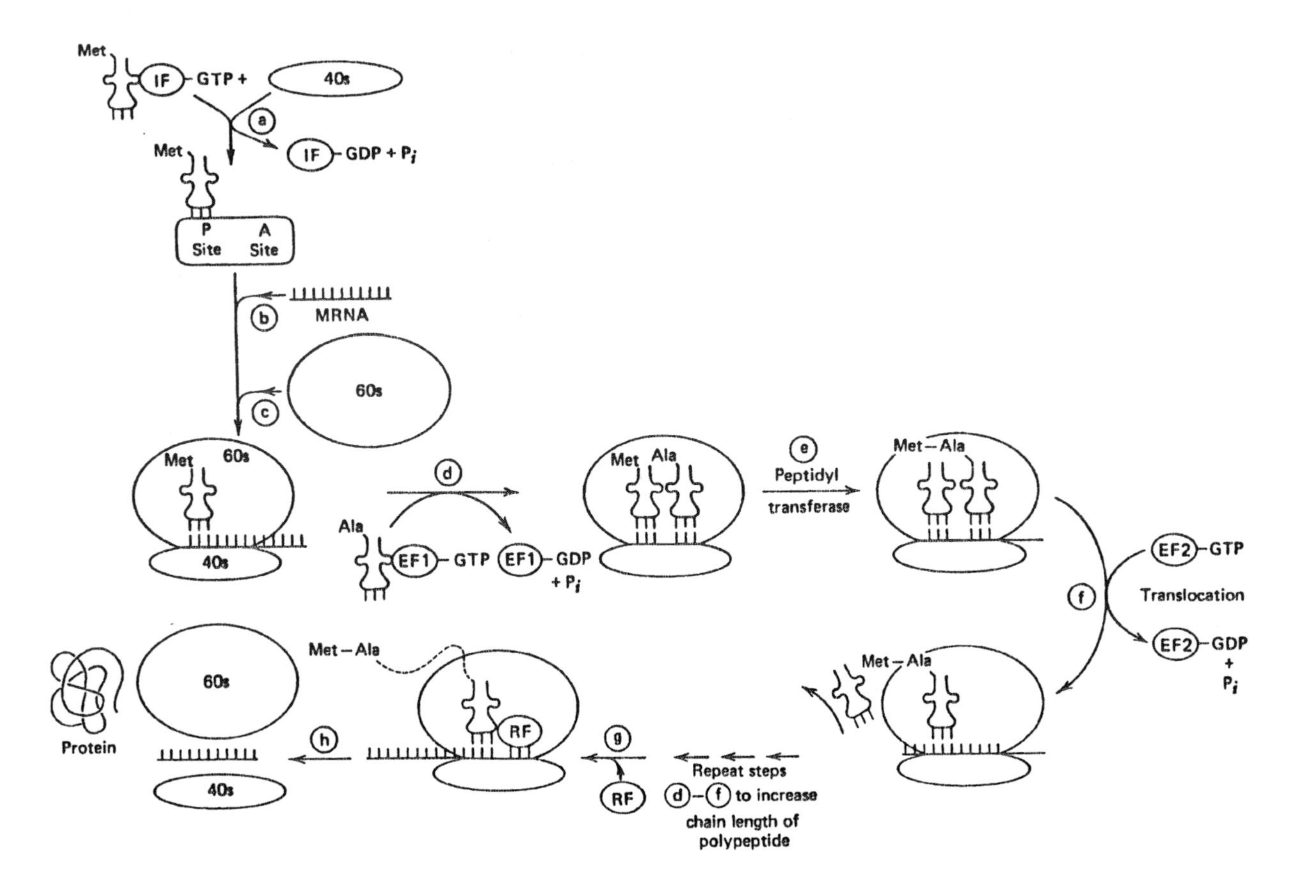

Figure 5.31. Outline of translation in *S. cerevisiae.* The figure represents a proposed scheme for initiation, elongation and termination of protein synthesis in *Saccharomyces cerevisiae.* From Garraway and Evans (1984) who provide further information on the different stages in yeast translation. Reproduced with permission. © John Wiley and Sons Ltd.

disposing of damaged or misfolded proteins and also for conferring short half-lives on certain proteins whose concentration rapidly changes with the physiological state of the cell (e.g. cyclins; see Chapter 4). Rapid and selective protein turnover takes place in proteasomes following covalent attachment of *ubiquitin* to the protein destined for proteolysis. Ubiquitinated proteins are thus specifically 'tagged' for destruction. The molecular biology of ubiquitin-dependent proteolytic pathways in eukaryotic cells have largely been elucidated using yeast cells (*S. cerevisiae*) as models (see Hershko and Ciehanover, 1992). Activation of this system in yeasts may occur in stressed cells. It is also noteworthy in the context of yeast biotechnology that heterologous proteins should not represent efficient substrates for recognition by ubiquitination for their successful production using recombinant DNA technology.

5.4 PHOSPHORUS AND SULPHUR METABOLISM

5.4.1 Phosphorus Metabolism

Phosphorus requirements of yeast cells are met by the uptake of inorganic phosphate ions from growth media. The phosphate taken up will even-

Table 5.13. Physiological roles of yeast proteases.

General function	Comments/Examples
Catabolite inactivation	Proteolytic inactivation of gluconeogenic enzymes such as fructose 1,6-bisphosphatase
Secretory protein processing	Proteolytic maturation of α-factor and killer toxin precursors
Degradation of aberrant/ harmful proteins	Proteolysis triggered by ubiquitination
Mitochondrial biogenesis	Proteolytic processing of precursors destined for the mitochondrial inner membrane
DNA repair processes	Protease induction follows UV irradiation and may trigger repair of mutagenic DNA
Nitrogen metabolism	Responses to nutritional stress (e.g. inactivation of NADP-dependent glutamate dehydrogenase during nitrogen-starvation) and sporulation (proteolysis induced by nitrogen-starvation)
Utilization of exogenous peptides	Periplasmic aminopeptidase activity
Autophagocytosis/ endocytosis	Proteolysis of endocytosed mating pheromone receptor
Protein biosynthesis	Removal of N-terminal methionine during translation
Cell growth	Activation of chitin synthetase zymogen during septum formation; proteolysis of cyclin subunits to permit cell cycle progression through S and M phases.

Information from Rendueles and Wolf, 1988; Van Den Hazel, Kielland-Brandt and Winther, 1996; Yamano *et al.*, 1996.

tually be incorporated into major cell constituents (phospholipids, nucleic acids, proteins) or may be employed in the numerous transphosphorylation reactions of intermediary metabolism. Phosphate transport into the cell is dependent on energy metabolism and is primarily regulated by the intracellular orthophosphate concentration (Borst-Pauwells, 1981). Thus, no net phosphate uptake will occur if the cellular orthophosphate levels are high. At the onset of active sugar catabolism by yeasts, orthophosphate levels decline followed by assimilation of extracellular phosphate. Johnston and Carlson (1992) have discussed several similarities which exist between phosphate and carbon utilization in *S. cerevisiae* (Table 5.14).

The intracellular concentration of free phosphate is generally maintained at very low levels in yeast cells (at around 20 μmol per gram dry

Table 5.14. Similarities between phosphate and carbon utilization in yeasts.

- Both carbon and phosphate are required in large amounts; carbon for energy and biosynthesis, phosphate for nucleic acids and phospholipids (primarily).
- Multiple sources of both carbon and phosphate exist.
- Both carbon and phosphate can be stored in intracellular polymeric forms (e.g. glycogen and polyphosphate, respectively).
- Multiple genes encode enzymes for carbon and phosphate utilization.
- Glucose represses genes for carbon utilization while phosphate similarly represses genes for phosphate utilization.
- Gene expression for both carbon and phosphate utilization is regulated by post-transcriptional cascades.

Information from Johnston and Carlson (1992).

weight in *S. cerevisiae*; Theobald *et al.*, 1986a). However, when *S. cerevisiae* cells switch from respiratory to fermentative metabolism following a glucose pulse, dynamic fluctuations in cellular phosphate have been observed (Theobald *et al.*, 1996b). The bulk of phosphate in yeast cells is in organic linkage (e.g. via ortho-bonds in phosphate esters of sugars, glyceraldehyde and via pyrophosphate-bonds in coenzymes) and in the form of polyphosphate. Polyphosphates are linear polymers of orthophosphate in anhydrous linkage. They are present in large amounts in yeast vacuoles (often exceeding 20% of cell dry weight) and because hydrolysis of polyphosphates yields the same amount of free energy as the splitting of the terminal ATP phosphate, they are very important for both phosphorus and energy metabolism in yeasts. The synthesis of polyphosphates in *S. cerevisiae* and *Kluyveromyces marxianus* in relation to the external phosphate availability has been studied by Schuddemat *et al.* (1989).

Polyphosphates play several roles in yeast physiology including: phosphate reserve, ATP synthesis reserve, metal chelation, glucose uptake, sporulation and neutralization of vacuolar amino acids.

The binding of phosphate to metabolic intermediates and to proteins is catalysed by phosphotransferase enzymes and the removal of phosphate by phosphatases. All of these enzymes require Mg^{2+} ions as activating cofactors. These processes usually involve ATP as the donor (energy producer) and ADP acceptor (energy consumer) of phosphate residues.

Plasma membrane-associated ATPases pump protons out of the cell at the expense of ATP hydrolysis (see Chapter 3). Mitochondrial ATPase acts as an ATP synthetase while vacuolar ATPase acidifies the internal space of yeast vacuoles. Kane (1995) has discussed the regulation of *S. cerevisiae* vacuolar proton-pumping ATPase in response to altered growth conditions.

Besides the ATP hydrolases and synthetases, yeast cells also contain other important enzymes involved in phosphorylation and dephosphorylation processes. These kinases and phosphatases are crucial in governing many cellular processes, not just in yeast (Rossou and Draetta, 1993; Stark, 1996) but in higher eukaryotic cells (Cohen, 1992). Some important yeast phosphatases are listed in Table 5.15. The regulation of phosphatase biosynthesis in *S. cerevisiae* has been discussed by Oshima *et al.* (1996) and by Stark (1996).

Phosphorylation reactions, on the other hand, are catalysed by many types of kinases which are integral to metabolic and growth processes of yeast cells. For example, sugars to be fermented or respired by yeast are firstly phosphorylated through kinase action. In glycolysis, phosphorylation reactions produce phosphorylated hexose which is converted into two trioses before their final conversion to pyruvate. Protein kinases belong to a family of proteins which transfer phosphate to the hydroxyl groups of serine, threonine or tyrosine on regulatory proteins. Table 5.16 summarizes some of the important protein kinase activities in yeast. As a consequence of the *S. cerevisiae* genome project, around 120 encoded protein kinases are now recognized, with over 60% of these having known, or suspected, functions (Hunter and Plowman, 1997). Protein kinases in yeast have been delimited into five *superfamilies* by Hunter and Plowman (1997) based on structural and functional relatedness.

Additional aspects of phosphorus metabolism in yeasts include the biosynthesis and breakdown of nucleotides and phospholipids. Biochemical and genetic aspects of nucleotide metabolism in *S. cerevisiae* have been reviewed by Jones and Fink (1982). Aspects relating to the catabolism of purines in *S. cerevisiae*, particularly control of allantoin metabolism, have been discussed by Cooper (1996). The synthesis and degradation of cyclic AMP (by adenylate cyclase and cAMP phosphodiesterase, respectively) is known to play important regulatory roles in yeast physiology, including: cell division, sporulation, mating and carbon metabolism. The action of cAMP via cAMP-dependent protein kinases (see Table 5.17) in signal transduction in yeasts has been discussed by Dickinson (1991), Gadd (1994) and

Table 5.15. Some important yeast phosphatases.

Enzyme	Comments	References
Alkaline and acid phosphatases	These periplasmic enzymes act non-specifically on several phosphate esters of sugars, alcohols and nucleosides.	Vogel and Hinnen (1990)
Inorganic pyrophosphatase	The cytoplasmic enzyme hydrolyses pyrophosphate into two orthophosphates; the vacuolar pyrophosphatase appears to act as a proton pump.	Lichko (1995)
Phosphatidate phosphatases	Membrane, cytosolic, microsomal and mitochondrial forms of this enzyme catalyse the hydrolysis of phosphatidate to diacylglycerol and inorganic phosphate. They are involved in lipid metabolism and in signal transduction.	Kocsis and Weselake (1996); Wu and Carman (1996)
Protein phosphatases	These enzymes are involved in regulating:	
	glucose repression;	Tu and Carlson (1995);
	glycogen accumulation;	Feng *et al.* (1991);
	protein translation;	Wek *et al.* (1992);
	cell cycle progression	Hisamoto *et al.* (1994);
	and mitosis.	Kinoshita *et al.* (1990)

Thevelein (1994). In addition Boy-Marcotte *et al.* (1996) have analysed the involvement of cAMP in regulating gene expression in glucose-exhausted cultures of *S. cerevisiae*.

Carman and Zeimetz (1996) and Greenberg and Lopes (1996) have reviewed the biosynthesis of phospholipids in *S. cerevisiae* (which is summarized in Figure 5.32). Phospholipids in yeast cells, comprising mainly phosphatidylcholine, phosphatidylethanolamine, phosphatidylinositol and phosphatidylserine, are integral structural components of cellular membranes. In addition, phosphatidylinositol and its phosphorylated derivatives have been proposed to play key roles in the control of intracellular responses to external stimuli (signal transduction). For example, Gadd and Foster (1997) have discussed the role of inositol 1,4,5-triphosphate in signal transduction during yeast cell–germ tube dimorphic transitions in *Candida albicans*. Phospholipid breakdown, through the activities of various phospholipases, plays important roles in the production of lipid-based messenger species that are part of signal transduction cascades in yeast cells. Ella *et al.* (1996) and Kohlwein *et al.* (1996) have discussed the involvement of *S. cerevisiae* phospholipases in the regulation of the cell cycle, stress response pathways and sporulation.

5.4.2 Sulphur Metabolism

The sulphur requirements of almost all yeasts can be met through assimilatory sulphate reduction and subsequent incorporation into sulphur amino acids (Figure 5.33). Sulphate metabolism in yeasts have been reviewed by Slaughter

Table 5.16. Some roles of protein kinases in yeasts.

General function	Comments
Signal transduction	The recognition of external signals by cell surface receptors is directly or indirectly coupled to protein kinase activation, which finally elicits the intracellular responses. For example, the generation of second messengers like cAMP (which activates protein kinase A) or inositol phosphate and diacylglycerol (which activates protein kinase C) and the use of G-proteins to transfer and amplify the signal.
Ion channel activity	The regulation of ion channel activity by activation and inhibition involves phosphorylation by protein kinases.
Cell cycle progression	The activation of tyrosine, threonine and serine kinases are key to the regulation of cell cycle checkpoints at G1–S and G2–M. (e.g. Cdc2 kinase, wee kinase, cyclin-dependent kinases – see Chapter 4).
Pseudohyphal growth	Some mitogen-activated protein (MAP) kinases of *S. cerevisiae* may be required for pseudohyphal development (e.g. Ime2).
Mating pheromone action	Several pheromone response kinases have been identified in *S. cerevisiae* to be involved in initial receptor binding, differentiation and G1 arrest.

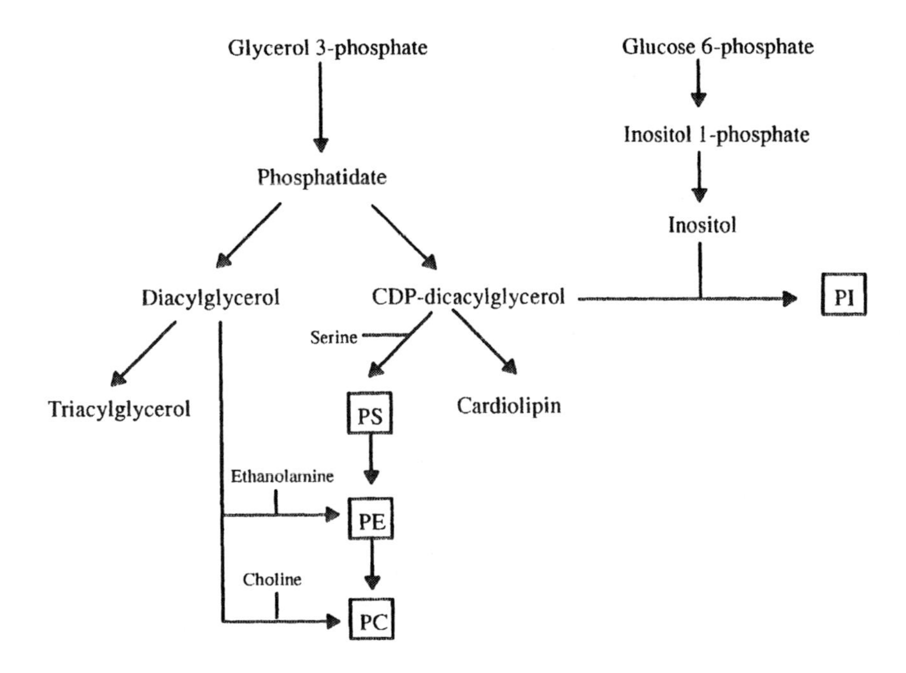

Figure 5.32. Summary of phospholipid biosynthesis in *S. cerevisiae*. CDP-diacylglycerol is the liponucleotide intermediate from which the branches of phospholipid synthesis diverge. Multiple steps in the pathway have been omitted in this summary. PI, phosphatidylinositol; PS, phosphatidylserine; PE, phosphatidylethanolamine; PC, phosphatidylcholine. Information from Carman and Zeimetz (1996) and Greenberg and Lopes (1996).

Table 5.17. Examples of complex carbon compound metabolism by yeasts.

Compounds	Comments	References
Phenols	Various *Candida* spp. (e.g. *C. maltosa*) and *Trichosporon cutaneum* can metabolize phenols, cresols, alkylphenols and monochlorophenol.	Hofmann and Schauer (1988); Neujahr (1990); Polnisch *et al.* (1992); Corti *et al.* (1995)
Polycyclic hydrocarbons	Some yeasts isolated from natural environments can biotransform polycyclic aromatic hydrocarbons.	MacGillivray and Shiaris (1993)
Benzene compounds	Yeasts from certain acidic soils can degrade benzene compounds. *Rhodotorula* spp. can convert benzaldehyde to less toxic benzyl alcohol.	Middelhoven *et al.* (1992); Wainwright (1992)
Biocides	Herbicides like paraquat/diquat can be degraded by *Lipomyces starkeyi*.	Hata *et al.* (1986)
	Sulphur-containing pesticides may be metabolized by wine yeasts.	Cabras *et al.* (1988)
	Yeast lipases can hydrolyse insecticidal pyrethroids.	Hirohara *et al.* (1985)
Aromatic and heterocyclic nitriles	Several *Candida* spp. (e.g. *C. fabianii*) can utilize these compounds as nitrogen source.	Brewis *et al.* (1995)
Purines, etc.	Uric acid, adenine, *n*-alkylamines and diamines may be utilized by *Trichosporon cutaneum* and *Candida famata*	Neujahr (1990)

(1989) and aspects of the regulation of sulphate assimilation and sulphur amino acid biosynthesis in *S. cerevisiae* has been discussed by Thomas *et al.* (1990) and Ono *et al.* (1996). Many yeasts, including most *Saccharomyces* spp., can also grow on sulphite or methionine. The metabolism of sulphur dioxide in yeasts has been reviewed by Rose (1989). Although some yeasts, notably *Candida utilis*, can grow on cysteine and cystine, most *Saccharomyces* spp. cannot utilize these compounds as sulphur sources. *S. cerevisiae* may form endogenous reserves of reduced sulphur in the form of glutathione (Elskens *et al.*, 1991).

Sulphur metabolism is very important in industrial yeast fermentations leading to the production of alcoholic beverages. In wines and beers, numerous volatile sulphur compounds are produced by yeast which can give rise to off-flavours and odours if their levels exceed certain threshold values. For example, residual H_2S which is not further incorporated into sulphur-amino acids is excreted by yeasts at levels which are highly dependent on yeast strain, concentrations of exogenous sulphate and growth conditions (Romano and Suzzi, 1992). Obnoxious sulphurous odours in wines may result if H_2S levels exceed around 50 μg/l (Rauhut, 1993). Brewing yeasts are known to produce H_2S when they are deficient in the vitamin pantothenate. This vitamin is a precursor of coenzyme A which is required for metabolism of sulphate into methionine. Pantothenate deficiency may therefore result in an imbalance in sulphur-amino acid biosynthesis, leading to excess sulphate uptake and excretion of H_2S (Slaughter and Jordan, 1986). In lager beers, dimethyl sulphide (CH_3SCH_3) is a common flavour-active

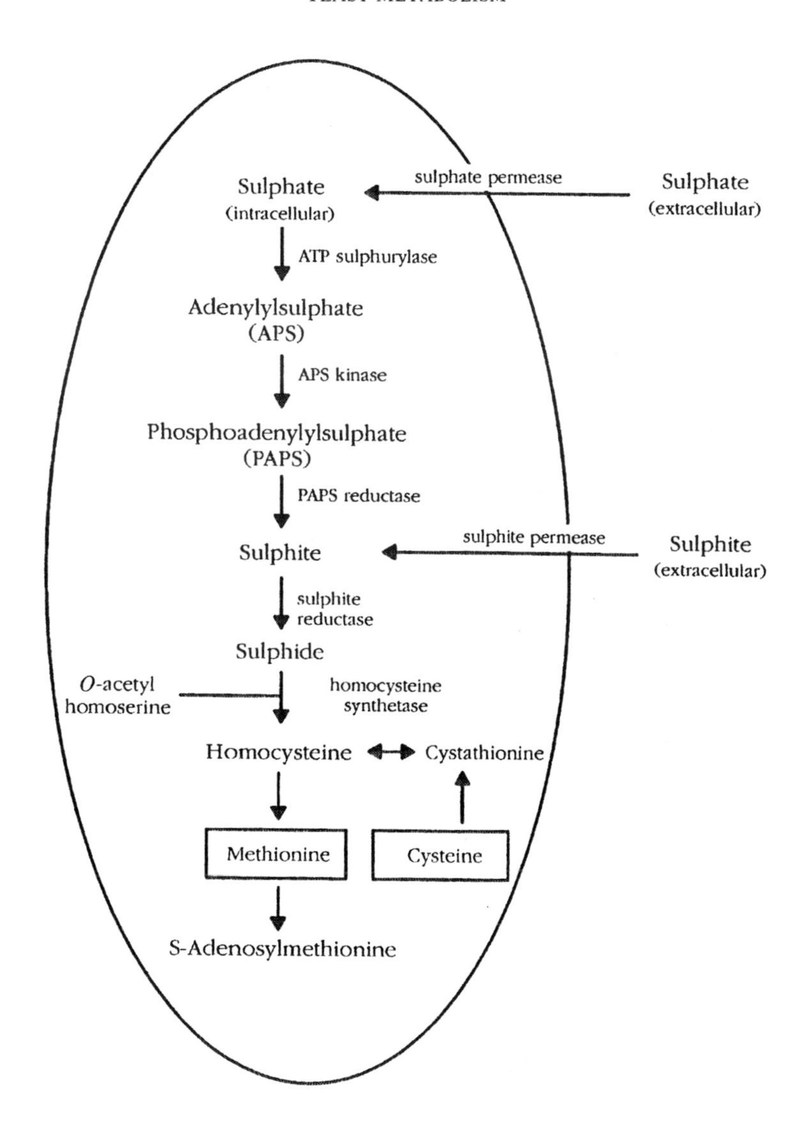

Figure 5.33. Inorganic sulphur assimilation and biosynthesis of sulphur-amino acids in *S. cerevisiae*.

compound which is formed when dimethyl sulphoxide present in malt wort is reduced by yeast. In wines, the origins of other organoleptically significant sulphur compounds (e.g. mercaptans, thioesters, carbon disulphide) have been discussed by Rauhut (1993).

Genetic manipulation of industrial yeasts has been employed to alter certain aspects of sulphur metabolism. For example, elimination of the gene encoding sulphite reductase *MET10* in brewing strains of *Saccharomyces* results in increased accumulation of SO_2 in beer (Hansen and Kielland-Brandt, 1996a). Similar inactivation of *MET2* led to increased sulphite levels in beer (Hansen and Kielland-Brandt, 1996b). Since sulphite is an antioxidant, beers produced with enhanced levels of sulphite possess increased flavour stability.

5.5 SPECIALIZED METABOLISM

Yeasts exhibit remarkable diversity in terms of their enzyme-catalysed reactions. For example, many complex carbon compounds, which are not generally regarded as metabolizable substrates, can be degraded and utilized by certain yeasts (Table 5.17). The biodegradation of several of these organic compounds by yeasts in the environment may be significant in the alleviation of chemical pollution.

Yeast enzyme catalysis may also be applied to organic chemical syntheses. For example, organic chemists frequently use commercial baker's yeast as a 'reagent' in the production of pure chemical compounds and as such it can be employed without microbiological or biochemical training. In organic chemistry, *S. cerevisiae* (baker's yeast) is able to stereospecifically reduce a wide variety of ketones and aldehydes in the preparation of chiral building blocks (Servi, 1990). Genetically modified yeasts have also been employed in chemical transformations. For example, bacterial genes (from *Acinetobacter* spp.) which encode cyclohexanone monooxygenase have been expressed in *S. cerevisiae* in order efficiently to catalyse Baeyer–Villiger oxidations (Stewart *et al.*, 1996). These reactions convert ketones to esters and are useful in the production of pharmaceuticals. Stewart *et al.* (1996) found that yeasts transformed in this way could react exclusively with only one mirror-image of particular racemic lactones. The enantioselectivity of yeasts in such reactions is regarded as being particularly advantageous in chemical syntheses.

5.6 SUMMARY

Yeasts can metabolize a remarkable array of organic substrates which provide cells with essential carbon and energy. These include hexose and pentose sugars, disaccharides, alcohols, polyols, alkanes, organic and fatty acids and several polymeric substrates. Even complex aromatic, heterocyclic and polycyclic carbon compounds can be degraded and utilized by certain yeasts. Knowledge of energy-transducing mechanisms and their regulation in yeasts holds the key to the exploitation of yeast cell physiology in biotechnology.

For historical reasons, glucose metabolism by *S. cerevisiae* is the best understood metabolic process due to its central importance in traditional fermentation industries. Growth conditions, in particular the availability of glucose and oxygen, are integral to the control of carbohydrate metabolism in *S. cerevisiae*. The competition between fermentation and respiration in this yeast is linked to the expression of regulatory phenomena such as the Pasteur and Crabtree effects. It is now recognized that the Pasteur effect is inoperative when *S. cerevisiae* grows in carbohydrate-rich media, as is the case during industrial ethanol fermentations. Only when cells are resting, starved of nitrogen or grown in glucose-limited chemostats are sugars aerobically respired and a Pasteur effect demonstrated. This is thought to be due to an irreversible inactivation of sugar transport systems triggered by the prevailing conditions of curtailed growth. Nevertheless, the underlying molecular mechanisms which dictate the expression of the Pasteur effect remain to be resolved.

The production of ethanol by *S. cerevisiae* is achieved through fermentative metabolism, irrespective of oxygen availability. In aerobic conditions, only a very small proportion of consumed glucose is respired due to glucose repression (mediated by an unidentified signal) of respiratory genes and inactivation (mediated by proteolysis) of respiratory enzymes. This is also due to the inherent limited respiratory capacity of glucose-sensitive yeasts like *S. cerevisiae*. Such limitation occurs at the pyruvate 'bottleneck'. The expression of such physiological behaviour in yeasts defines the Crabtree effect which has a corollary with industrial practice. Thus, in order to maximize respiratory growth and yeast biomass yield, while at the same time suppressing fermentation, good aeration together with controlled sugar feeding regimes (as in fed-batch cultivation) are required.

Although intermediary carbon metabolism and its regulation are relatively well understood in *S.*

cerevisiae, the same cannot be said for non-*Saccharomyces* yeasts growing on non-conventional carbon sources. Further research is especially required into metabolic control processes in methylotrophic and oleaginous yeasts and in yeasts capable of fermenting lactose and xylose.

In addition to carbon and energy metabolism, this chapter has also discussed the assimilation and metabolism of essential inorganic nutrients: nitrogen, phosphorus and sulphur. Again, the majority of fundamental studies have been conducted in *S. cerevisiae* and a lot of information has now accrued from laboratory experiments conducted in chemically defined media. However, yeasts in industry are rarely propagated under such 'clean' conditions and research is required into metabolism of organic and inorganic nutrients during yeast fermentations of complex feedstocks.

5.7 REFERENCES

Abel, C., Linz, F., Scheper, T. and Schügerl, K. (1994) Transient behaviour of continuously cultivated baker's yeast during enforced variations of dissolved oxygen and glucose concentrations. *Journal of Biotechnology*, **33**, 183–193.

Albers, E., Larsson, C., Liden, G., Niklasson, C. and Gustafsson, L. (1996) Influence of the nitrogen source on *Saccharomyces cerevisiae* anaerobic growth and product formation. *Applied and Environmental Microbiology*, **62**, 3187–3195.

Alberts, B., Bray, D., Lewis, J., Raff, M., Roberts, K. and Watson, J.D. (1994) *The Molecular Biology of the Cell*. 3rd edn. Garland Publishing Co., New York.

Aldridge, J. and Pye, E.K. (1976) Cell density dependence of oscillatory metabolism. *Nature*, **259**, 670–671.

Alexander, M.A. and Jeffries, T.W. (1990) Respiratory activity and metabolic partitioning as regulatory phenomena in yeasts. *Enzyme and Microbial Technology*, **12**, 2–19.

Amore, R., Wilhelm, M. and Hollenberg, C.P. (1989) The fermentation of xylose – an analysis of the expression of *Bacillus* and *Actinoplanes* xylose isomerase genes in yeast. *Applied Microbiology and Biotechnology*, **30**, 351–357.

Anderson, C.M. and Tatchell, K. (1996) The role of glycogen in *Saccharomyces cerevisiae*. *Molecular Biology of the Cell*, **7**, 1061.

Ansanay, V., Dequin, S., Camarasa, C., Schaeffer, V., Grivet, J.-P., Blondin, B., Salmon, J.-M. and Barre, P. (1996) Malolactic fermentation by engineered *Saccharomyces cerevisiae* as compared with engineered *Schizosaccharomyces pombe*. *Yeast*, **12**, 215–225.

Aon, M.A., Cortassa, S., Westerhoff, H.V. Berden, J.A., van Spronsen, E. and van Dam, K. (1991) Dynamic regulation of yeast glycolytic oscillations by mitochondrial functions. *Journal of Cell Science*, **99**, 325–334.

Barnett, J.A. (1981) The utilization of disaccharides and some other sugars by yeasts. *Advances in Carbohydrate Chemistry and Biochemistry*, **39**, 347–404.

Barnett, J.A. (1992) Some controls on oligosaccharide utilization by yeasts. The physiological basis of the Kluyver effect. *FEMS Microbiology Letters*, **100**, 371–378.

Barnett, J.A., Payne, R.W. and Yarrow, D. (1979) *A Guide to Identifying and Classifying Yeasts*. Cambridge University Press, London.

Biely, P. (1985) Microbial xylanolytic systems. *Trends in Biotechnology*, **3**, 286–290.

Biely, P. and Slavikova, E. (1994) New search for pectolytic yeasts. *Folia Microbiologica*, **39**, 485–488.

Bier, M., Teusink, B., Khlodenko, B.N. and Westerhoff, H.V. (1996) Control analysis of glycolytic oscillations. *Biophysical Chemistry*, **62**, 15–24.

Bilinski, C.A. and Stewart, G.G. (1990) Yeast proteases and brewing. In *Yeast: Biotechnology and Biocatlaysis* (eds H. Verachtert and R. DeMot), pp. 147–162. Marcel Dekker Inc., New York and Basel.

Boles, E., Müller, S. and Zimmerman, F.K. (1996) A multi-layered sensory system controls yeast glycolytic gene expression. *Molecular Microbiology*, **19**, 641–642.

Borst-Pauwels, G.W.F.H. (1981) Ion transport in yeast. *Biochimica et Biophysica Acta*, **650**, 88–127.

Boulton, C.A. and Ratledge, C. (1984) The physiology of hydrocarbon-utilizing microorganisms. In *Topics in Enzyme and Fermentation Biotechnology*. Vol. 19 (ed. A. Wiseman), pp. 11–77. Ellis Horwood, Chichester.

Boy-Marcotte, E., Tadi, D., Perrot, M., Boucherie, H. and Jacquet, M. (1996) High cAMP levels antagonize the reprogramming of gene expression that occurs at the diauxic shift in *Saccharomyces cerevisiae*. *Microbiology (UK)*, **142**, 459–467.

Brewis, E.A., Van Der Walt, J.P. and Prior, B.A. (1995) The utilization of aromatic, cyclic and heterocyclic nitriles by yeast. *Systematic and Applied Microbiology*, **18**, 338–342.

Brindle, K.M. (1996) Analysis of metabolic control *in vivo* using molecular genetics. *Cell Biochemistry and Function*, **14**, 269–276.

Bruinberg, P.M., de Bot, P.H.M., van Dijken, J.P. and Scheffers, W.A. (1983) The role of redox balances in anaerobic fermentation by yeasts. *European Journal of Applied Microbiology and Biotechnology*, **18**, 287–292.

Busturia, A. and Lagunas, R. (1986) Catabolite inactivation of the glucose transport system in *Saccharomyces cerevisiae*. *Journal of General Microbiology*, **132**, 379–385.

Cabib, E., Silverman, S.J., Sburlati, A. and Slater, M.L. (1990) Chitin synthesis in yeast (*Saccharomyces cerevisiae*). In *Biochemistry of Cell Walls and Membranes in Fungi* (eds P.J. Kuhn, A.P.J. Trinci, M.J. Jung, M.W. Goosey and L.G. Coping), pp. 31–41. Springer-Verlag, Berlin.

Cabib, E., Shaw, J.A., Mol, P.C., Bowers, B. and Choi, W.-J. (1996) Chitin biosynthesis and morphogenetic processes. In *The Mycota III. Biochemistry and Molecular Biology* (eds R. Brambl and G.A. Marzluf), pp. 243–267. Springer-Verlag, Berlin and Heidelberg.

Cabras, P., Meloni, M., Pirisi, F.M., Farris, G.A. and Fatichenti, F. (1988) Yeast and pesticide interaction during aerobic fermentation. *Applied Microbiology and Biotechnology*, **29**, 298–301.

Carman, G.M. and Zeimetz, G.M. (1996) Regulation of phospholipid biosynthesis in the yeast *Saccharomyces cerevisiae*. *Journal of Biological Chemistry*, **271**, 3293–3296.

Carlson, M. (1987) Regulation of sugar utilization in *Saccharomyces* species. *Journal of Bacteriology*, **169**, 4873–4877.

Cartledge, T.G. (1987) Substrate utilization, non-carbohydrate substrates. In *Yeast Biotechnology* (eds D.R. Berry, I. Russell and G.G. Stewart), pp. 311–342. Allen and Unwin, London.

Castrillo, J.I., Kaliterna, J., Weusthuis, R.A., van Dijken, J.P. and Pronk, J.T. (1996) High-cell density cultivation of yeasts on disaccharides in oxygen-limited batch cultures. *Biotechnology and Bioengineering*, **49**, 621–628.

Chambers, A., Packham, E.A. and Graham, I.R. (1995) Control of glycolytic gene expression in the budding yeast *Saccharomyces cerevisiae*. *Current Genetics*, **29**, 1–9.

Chance, B., Estabrook, R. and Ghosh, A. (1964) Damped sinusoidal oscillations of cytoplasmic reduced pyridine nucleotides in yeast cells. *Proceedings of the National Academy of Sciences* (*USA*), **51**, 1244–1251.

Chance, B., Keyhani, E. and Saronio, C. (1978) Oxygen metabolism in yeast cells. In *Biochemistry and Genetics of Yeast* (eds M. Bacila, B.L. Horecker and A.O.M. Stopanni), pp. 17–48. Academic Press, New York.

Chen, C.-I. and McDonald, K.A. (1990) Oscillatory behaviour of *Saccharomyces cerevisiae* in continuous culture III. Analysis of cell synchronization and metabolism. *Biotechnology and Bioengineering*, **36**, 28–38.

Cohen, P. (1992) Signal integration at the level of protein kinases, protein phosphatases and their substrates. *Trends in Biochemical Sciences*, **17**, 408–413.

Cooper, T.G. (1996) Regulation of allantoin catabolism in *Saccharomyces cerevisiae*. In *The Mycota III. Biochemistry and Molecular Biology* (eds R. Brambl and G.A. Marzluf), pp. 139–169. Springer-Verlag, Berlin and Heidelberg.

Cooper, D.G. and Paddock, D.A. (1984) Production of a biosurfactant from *Torulopsis bombicola*. *Applied and Environmental Microbiology*, **47**, 173–176.

Corti, A., Frassinetti, S., Vallini, G., D'Antone, S., Fichi, C. and Solaro, R. (1995) Biodegradation of nonionic surfactants I. Biotransformation of 4-(1-nonyl) phenol by a *Candida maltosa* isolate. *Environmental Pollution*, **90**, 83–87.

Davies, J. (1988) Yeast oil from cheese whey-process development. In *Single Cell Oil* (ed. R.S. Moreton), pp. 99–145. Longman, Harlow, UK.

Davies, S.E.C. and Brindle, K.H. (1992) Effects of overexpression of phosphofructokinase on glycolysis in the yeast *Saccharomyces cerevisiae*. *Biochemistry*, **31**, 4729–4735.

Davis, R.H. (1996) Polyamines in fungi. In *The Mycota III. Biochemistry and Molecular Biology* (eds R. Brambl and G.A. Marzluf), pp. 347–356. Springer-Verlag, Berlin and Heidelberg.

Dawes, E.A. (1986) *Microbial Energetics*. Blackie and Son, Glasgow.

De Deken, R.H. (1966) The Crabtree effect: a regulatory system in yeast. *Journal of General Microbiology*, **44**, 149–156.

DeJong-Gubbels, P., Vanrolleghem, P., Heijnen, S., Van Dijken, J.P. and Pronk, J.T. (1995) Regulation of carbon metabolism in chemostat cultures of *Saccharomyces cerevisiae* grown on mixtures of glucose and ethanol. *Yeast*, **11**, 407–418.

De Koning, W. and Harder, W. (1992) Methanol-utilizing yeasts. In *Methane and Methanol Utilizers* (eds J.C. Murrell and H. Dalton), pp. 207–244. Plenum Press, New York.

De Mot, R. (1990) Conversion of starch by yeasts. In *Yeast: Biotechnology and Biocatalysis* (eds H. Verachtert and R. De Mot), pp. 163–222. Marcel Dekker Inc., New York and Basel.

Dickinson, J.R. (1991) Metabolism and biosynthesis. In *Saccharomyces* (eds M.F. Tuite and S.G. Oliver), pp. 59–100. Plenum Press, New York and London.

Dombek, K.N. and Ingram, L.O. (1988) Intracellular accumulation of AMP as a cause for the decline in the rate of ethanol production by *Saccharomyces cerevisiae* during batch fermentation. *Applied and Environmental Microbiology*, **54**, 98–104.

Duboc, P., Marison, I. and von Stockar, U. (1996) Physiology of *Saccharomyces cerevisiae* during cell cycle oscillations. *Journal of Biotechnology*, **51**, 57–72.

Eisenberg, A., Seip, J.E., Gavagan, J.E., Payne, M.S., Anton, D.L. and DiCosimo, R. (1997) Pyruvic acid production using methylotrophic yeast transformants as catalyst. *Journal of Molecular Catalysis B: Enzymatic*, **2**, 223–232.

Ella, K.M., Dolan, J.W., Qi, C. and Meier, K.E. (1996) Characterization of *Saccharomyces cerevisiae* deficient in expression of phospholipase D. *Biochemical Journal*, **314**, 15–19.

Elskens, M.T., Jaspers, C.J. and Penninckx, M.J. (1991) Glutathione as an endogenous sulphur source in the yeast *Saccharomyces cerevisiae*. *Journal of General Microbiology*, **137**, 637–644.

Entian, K.-D. and Barnett, J.A. (1992) Regulation of sugar utilization by *Saccharomyces cerevisiae*. *Trends in Biochemical Sciences*, **17**, 506–510.

Evans, C.T. and Ratledge, C. (1984) Induction of xylulose-5-phosphate phosphoketolase in a variety of yeasts grown on D-xylose: the key to efficient xylose metabolism. *Archives of Microbiology*, **139**, 48–52.

Faber, K.N., Harder, W., Ab, G. and Veenhuis, M. (1995) Review – methylotrophic yeasts as factories for the production of foreign proteins. *Yeast*, **11**, 1331–1344.

Fell, D. (1997) *Understanding the Control of Metabolism*. Portland Press, London and Miami.

Fell, D.A. and Thomas, S. (1995) Physiological control of metabolic flux – the requirement for multisite modulation. *Biochemical Journal*, **311**, 35–39.

Feng, Z., Wilson, D.E., Peng, Z.-Y., Schlender, K.K., Reimann, E.M. and Trumbly, R.J. (1991) The yeast *GLC7* gene required for glycogen accumulation encodes a type 1 protein phosphatase. *Journal of Biological Chemistry*, **266**, 23796–23801.

Fiechter, A. and Seghezzi, W. (1992) Regulation of glucose metabolism in growing yeast cells. *Journal of Biotechnology*, **27**, 17–45.

Fiechter, A., Fuhrmann, G.F. and Käppeli, O. (1981) Regulation of glucose metabolism in growing yeast cells. *Advances in Microbial Physiology*, **22**, 123–183.

Filipini, C., Sonnleitner, B., Fiechter, A., Bradley, J. and Schmid, R. (1991) On-line determination of glucose in biotechnological processes: comparison between FIA and *in situ* enzyme electrode. *Journal of Biotechnology*, **18**, 153–160.

Fogarty, W.M. and Kelly, C.T. (1990) Recent advances in microbial amylases. In *Microbial Enzymes and Biotechnology*, 2nd edn (ed. W.M. Fogarty and C. T. Kelly), pp. 71–132. Elsevier Applied Science, London and New York.

François, J., van Schaftingen, E. and Hers, H-G. (1984) The mechanism by which glucose increases fructose-2,6-bisphosphate concentration in *Saccharomyces cerevisiae*. *European Journal of Biochemistry*, **145**, 187–193.

Gadd, G.M. (1988) Carbon nutrition and metabolism. In *Physiology of Industrial Fungi* (ed. D.R. Berry), pp. 21–57. Blackwell Scientific Publishers, Oxford.

Gadd, G.M. (1994) Signal transduction in fungi. In *The Growing Fungus* (eds N.A.R. Gow and G.M. Gadd), pp. 183–210. Chapman & Hall, London.

Gadd, G.M. and Foster, S.A. (1997) Metabolism of inositol 1,4,5-triphosphate in *Candida albicans*: significance as a precursor of inositol polyphosphates and in signal transduction during the dimorphic transition from yeast cells to germ tubes. *Microbiology (UK)*, **143**, 437–48.

Galons, J.P., Tanida, I., Ohya,Y. (1990) A multinuclear magnetic resonance study of a clz 11 mutant showing the Pet-phenotype of *Saccharomyces cerevisiae*. *European Journal of Biochemistry*, **193**, 111–119.

Gainvors, A. and Belarbi, A. (1995) Detection method for polygalacturonase-producing strains of *Saccharomyces cerevisiae*. *Yeast*, **11**, 1493–1499.

Gancedo, C. (1992) Carbon catabolite repression in yeast. *European Journal of Biochemistry*, **206**, 297–313.

Gancedo, C., Gancedo, J.M. and Sols, A. (1968) Glycerol metabolism in yeasts. Pathways of utilization and production. *European Journal of Biochemistry*, **5**, 165–172.

Gancedo, C. and Serrano, R. (1989) Energy-yielding metabolism. In *The Yeasts, 2nd edn, Vol. 3* (eds A.H. Rose and J.S. Harrison), pp. 205–259. Academic Press, London.

Garraway, M.O. and Evans, R.C. (1984) *Fungal Nutrition and Physiology*. J. Wiley and Son, New York.

Gellissen, G. and Melber, K. (1996) Methylotrophic yeast *Hansenula polymorpha* as production organism for recombinant pharmaceuticals. *Drug Research*, **46**, 943–948.

Gellissen, G., Hollenberg, C.P. and Janowicz, Z.A. (1995) Gene expression in methylotrophic yeasts. In *Gene Expression in Recombinant Microorganisms* (ed. A. Smith), pp. 195–239. Marcel Dekker, New York.

Gillies, R.J., Alger, J.R., den Hollander, J.A. and Shulman, R.G. (1982) Intracellular pH measured by NMR: methods and results. In *Intracellular pH: its Measurement, Regulation and Utilization in Cellular Functions*, pp. 79–104. Alan R. Liss. Inc., New York.

Godfredsen, S.E. (1990) Microbial lipases. In *Microbial*

Enzymes and Biotechnology, 2nd edn (eds W.M. Fogarty and C.T. Kelly), pp. 255–274. Elsevier Applied Science, London and New York.

Goffeau, A. and Crosby, B. (1978) A new type of cyanide-insensitive, azide-sensitive respiration in the yeasts *Schizosaccharomyces pombe* and *Saccharomyces cerevisiae*. In *Biochemistry and Genetics of Yeasts* (eds M. Bacila, B.L. Horecker and A.O.M. Stoppani), pp. 81–96. Academic Press, New York.

Gooday, G.W. (1994) Cell membrane. In *The Growing Fungus* (eds N.A.R. Gow and G.M. Gadd), pp. 63–74. Chapman & Hall, London.

Greenberg, M.L. and Lopes, J.M. (1996) Genetic regulation of phospholipid biosynthesis in *Saccharomyces cerevisiae*. *Microbiological Reviews*, **60**, 1–20.

Guiraud, J.-P. and Galzy, P. (1990) Inulin conversion by yeasts. In *Yeasts: Biotechnology and Biocatalysis* (eds H. Verachtert and R. De Mot), pp. 255–296. Marcel Dekker Inc., New York and Basel.

Hahn-Hägerdal, B., Jeppsson, J., Skoog, K. and Prior, B.A. (1994) Biochemistry and physiology of xylose fermentation by yeasts. *Enzyme and Microbial Technology*, **16**, 933–943.

Hansen, J. and Keilland-Brandt, M.C. (1996a) Inactivation of *MET10* in brewers' yeast specifically increases SO_2 formation during beer production. *Nature Biotechnology*, **14**, 1587–1591.

Hansen, J. and Kielland-Brandt, M.C. (1996b) Inactivation of *MET2* in brewer's yeast increases the level of sulphite in beer. *Journal of Biotechnology*, **50**, 75–87.

Harder, W. and Brooke, A.G. (1990) Methylotrophic yeasts. In *Yeast: Biotechnology and Biocatalysis* (eds H. Verachtert and R. DeMot), pp. 395–428. Marcel Dekker Inc., New York and Basel.

Harder, W. and Veenhuis, M. (1989) Metabolism of one-carbon compounds. In *The Yeasts, 2nd edn, Vol. 3* (eds A.H. Rose and J.S. Harrison), pp. 289–316. Academic Press, London.

Hassan, M., Blanc, P.J., Pareilleux, A. and Goma, G. (1994) Selection of fatty acid auxotrophs from oleaginous yeast *Cryptococcus curvatus* and production of cocoa butter equivalents in batch culture. *Biotechnology Letters*, **16**, 819–824.

Hata, S., Shirata, K. and Takagishi, H. (1986) Degradation of paraquat and diquat by the yeast *Lipomyces starkeyi*. *Journal of General and Applied Microbiology*, **32**, 193–202.

Heinritz, B., Ghildyal, N.P., Rogge, G., Hanschmann, G. and Ringpfeil, M. (1985) Influence of yeast-cell state on efficiency of ethanol production. *Acta Biotechnology*, **5**, 101–103.

Henschke, P.A. and Jiranek, V. (1993) Yeasts – metabolism of nitrogen compounds. In *Wine Microbiology and Biotechnology* (ed. G.H. Fleet), pp. 77–164. Harwood Academic Publishers, Chur, Switzerland.

Hershko, A. and Ciechanover, A. (1992) The ubiquitin system for protein degradation. *Annual Review of Biochemistry*, **61**, 761–807.

Hinkle, P.C., Kumar, M.A., Resetar, A. and Harris, D.L. (1991) Mechanistic stoichiometry of mitochondrial oxidative phosphorylation. *Biochemistry*, **30**, 3576–3582.

Hinnebusch, A.G. (1990) Transcriptional and translational regulation of gene expression in the general control of amino acid biosynthesis in *Saccharomyces cerevisiae*. *Progress in Nucleic Acid Research and Molecular Biology*, **38**, 195–240.

Hinnebusch, A.G. (1992) General and pathway-specific regulatory mechanisms controlling the synthesis of amino acid biosynthetic enzymes in *Saccharomyces cerevisiae*. In *The Molecular and Cellular Biology of the Yeast* Saccharomyces. *Vol. II. Gene expression* (eds E.W. Jones, J.R. Pringle and J.R. Broach), pp. 319–414. Cold Spring Harbor Laboratory Press, Cold Spring Harbor, New York.

Hinnebusch, A.G. and Liebman, S.W. (1991) Protein synthesis and translational control in *Saccharomyces cerevisiae*. In *The Molecular and Cellular Biology of the Yeast* Saccharomyces. *Vol. I. Genome Dynamics, Protein Synthesis and Energetics* (eds J.R. Broach, J.R. Pringle and E.W. Jones), pp. 627–735. Cold Spring Harbor Laboratory Press, Cold Spring Harbor, New York.

Hipkin, C.R. (1989) Nitrate assimilation in yeasts. In *Molecular and Genetic Aspects of Nitrate Assimilation* (eds J.L. Wray and J.R. Kinghorn), pp. 51–68. Oxford Science Publications, Oxford.

Hirohara, H., Mitsuda, S., Ando, E. and Komaki, R. (1985) Enzymatic preparation of optically active alcohols related to synthetic pyrethroid insecticides. In *Biocatalysis in Organic Syntheses* (eds J. Tramper, H.C. van der Plas and P. Linko), pp. 119–134. Elsevier Science Publishers, Amsterdam.

Hisamoto, N., Sugimoto, K. and Matsumoto, K. (1994) The Glc7 type 1 protein phosphatase of *Saccharomyces cerevisiae* is required for cell cycle progression in G2/M. *Molecular and Cellular Biology*, **14**, 3158–3165.

Hofmann, K.H. and Schauer, F. (1988) Utilization of phenol by hydrocarbon assimilating yeasts. *Antonie van Leeuwenhoek*, **54**, 179–188.

Hoffman, C.S. and Winston, F. (1991) Glucose repression of transcription of *Schizosaccharomyces pombe fbp1* gene occurs by a cAMP signalling pathway. *Genes and Development*, **5**, 561–571.

Hofmann, M., Boles, E. and Zimmerman, F.K. (1994) Characterization of the essential yeast gene encoding *N*-acetylglucosamine-phosphate mutase. *European Journal of Biochemistry*, **221**, 741–747.

Holmes, A.R., Collings, A., Farnden, K.J.F. and Shepherd, M.G. (1989) Ammonium assimilation by *Candida albicans* and other yeasts: evidence for activity of glutamate synthase. *Journal of General Microbiology*, **135**, 1423–1430.

Holzer, H. (1961) Regulation of carbohydrate metabolism by competition. *Cold Spring Harbor Symposia on Quantitative Biology*, **26**, 277–288.

Holzer, H. (1976) Catabolite inactivation in yeast. *Trends in Biochemical Sciences*, **1**, 178–181.

Homma, M., Kanbe, T., Chibana, H. and Tanaka, K. (1992) Detection of intracellular forms of secretory aspartic proteinase in *Candida albicans*. *Journal of General Microbiology*, **138**, 627–633.

Hommes, F.A. (1964) Oscillatory reductions of pyridine nucleotides during anaerobic glycolysis in brewer's yeast. *Archives of Biochemistry and Biophysics*, **108**, 36–46.

Horecker, H., Rosen, O.M., Kowal, J., Rosen, S., Scher, B., Lai, C.Y., Hoffee, P. and Cremona, T. (1968) Comparative studies of aldolases and fructose diphosphatases. In *Aspects of Yeast Metabolism* (eds A.K. Mills and H. Krebs), pp. 71–103. Blackwell Scientific Publications, Oxford and Edinburgh.

Hough, J.S. (1985) *The Biotechnology of Malting and Brewing*. Cambridge University Press, Cambridge.

Hough, J.S., Briggs, D., Stevens, R. and Young, T.W. (1982) *Malting and Brewing Science. Vol. II. Hopped Wort and Beer*. Chapman & Hall, London.

Hunter, T. and Plowman, G.D. (1997) The protein kinases of budding yeasts: six score and more. *Trends in Biochemical Sciences*, **22**, 15–22.

Jeffries, T.W. (1990) Fermentation of D-xylose and cellobiose. In *Yeast: Biotechnology and Biocatalysis* (eds H. Verachtert and R. DeMot), pp. 349–394. Marcel Dekker Inc., New York and Basel.

Jennings, D.M. (1995) *The Physiology of Fungal Nutrition*. Cambridge University Press, Cambridge.

Jeppsson, H., Yu, S. and Hahn-Hägerdahl, B. (1996) Xylulose and glucose fermentation by *Saccharomyces cerevisiae* in chemostat culture. *Applied and Environmental Microbiology*, **62**, 1705–1709.

Johnson, V., Singh, M. and Saini, V.S. (1992) Bioemulsifier production by an oleaginous yeast *Rhodotorula glutinis* 11P–30. *Biotechnology Letters*, **6**, 487–490.

Johnston, M. and Carlson, M. (1992) Regulation of carbon and phosphate utilization. In *The Molecular and Cellular Biology of the Yeast* Saccharomyces. *Gene Expression* (eds E.W. Jones, J.R. Pringle and J.R. Broach), pp. 193–281. Cold Spring Harbor Laboratory Press, Cold Spring Harbor, New York.

Jones, E.W. (1991) Three proteolytic systems in the yeast *Saccharomyces cerevisiae*. *Journal of Biological Chemistry*, **266**, 7963–7966.

Jones, E.W. and Fink, G.R. (1982) Regulation of amino acid and nucleotide biosynthesis in yeast. In *The Molecular Biology of the Yeast* Saccharomyces. *Metabolism and Gene Expression* (eds J.N. Strathern, E.W. Jones and J.R. Broach), pp. 181–299. Cold Spring Harbor Laboratory Press, Cold Spring Harbor, New York.

Kaliterna, J., Weusthuis, R.A. Castrillo, J.I., van Dijken, J.P. and Pronk, J.T. (1995a) Coordination of sucrose uptake and respiration in the yeast *Debaryomyces yamadae*. *Microbiology* (*UK*), **141**, 1567–1574.

Kaliterna, J., Weusthuis, R.A., Castrillo, J.I., Van Dijken, J.P. and Pronk, J.T. (1995b) Transient responses of *Candida utilis* to oxygen limitation: regulation of the Kluyver effect for maltose. *Yeast*, **11**, 317–325.

Kane, P.M. (1995) Disassembly and reassembly of the yeast vacuolar H^+ ATPase *in vivo*. *Journal of Biological Chemistry*, **270**, 17025–17032.

Käppeli, O. (1986) Cytochomes P-450 in yeasts. *Microbiological Reviews*, **50**, 244–258.

Käppeli, O. and Sonnleitner, B. (1986) Regulation of sugar metabolism in *Saccharomyces*-type yeast: experimental and conceptual considerations. *Critical Reviews in Biotechnology*, **4**, 299–325.

Käppeli, O., Arreguin, M. and Rieger, M. (1985) The respirative breakdown of glucose by *Saccharomyces cerevisiae*: an assessment of a physiological state. *Journal of General Microbiology*, **131**, 1411–1416.

Keulers, M., Satroudinov, A.D., Suzuki, T. and Kuriyama, H. (1996) Synchronization affector of autonomous short-period oscillation of *Saccharomyces cerevisiae*. *Yeast*, **12**, 673–682.

Kinoshita, N., Ohkura, H. and Yanagida, M. (1990) Distinct, essential roles of type 1 and 2A protein phosphatases in the control of the fission yeast cell division cycle. *Cell*, **63**, 405–415.

Kocsis, M.G. and Weselake, R.J. (1996) Phosphatidate phosphatases of mammals, yeast and higher plants. *Lipids*, **31**, 785–802.

Kohlwein, S.D., Daum, G., Schneiter, R. and Paltauf, F. (1996) Phospholipids: synthesis, sorting, subcellular traffic – the yeast approach. *Trends in Cell Biology*, **6**, 260–266.

Köller, W. (1992) Antifungal agents with target sites in sterol functions and biosynthesis. In *Target Sites of Fungicide Action* (ed. W. Köller), pp. 119–206. CRC Press, Boca Raton, Florida.

Kötter, P. and Ciriacy, M. (1993) Xylose fermentation by *Saccharomyces cerevisiae*. *Applied Microbiology and Biotechnology*, **38**, 776–783.

Kovac, L., Stella, C.A. and Ramos, E.H. (1996) Why *Saccharomyces cerevisiae* can oxidize but not decarboxylate external pyruvate. *Biochimica et Biophysica Acta*, **1289**, 79–82.

Lacerda, V., Marsden, A., Buzato, J.B. and Ledingham, W.M. (1990) Studies on ammonium assimilation in continuous cultures of *S. cerevisiae* under carbon and nitrogen limitation. In *Proceedings of the 5th European Congress on Biotechnology* (eds C. Christiansen, L. Munck and J. Villadsen), pp. 1075–1078. Munksgaard International Publishers, Copenhagen.

Lagunas, R. (1979) Energetic irrelevance of aerobiosis for *S. cerevisiae* growing on sugars. *Molecular and Cellular Biochemistry*, **27**, 139–146.

Lagunas, R. (1981) Is *Saccharomyces cerevisiae* a typical facultative anaerobe? *Trends in Biochemical Science*, **6**, 201–202.

Lagunas, R. (1986) Misconceptions about energy metabolism of *Saccharomyces cerevisiae. Yeast*, **2**, 221–228.

Lagunas, R. (1993) Sugar transport in *Saccharomyces cerevisiae. FEMS Microbiology Reviews*, **104**, 229–242.

Lagunas, R. and Gancedo, C. (1983) Role of phosphate in the regulation of the Pasteur effect in *Saccharomyces cerevisiae. European Journal of Biochemistry*, **137**, 479–483.

Lagunas, R., Dominquez, C., Busturia, A. and Saez, M.J. (1982) Mechanisms of appearance of the Pasteur effect in *Saccharomyces cerevisiae*: inactivation of sugar transport systems. *Journal of Bacteriology*, **152**, 19–25.

Large, P.J. (1986) Degradation of organic nitrogen compounds by yeasts. *Yeast*, **2**, 1–34.

Leman, J., Bednarksi, W. and Tomasik, J. (1990) Influence of cultivation conditions on the composition of oil produced by *Candida curvata. Biological Wastes*, **31**, 1–15.

Lichko, L.P. (1995) H^+ATPase and H^+pyrophosphatase of the yeast vacuolar membrane. *Biochemistry (Moscow)*, **60**, 635:643.

Lloyd, D., Kristensen, B. and Degn, H. (1983) Glycolysis and respiration in yeasts: the effect of ammonium ions studied by mass spectrometry. *Journal of General Microbiology*, **129**, 2125–2127.

Lodolo, E.J., O'Connor-Cox, E.S.C. and Axcell, B.C. (1995) Novel application of glucagon and insulin to study yeast glycogen concentrations. *Journal of the American Society of Brewing Chemists*, **53**, 145–151.

MacGillivray, A.R. and Shiaris, M.P. (1993) Biotransformation of polycyclic aromatic hydrocarbons by yeasts isolated from coastal sediments. *Applied and Environmental Microbiology*, **59**, 1613–1618.

Magasanik, B. (1992) Regulation of nitrogen utilization. In *The Molecular and Cellular Biology of the Yeast* Saccharomyces. *Vol. II. Gene Expression* (eds E.W. Jones, J.R. Pringle and J.R. Broach), pp. 283–317. Cold Spring Harbor Laboratory Press, Cold Spring Harbor, New York.

Martegani, E., Porro, D., Ranzi, B.M. and Alberghina, L. (1990) Involvement of a cell size control mechanism in the induction and maintenance of oscillations in continuous cultures of budding yeast. *Biotechnology and Bioengineering*, **36**, 453–459.

Middelhoven, W.J., Koorevaar, M. and Schuur, G.W. (1992) Degradation of benzene compounds by yeasts in acidic soils. *Plant Soil*, **145**, 37–43.

Müller, S., Boles, E., May, M. and Zimmerman, F.K. (1995) Different internal metabolites trigger the induction of glycolytic gene expression in *Saccharomyces cerevisiae. Journal of Bacteriology*, **177**, 4517–4519.

Münch, T., Sonnleitner, B. and Fiechter, A. (1992) New insights into the synchronization mechanism with forced synchronous cultures of *Saccharomyces cerevisiae. Journal of Biotechnology*, **24**, 299–314.

Neklyudov, A.D., Ilyukhina, V.P., Mosina, G.I. Petrakova, A.N., Federova, N.F. and Kuznetsov, V.D. (1996) Hydrolysis of protein substrates by yeast proteases. *Applied Biochemistry and Microbiology*, **32**, 212–314.

Neujahr, H.Y. (1990) Yeasts in biodegradation and biodeterioration. In *Yeasts: Biotechnology and Biocatalysis* (eds H. Verachtert and R. DeMot), pp. 321–348. Marcel Dekker Inc., New York and Basel.

Niederberger, P. (1989) Amino acid production in microbial eukaryotes and prokaryotes other than Coryneforms. In *Microbial Products: New Approaches* (eds S. Baumberg, I. Hunter and M. Rhodes), pp. 1–24. Cambridge University Press, Cambridge.

Nissen, T.L., Schulze, U., Nielsen, J. and Villadsen, J. (1997) Flux distributions in anaerobic, glucose-limited continuous cultures of *Saccharomyces cerevisiae. Microbiology (UK)*, **143**, 203–218.

O'Connor-Cox, E.S.C., Lodolo, E.J. and Axcell, B.C. (1996) Mitochondrial relevance to yeast fermentative performance: a review. *Journal of the Institute of Brewing*, **102**, 19–25.

Ogata, K., Nishikawa, H. and Ohsugi, M. (1969) A yeast capable of utilizing methanol. *Agricultural and Biological Chemistry*, **33**, 1519–1520.

Olama, Z.A., Shaban, N.Z. and Temsah, S.A. (1990) Lipid and protein production by *Candida guilliermondii* strain 1 grown on the solar fraction of crude oil. *Microbios*, **64**, 103–109.

Olson, L. and Hahn-Hägerdal, B. (1996) Fermentation of lignocellulosic hydrolysates for ethanol production. *Enzyme and Microbial Technology*, **18**, 312–331.

Ono, B.-I., Kijima, K., Ishii, N. Kawato, T., Matsuda, A., Paezewski, A. and Shimoda, S. (1996) Regulation of sulphate assimilation in *Saccharomyces cerevisiae. Yeast*, **12**, 1153–1162.

Oshima, Y., Ogawa, N. and Harashima, S. (1996) Regulation of phosphatase synthesis in *Saccharomyces cerevisiae* – a review. *Gene*, **179**, 171–177.

Oshita, K. Kubota, M., Uchida, M. and Ono, M. (1995) Clarification of the relationship between fusel alcohol formation and amino acid assimilation by brewing yeast using ^{13}C-labelled amino acid. European Brewery Convention Proceedings. 25th Convention, Brussels 1995. pp. 387–394. IRL Press at Oxford University Press, Oxford.

Paalme, T., Elken, R., Vilu, R. and Korhola, M. (1997) Growth efficiency of *Saccharomyces cerevisiae* on glucose/ethanol media with a smooth change in the dilution rate (A-stat). *Enzyme and Microbial Technology*, **20**, 174–181.

Paltauf, F., Kohlwein, S.D. and Henry, S.A. (1992) Regulation and compartmentalization of lipid synthesis in yeast. In *The Molecular and Cellular Biology of the Yeast* Saccharomyces. *Vol. II. Gene Expression* (eds E.W. Jones, J.R. Pringle and J.R. Broach), pp. 415–500. Cold Spring Harbor Laboratory Press, Cold Spring Harbor, New York.

Panek, A.D. (1991) Storage carbohydrates. In *The Yeasts, 2nd edn, Vol. 4. Yeast Organelles* (eds A.H. Rose and J.S. Harrison), pp. 655–678. Academic Press, London.

Parks, L.W. and Casey, W.M. (1995) Physiological implications of sterol biosynthesis in yeast. *Annual Review of Microbiology*, **49**, 95–116.

Parulekar, S.J., Semones, G.B., Rolf, M.J., Lievense, J.C. and Lim, H.C. (1986) Induction and elimination in continuous cultures of *Saccharomyces cerevisiae*. *Biotechnology and Bioengineering*, **28**, 700–710.

Phaff, H.J., Miller, M.W. and Mrak, E.M. (1978) *The Life of Yeasts*. Harvard University Press, Cambridge (USA) and London (UK).

Picataggio, S., Rohrer, T., Deanda, K., Lanning, D., Reynolds, R., Mielenz, J. and Eirich, L.D. (1992) Metabolic engineering of *Candida tropicalis* for the production of long-chain dicarboxylic acids. *Bio/technology*, **10**, 894–898.

Polnisch, E., Kneifel, H., Franzke, H. and Hofman, K.H. (1992) Degradation and dehalogenation of monochlorophenols by the phenol-assimilating yeast *Candida maltosa*. *Biodegradation*, **2**, 193–199.

Porro, D., Martegani, E., Ranzi, B.M. and Alberghina, L. (1988) Oscillation in continuous cultures of budding yeast: a segregated parameter analysis. *Biotechnology and Bioengineering*, **32**, 411–417.

Postma, E., Scheffers, W.A. and van Dijken, J.P. (1988) Adaptation of the kinetics of glucose transport to environmental conditions in the yeast *Candida utilis* CBS621: a continuous culture study. *Journal of General Microbiology*, **134**, 1109–1116.

Prior, B.A., Kilian, S.G. and du Preez, J.C. (1989) Fermentation of D-xylose by the yeasts *Candida shehatae* and *Pichia stipitis*: prospects and problems. *Process Biochemistry*, **24**, 21–32.

Pronk, J.T., Steensma, H.Y. and van Dijken, J.P. (1996) Pyruvate metabolism in *Saccharomyces cerevisiae*. *Yeast*, **12**, 1607–1633.

Prusiner, S. and Stadtman, E.R. (1973) *The Enzymes of Glutamine Metabolism*. Academic Press, New York and London.

Quain, D.E. (1988) Studies on yeast physiology – impact on fermentation performance and product quality. *Journal of the Institute of Brewing*, **95**, 315–323.

Quain, D.E. and Boulton, C.A. (1987) Growth and metabolism of mannitol by strains of *Saccharomyces cerevisiae*. *Journal of General Microbiology*, **133**, 1675–1684.

Ratledge, C. (1991) Yeast physiology – a micro-synopsis. *Bioprocess Engineering*, **6**, 195–203.

Ratledge, C. and Evans, C.T. (1989) Lipids and their metabolism. In *The Yeasts, 2nd edn, Vol. 3* (eds A.H. Rose and J.S. Harrison), pp. 367–455. Academic Press, London.

Ratledge, C. and Tan, K.-H. (1990) Oils and fats: production, degradation and utilization by yeasts. In *Yeast: Biotechnology and Biocatalysis* (eds H. Verachtert and R. DeMot), pp. 223–253. Marcel Dekker Inc., New York and Basel.

Rauhut, D. (1993) Yeasts – production of sulphur compounds. In *Wine Microbiology and Biotechnology* (ed. G.H. Fleet), pp. 183–223. Harwood Academic Publishers, Chur, Switzerland.

Ray, R.R. and Nanda, G. (1996) Microbial β-amylases: biosynthesis characteristics and industrial applications. *Critical Reviews in Microbiology*, **22**, 181–199.

Rehm, H.J. and Reiff, I. (1981) Mechanisms and occurrence of microbial oxidation of long-chain alkanes. *Advances in Biochemical Engineering*, **19**, 176–215.

Reid, M.F. and Fewson, C.A. (1994) Molecular characterization of microbial alcohol dehydrogenases. *Critical Reviews in Microbiology*, **20**, 13–56.

Rendueles, P.S. and Wolf, D.H. (1988) Proteinase function in yeast: biochemical and genetic approaches to a central mechanism of post-translation control in the eukaryotic cell. *FEMS Microbiology Reviews*, **54**, 17–46.

Rep, M. and Grivell L.A. (1996) The role of protein degradation in mitochondrial function and biogenesis. *Current Genetics*, **30**, 367–380.

Richard, P., Bakker, B.M., Teusink, B., van Dam, K. and Westerhoff, H.V. (1996a) Acetaldehyde mediates the synchronization of sustained glycolytic oscillations in populations of yeast cells. *European Journal of Biochemistry*, **235**, 238–241.

Richard, P., Teusink, B., Hemker, B., van Dam, K. and Westerhoff, H.V. (1996b) Sustained oscillations in free-energy state and hexose phosphates in yeast. *Yeast*, **12**, 731–740.

Romano, P. and Suzzi, G. (1992) Production of H_2S by different yeast strains during fermentation. *Proceedings of the 22nd Convention of the Institute of Brewing (Australia and New Zealand Section)*, Melbourne 1991. pp. 96–98. Institute of Brewing, Adelaide, Australia.

Rose, A.H. (1989) Transport and metabolism of sulphur dioxide in yeasts and filamentous fungi. In *Nitrogen, Phosphorus and Sulphur Utilization* by Fungi (eds L. Boddy, R. Marchant and D.J. Read), pp. 58–70. Cambridge University Press, Cambridge.

Rossou, I. and Draetta, G. (1993) Phosphorylation in yeast cell processes. *Cellular Signalling*, **5**, 381–387.

Rowe, S.M., Simpson, W.J. and Hammond, J.R.M. (1994) Intracellular pH of yeast during brewery fermentation. *Letters in Applied Microbiology*, **18**, 135–137.

Rowen, D.W., Meinke, M. and LaPorte, D.C. (1992) *GLC3* and *GHA1* of *Saccharomyces cerevisiae* are allelic and encode the glycogen branching enzyme. *Molecular and Cellular Biology*, **12**, 22–29.

Sachs, M.S. (1996) General and cross-pathway controls of amino acid biosynthesis. In *The Mycota III. Biochemistry and Molecular Biology* (eds R. Brambl and G.A. Marzluf), pp. 315–345. Springer-Verlag, Berlin and Heidelberg.

Satroudinov, A.D., Kuriyama, H. and Kobayashi, H. (1992) Oscillatory metabolism of *Saccharomyces cerevisiae*. *FEMS Microbiology Letters*, **98**, 261–268.

Scheffers, W.A. (1987) Alcoholic fermentation. *Studies in Mycology*, **30**, 321–332.

Scheffers, W.A. (1966) Stimulation of fermentation in yeasts by acetoin and oxygen. *Nature*, **210**, 533–534.

Schneider, H. (1988) Conversion of pentoses to ethanol by yeasts and fungi. *Critical Reviews in Biotechnology*, **9**, 1–40.

Schuddemat, J., deBoo, R., van Leeuwen, C.C.M., van den Broek, P.J.A. and van Stevenick, J. (1989) Polyphosphate synthesis in yeast. *Biochimica et Biophysica Acta*, **1010**, 191–198.

Schwan, R. and Rose, A.H. (1994) Polygalacturonase production by *Kluyveromyces marxianus*: effect of medium composition. *Journal of Applied Bacteriology*, **76**, 62–67.

Servi, S. (1990) Baker's yeast as a reagent in organic synthesis. *Synthesis*, **January 1990**, 1–25.

Shennan, J.L. and Levi, J.D. (1974) The growth of yeasts on hydrocarbons. In *Progress in Industrial Microbiology*. Vol. 13 (ed. D.J.D. Hockenhull), pp. 1–57. Churchill Livingstone, Edinburgh.

Sills, A.M. and Stewart, G.G. (1982) Production of amylolytic enzymes by several yeast species. *Journal of the Institute of Brewing*, **88**, 313–316.

Sims, AP. and Barnett, J.A. (1978) The requirement of oxygen for the utilization of maltose, cellobiose and D-galactose by certain anerobically fermenting yeasts (Kluyver effect). *Journal of General Microbiology*, **106**, 277–288.

Sims, A.P. and Barnett, J.A. (1991) Levels of activity of several enzymes involved in aerobic utilization of sugars by six yeast species: observations towards understanding the Kluyver effect. *FEMS Microbiology Letters*, **77**, 295–298.

Singh, A. and Mishra, P. (1995) Microbial pentose utilization. In *Progress in Industrial Microbiology. Vol. 33*. Elsevier, Amsterdam.

Singh, N., Wakil, S.J. and Stoops, J.K. (1985) Yeast fatty acid synthase: structure to function relationship. *Biochemistry*, **24**, 6598–6602.

Slaughter, J.C. (1988) Nitrogen metabolism. In *Physiology of Industrial Fungi* (ed. D.R. Berry), pp. 58–76. Blackwell Scientific Publications, Oxford.

Slaughter, J.C. (1989) Sulphur compounds in fungi. In *Nitrogen, Phosphorus and Sulphur Utilization by Fungi* (eds L. Boddy, R. Marchant and D.J. Reed), pp. 91–105. Cambridge University Press, Cambridge.

Slaughter, J.C. and Jordan, B. (1986) The production of hydrogen sulphide by yeast. In *Proceedings of the Second Aviemore Conference on Malting, Brewing and Distilling* (eds I. Campbell and F.G. Priest), pp. 308–310. Institute of Brewing, London.

Sonnleitner, B. and Hahnemann, U. (1994) Dynamics of the respiratory bottleneck of *Saccharomyces cerevisiae*. *Journal of Biotechnology*, **38**, 63–79.

Stark, M.J.R. (1996) Yeast protein serine/threonine phosphatases – multiple roles and diverse regulation. *Yeast*, **12**, 1647–1675.

Stewart, J.D., Reed, K.W., Zhu, J., Chen, G. and Kayser, M.M. (1996) A 'designer yeast' that catalyses the kinetic resolution of 2-alkyl-substituted cyclohexanones by enantioselective Baeyer-Villiger oxidations. *Journal of Organic Chemistry*, **61**, 7652–7653.

Steyn, A.J.C. and Pretorius, I.S. (1995) Characterization of a novel α-amylase from *Lipomyces kononenkoae* and expression of its gene *LKA1* in *Saccharomyces cerevisiae*. *Current Genetics*, **28**, 526–533.

Subden, R.E (1990) Wine yeast: selection and modification. In *Yeast Strain Selection* (ed. C.J. Panchal), pp. 113–137. Marcel Dekker Inc., New York.

Subden, R.E. and Osothsilp, C. (1987) Malic acid metabolism in wine yeasts. In *Biological Research in Industrial Yeasts* (eds G.G. Stewart, I. Russell, R.D. Klein and R.R. Hiebsch), pp. 67–76. CRC Press, Boca Raton, Florida.

Sudbery, P.E. (1995) Genetics of industrial yeasts other than *Saccharomyces cerevisiae* and *Schizosaccharomyces pombe*. In *The Yeasts, 2nd edn, Vol. 6* (eds

A.E. Wheals, A.H. Rose and J.S. Harrison), pp. 255–283. Academic Press, London.
Tabor, C.W. and Tabor, H. (1985) Polyamines in microorganisms. *Microbiological Reviews*, **49**, 81–99.
Teusink, B., Larsson, C., Diderich, J., Richard, P., van Dam, K., Gustafsson, L. and Westerhoff, H.V. (1996) Synchronized heat flux oscillations in yeast cell populations. *Journal of Biological Chemistry*, **271**, 24442–24448.
Theobald, U., Mohns, J. and Rizzi, M. (1996a) Determination of *in vivo* cytoplasmic orthophosphate concentration in yeast. *Biotechnology Techniques*, **10**, 297–302.
Theobald, U., Mohns, J. and Rizzi, M. (1996b) Dynamics of orthophosphate in yeast cytoplasm. *Biotechnology Techniques*, **10**, 461–466.
Thevelein, J.M. (1994) Signal transduction in yeast. *Yeast*, **10**, 1753–1790.
Thevelein, J.M. (1996) Regulation of trehalose metabolism and its relevance to cell growth and function. In *The Mycota III. Biochemistry and Molecular Biology* (eds R. Brambl and G.A. Marzluf), pp. 395–420. Springer-Verlag, Berlin and Heidelberg.
Thomas, K.C. and Ingledew, W.M. (1990) Fuel alcohol production: effects of free amino nitrogen on fermentation of very-high-gravity wheat mashes. *Applied and Environmental Microbiology*, **56**, 2046–2050.
Thomas, D., Barbey, R. and Surdin-Kerjan, Y. (1990) Gene-enzyme relationship in the sulphate assimilation pathway of *Saccharomyces cerevisiae*. *Journal of Biological Chemistry*, **265**, 15518–15524.
Toivola, A., Yarrow, D., van den Bosch, E., van Dijken, J.P. and Scheffers, W.A. (1984) Alcoholic fermentation of D-xylose by yeasts. *Applied and Environmental Microbiology*, **47**, 1221–1223.
Trumbly, R.J. (1992) Glucose repression in the yeast *Saccharomyces cerevisiae*. *Molecular Microbiology*, **6**, 15–21.
Tsai, C.S., Mitton, K.P. and Johnson, B.F. (1989) Acetate assimilation by the fission yeast, *Schizosaccharomyces pombe*. *Biochemistry and Cell Biology*, **67**, 464–467.
Tubb, R.S. (1986) Amylolytic yeasts for commercial applications. *Trends in Biotechnology*, **4**, 98–104.
Tu, J.L. and Carlson, M. (1995) REG1 binds to protein phosphatase type-1 and regulates glucose repression in *Saccharomyces cerevisiae*. *EMBO Journal*, **14**, 5939–5946.
Tuite, M.F. (1989) Protein synthesis. In *The Yeasts, 2nd edn, Vol. 3* (eds A.H. Rose and J.S. Harrison), pp. 162–204. Academic Press, London.
Tyler, B.M. and Holland, M.J. (1996) RNA polymerases and transcription factors. In *The Mycota III. Biochemistry and Molecular Biology* (eds R. Brambl and G.A. Marzluf), pp. 111–138. Springer-Verlag, Berlin and Heidelberg.
Van Dam, K. (1996) Role of glucose signalling in yeast metabolism. *Biotechnology and Bioengineering*, **52**, 161–165.
Van Den Hazel, H.B., Kielland-Brandt, M.C. and Winther, J.R. (1996) Review: biosynthesis and function of yeast vacuolar proteases. *Yeast*, **12**, 1–16.
Van Dijken, J.P. and Bos, P. (1981) Utilization of amines by yeasts. *Archives of Microbiology*, **128**, 320–324.
Van Dijken, J.P. and Scheffers, W.A. (1986) Redox balances in the metabolism of sugars by yeasts. *FEMS Microbiology Reviews*, **32**, 199–224.
Van Laere, A. (1994) Intermediary metabolism. In *The Growing Fungus* (eds N.A.R. Gow and G.M. Gadd), pp. 211–238. Chapman & Hall, London.
Visser, W., Van Spronsen, E.A., Nanninga, N., Pronk, J.T., Keunen, J.G. and Van Dijken, J.P. (1995) Effects of growth conditions on mitochondrial morphology in *Saccharomyces cerevisiae*. *Antonie van Leeuwenhoek*, **67**, 2013–2107.
Vogel, K. and Hinnen, A. (1990) The yeast phosphatase system. *Molecular Microbiology*, **4**, 2013–2017.
Wainwright, M. (1992) *An Introduction to Fungal Biotechnology*. J. Wiley and Sons, Chichester.
Wainwright, M. and Falih, A.M.K. (1996) Involvement of yeasts in urea hydrolysis and nitrification in soil amended with a natural source of sucrose. *Mycological Research*, **100**, 307–310.
Walker, G.M. (1994) The roles of magnesium in biotechnology. *Critical Reviews in Biotechnology*, **14**, 311–354.
Webb, S.R. and Lee, H. (1990) Regulation of D-xylose utilization by hexoses in pentose-fermenting yeasts. *Biotechnology Advances*, **8**, 685–698.
Wek, R.C., Cannon, J.F., Dever, T.E. and Hinnebusch, A.G. (1992) Truncated protein phosphatase GLC7 restores translational activity of *GCN4* expression in yeast mutants defective for the eIF-2α kinase GCN2. *Molecular and Cellular Biology*, **12**, 5700–5710.
Weusthuis, R.A., Visser, W., Pronk, J.T., Scheffers, W.A. and van Dijken, J.P. (1994) Effects of oxygen limitation on sugar metabolism in yeasts: a continuous-culture study of the Kluyver effect. *Microbiology (UK)*, **140**, 703–715.
Whitaker, J.R. (1990) Microbial pectolytic enzymes. In *Microbial Enzymes and Biotechnology. 2nd edn* (eds W.M. Fogarty and C.T. Kelly), pp. 133–176. Elsevier Applied Science, London and New York.
White, T., Miyasaki, S.H. and Agabian, N. (1993) Three distinct secreted aspartyl proteinases in

Candida albicans. *Journal of Bacteriology*, **175**, 6126–6132.

Wijsman, M.R., Van Dijken, J.P., Van Kleef, B.H.A. and Scheffers, W.A. (1984) Inhibition of fermentation and growth in batch cultures of the yeast *Brettanomyces intermedius* upon a shift from aerobic to anaerobic conditions (Custers effect). *Antonie van Leeuwenhoek*, **50**, 183–192.

Wills, C. (1990) Regulation of sugar and ethanol metabolism in *Saccharomyces cerevisiae*. *Critical Reviews in Biochemistry and Molecular Biology*, **25**, 245–280.

Wills, C. (1996) Some puzzles about carbon catabolite repression in yeast. *Research in Microbiology*, **147**, 556–572.

Wolf, H.J. and Hanson, R.S. (1979) Isolation and characterization of methane-utilizing yeasts. *Journal of General Microbiology*, **114**, 187–194.

Wu, W.-I. and Carman, G.M. (1996) Regulation of phosphatidate phosphatase activity from the yeast *Saccharomyces cerevisiae* by phospholipids. *Biochemistry*, **35**, 3790–3796.

Yamano, H., Gannon, J. and Hunt, T. (1996) The role of proteolysis in cell cycle progression in *Schizosaccharomyces pombe*. *EMBO Journal*, **15**, 5268–5279.

Zimmerman, F.K. (1992) Glycolytic enzymes as regulatory factors. *Journal of Biotechnology*, **27**, 17–26.

6

YEAST TECHNOLOGY

6.1 INTRODUCTION
6.2 APPLIED MOLECULAR GENETICS OF YEASTS
- 6.2.1 Genetics of 'Industrial' *Saccharomyces* Yeasts
- 6.2.2 Recombinant DNA Technology in Yeasts
  - *6.2.2.1 Molecular Genetic Aspects*
  - *6.2.2.2 Cellular Aspects*
  - *6.2.2.3 Technical Aspects*
  - *6.2.2.4 Commercial Aspects*

6.3 DEVELOPMENTS IN YEAST TECHNOLOGIES
- 6.3.1 Alcoholic Beverages
  - *6.3.1.1 Developments in Brewing Yeasts and Fermentation*
  - *6.3.1.2 The Brewing Process*
  - *6.3.1.3 Brewing Yeasts*
  - *6.3.1.4 Yeast Fermentation Technology Developments*
  - *6.3.1.5 Developments in Yeasts Producing other Alcoholic Beverages*
- 6.3.2 Industrial Alcohols
  - *6.3.2.1 Bioethanol*
  - *6.3.2.2 Other Alcohols from Yeast*
- 6.3.3 Yeast Biomass-Derived Products
  - *6.3.3.1 Baker's Yeast*
  - *6.3.3.2 Other Types of Whole-Cell Yeast Biomass*
  - *6.3.3.3 Extracted Yeast Cell Products*
- 6.3.4 Industrial Enzymes and Chemicals
- 6.3.5 Therapeutic Proteins

6.4 YEASTS IN BIOMEDICAL RESEARCH
6.5 SUMMARY
6.6 REFERENCES

6.1 INTRODUCTION

Yeast technology encompasses all industrial processes which exploit the activities of yeast cells. The products of yeast biotechnologies impinge on many commercially important sectors, including: food, beverages, chemicals, industrial enzymes, pharmaceuticals, agriculture and the environment. The diversity of yeast metabolic processes (described in Chapter 5) is thus reflected in the diversity of yeast technologies.

This chapter focuses on the exploitation of yeast growth, gene expression and metabolism in several areas of industrial, medical and environmental biotechnology and also discusses avenues of potential future development in these areas. The technological exploitation of *S. cerevisiae* has been extensively covered in Volume 5 of *The Yeasts* (Rose and Harrison, 1993), by Reed and Nagodawithana (1991) and by Berry *et al.* (1987). Non-*Saccharomyces* yeasts are becoming increasingly significant as industrial organisms and basic and applied aspects of these yeasts have been discussed by Klein and Zaworski (1990), Sudbery (1994) and by several authors in *Nonconventional Yeasts in Biotechnology*, edited by Wolf (1996).

Yeasts have been exploited by Mankind for thousands of years, albeit fortuitously, in food and fermentation processes. For most of the present century, scientific and technological advances have mirrored developments in the brewing and allied industries. In recent years, however, the industrial importance of yeasts has extended beyond traditional fermentation uses into, for example, the health-care sector. Nevertheless, it should be stressed that ethyl alcohol produced by yeast fermentation is likely to remain the premier world-wide biotechnological commodity for many years to come. Figure 6.1 outlines some of the principal traditional and modern industrial applications of yeasts. 'Traditional' means at least around 100 years old, and by no means obsolete in modern times.

Although *S. cerevisiae* has been described as Mankind's most domesticated organism, and is still the most widely exploited yeast species in industry, other yeasts are playing important roles in modern biotechnology. Table 6.1 summarizes the applications of *S. cerevisiae* and lists some of the other biotechnologically important yeast species. It should also be stressed that yeast biodiversity is almost untapped from an industrial viewpoint and many yeast products have yet to be harnessed and commercialized. As Fogel and Welch (1987) state: 'The gene pool in industrial and wild yeasts is wide, deep and virtually unplumbed. To this we must add the gene pools of 500 other yeast species'.

The focus of this chapter will be the exploitation of yeast growth and metabolic process in industry. Emphasis will be placed on the relationship between yeast cell physiology and biotechnology, rather than providing comprehensive coverage of recombinant DNA and fermentation technologies with yeasts. Due recognition will also be given to the increasing industrial importance of non-*Saccharomyces* yeasts, particularly in the production of biopharmaceuticals. Finally, the importance of yeasts as experimental models in fundamental studies of biomedical science will be discussed.

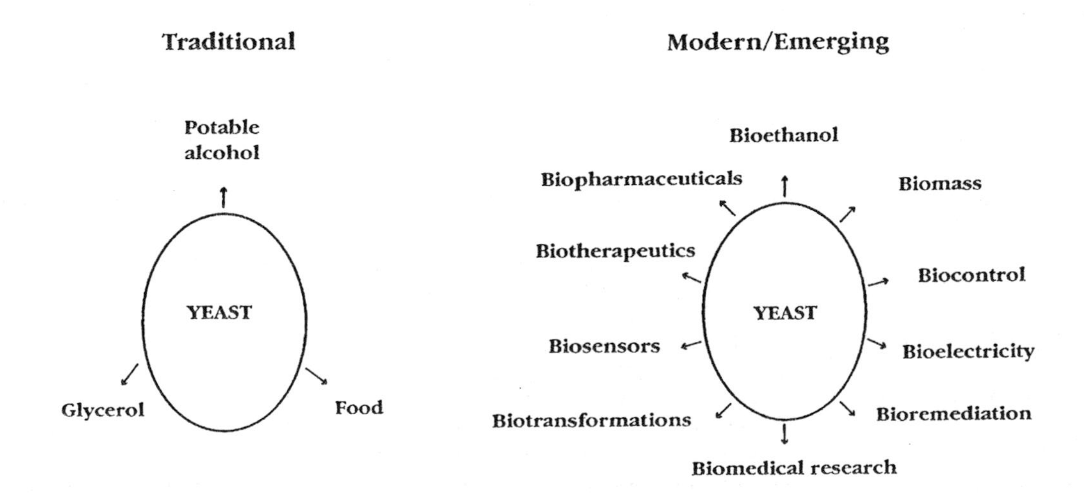

Figure 6.1. Summary of yeast technologies.

Table 6.1. Some biotechnologically important yeasts.

Yeasts (alphabetical)	Uses and potential uses in industry
Arxula adeninivorans	Assimilates nitrate and amines; grows at temperatures >45°C, secretes several hydrolases e.g. proteases, glycosidases and pectinases (see Kunze and Kunze, 1996).
Candida spp.	Widespread uses of different species, for example: *C. albicans* (6-aminopenicillanic acid, vitamin B_6 from hydrocarbons); *C. boidinii* (NAD, FAD, methyl ketones and citric acid); *C. famata* (riboflavin); *C. maltosa* (fatty acid and alkane utilization for biomass protein); *C. tropicalis* (tryptophan); *C. pelliculosa* (biomass protein from cellulosic material); *C. utilis* (range of products, grows on xylose, candidate for cloning technology); *C. shehatae* (xylose fermentations).
Hansenula polymorpha	Methylotrophic yeast useful in the expression of heterologous genes (see Sudbery, 1995).
Kluyveromyces marxianus, *K. lactis*	Lactose and polyfructosan fermented. Natural cocoa fermentations. Several enzymes (lactase, pectinase, etc.). Candidates for cloning technology.
Pachysolen tannophilus	Fermentation of pentose sugars derived from plant/wood lignocellulosic hydrolysates.
Phaffia rhodozyma	Astaxanthin pigment (food colorant).
Pichia spp.	*P. guilliermondii* (oversynthesizes riboflavin; biomass protein from hydrocarbons – see Sibirny, 1996a). *P. methanolica* produces alcohol oxidase of use in ethanol biosensors (Sibirny, 1996b). *P. pastoris* is a methylotrophic yeast which produces biomass protein from methanol and is useful in expression of heterologous genes and in production of human therapeutic proteins (see Sudbery, 1995).
Rhodosporidium toruloides	Source of phenylalanine ammonium lyase (PAL) which has potential in diagnosis and treatment of phenylketonuria.
Saccharomyces spp.	*S. cerevisiae* is used in classical food fermentation applications (beer, bread, yeast extracts/vitamins, wine, saké, distilled spirits). Also produces fuel alcohol, glycerol, invertase and animal feed. Numerous proteins (e.g. biopharmaceuticals) via recombinant DNA technology. *S. cerevisiae* var. *diastaticus* utilizes starch. *S. boulardii* used as a biotherapeutic agent. *S. pyriformis* is the starter culture for traditional ginger beer and *S. bayanus* is used in champagne fermentations.
Saccharomycopsis spp.	*S. fibuligera* is an amylolytic yeast.
Schizosaccharomyces pombe	Traditional African alcoholic beverages (millet beer or 'Pombe' and palm juice gin or 'Ogogoro'). Rum fermentations and de-acidification of wines by malo-ethanolic fermentations. High ethanol/osmotic tolerance and xylose fermentation (candidate for fuel ethanol). Candidate for biomass protein and expression of heterologous genes. Genotoxicity/mutagenesis testing.
Schwanniomyces spp.	*S. castellii* and *S. occidentalis* are amylolytic yeasts. *S. occidentalis* can efficiently convert starch and inulin to ethanol (see Ingledew, 1987) and it also has potential in heterologous gene expression.
Trichosporon cutaneum	Biosensor for monitoring phenol (Neujahr, 1990) and B.O.D. (Reiser *et al.*, 1996).
Yarrowia lipolytica	Biomass protein from lipids and hydrocarbons. Production of citric acid and secretion of extracellular enzymes (e.g. lipase). Potential in heterologous gene expression (Barth and Gaillardin, 1996).
Zygosaccharomyces rouxii	Osmotolerant and halophilic yeast used to provide characteristic aromas to Japanese soy sauce and miso.

6.2 APPLIED MOLECULAR GENETICS OF YEASTS

6.2.1 Genetics of 'Industrial' *Saccharomyces* Yeasts

Saccharomyces cerevisiae has long been a favoured organism for the study of genetics (see Hall and Linder, 1993). Particular genetic attributes of this yeast are: its ease and rapidity of growth; the ready isolation and selection of mutants; its small genome and the existence of both haploid and diploid life cycles. Such facilities have enabled *S. cerevisiae* to be used not only for basic research into genetic phenomena, but have also provided a means to improve strains of this important yeast for biotechnology.

'Classical' genetic approaches in *S. cerevisiae* involve mating of haploids of opposite mating-type (**a**, and α cells). Subsequent meiosis and sporulation results in the production of a tetrad ascus with four spores which can be isolated, propagated and genetically analysed (i.e. tetrad analysis). Such procedures are generally very feasible with so-called *laboratory reference* or *academic* strains of *S. cerevisiae* which can be propagated vegetatively as stable haploids or diploids. *Industrial* strains on the other hand, such as those employed in the fermented beverage industries are genetically quite complex and do not exhibit stable haploidy. For example, brewing strains of *Saccharomyces* are polyploid or aneuploid, ranging from diploid to heptaploid (Hinchliffe, 1991; Hansen and Kielland-Brandt, 1996a), they are reticent to mate, and exhibit poor sporulation with low spore viability. As a consequence, it is generally fruitless to perform tetrad analysis and 'breeding' with brewing yeasts (Kielland-Brandt *et al.*, 1995). Rather than systematically breeding yeast strains, brewers have traditionally relied upon natural selection of those yeasts which characteristically ferment well and which impart desirable organoleptic properties to the beer. Nevertheless, several genetic opportunities exist to improve fermentation processes in a consistent, reproducible manner with 'new' brewing yeast strains (Table 6.2).

Several workers have discussed strategies for circumventing the sexual reproductive deficiencies of brewing yeasts which have total disregard for ploidy and mating type (e.g. Bilinski *et al.*, 1987; Russell *et al.*, 1987; Kielland-Brandt *et al.*, 1995; Hansen and Kielland-Brandt, 1996a). Undoubtedly, the most significant contributions to strain improvement of industrial yeasts are made by introducing novel genetic characteristics using recombinant DNA technology.

Other molecular genetic techniques have also proved beneficial in the use of *Saccharomyces* yeasts in biotechnology. For example, in the differentiation of brewing yeasts with similar physiological properties using *genetic fingerprinting* techniques. Such procedures encompass various methods of DNA analysis including restriction fragment length polymorphisms (RFLPs) electrophoretic karyotyping and the polymerase chain reaction (PCR) – see section 6.3.

Electrophoretic karyotype analysis has also been applied to baking yeast strains (Codon and Benitez, 1995), to non-*Saccharomyces* yeasts (Zimmerman and Fournier, 1996) and to pathogenic yeasts (Boekhout *et al.*, 1997). Differentiation between wine strains of *S. cerevisiae* has been achieved using mitochondrial DNA restriction analysis (Querol and Ramon, 1996).

6.2.2 Recombinant DNA Technology in Yeasts

'Genetic engineering' – or more strictly speaking, recombinant DNA technology – has rapidly revolutionized several areas of basic and applied biology. Yeast cells are at the forefront of many of these developments in modern biotechnology. Particular attributes of yeasts as hosts for expressing foreign genes are listed in Table 6.3. Many of these attributes are not shared by prokaryotic cells and this often makes yeasts the preferred microbial hosts for production of many eukaryotic proteins.

Although yeast cells, particularly those of *S. cerevisiae*, have often been described as the *ideal* hosts for expressing heterologous genes, there

Table 6.2. Some methods employed in genetic research and development of brewer's yeasts.

Method	Comments/References
Hybridization	Cannot generally be used directly, but method is not entirely obsolete. Has been used to study the genetic control of flocculation, sugar uptake and flavour production (Bilinski *et al.*, 1987). Cross-breeding of *S. carlsbergensis* lager yeast may be applied to combine known desired features of parental strains (Kielland-Brandt *et al.*, 1995). Hybridization of spore-derived clones of distiller's strains of *S. cerevisiae* has also been accomplished (Christensen, 1987).
Mutation and selection	For example, to induce auxotrophic and derepressed mutants for efficient maltose fermentation in the presence of glucose (Russell *et al.*, 1987).
Rare mating	Mixing of non-mating strains at high cell density (*c.* 10^8 cells/ml) results in a few true hybrids with fused nuclei. Cytoduction (introduction of cytoplasmic elements without nuclear fusion) can also be used to impart killer activity (using karyogamy deficient, kar$^-$, mutants). See Russell *et al.* (1987).
Spheroplast fusion	Spheroplasts from yeast strains of one species, the same genus, or different genera can be fused to produce intraspecific, interspecific or intergeneric fusants, respectively (see Kavanagh and Whittaker, 1996; Zimmerman and Sipiczki, 1996). The possibility exists to introduce novel characteristics into brewing strains which are incapable of mating (Russell and Stewart, 1979).
Single-chromosome transfer	Transfer of whole chromosomes from brewing strains (using the kar$^-$ mutation) into genetically defined strains of *S. cerevisiae.* (Kielland-Brandt *et al.*, 1995).
Transformation	Introduction of genes from other yeasts and other organisms (Hinchliffe, 1991; Hammond, 1995; and section 6.3.1).

are potential drawbacks worth considering in *S. cerevisiae* cloning technology. For example, the glycosylation patterns of proteins processed by this yeast may not match those of human glycoproteins for therapeutic use. This is due to the presence of antigenic α-1,3 mannose linkages which *S. cerevisiae* adds to some heterologous proteins. In addition, if heterologous proteins are intracellularly accumulated (rather than extracellularly secreted) they may prove deleterious to yeast growth and costly to extract from disrupted cells. Genetic drawbacks may also arise if considering problems of plasmid instability and economic costs of providing growth media minus an auxotrophic nutrient which may be required to place selective pressure on recombinant plasmids. Nevertheless, these problems can largely be overcome using a variety of cell physiological and molecular genetic approaches and by using alternative yeasts to *S. cerevisiae* (see below). This means that yeasts are likely to remain the host organisms of choice for many biotechnology companies in the future production of recombinant human proteins.

Recombinant DNA technology in *S. cerevisiae* has been extensively reviewed in recent years (e.g. Barr *et al.*, 1989; Romanos *et al.*, 1992; Curran and Bugeja, 1993; Hadfield *et al.*, 1993; Hinnen *et al.*, 1995). Although the majority of research and development into recombinant protein synthesis in yeasts has been conducted using *S. cerevisiae*, several non-*Saccharomyces* species are now being studied and exploited in this sector of biotechnology. Some of these so-called (but unsuitably named) *non-conventional* yeasts exhibit particular advantages in cloning technology com-

Table 6.3. Attributes of yeasts as foreign gene expression hosts.

General attributes	Examples
Historical	• Yeasts and yeast products have always held a generally favourable public acceptability. • Industrial yeasts are non-pathogenic and *S. cerevisiae* and *K. lactis* possess GRAS (generally regarded as safe) status (no endotoxins, pyrogens, viruses, etc.). • *S. cerevisiae* is the most studied of simple eukaryotes in terms of biochemistry and genetics.
Technological	• There is simplicity and safety associated with large-scale yeast growth. • Fermentation and downstream processing technologies are well understood and developed for yeasts. • Research-to-production scale-up yeast technology is relatively straightforward. • A wide range of readily utilizable and inexpensive carbon sources for yeast growth and fermentation exist.
Genetical	• Yeasts are eukaryotes and can efficiently express heterologous eukaryotic genes. • Industrial yeasts (e.g. brewing strains) are well-adapted for high yields of metabolic products. They are also fairly 'robust' in their ability to withstand environmental stress. • The *S. cerevisiae* genome is relatively small and can be modified by both classical genetic hybridization and modern recombinant DNA techniques. • Stable mutants are achievable that enhance productivity. • Preferred codon usage known. • Selectable markers available for polyploid strains. • Contains 'natural' plasmids (e.g. 2μm DNA) but no viruses to kill cells in large-scale propagations.
Molecular biological	• Both autonomous and integrative transformation possible. • Post-translational modification (e.g. glycosylation) occurs together with proteolytic maturation and multimeric particle assembly. • Protein secretion is quite efficient and controllable using endogenous signal sequences. • Yeasts can often splice introns from animal genes. • Yeast RNA polymerases recognize many animal promoters.

pared with *S. cerevisiae* and these are outlined in Table 6.4.

The next sections will provide an overview of the following aspects of recombinant DNA technology in yeasts: molecular genetic aspects; cell physiological aspects; technical aspects and commercial aspects.

6.2.2.1 Molecular Genetic Aspects

The transformation of *S. cerevisiae* cells by 'foreign' DNA was first described around 20 years ago (Beggs, 1978; Hinnen *et al.*, 1978). Note, however, that about 20 years before these classic papers, Oppenoorth (1959) claimed the *transformation* of brewing yeast with 'crude' DNA preparations from a wild yeast (*S. chevalieri*). In recent years, *S. cerevisiae* has arguably established itself as *the* experimental model organism in which to study the expression of genes from higher eukaryotes. Other species of yeast have also been studied with regard to both fundamental and applied aspects of recombinant DNA technology, some of which are listed in Table 6.4. Nevertheless, the overall strategies of

Table 6.4. Attributes of some non-*Saccharomyces* yeasts in cloning technology.

Yeast	Attributes
Candida maltosa	This yeast can readily assimilate *n*-alkanes and fatty acids as carbon sources. *n*-Alkane induction of endogenous P450 promoters can be used to control expression of heterologous genes. Can target the synthesis of foreign proteins to specific organelles (ER or peroxisomes). This yeast can be used to optimize the biotransformation and intracellular transport of hydrophobic organic compounds. **Reference:** Mauersberger *et al.* (1996)
Hansenula polymorpha	Stringently promoted gene expression (using *MOX*, methanol oxidase, or formate dehydrogenase). Stable, multi-copy integration of foreign DNA. Growth up to 45°C. Crabtree effect absent (see Chapter 5). Range of auxotrophic host strains (*ura*$^-$, *leu*$^-$, etc.). Proteins not hyperglycosylated. Can either secrete proteins or accumulate potentially toxic proteins in peroxisomes. **References:** Sudbery (1994); Gellissen and Melber (1996); Hansen and Hollenberg (1996)
Kluyveromyces spp. (*K. lactis, K. marxianus*)	*K. lactis* has GRAS status. Lactose-fermenting and can be grown on cheap carbon sources such as cheese whey. Very high cell density propagations ($>$ 100 g cell dry wt/l). Contains endogenous plasmids which can be used as vectors (e.g. *K. lactis* linear killer plasmids and circular pKD1 plasmids which are similar to 2 μm DNA fragments). **Reference:** Wésolowski-Louvel *et al.* (1996)
Pichia pastoris	Methanol-inducible *AOX* (alcohol oxidase) promoter (tightly regulated). Very high cell density propagations ($>$ 100 g dry cell wt/l) on simple growth media. High heterologous protein expression levels (5–40% of cell protein) with efficient secretion ($>$ 1 g/l). Proteins not hyperglycosylated. G418^R selection simplifies rapid isolation of multicopy transformants. **References:** Romanos (1995); Sreekrishna and Kropp (1996)
Pichia methanolica	Attributes similar to *P. pastoris* but *P. methanolica* is the only methylotrophic yeast which can be analysed by classical genetics, including gene mapping. **Reference:** Sibirny (1996b)
Schizosaccharomyces pombe	Can correctly splice mammalian introns. Galactosylates proteins (as in higher eukaryotes). Distance from TATAA box to transcription start site resembles that in higher eukaryotes. Glucose-dependent lysis of cells at end of growth cycle may facilitate harvesting of heterologous proteins. **References:** Evans and McAthey (1991); Bligh and Kelly (1990)
Schwanniomyces occidentalis	Starch utilizer. Does not hyperglycosylate. Secretes proteins greater than 140 kDa. **Reference:** Dohmen and Hollenberg (1996)
Yarrowia lipolytica	Efficient and precise integrative transformation Efficient secretion–signal recognition resembles that of higher eukaryotes. Grows on *n*-paraffins as sole carbon source. **Reference:** Barth and Gaillardin (1996)

Table 6.5. Main stages in heterologous gene expression in yeast cells.

Stage	Brief description
Gene isolation	A structural gene (or, more likely, complementary DNA) encoding the protein of interest is isolated from restriction fragments of DNA from a donor organism.
Cloning (vectorization)	The *in vitro*-manipulated heterologous gene is recombined into an appropriate expression vector.
Transformation	The vector is introduced into yeast cells or spheroplasts by autonomous or integrative transformation.
Selection	Selectable genetic markers, carried by plasmid vectors, are used to identify transformants (e.g. by complementation of a recessive mutation in the host strain). Empirical selection (after mutagenesis) may be employed for supersecretor strains.
Expression	Heterologous gene expression involves: transcription using promoter sequences fused upstream from the structural gene; termination of transcription using terminator sequences fused downstream from the structural gene; mRNA transport from nucleoplasm to cytoplasm; initiation of translation and elongation of the polypeptide chain.
Modification	Heterologous proteins are subjected to post-translational chemical modification. (e.g. glycosylation, acetylation, etc.)
Signalling	Specific secretion signal peptides are removed for secretory proteins, or proteins may be allowed to accumulate in cells or specific organelles.
Secretion	Certain proteins, if desired, are exported from the cell into the surrounding medium.

yeast cloning are common to all species and involve several basic stages (Table 6.5).

Figure 6.2 schematically summarizes some of the basic procedures undertaken in yeast recombinant DNA technology. Molecular genetic aspects involved in choice of vectors, promoters, selectable markers and secretion signals will now be discussed with particular reference to 'industrial' yeast strains. Depending on the biotechnological application, heterologous genes for yeast cloning may be isolated from viral, prokaryotic, or eukaryotic DNA (see Table 6.35). The latter may include genes from other yeasts.

Numerous plasmid **vectors** have been constructed for genetically transforming yeast cells (Table 6.6). Generally speaking, the vectors for the most efficient transformation of yeast cells would contain the following:

- Sequences of bacterial (*Escherichia coli*) plasmids which serve as a 'cloning and amplification device' (Hinnen *et al.*, 1995) and which permit DNA to be shuttled back and forth between yeast and bacteria. Many yeast vectors are therefore *shuttle plasmids*.
- An autonomously replicating sequence (ARS).
- A centromeric (CEN) sequence for stable mitotic segregation.
- A suitable and convenient selectable marker.
- One or several unique restriction enzyme sites to allow for insertion of heterologous DNA.

Promoters are employed in yeast cloning vectors to enhance transcription of the gene of interest. Promoter sequences may be yeast-derived (homologous) or heterologous. The former are generally more efficient and can either be constitutive or regulated (Table 6.7). Frequently used constitutive yeast promoters are glycolytic genes which lead to high-level transcriptional expression. Glycolytic enzymes in fermenting yeast cells can account for between 1–5% of total soluble protein and so promoters like *ADH*, *PGK* and *ENO* are popular choices

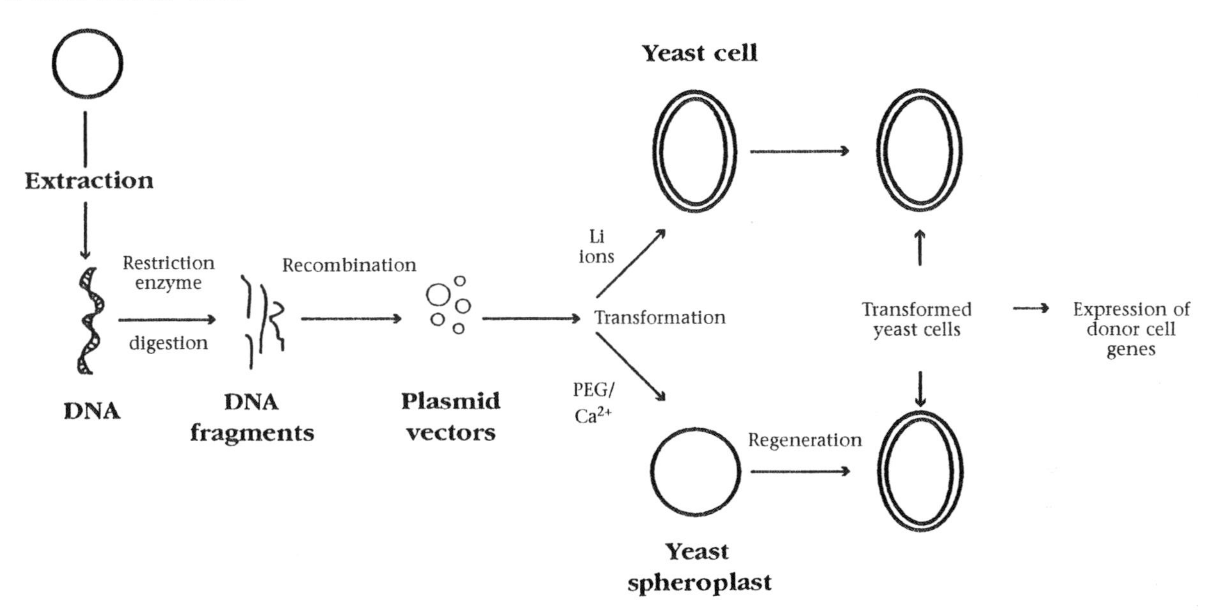

Figure 6.2. Basic procedures in yeast genetic engineering.

for augmenting transcriptional activity of heterologous genes. However, glycolytic promoters are not tightly regulated. Regulated promoters on the other hand facilitate yeast cell growth in the absence of heterologous gene expression. In other words, biomass increase can be temporally separated from gene expression by controlling the availability of certain nutrients (e.g. galactose in the case of *GAL1*, phosphate in the case of *PHO5*, methionine in the case of *MET25*, or copper ions in the case of *CUP1*; see Table 6.11). In this way, the yeast cell population can be maximized before nutrient-regulated induction of foreign gene expression. Temperature shifts can also be used to regulate heterologous gene expression in yeasts (Piper and Kirk, 1991).

Particularly strong transcriptional activation of heterologous genes can be achieved in methylotrophic yeasts such as *H. polymorpha* and *P. pastoris*. This is because of the stringent derepression of methanol-utilizing genes when nutrient feed is changed from glycerol to methanol as sole carbon source (see Faber *et al.*, 1995).

Another molecular genetic aspect relevant to the success of recombinant DNA technology in yeasts is the choice of a suitable **selectable marker** which is used to isolate and identify transformant cells. Table 6.8 lists some of the recessive and dominant selectable markers which have been used in yeast cell transformation. Recessive markers are genes which complement a specific auxotrophic mutation in the host yeast strain. Many of these markers carried by vectors are genes complementing particular amino acid or nucleotide auxotrophy. Specific dominant markers are required for selecting transformants of industrial (brewing) strains of *S. cerevisiae* due to their polyploid genetic make-up and absence of available auxotrophic mutants.

Secretion of heterologous proteins by yeast is generally desirable from the cells' viewpoint since it avoids potential toxicity of intracellularly accumulated protein, and also from the yeast technologists' viewpoint since it simplifies protein purification procedures. Additional advantages lie in post-translational modification (e.g. glyco-

Table 6.6. Plasmid vectors for transforming yeast cells.

Vector type	Examples	Characteristics
Integrating plasmids	Yeast Integrative plasmid (YIp)	Low-frequency integration into chromosomal loci occurs by homologous recombination, since these vectors do not have functional replication origins. YIps are integrated in single copies, but transformants are stable – even in absence of selective pressure.
Independently replicating plasmids	Yeast Replicating plasmid (YRp)	High-frequency transformation (500–2000 transformants per μg DNA), due to ARS elements which maintain extra-chromosomal replication. Plasmids are unstable during yeast growth due to asymmetric segregation at cell division. YRps are rarely used.
	Yeast Episomal plasmid (YEp)	These are derived from *S. cerevisiae* 2μm plasmids with origin of replication (ori) sequences. YEps are relatively stably maintained extrachromosomally at 50–100 copies per cell.
	Yeast Centromeric plasmid (YCp)	Inclusion of functional yeast centromeric (CEN) sequences leads to stable segregation at mitosis. Plasmids are maintained at 1–3 copies per cell but are stable in dividing cells.
Specialized plasmids	Yeast Linear plasmid (YLp)	These artificial 'chromosomal' plasmids contain telomeric sequences for maintaining a linear state.
	Yeast Artificial Chromosomes (YACs)	Large artificial linear chromosomes very stably maintained (contain CEN sequences). Useful for accepting very large DNA inserts for long-range physical mapping techniques currently used in human genome analysis.
	Yeast Expression plasmid (YXp)	Contain transcriptional promoter and terminator sequences to which heterologous genes can be fused. Many also contain coding sequences for modification and secretion of foreign proteins.
	Yeast Retrotransposons (Ty elements)	Ty fusion proteins can produce hybrid Ty virus like particles which are potentially usefully in production of vaccines.
	Yeast Killer Plasmids	e.g. pGKL1 and 2 from *K. lactis* are linear killer plasmids that have been developed as cloning vectors.
	Yeast Disintegration plasmids (YDp)	Useful for totally losing bacterial vector sequences.

Information mainly from Curran and Bugeja (1993) and Parent and Bostian (1995).

sylation), proteolytic maturation and folding that occurs as proteins pass through the yeasts' secretory pathway (see Chapter 2). The molecular elements which direct protein secretion are the *signal sequences* at the N-terminal of the protein destined for export. These signal peptides (of approx. 15–30 amino acids), which are removed by specific proteases, are encoded by yeast (native) or heterologous (foreign) genes which are inserted at the junction between the 5′ end of the structural gene and the 3′ end of the promoter. Examples of yeast secretion signals are given in Table 6.9.

Protein secretion in yeast is a complex process and there is no universally accepted signal sequence which directs secretion in all yeast species. Each system requires considered analysis on a case-by-case basis. Although several foreign proteins can be secreted under the direction of their own signals (Hadfield *et al.*, 1993), generally speaking, homologous signal sequences are more likely to be successful and can result in highly

Table 6.7. Transcriptional promoters used in yeast recombinant DNA technology.

Promoter type		Gene	Encoded protein
Constitutive:		*ADH1*	Alcohol dehydrogenase 1
		PGK1	Phosphoglycerate kinase
		GAP 491	Glyceraldehyde 3-phosphate dehydrogenase
		TPI	Triose phosphate isomerase
		PYK	Pyruvate kinase
		ENO	Enolase
		PMA1	H^+-pumping ATPase (*H. polymorpha*)
Regulated:		*PHO5*	Acid phosphatase
		GAL1, 7, 10	Products of the Leloir pathway
		MET25	*O*-acetyl homoserine sulphydrylase
		ADH2	Alcohol dehydrogenase 2
		MEL1	Melibiase
		CUP1	Copper metallothionein
		HSE	Heat shock element[a]
		*MFα1/MF***a***1*	Mating pheromones α and **a**
		AOX	Alcohol oxidase (*P. pastoris*)
		MOX	Methanol oxidase (*H. polymorpha*)
Heterologous:	viral	*SV40*	Simian virus early promoter[a] (*Sch. pombe*)
		CaMV	Cauliflower mosaic virus 35S promoter[a]
	plant	*Opaque-2*	Maize zein protein
	animal	*GRE*	Glucocorticoid response element[a]
		ARE	Androgen response element[a]

[a] These are not encoded proteins, but DNA 'modules'.

expressed proteins comprising over 80% of the total protein recoverable from the extracellular medium. Examples of frequently used signal sequences in *S. cerevisiae* include those derived from invertase (encoded by *SUC 2*), acid phosphatase (*PHO 5*) and α-factor mating pheromone (*MFα1*). It is noteworthy that since the specificity of the KEX2 and STE13 exopeptidases is for sequences on the N-terminal of the cleavage point, secretion using the α-factor system can provide products with authentic N-termini. The composition and processing of the latter signal sequence is outlined in Figure 6.3. The α-factor leader has been used to successfully direct the secretion of numerous heterologous proteins of use as human therapeutic agents (see Brake, 1989). Other attributes are that it can also be used to efficiently secrete proteins in non-*Saccharomyces* yeast such as *P. pastoris*.

Post-translational processing and modification of recombinant yeast proteins is another important molecular aspect which requires consideration, particularly with regard to the production of human therapeutic proteins. The first post-translational processing step is the removal of the initiator methionine from newly synthesized protein by a methionyl aminopeptidase, a process common to higher eukaryotes. The chemical modifications which yeasts may then carry out on heterologous proteins include: glycosylation, phosphorylation, acetylation, methylation, myristylation and isoprenylation.

N- and O-linked glycosylation patterns may prove to be problematic since some yeasts glycosylate secretory proteins resulting in the production of non-identical native glycoproteins. For example, *S. cerevisiae* adds mannose units to threonine or serine residues (i.e. O-linked) whereas many higher eukaryotes have sialic acid O-linked side chains. Such differences may affect

Table 6.8. Selectable genetic markers used in yeast recombinant DNA technology.

Marker type	Gene	Comments
Recessive	*LEU 2* *TRP 1* *HIS 3* *LYS 2*	Genes which complement auxotrophic mutations for amino acid biosynthesis
	URA 3 *ADE 2*	Genes which complement autotrophy for nucleotides
	Amyloglucosidase β-Lactamase	Genes conferring enzyme activity (For screening, rather than selecting)
Dominant	*CUP 1*	Copper resistance
	G418^R	Aminoglycoside antibiotic resistance
	TUN^R	Tunicamycin resistance
	KILk1	Killer toxin immunity
	C230	Chromogenic marker
	SMR1	Sulphometuron methyl resistance
	SFA	Encodes formate dehydrogenase (cells can grow in 6 mM formaldehyde)
	Hygromycin^R Methotrexate^R Chloramphenicol^R Diuron^R Zeocin^R Canavanine^R	Genes which confer resistance to drugs

the folding, stability, activity and immunogenicity of heterologous proteins produced by *S. cerevisiae*. N-linked glycosylation in *S. cerevisiae* appears to resemble processes which occur in higher eukaryotes.

Problems caused by over-glycosylation (e.g. mannosylation) of proteins by *S. cerevisiae* (and *K. lactis*) may be circumvented by using other yeast species such as *Pichia pastoris* which do not have a tendency to hyperglycosylate. Further aspects of the molecular biology of protein glycosylation in relation to yeast recombinant DNA technology have been discussed by Kukuruzinska *et al.*, (1987); Hadfield *et al.*, (1993) and Eckart and Bussineau (1996).

Once synthesized and modified, heterologous proteins produced by yeast may undergo intracellular proteolytic degradation before their purification. In *S. cerevisiae*, such proteolysis may be unspecific and associated with the vacuole (Hirsch *et al.*, 1989), or it may be specific and coupled to the ubiquitin system (Hilt and Wolf, 1992). Proteolysis can be circumvented by exploiting the secretory pathway (e.g. use of 'supersecreters') or by using protease-deficient host strains. Hinnen *et al.*, (1995) have discussed the problem of proteolysis in *S. cerevisiae* with specific reference to the production of recombinant hirudin, a thrombin inhibitor from the leech *Hirudo medicinalis*.

6.2.2.2 Cellular Aspects

Several cell physiological aspects require consideration before production of recombinant proteins from yeasts. These will be discussed in turn and include: intact cell and spheroplast transfor-

Table 6.9. Signal sequences used for secretion of heterologous proteins from yeasts.

Type	Signal sequence gene	Comments
Homologous	*SUC 2*	Encodes periplasmic invertase
	PHO 5	Encodes periplasmic acid phosphatase
	Killer Toxin	e.g. from *K. lactis*
	α Factor	The prepro-MFa1 leader is most frequently used
	MEL 1	Encodes melibiase, a secreted glycoprotein
Heterologous	*HSA*	Encodes human serum albumin (*HSA* prepro has been used in *P. pastoris*)
	Mucor pusillus rennin	see Hadfield *et al.* (1993) for references
	Chicken lysozyme	
	Bovine prolactin	
	Human gastrin	

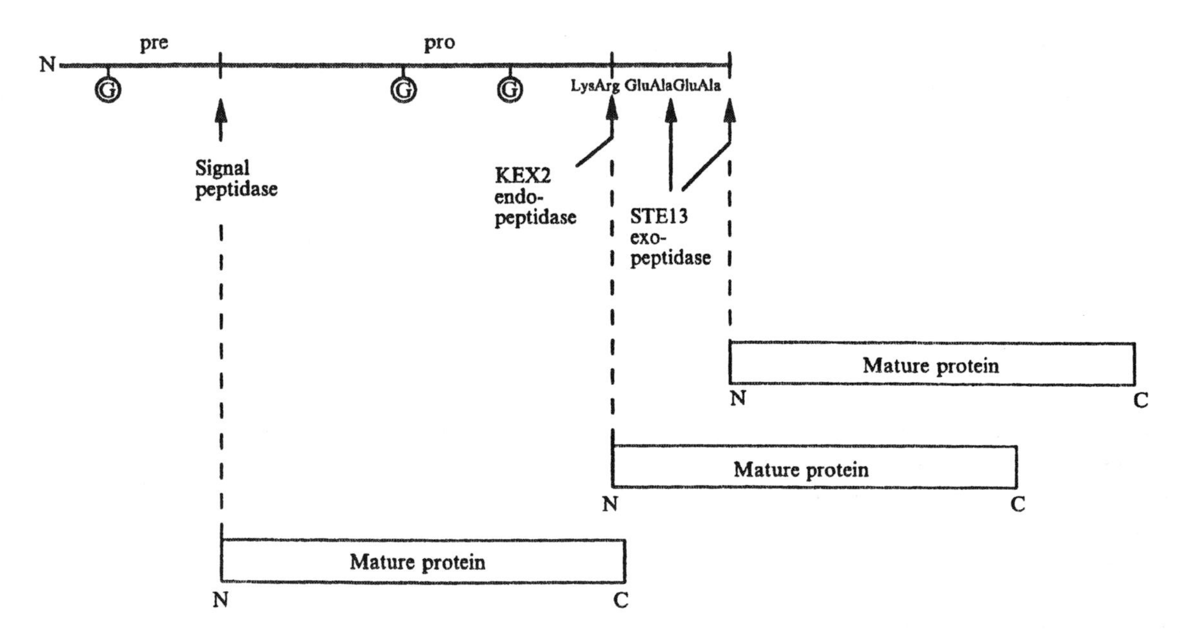

Figure 6.3. Composition and processing of the α-factor leader of *S. cerevisiae.* The pre-peptide directs secretion into the ER and is removed by signal peptidase. The pro-peptide is cleaved in the Golgi by KEX2 endopeptidase, which cuts C-terminally to LysArg. This leaves the two GluAla dipeptides attached to the N-terminus of the mature protein. These are serially removed by STE13 exopeptidase digestion. The possible sites for heterologous protein fusion to the MFα1 leader are indicated. The three glycosylation sites are indicated (G). Reproduced with permission from Hadfield *et al.* (1993) and the British Mycological Society. Further details on the molecular biology of the α-factor leader and its use in recombinant DNA technology can be found in Brake (1989) and Hadfield *et al.* (1993).

mation strategies; regulation of heterologous genes by growth conditions; stabilization of plasmids and the cellular or extracellular destination of mature foreign proteins.

The yeast cell wall presents a formidable barrier to the entry of exogenous macromolecules (see Chapter 2). Transformation of *S. cerevisiae* cells with foreign DNA was originally accomplished using osmotically stabilized spheroplasts in which the cell wall had been partly digested by lytic enzymes. In the presence of polyethylene glycol (PEG, which co-precipitates cells) and Ca^{2+} ions, exogenously added DNA is able to transfect the spheroplasts which are then allowed to regenerate their cell walls in soft agar. Problems encountered with this spheroplast transformation technique include: strain variation in transformation efficiency, poor transformant recovery and high frequency of diploidization (see Rose, 1995). Many of these problems can be overcome by directly transforming intact (or *competent*) yeast cells. This circumvents the laborious preparation of spheroplasts and avoids their time-consuming regeneration. Methods employed are outlined in Table 6.10, and further details can be found in Watts and Stacey (1991) and Rose (1995).

Once yeast cells have been transformed by foreign DNA, either by genomic chromosomal integration or by autonomously replicating plasmids, the heterologous genes need to be expressed efficiently. In addition to the molecular genetic aspects discussed previously, several physiological factors are important in governing the expression of foreign genes and the subsequent modification of foreign proteins in yeast cells. With regard to nutrient and temperature inducible transcription, several examples of physiologically regulated promoters operate in yeasts and these are summarized in Table 6.11. Piper and Kirk (1991) have discussed the benefits of inducible over constitutive gene expression in recombinant *S. cerevisiae* cells.

It is essential that due consideration is given to growth media constituents and physical growth conditions when attempting to maximize foreign gene expression in yeasts using regulated promoters. Regarding media, chemically defined nutrients are preferred due to the inconsistency and impurity of complex industrial fermentation feedstocks such as molasses, sulphite waste liquor, etc. Much tighter physiological control over transcriptional promoters can be achieved using growth media whose make-up is precisely known and which can be finely tuned to control growth, metabolism and foreign gene expression in recombinant yeasts propagated in large-scale bioreactors (see Vasavada, 1995).

In yeast recombinant DNA technology, it is often desirable to 'switch on' the synthesis of heterologous proteins only when cells have grown to a sufficient culture density in bioreactors. Methods employed to achieve high cell densities by recombinant yeasts are described below. 'Switching on' can be relatively simply accomplished in dense cell populations by altering the availability of a particular nutrient (preferably in defined media) or by changing the fermentation temperature. Such strategies which temporally separate biomass growth from foreign protein production may prevent product toxicity and plasmid instability.

Table 6.10. Transformation strategies for intact yeast cells.

Method	Description	Comments
Chemical	Cells are permeabilized with lithium salts (e.g. acetate), DNA and PEG added and cells briefly heat shocked	Generally less efficient compared with the spheroplast method, but refinements can improve transformation frequencies
Electrical	Cells and DNA are electrically pulsed in an electroporation device (commercially available)	Efficient and convenient method for transforming yeast cells, but relatively expensive equipment needed
Biolistic	DNA-coated tungsten microprojectile particles are propelled in a 'gun' at high speed into the cell	Can also be used to transform yeast mitochondria

Table 6.11. Physiological regulation of transcriptional promoters in yeast.[a]

Promoter[b]	Physiological regulation
GAL 1	Repressed by glucose, induced by galactose
PHO 5	Repressed by high, and induced by low, inorganic phosphate levels
ADH 2	Repressed by glucose, derepressed naturally towards the end of growth on glucose
MFα1	Active in MATα cells, inactive in MAT**a** or **a**/α cells
MFα1	Induced by low temperature shift
CUP 1	Induced by copper ions[c]
MEL 1	Induced by galactose, repressed by glucose
MET25	Repressed by methionine
PGK/ARE hybrid	Induced by dihydrotestosterone
CYC/GRE hybrid	Induced by deoxycorticosterone
PGK or *TPI/α2* operator	Induced by low temperature shift
HSE	Induced by heat shock (e.g. 39°C)
AOX 1	Repressed by glycerol, induced by methanol (in *P. pastoris*)
MOX 1	Repressed by glycerol, induced by methanol (in *H. polymorpha*)

[a] Yeast refers to *S. cerevisiae*, unless mentioned otherwise.
[b] See Table 6.7 for list of proteins encoded by these promoters.
[c] High expression levels may also be achieved with Mg^{2+} ions (Macreadie *et al.*, 1995).

The stability of hybrid plasmids in recombinant yeasts is a very important consideration which directly impinges on the successful commercial production of heterologous proteins from yeasts. This is mainly due to the fact that the majority of plasmid vectors used are unstable shuttle vectors (see Table 6.6). Problems arise due to the structural and segregational instability of plasmid vectors in yeast which may lead to the entire loss of plasmids from transformed cells. The inherent instabilities of plasmids, together with external physiological factors, may result in plasmid-free cells outgrowing their plasmid-containing counterparts and this would have serious consequences for large-scale production of recombinant proteins (Zhang et al., 1996).

Plasmid stabilization techniques are therefore required to prevent such losses and the main strategies (outlined in Table 6.12) are molecular genetic (e.g. novel plasmid construction) and bioprocess-related (e.g. nutrient availability and physical environment). Vasavada (1995) and Zhang *et al.*, (1996) have further discussed strategies for improving plasmid stability in recombinant yeast.

The importance of yeast cell physiology also needs to be recognized when considering the cellular destiny of synthesized heterologous proteins. For example, are proteins to be recovered from the cytoplasm, organelles, membranes, periplasmic space, cell wall, or from the extracellular medium?

Location of proteins intracellularly would be desired for heterologous genes which are *normally* expressed in the cytoplasm and also for secreted proteins which have no or few disulphide bonds (Eckart and Bussineau, 1996). The successful targeting of proteins to specific yeast membranes has been reported for several heterologous membrane-associated proteins from plants (e.g. nutrient and ion-transporters; Frommer and Ninnemann, 1996). Specific organelle-targeting of heterologous proteins can also be accomplished in yeasts and the accumulation of potentially toxic proteins into the peroxisomes of methylotrophic yeasts would be particularly advantageous in this regard (see Chapter 2). The specific directing of expressed heterologous proteins to the yeast cell wall is generating significant interest in yeast biotechnology – for example, in the production of whole-cell oral vaccines (e.g. against hepatitis B; see Schreuder *et al.*, 1996) and for

Table 6.12. Factors affecting plasmid stability in recombinant yeasts.

Factor	Examples which affect stability	Approaches to improve stability
Genetic	Plasmid make-up and copy number. Expression level. Selectable markers. Host cell properties (including ploidy).	Inclusion of stabilization, partitioning (e.g. CEN) and replication sequences. Optimization of copy number. Suitable promoters and selectable markers (e.g. autoselection).
Physiological	Phase of growth (e.g. exponential or stationary) and growth rate (e.g. changing as in batch culture or constant as in chemostat culture).	Data on cell growth physiology in relation to plasmid stability is conflicting (see Vasavada, 1995).
Environmental	Medium formulation, dissolved oxygen tension, temperature, pH, dilution rate, bioreactor operation modes.	Addition of antibiotics/removal of amino acids. Optimizing and changing dissolved oxygen and temperature. Oscillating dilution rate and substrate concentrations. Use of fed-batch and immobilized bioreactors.

the immobilization of industrial enzymes (e.g. α-galactosidase; see Van der Vaart *et al.*, 1997). Several recombinant proteins have been shown to be toxic to host yeast cells or inhibitory to their growth (Hadfield *et al.*, 1993) and so their extracellular secretion is desirable in such circumstances. Other molecular and technological advantages of secretion in recombinant yeasts have been discussed previously.

6.2.2.3 Technical Aspects

Two major approaches to increasing the yield of biologically active recombinant proteins produced by yeasts are: to use molecular genetics and to use fermentation technology. The two should **not** be mutually exclusive. Regarding the latter, understanding the nutrition, growth and metabolism of recombinant yeasts (i.e. their cell physiology) in fermenters is crucial to the development of commercial products by genetic engineering. Concerning yeast growth, several options are available to cultivate recombinant yeasts in large-scale bioreactors in which the nutritional growth environment can be manipulated and controlled in various ways (Table 6.13).

The rate and extent of host cell growth is a very important aspect of recombinant DNA technology in yeasts. Active cell growth and high biomass levels will generally be associated with higher rates and levels of recombinant protein biosynthesis and this, in turn, leads to more economical fermentation. **High cell density** fermentations are therefore desirable and can be achieved in *S. cerevisiae* using fed-batch technology originally developed for the production of baker's yeast using molasses as the growth substrate. The fed-batch propagation mode is particularly advantageous since it overcomes such effects as substrate and product inhibition and catabolite repression of respiration. It is desirable to grow recombinant *S. cerevisiae* to high cell densities (e.g. >50 g dry cell wt/l) using fed-batch technology, or using modifications such as the fill-and-draw system (where a small residual inoculum is kept in the vessel after removal of the bulk of the culture at the end of a fed-batch run).

Very high cell density fermentations (e.g. >150 g dry cell wt/l) can successfully be achieved with several non-*Saccharomyces* yeasts including *Hansenula polymorpha* and *Pichia pastoris* (Sudbery, 1994). These methylotrophic yeasts are generally propagated in a batch mode initially with glycerol as carbon source, before a controlled methanol feeding regime to induce gene

Table 6.13. Growth systems for recombinant yeasts.

System	Comments
Batch culture	The ideal system would entail biomass accumulation in the absence of heterologous protein synthesis (to avoid toxicity/growth inhibition) followed by production towards the end of growth.
Fed-batch culture	Method chosen for many heterologous proteins (see Mendoza-Vega *et al.*, 1994). Especially important in controlled glucose-feeding regimes to avoid the Crabtree effect (Chapter 5) and to maximize respiratory growth.
Continuous culture	May be advantageous when constitutive (and continuous) gene expression is allied to secretion (Piper and Kirk, 1991). However, continuous culture may select for faster-growing non-producing (e.g. plasmid-free) cells. Plasmid stability is not only strongly influenced by the dilution rate, but also by the type of nutrient (C, N, P, Mg, etc.) limitation in the chemostat (e.g. O'Kennedy *et al.*, 1995).
Immobilized culture	High cell densities, enhanced productivity and improved plasmid stability of immobilized recombinant yeast systems has been reported (Zhang *et al.*, 1996).

expression. Limitation or complete removal of glycerol is required before induction by methanol because glycerol will repress the appropriate promoters (i.e. *AOX* in the case of *P. pastoris* and *MOX* in the case of *H. polymorpha*). Both of these methylotrophic yeasts are Crabtree-negative (see Chapter 5) and will not accumulate growth inhibitory levels of ethanol during aerobic propagation in the presence of glucose. *H. polymorpha* has an additional advantage in being thermotolerant (growth optimum 42°C), which simplifies heat loss problems in large-scale bioreactors.

A possible technological constraint in achieving very high cell densities with *H. polymorpha* and *P. pastoris* is the avoidance of oxygen limitation. Nevertheless, these problems can largely be overcome by careful control of dissolved oxygen and agitation. For example, Chen *et al.* (1997) used such techniques to achieve very high cell densities of *P. pastoris* (420 g wet cell/l!) in the production of recombinant thrombomodulin fragment.

Downstream processing of recombinant proteins and their subsequent purification from either yeast cells or fermentation media are crucial stages in the commercialization of yeast genetic engineering. In order to obtain regulatory approval and for the continued economic viability of any recombinant fermentation, downstream processes need to achieve acceptable consistency, purity and yield of the product. Recovery of intracellularly expressed proteins is fraught with difficulties due to the necessity of rupturing or hydrolysing the tough yeast cell wall. Such problems can be circumvented by secretion of the recombinant proteins into the culture medium. Thenceforth, various procedures can be employed for purification. These generally involve centrifugation, ultrafiltration, precipitation and chromatographic procedures. As a rule of thumb, biochemical engineers can expect to lose around 10% of the protein yield following each individual purification step. Examples of specific methods for the purification from yeast of an intracellular recombinant protein (e.g. interferon-α8) and an extracellular recombinant protein (e.g. α1-antitrypsin) can be found in, respectively, Di Marco *et al.* (1996) and Kwon, Song and Yu (1995).

6.2.2.4 Commercial Aspects

Perhaps the most important aspects to be considered in yeast recombinant biotechnology relate to the commercialization of product. This covers a multitude of scientific, technical, economic, marketing, safety, regulatory, legal and ethical issues which need to be addressed by all biotechnology companies, not just those which exploit yeasts for commercial gain. General legal, economic and social issues confronting modern biotechnology have been reviewed (e.g. Brauer, 1995; Moses and Moses, 1995). Although it is outwith the scope of this book to consider these issues, some brief points are worth noting with reference to therapeutic and food uses of recombinant yeasts.

Several recombinant proteins from yeasts have already been successfully marketed and many more are under development. Most of the products from the 'new' yeast biotechnologies are biopharmaceutical agents of use in the prevention and treatment of human disease (see section 6.3). The approval from regulatory bodies (e.g. the US Food and Drug Administration, FDA) for the clinical use of new drugs is an expensive and lengthy exercise. Particular drawbacks may be experienced with products which are manufactured from yeasts which currently do not enjoy GRAS status (e.g. methylotrophic yeasts).

For food use, the sale and usage of recombinant yeasts generally requires approval from various Governmental bodies. Safety issues relating to genetically 'engineered' foods are the remit of organizations such as the US FDA and the International Food Biotechnology Council. In the UK, foods which are obtained from genetically modified yeasts are regarded as *novel* foods and are evaluated by various committees including the Food Advisory Committee (FAC), the Advisory Committee for Genetic Manipulation (ACGM) and the Advisory Committee on Novel Foods and Processes (ACNFP). This latter committee is an independent panel of experts which report to the UK Ministry of Agriculture Fisheries and Food (MAFF) on the safety of individual novel foods and on more general topics. Reports are made available to the general public. Some of the procedures for gaining approval of recombinant yeasts for food use are outlined in Figure 6.4.

To date, the only 'genetically modified' yeast cells (as opposed to products) cleared for food use in the UK are a baking and a brewing strain of *S. cerevisiae* (see Table 6.14). The former was, to this author's knowledge, the world's first genetically engineered organism to gain national Government clearance for food use. This represented a milestone in modern biotechnology (see Walker and Gingold, 1993), but was not met with universal public approval (e.g. Erlichman, 1990).

Further information on these two yeasts can be found in Hodgson (1990) and Newswatch (1990) for the baking strain, and Hammond (1995) for the brewing strain.

Several important questions regarding safety aspects of recombinant yeasts employed in food-related fermentation processes have been raised

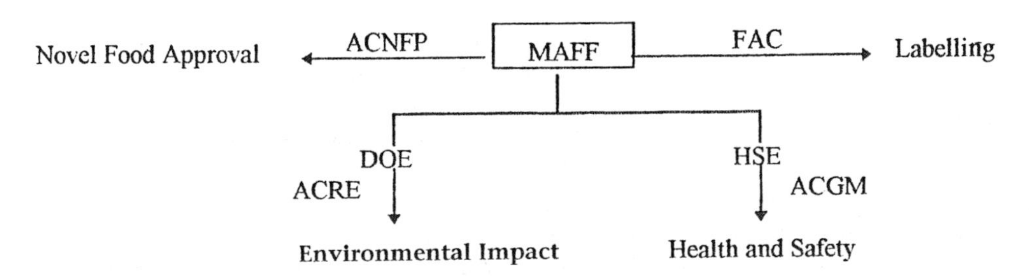

Figure 6.4. UK Government approval routes for commercial food use of genetically modified yeasts. MAFF, Ministry for Agriculture, food and Fisheries; ACNFP, Advisory Committee on Novel Foods and Processes; FAC, Food Advisory Committee; HSE, Health and Safety Executive; ACGM, Advisory Committee for Genetic Manipulation; DoE, Department of the Environment; ACRE, Advisory Committee on Release to the Environment.

Table 6.14. Genetically modified yeasts approved for food use (UK).

Yeast	Description	Approval	Granted to
Baking strain of *S. cerevisiae*	Altered promotion (with genes from another *S. cerevisiae* strain) of maltose-utilizing genes to avoid glucose repression and speed up dough fermentation.	ACNFP recommendations: 1989 Press release: 1990 Government approval: 1991	Gist-Brocades (UK subsidiary: British Fermentation Products)
Brewing strain of *S. cerevisiae* (No. BRG 6050)	Modified by incorporation of plasmid pDVK2 containing the *STA2* gene from *S. cerevisiae* var. *diastaticus* to enable brewing yeast to partially hydrolyze maltodextrins during wort fermentation.	ACNFP recommendations: 1992 Press release: 1993 Government approval: 1994	Brewing Research Foundation International

by Inose and Murata (1995). They found that in *S. cerevisiae* cells transformed with some glycolytic enzyme genes, levels of intracellular methylglyoxal (MG, a highly toxic 2-oxoaldehyde) were formed from a glycolytic by pass reaction during glucose fermentations. The concentrations of MG were sufficient to induce mutagenicity in a *Salmonella* tester strain. Inose and Murata (1995) suggested future monitoring of the safety levels of MG in recombinant yeasts employed in food fermentations. This report serves to emphasize the importance of studies on yeast cell physiology in relation to recombinant DNA technology.

6.3 DEVELOPMENTS IN YEAST TECHNOLOGIES

6.3.1 Alcoholic Beverages

The production of alcoholic beverages is as old as history. Industrial yeast fermentations today represent a significant contribution to the economies of many countries. The alcoholic beverages produced throughout the world are primarily beer, wine, distilled spirits, cider, saké and liqueurs (see Lea and Piggott, 1995). Table 6.15 summarizes some of the production differences and similarities among some alcoholic beverages.

6.3.1.1 Developments in Brewing Yeasts and Fermentation

Since the brewing of beer is the oldest biotechnology known (e.g. Samuel, 1996), it will be covered first in relation to yeast scientific and technological developments. Brewing science in general has been covered in several textbooks (e.g. Hough *et al.*, 1982; Lewis and Young, 1995; Priest and Campbell, 1996) and reviews specifically dealing with brewing yeasts have been published (e.g. Kirsop, 1982; Stewart and Russell, 1986; Johnston, 1990; Hammond, 1993; Stewart and Russell, 1995). The following represents a brief outline of brewing and brewer's yeast, together with a discussion of new developments in yeast 'biotechnology' and yeast fermentation technology.

6.3.1.2 The Brewing Process

Beer is a malt beverage resulting from an alcoholic fermentation of the aqueous extract of malted barley with hops. Brewing is therefore a

Table 6.15. Major alcoholic beverages – summary of production.

	Beer	Whisky	Wine	Spirits, liqueurs
Raw ingredients	Barley, adjuncts (rice, wheat, maize, etc.)	Barley (Malt whisky) Barley, wheat, etc. (Grain Whisky)	Grapes	Barley, maize, molasses, grapes, whey, etc.
Pre-treatment	Malting, mashing	Malting, mashing	Crushing, maceration	Variable dependent on substrate
Boiling	Yes (hops)	No	No	No
Fermentation	*S. cerevisiae* (ale) *S. carlsbergensis* (lager)	*S. cerevisiae*	*S. cerevisiae* (starters or 'natural' yeasts)	*S. cerevisiae* *K. marxianus* (whey)
Yeast recycling	Yes	No	No	No
Distillation	No	Yes	No	Yes
Maturation	Short (weeks)	Long (Years)	Long (Years)	Varies. Gin/Vodka: none Cognac: years
Final alcohol content (%,v/v)	3–6	40–45	8–12	35–45

multistage process involving biological conversion of raw materials to final product (see Figure 6.5).

Brewing yeast fermentation performance; that is, the ability of yeasts to 'consistently metabolize wort constituents into ethanol and other fermentation products in order to produce beer with satisfactory quality and stability' (Russell and Stewart, 1995), is influenced and controlled by a number of factors:

- **Genetic characteristics**, i.e. the choice of yeast strain employed.
- **Cell physiology**, e.g. the stress tolerance of yeast cells, the viability and vitality of the cells and the inoculum cell density (or *pitching rate*).
- **Nutritional availability**, e.g. the concentration and category of assimilable nitrogen, the spectrum of wort sugars and the concentration and availability of metal ions.
- **Physical environment**, e.g. temperature, pH, dissolved oxygen and wort gravity.

6.3.1.3 Brewing Yeasts

Traditionally used ale brewing yeast are all strains of *Saccharomyces cerevisiae* that represent quite a diverse group of microorganisms. Lager yeasts are also strains of *S. cerevisiae* (Dr Ann Vaughan-Martini, University of Perugia, Italy, personal communication), but for historical and practical, rather than definitively taxonomic reasons, they will be referred to as *S. carlsbergensis* (more strictly, *S. cerevisiae* var. *carlsbergensis*). Kielland-Brandt *et al.* (1995) have discussed the scientific reasons for distinguishing between *S. cerevisiae* and *S. carlsbergensis* and Table 6.16 compares and contrasts some of the genetic and physiological differences between ale and lager

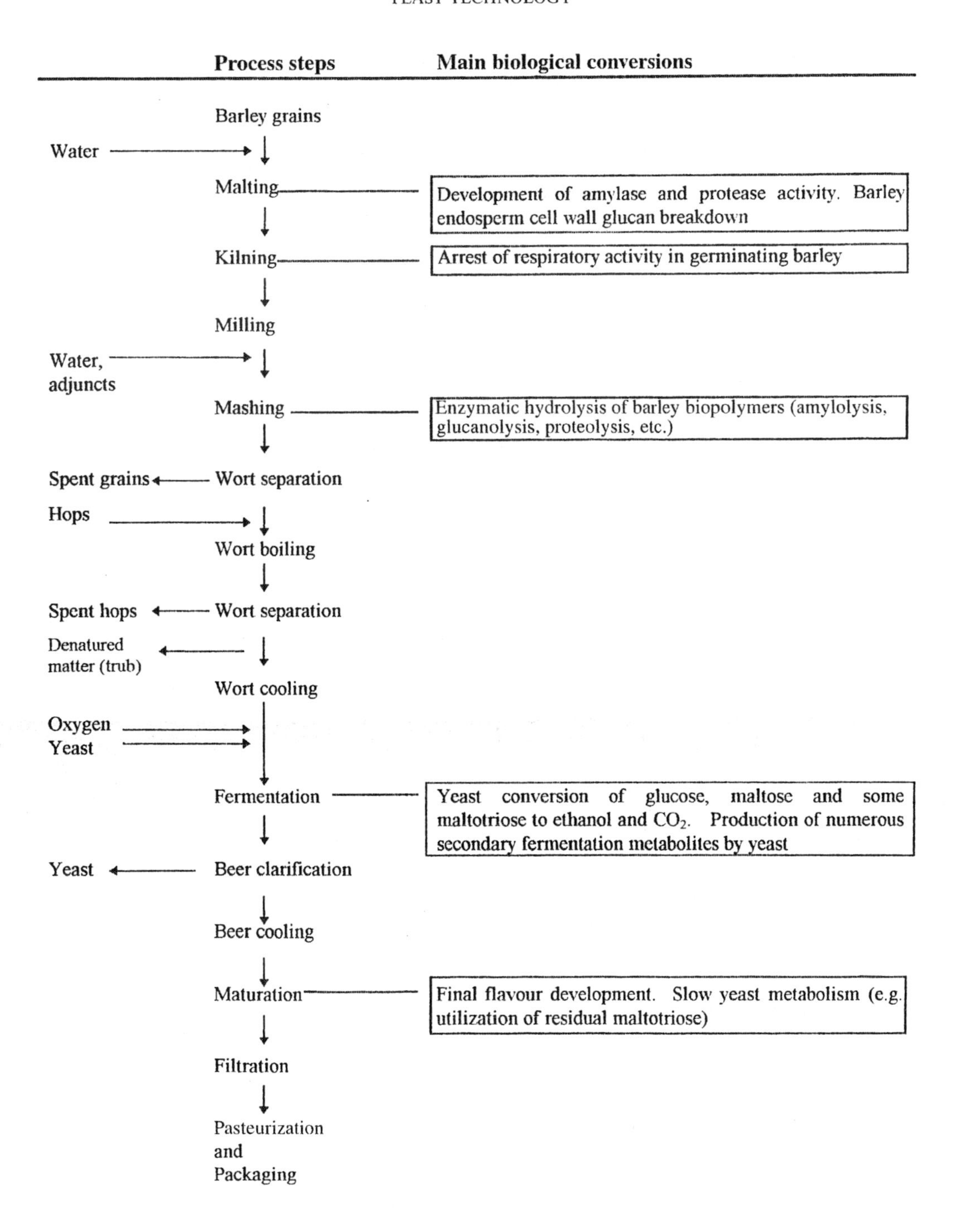

Figure 6.5. Flow diagram of the brewing process.

Table 6.16. Some genetic and physiological differences between laboratory and brewing strains of *S. cerevisiae*.

Difference parameter		Laboratory strains	Ale strains	Lager strains
Genetic	Selection	Selected for genetic studies	Selected for alcohol production	Selected for alcohol production
	Auxotrophy	Auxotrophic mutants available	Prototrophic	Prototrophic
	Ploidy	Stable haploids or diploids	Polyploid or aneuploid	Allotetraploid[a]
	Karyotype	16 chromosomes	Variable	Variable (two types of same chromosome present)
	Sporulation	Good	Poor	Poor
Growth	Rate	Fast growth rate	Slow growth rate	Slow growth rate
	Flocculation	Generally non-flocculent (freely suspended behaviour)	Weakly flocculent (flotation behaviour)	Strongly flocculent (sedimentary behaviour)
	Killer character	Several killer phenotypes	Non-killer	Non-killer
Metabolic	Fermentation	Low rate	High rate	High rate
	Sugar utilization	Melibiose not utilized	Melibiose not utilized	Melibiose cleaved (α-galactosidase) and fermented
		Single maltose transporter	Glucose inhibits high-affinity maltose uptake	Glucose does not inhibit high-affinity maltose uptake
		Variable maltotriose transport	Poor maltotriose transport	Good maltotriose transport
	Amino acid transport	Valine transported	Poor valine uptake	Poor valine uptake
	Flavour compounds	Ferrulic acid decarboxylated (to produce phenolic off-flavour – *POF1* allele)	Lack *POF1* – no ferrulic acid decarboxylation	Lack *POF1* – no ferrulic acid decarboxylation
	Ferment at 37°C	?	Yes	No

[a] The lager yeast strain, *S. carlsbergensis*, possesses two divergent genomes – one from *S. cerevisiae* and one from *S. monacensis*. *S. carlsbergensis* may therefore be a natural hybrid between *S. cerevisiae* and *S. monacensis* (see Kielland-Brandt *et al.* 1995).

brewing yeast. Laboratory (or 'scientific') strains of *S. cerevisiae* are also included in Table 6.16 to differentiate between their properties and those of the industrial strains. Of course, the applications of these yeasts is different: laboratory yeasts are employed for fundamental research and are generally grown on solid media, in shake-flasks or in chemostats at temperatures round 25–30°C; brewing yeasts are employed for industrial beverage production in large non-agitated fermenters below ambient temperature. Within the brewing yeasts, there are also significantly different growth conditions. For example, most ale fermentations are generally completed faster (e.g. 5 days versus 14 days) and conducted at higher temperatures (e.g. 20°C versus 8°C) compared with lager fermentations. Traditionally, ale yeasts were described as 'top fermenters' which rose to the top of vessels at the end of fermentation and were harvested by *skimming* from thick floating films of yeast biomass. Lager yeasts, on the other hand, were 'bottom fermenters' which flocculated to the base of vessels at the end of fermentation and were harvested from a sedimented pellet of yeast biomass. With the advent of tall (approx. 20 m) cylindroconical fermentation vessels, bottom-fermenting yeasts can be employed for ale brewing and processing differences between the buoyant-density ale yeasts and the flocculent lager yeasts are no longer strictly discriminatory.

In recent years, several molecular methods have been utilized to characterize and differentiate brewing yeast stains (Table 6.17). Genetic fingerprinting techniques, in particular, have proved useful in strain management, authentication and verification, especially when multiple strains of *S. cerevisiae* are stored and distributed for brewing under licence, or in simultaneous multi-beverage production.

Molecular genetic techniques can easily discriminate between brewing strains with similar physiological properties. For example, pulse-field electrophoretic karyotype analysis can be used to detect chromosome length polymorphisms (CLPs) in industrial yeasts because each individual strain possesses its own, and characteristic, banding pattern (e.g. Naumov *et al.*, 1992). Nevertheless, although such molecular identification approaches for brewing yeasts are considered to be more objective, sensitive and reproducible compared with traditional morphological and biochemical tests, they are still not

Table 6.17. Molecular methods for brewing yeast strain differentiation.

Method	Description	References
Restriction enzyme analysis (DNA fingerprinting)	Total, ribosomal or mitochondrial DNA is digested with restriction endonucleases and specific fragments hybridized after electrophoretic separation with multi-locus DNA probes such as the Ty1 retrotransposon. Restriction fragment length polymorphisms (RFLPs) are detected.	Pedersen (1986); Walmsley *et al.* (1989) Wightman *et al.* (1996)
Electrophoretic karyotyping (chromosome fingerprinting)	Whole yeast chromosomes are separated electrophoretically using pulse-field techniques.	Pedersen (1987); Oakley Gutowski *et al.* (1992); Vaughan-Martini *et al.* (1993)
Polymerase chain reaction (PCR)	Specific DNA sequences are exponentially propagated *in vitro* and the amplified products analysed after electrophoretic separation. Random amplified polymorphic DNA can also be analysed by PCR (RAPD–PCR).	Ness *et al.* (1993)
Genetic tagging	Specific genetic sequences, including selectable markers, are introduced into yeasts to facilitate their recognition. (e.g. replacement of chloramphenicol resistance sequences with a 'tag' which confers sensitivity to the antibiotic).	Lancashire and Hadfield (1986)
Chromatography	Pyrolysis-gas chromatography or gas chromatography (of long-chain fatty acid methyl esters).	Hammond (1993)
Polyacrylamide gel electrophoresis (PAGE)	Total soluble yeast proteins are electrophoresed and banding patterns analysed by computer.	Hammond (1993)

particularly *rapid* for the needs of modern large-scale brewery operations (Meaden, 1996).

Additional information on genetic fingerprinting of brewing yeast strains can be found in Meaden (1990), Walmsley (1994), Pedersen (1994) and Schofield *et al.* (1995).

Molecular genetic improvement of brewing yeasts

Priest and Campbell (1996) have defined the properties of 'good' brewing yeasts as follows:

- Rapid fermentation rate without excessive yeast growth
- Efficient utilization of maltose and maltotriose with good conversion to ethanol
- Ability to withstand the stresses imposed by the alcohol concentrations and osmotic pressures encountered in brewing
- Reproducible production of correct levels of flavour and aroma compounds
- Ideal flocculation character for the process employed
- Good handling characteristics (e.g. retention of viability during storage, genetic stability).

Unfortunately, brewing yeast strains are limited in several of these characteristics and in reality are quite inefficient at fermenting the available sugars in brewers wort. As Hammond (1995) states: 'Brewing yeasts are far from optimized for the task which they are set by the brewer'.

A variety of approaches exist to genetically modify, and hopefully improve, the characteristics of brewing yeasts. Due to their lack of a 'sex-life', classical genetic breeding approaches are impracticable, but several strategies have been devised to circumvent this (see Table 6.2). The most exciting developments have been achieved using recombinant DNA technology. Transformation strategies, in particular, have opened up the possibility of using brewing yeasts which: ferment a greater array of (otherwise unfermentable) sugars; flocculate appropriately and sufficiently early; tolerate better the chemical and physical stresses in brewing and produce more stable, flavoursome beer.

Table 6.18 highlights some of the improvements to the brewing process which are achievable using recombinant yeasts. Generally, the benefits to the industry relate to reduced raw material costs, increased efficiencies and productivities of fermentation, improved beer quality and the development of 'new' beers.

Nevertheless, several factors have hindered, and may continue to hinder in the future, the usage of recombinant DNA technology in brewing yeast strain selection programmes. Casey (1990) has outlined these factors which include: uncertain and lengthy government regulatory approval, availability of alternative (traditional) solutions, patent applications and concerns over consumer acceptance. Added to these concerns should be the generally inadequate knowledge of the cell physiology of recombinant brewing yeasts.

As discussed previously (see Table 6.14), one recombinant brewing yeast has been cleared for commercial use in the UK (Hammond, 1995), but it has yet to receive widespread industrial acceptance The important point to make is that although scientific, technical and regulatory hurdles with recombinant brewing yeasts can be overcome, public acceptability remains the most challenging hurdle.

6.3.1.4 Yeast Fermentation Technology Developments

Brewing improvements centred on the activities of yeast can be made not only through genetic modifications, but also through process-related modifications. These latter considerations relate to the design, operation, monitoring and control of yeast fermentations. Examples of process control developments include on-line measurement and modulation of parameters such as: yeast biomass (using optical density or impedance-based biosensors); ethanol concentrations (potentially with an alcohol-electrode); specific wort gravities and CO_2 evolution. Fermenter

Table 6.18. Aspects relating to practical applications and potential of recombinant brewing yeasts.

Current strain limitations	Desirable strain properties	Improvements using recombinant DNA technology
Limited ethanol tolerance	Higher ethanol tolerance for fermentation of very high gravity wort	Transfer of acetoacetyl CoA thiolase (*ERG 10*) gene from tolerant yeast may increase ethanol tolerance (see Cantwell and McConnell, 1987) but this trait is difficult to confer due to its polygenic nature.
Narrow range of fermentable carbohydrates	Utilization and fermentation of maltodextrins to produce dry and light (low carbohydrate) beer	Achieved by transfer of glucoamylase (*STA*) genes from *S. cerevisiae* var. *diastaticus* or from *Aspergillus* spp. (e.g. Tubb, 1986; Perry and Meaden, 1988; Hammond, 1995).
Susceptible to contamination	Antimicrobial (against wild yeasts and bacteria) properties so that brewing strains could 'self-cleanse' contaminated fermentations	Killer toxin plasmids (from laboratory killers) transferred by cytoduction (Young, 1981) or electro-transformation (Salek *et al.*, 1992) to brewing strains. Flocculent killers also available through protoplast fusion (Javadekav *et al.*, 1995).
Limited stress tolerance	Osmo-, thermo- and barotolerance required to levels above those encountered in brewing	Changes to cell physiology brought about by nutrients and physical conditions are more likely to succeed than cloning technology.
No β-glucan hydrolysis'	Degradation of viscous β-glucan derived from malted barley desirable to improve wort filtration and eliminate beer haze	β-Glucanase genes have been successfully cloned into brewing yeast from bacteria, fungi and barley (e.g. Cantwell *et al.*, 1985).
Limited protein hydrolysis	Proteolysis would improve wort nitrogen utilization and beer haze prevention (through 'chill proofing')	*S. cerevisiae* has been transformed to successfully secrete a protease (Young and Hosford, 1987)
Beer produced requires lengthy maturation	Reduced levels, or metabolism of, undesired off-flavours (e.g. diacetyl and H_2S) which would result in reduced maturation times	Genes encoding enzymes from the valine synthetic pathway (which can give rise to diacetyl) have been cloned from various organisms into brewing yeasts (e.g. Tada *et al.*, 1995). Reduced H_2S production also possible by transfer of cystathionine synthase genes (Tezuka *et al.*, 1992).
Uncontrolled flocculation	Correct timing of yeast flocculation would assist downstream processing of beer	Flocculation genes (e.g. *FLO1*) have been cloned into brewing yeasts and are only expressed at end of fermentation (Watari *et al.*, 1994).
Limited ester synthesis	Certain esters (e.g. ethyl and isoamyl acetate) provide desirable flavour and aroma notes to beer	The gene encoding alcohol acetyl transferase (*AFTI*) which is responsible for acetate ester synthesis has been cloned into a (as yet non-brewing) strain of *S. cerevisiae* (Fujii *et al.*, 1994).
Flavour instability of produced beer	Increased SO_2 production to act as flavour stabilizer and antioxidant	Sulphite reductase genes (*MET10*) have been deleted resulting in increased SO_2 levels in beer (Hansen and Kielland-Brandt, 1996b).
Maltose utilization repressed	Avoidance of glucose repression would improve maltose fermentation and permit greater usage of glucose-based adjuncts	Transfer of maltose permease genes (e.g. *MAL61*) results in improved maltose utilization (Kodama *et al.*, 1995).

design developments include the use of immobilized yeast bioreactors for primary fermentations and for beer maturation (Masschelein *et al.*, 1994; Norton and D'Amore, 1994; Cashin, 1996). Further discussion of yeast cell immobilization for biotechnological applications, including brewing, was provided in Chapter 4 (see Tables 4.6 and 4.15).

6.3.1.5 Developments in Yeasts Producing other Alcoholic Beverages

Strains of *S. cerevisiae* are also exploited in the manufacture of other alcoholic beverages, including wine, cider, saké and distilled spirits (e.g. whisky). Some of the special desired properties of some of these yeasts are summarized in Table 6.19.

Like brewing, winemaking is a large, global yeast-based technology which significantly impacts on the economic well-being of many countries. An outline of the winemaking process is shown in Figure 6.6. Traditional wine fermentations are carried out using the wild microflora found on the surface of grape skins and indigenous yeast associated with winery surfaces (which are predominantly strains of *S. cerevisiae* – see Martini, 1993). Various yeasts and bacteria participate in natural wine fermentations and the reader is directed to Fleet (1993) and Jackson (1994) for more detailed information on wine microbiology. Important non-*Saccharomyces* species are the apiculate yeasts *Kloeckera apiculata* and *Hansensiaspora uvarum* which predominate in the early stages, followed by several other yeasts (e.g. *Candida stellata, Torulaspora delbrueckii, Kluyveromyces and Pichia* spp.) in the middle stages when ethanol levels rise to 3–4% (v/v). The latter stages of wine fermentation are predominated by *S. cerevisiae*.

The rule, rather than the exception, for modern wineries is the use of specially selected

Table 6.19. Characteristics of some alcoholic beverage-producing yeasts.

Beverage	Desirable yeast traits
Whisky	Some whisky yeasts are hybrids between *S. cerevisiae* and *S. cerevisiae* var. *diastaticus* which possess glucoamylase activity enabling them to hydrolyse and ferment maltotetraose and low molecular weight maltodextrins. Strains must be stress-tolerant (especially to ethanol and temperature) and produce desirable congeners (esters, organic acids, aldehydes, higher alcohols etc.) Note that some Scotch whisky is made using mixed cultures of distiller's and brewer's yeast strains and there also is an important contribution to flavour made by lactic acid bacteria (also in American 'Sour mash' whiskey).
Wine	Wine yeasts are generally homothallic diploids. Wine yeast desirable characteristics include: correct volatile acidity (in relation to ethanol produced); fermentation vigour; aromatic character (esters, terpenes, succinic acid, glycerol etc.); SO_2 tolerance; correct balance of sulphur compound production; low acetaldehyde; killer character; low urea excretion (in view of ethyl carbamate production).
Cider	Desirable traits include: rapid, non-foaming fermentation; polygalacturonase activity; tolerant to SO_2, low pH and high ethanol; low vitamin, fatty acid and O_2 requirements; compatibility with chosen malolactic bacterial strains; non-production of H_2S or acetic acid; correct balance of aromatic compounds.
Saké	Desirable traits include; high ethanol tolerance; non-foaming fermentations; auxotrophic (e.g. adenine) mutants (pink saké); killer character; arginase-deficiency (low urea production)
Whey beverages	Lactose fermentation necessary – *Kluyveromyces marxianus* used (see Mawson, 1994; Siso, 1996).

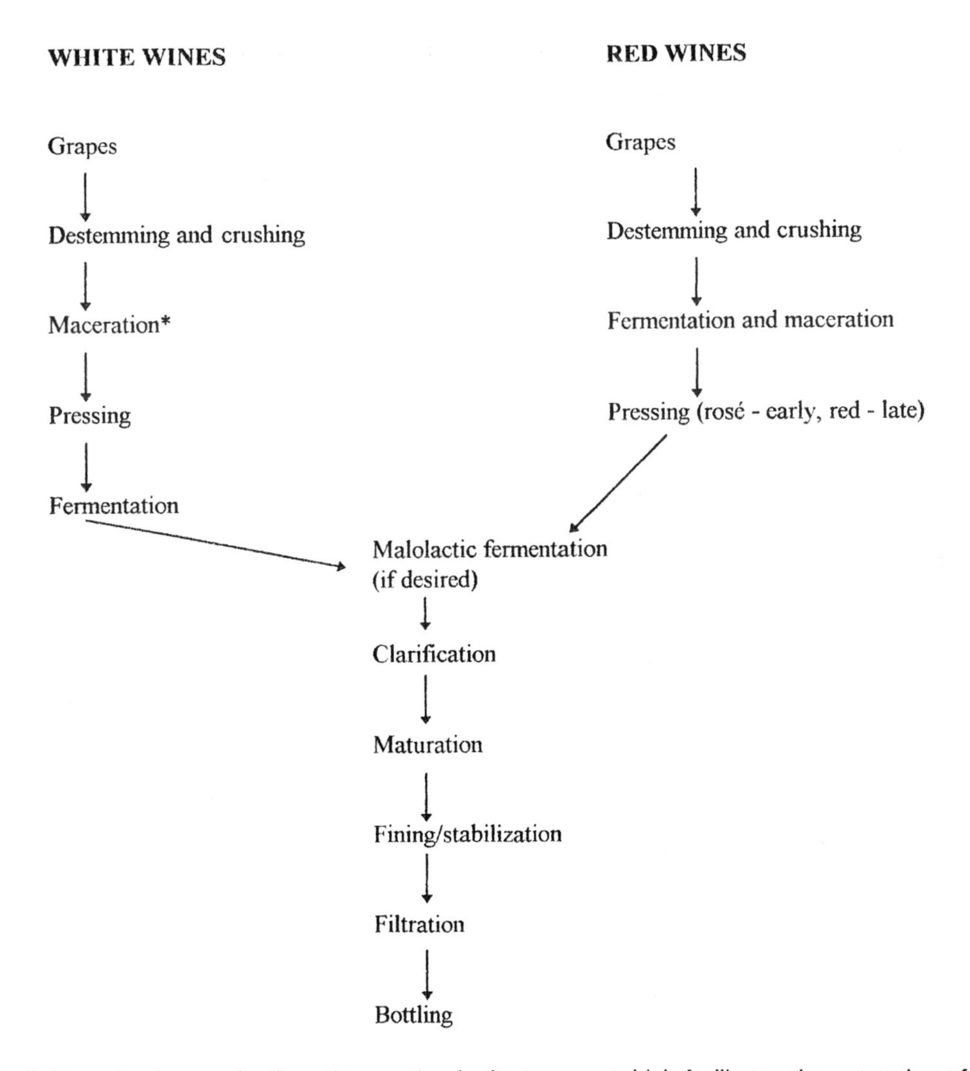

Figure 6.6. Outline of wine production. *Maceration is the process which facilitates the extraction of compounds from the seeds and skins and is initiated by the action of hydrolytic enzymes released from cells ruptured during crushing.

starter cultures of *S. cerevisiae* (Subden, 1990). These cultures, which are inoculated into grape must following SO_2 applications to suppress natural microflora, have reduced lag phases and can complete sugar conversion to alcohol much faster than un-inoculated fermentations. The desired characteristics of selected wine yeast starters, or their 'enological aptitude', are outlined in Table 6.19. It should be noted that the use of such cultures may not necessarily prevent the growth and metabolic activity of indigenous, winery associated *S. cerevisiae* strains or other natural yeasts such as *K. apiculata* (see Fleet and Heard, 1993).

Modern molecular biotechnology and advances in fermentation technology are having an impact on wine yeast applications, as outlined in Table 6.20.

Table 6.20. Some biotechnological developments related to wine yeasts.

Developments	Comments	References
Genetic fingerprinting	RFLP and RAPD-PCR analysis, together with electrophoretic karyotyping have been used to characterize wine yeasts	Querol and Ramon (1996)
Recombinant DNA technology	Several genes from various organisms (including other yeasts) have been cloned into wine yeasts. KI killer toxin (to combat wild yeasts); pectinases (to increase wine filterability and fruity aroma); glucanases (releases bound terpenes to increase fruity aroma); lactate dehydrogenase (to promote mixed fermentation and acidification); malolactic enzyme (to promote malolactic fermentation); malic enzyme (to promote malo-ethanolic fermentation); DHAP reductase and glycerol phosphatase (to increase glycerol levels).	Subden (1990); Barre *et al.* (1993); Butzke and Bisson (1996); Querol and Ramon (1996)
Hybridization	Flocculation properties have been successfully introduced into wine yeasts strains of *S. cerevisiae* by hybridization.	Shinohara *et al.* (1997)
Fermentation technology	Production of sparkling wines using yeasts immobilized in natural gels	Godia *et al.* (1991)

In particular, recombinant DNA technology represents the greatest potential for improvement of wine yeast fermentations. As Barre *et al.* (1993) have stated: 'Cloned DNA transfer methods will predominate in the future construction of wine strains'. Two main aspects of enology relating to consistency and quality of the product are most likely to benefit from the use of recombinant wine yeasts (Table 6.21).

Another group of alcoholic beverages to be mentioned which are reliant on yeast metabolism for their production are the distilled spirits. These include whisky (Scotch Malt and Grain), whiskey (Irish and North American), wine spirits (Cognac, brandy, Armagnac, Grappa), neutral spirits (vodka), rum and various flavoured spirits (e.g. gin, aquavit, ouzo) and liqueurs. Figure 6.7 outlines the production of some cereal-based distilled beverages and Table 6.22 summarizes the raw materials used in the production of these spirits. Further information on their manufacture can be found in Watson (1993), and Lea and Piggott (1995).

Regarding biotechnological developments in whisky manufacture, it must be noted that, especially in Scotland, the product is associated with

Table 6.21. Potential winemaking improvements with recombinant yeasts.

Enological aspect	Examples of improvements
Fermentation	Control and reliability of fermentation through: expression of malolactic fermentation by yeast; control of flocculation and foaming; control of yeast population growth; development of cryotolerant strains for more controlled fermentations.
Flavour	Elimination of flavour defects due to sulphide compounds and introduction of 'novel' flavours.

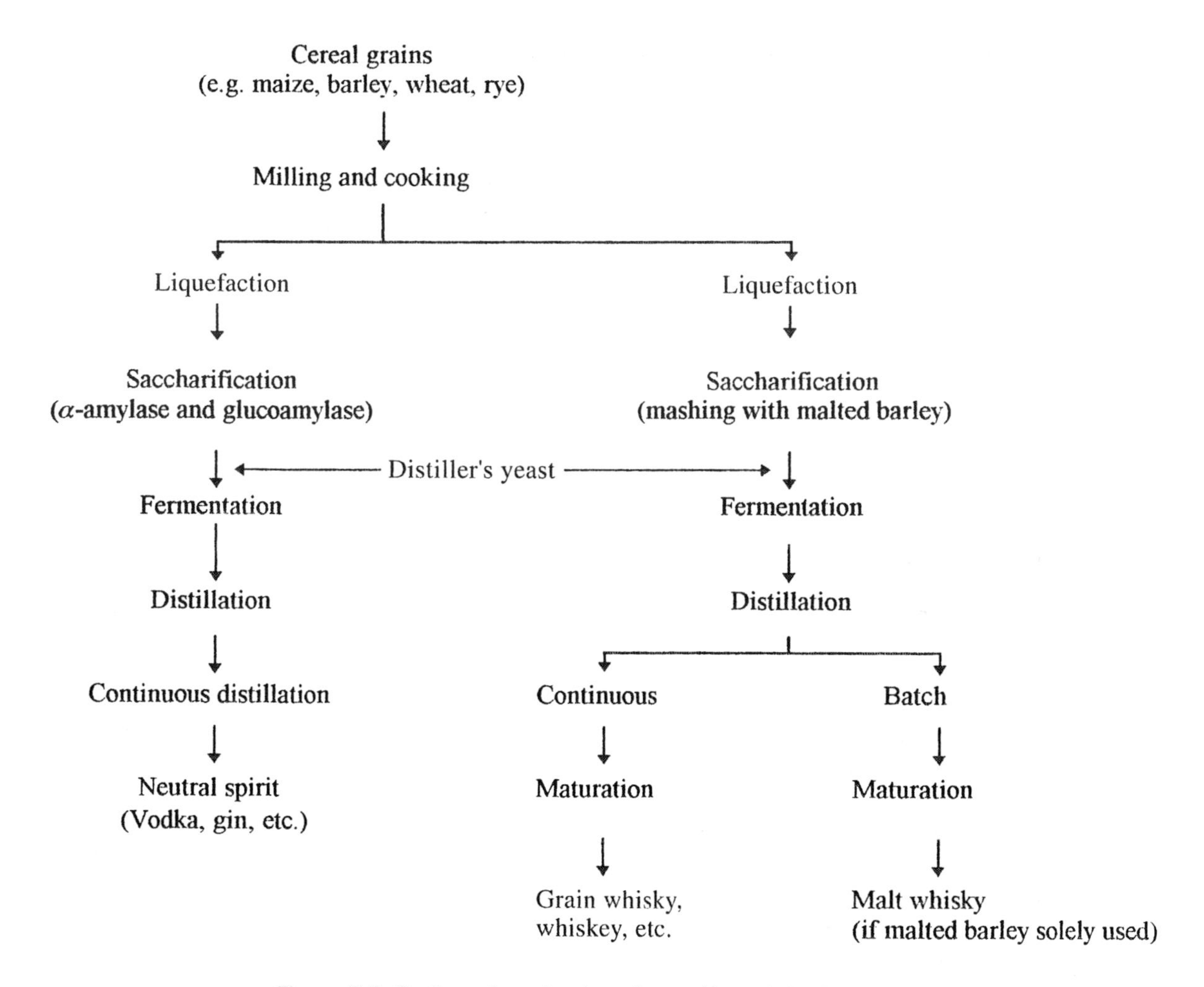

Figure 6.7. Outline of production of cereal-based distilled spirits.

tradition and purity and the industry is extremely reluctant to embrace new technology which may possibly tarnish this image in the eyes of the consumer. Nevertheless, there is nothing in the legal definition of Scotch whisky which precludes the use of recombinant strains of yeast. In fact, Watson (1993) made the following comment:

'Although there is still some reluctance to use genetically engineered yeasts in the distilling industry, because of the regulatory hurdles and concerns about adverse consumer response, it is felt that these fears will soon be overcome and their use will become widespread by the end of the century'.

However, it is the view of the present author that this situation is unlikely to arise in the near future. What is more likely is the use of recombinant yeasts for the production of neutral distilled spirits and fuel ethanol.

Table 6.22. Carbohydrate sources for distilled spirit fermentations.

Raw material	Fermentable carbohydrate	Product
Malted barley	Hydrolysed starch	Scotch malt whisky
Malt plus wheat or maize	Hydrolysed starch	Scotch grain whisky and neutral spirits (gin, after flavouring)
Malt plus rye	Hydrolysed starch	Bourbon whiskey (US)
Grapes	Grape juice sugars	Brandy, Cognac, Armagnac
Fermented (spent) grape must/ grape residues	Grape juice sugars	Grappa (Italy)
Sugar cane or beet molasses	Sucrose	Rum, neutral spirits
Potatoes, cereals, plus malt	Hydrolysed starch	Vodka, flavoured spirits
Rice + *Aspergillus oryzae*	Hydrolysed starch	Shochu (Japan)
Palm oil	Various sugars	Raki
Agave tequilana	Hydrolysed inulin	Tequila (Mexico)
Cheese whey	Lactose	Neutral spirits, 'cream' liqueurs

6.3.2 Industrial Alcohols

6.3.2.1 Bioethanol

World-wide interest in the production of fermentation ethanol (or 'bioethanol') as a source of renewable energy has oscillated during most of this century. Production was initiated in the 1930s but waned due to low oil prices. During the 1970s oil crisis interest was renewed but subsequently declined after the crisis. In the 1990s, bioethanol attracted more attention due to its application as an octane enhancer and replacer in petroleum (gasoline) for internal combustion engines. For some individual countries, notably Brazil, the interest and practical applications of bioethanol has followed a more stable picture. Major decisions were made by the Brazilian Government's Alcohol National Plan (PNA) in 1975 to commence production of fuel ethanol by yeast (mainly baker's strains) fermentation of sugar cane juice and molasses, and to a lesser extent, hydrolysed cassava. This subsequently resulted in the production of millions of litres of bioethanol being produced for fuel use by the mid-1980s. The majority of automobiles in Brazil are now fuelled by ethanol or ethanol–petroleum mixtures. The economic benefits for countries with vast supplies of fermentable sugar are significant.

Substrates for bioethanol

In addition to the sucrose derived from sugar cane, numerous other sources of potentially fermentable carbohydrates have been considered for production of bioethanol by yeasts (see Table 6.23).

The greatest potential for increasing supplies of bioethanol comes from plant biomass and the fermentation of lignocellulosic hydrolysates. Although presently not competitive with petroleum, biotechnology and bioprocess engineering may enhance the future competitiveness of lignocellulose-derived ethanol (see Bashir and Lee, 1994; Olsson and Hahn-Hägerdal, 1996; Szczodrak and Fiedurek, 1996). A summary of the lignocellulosic hydrolysate fermentation process is shown in Figure 6.8.

Olsson and Hahn-Hägerdal (1996) have evaluated the performance of several pentose-fermenting yeasts (including recombinant strains) in converting lignocellulosic hydrolysates to ethanol. *Candida shehatae* and *Pichia stipitis* were the most successful yeasts studied, although the

Table 6.23. Sources of potentially fermentable carbohydrates for fuel ethanol production.

Raw material	Pre-treatments required	Principal fermentable sugars	Yeasts considered
Sugar crops (sugar cane, sugar beet)	Physical and chemical extraction	Sucrose, fructose, glucose	*S. cerevisiae*
Cow's milk	Bacterial or acidic precipitation of casein and physical separation of whey	Lactose	*K. marxianus*
Cereal starch (maize, barley, wheat, rice, rye, oats, sorghum)	Gelatinization, liquefaction and enzymic (amylase) saccharification	Maltose, glucose, sucrose, fructose, maltotriose, maltodextrins	*S. cerevisiae* (hydrolysed starch) or *Schw. castellii* (raw starch)
Tuber plants (cassava, yams, Jerusalem artichokes, potatoes)	Hydrolysis with amylases (potatoes) or inulinases (artichokes)	Maltose, glucose etc. (from starch) Fructose (from inulin)	*S. cerevisiae* *K. marxianus*
Cellulose, lignocellulose and hemicellulose (e.g. paper, wood, straw, sawdust, stover)	Physical (e.g. steam explosion), chemical (e.g. acids, alkalis) and enzymic (cellulase) hydrolysis	Xylose, glucose cellobiose	*P. stipitis* and *C. shehatae* (xylose); *K. marxianus* and *C. wickerhamii* (cellobiose)
Industrial effluents (sulphite waste liquor, corn steep liquor, molasses, sweet sorghum syrup)	Dilution, pH adjustment	Xylose, sucrose	Pentose-fermenters (e.g. *P. stipitis*) and *S. cerevisiae*

presence of inhibitory chemicals in the hydrolysates curtailed their growth and metabolic activities. The economic viability of simultaneous saccharification and fermentation (SSF), separate hydrolysis and fermentation (SHF) and direct microbial conversion (DMC) in cellulose-to-ethanol biotransformations has been discussed by several authors (e.g. Bashir and Lee, 1994; Szczodrak and Fiedurek, 1996). SHF appears to offer the greatest potential at present. Barron *et al.* (1995) have, however, also favourably assessed the performance of *Kluyveromyces marxianus* SSF processes (at 45°C) in cellulose-to-ethanol conversions. Bacterial alternatives to yeasts have additionally been evaluated and Asghari *et al.* (1996) reported good conversions of hemicellulose (from agricultural wastes) to ethanol using a recombinant strain of *Escherichia coli*.

For the production of bioethanol from cellulosic biomass and other available substrates to be economically feasible, developments in both yeast biotechnology and yeast process technology will be required. Some of these developments will now be discussed.

Yeast genetic developments for bioethanol production

Several yeasts, bacteria and fungi are capable of producing ethanol through fermentative metabolism of carbohydrates (see Russell *et al.*, 1987). Although the bacterium *Zymomonas mobilis* (and some recombinant *E. coli* strains – see above) possesses some favourable properties in respect of ethanol-producing capabilities, the overwhelm-

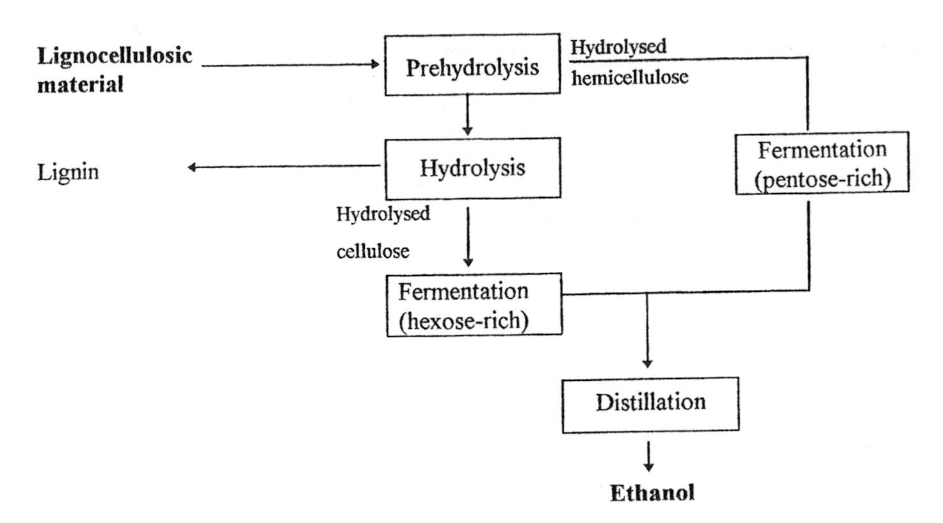

Figure 6.8. Outline of ethanol production from lignocellulosic materials. Adapted from Olsson and Hahn-Hägerdal (1996).

ing majority of fermentation alcohol produced world-wide uses *S. cerevisiae*. The lactose-fermenting yeast, *Kluyveromyces marxianus*, the amylolytic yeast *Schwanniomyces castellii* and some pentose-fermenting yeasts (e.g. *Pichia stipitis*, *Pachysolen tannophilus* and *Candida shehatae*) play relatively minor roles in global bioethanol production at present. This situation may change in the future.

The 'ideal' fuel ethanol-producing yeast would possess fermentation and growth properties as outlined in Table 6.24. Unfortunately, no single yeast exists which possesses all of these characteristics and so there is great scope in researching yeast physiology and biotechnology in relation to the industrial production of bioethanol.

Many producers of bioethanol presently use baker's yeast, primarily because of its convenience, wide availability and relative inexpense. Such strains, however, are not ideally suited for many types of fermentation and several genetic and physiological approaches have been undertaken aimed at improving *S. cerevisiae* and other species of yeast. Even small improvements in yeast fermentation performance would be economically significant due to the scale and future potential of the bioethanol industry.

With regard to genetic strategies for strain improvement, both classical genetics and recombinant DNA technology may be employed (Table 6.25) depending on the yeast species in question and the substrate to be fermented. Nevertheless, it should be noted that ethanol production by yeasts is a polygenic trait and the enhancement of fermentative metabolism is not a straightforward genetic task (Christensen, 1987; Tavares and Echeverrigaray, 1987; Panchal and Tavares, 1990).

Some success has been achieved through genetic 'breeding' by selection and self-crossings and Tavares and Echeverrigaray (1987) have argued that such approaches still present a viable and effective means of constructing fuel ethanol-producing yeasts, in spite of advances in recombinant DNA technology.

Where genetic engineering of yeasts may be particularly beneficial is in enabling cells, mainly of *S. cerevisiae*, to ferment otherwise recalcitrant materials. Table 6.26 outlines some of the donor cells and their genes employed to construct new yeast strains capable of fermenting 'unfermentable' substrates.

Other molecular genetic approaches aimed at improving the ethanol productivities of yeast

Table 6.24. Desired properties of fuel ethanol-producing yeasts.

General property	Examples
Fermentation	Fast fermentation rates and high and reproducible ethanol yields in very high-gravity media (to produce > 18% v/v ethanol). High ethanol tolerance. Low pH and high temperature optima for fermentation. Efficient utilization of varied substrates. Reduced levels of minor fermentation metabolites (organic acids, glycerol, higher alcohols, esters, aldehydes). Derepression for a variety of saccharides in presence of glucose. Amylolytic and cellulolytic activities.
Growth	High rate of yeast growth but lowered final growth yield. High cell viability and vitality. Tolerance to high sugar, toxic chemicals/inhibitors and to temperature fluctuations. Resistance to bacterial contamination (i.e. good 'competitor'). Genetic stability. Easy to propagate. Minimal heat generation during fermentation. Possession of appropriate flocculation characteristics depending on process requirements. Possession of killer character.

Table 6.25. Genetic strategies for yeast strain improvement for fuel ethanol production.

Strategies	Examples
• Natural crosses:	polybreeding; recurrent selections; selections after self-crossings
• Hybridizations	
• Rare matings	
• Mutagenesis:	chemical (e.g. NTG); radiation (UV)
• Spheroplast fusion:	e.g. intergeneric (*S. cerevisiae* var. *diastaticus*/*S. uvarum*)
• Transformation:	recombinant DNA technology mainly to increase the array of carbohydrates fermented (e.g. genes encoding cellulases, glucoamylases, etc.)

centre on 'metabolic engineering'. For example, numerous attempts to enhance the flux through yeast glycolysis by over-expressing individual glycolytic enzyme genes (e.g. phosphofructokinase) have been made (e.g. Schaaf *et al.*, 1989; Davies and Brindle, 1992). However, these attempts have generally met with little success in increasing the rate or extent of ethanolic fermentation by *S. cerevisiae*. This is because yeast cells have a strong homeostatic response tendency and also the control of a major pathway like glycolysis involves complex regulation of many different enzymes (see Fell, 1997 and Chapter 5). Clearly there is much to be learned about yeast physiology before we can effectively genetically engineer yeasts to improve metabolic fluxes and, consequentially, fermentation performance.

Yeast physiological developments for bioethanol production

The importance of yeast cell physiology in yeast alcohol fermentations has been emphasized by

Table 6.26. Recombinant yeasts with fuel alcohol potential.

Donor organism	Gene-encoded enzymes	Recombinant host yeast	Comments	References
Trichoderma reesei	Cellulase	*S. cerevisiae*	Cellulose hydrolysis difficult	Knowles *et al.* (1987)
Aspergillus awamori	Glucoamylase	*S. cerevisiae*	Ferments liquefied starch	Innis *et al.* (1987)
Saccharomyces cerevisiae var. *diastaticus*	Glucoamylase	*S. cerevisiae*	Ferments liquefied starch	Russell *et al.* (1987)
Escherichia coli	Xylose isomerase	*Sch. pombe*	Xylose fermentation	Chan *et al.* (1989)
Pichia stipitis	Xylose reductase and Xylitol dehydrogenase	*S. cerevisiae*	Xylose fermentation	Meinander *et al.* (1994)
Kluyveromyces lactis	β-Galactosidase and lactose permease	*S. cerevisiae*	Whey fermentation	Sreekrishna and Dickson (1985); Porro *et al.* (1992)

several authors. For example, by Olsson and Hahn-Hägerdal (1996) for pentose fermentations by *P. stipitis* and *C. shehatae*, by Thomas and Ingledew (1992) for high-gravity cereal fermentations by *S. cerevisiae* and by Walker *et al.* (1996) for sugar cane molasses fermentations. Aspects of cell physiology relating specifically to the phenomenon of ethanol tolerance in yeasts were discussed in Chapter 4 (e.g. section 4.4.3). Environmental factors which play important roles in dictating yeast fermentation performance are noted in Table 6.27.

Dramatic examples of how important it is to address aspects of yeast physiology and 'cell engineering', instead of focusing primarily on genetic engineering in relation to fermentation are to be found from the work of Ingledew and his colleagues on fuel alcohol from cereals (e.g. Thomas and Ingledew, 1992; Jones and Ingledew, 1994; Thomas *et al.*, 1996). Simply by addressing questions relating to yeast inoculum cell density, temperature and key nutritional requirements, it is possible greatly to enhance ethanol yields from commonly used industrial strains of *S. cerevisiae* in high-sugar fermentations. For example, ethanol yields of over 21% (v/v) in molasses, and 23% (v/v) in wheat mashes have been achieved by optimization of Mg^{2+} ion and assimilable nitrogen availability, respectively, in *S. cerevisiae* fermentations (G.M. Walker, unpublished results and W.M. Ingledew, University of Saskatchewan, Canada, personal communication).

The important point to be made is that in addition to the yeasts' genotype, its growth environment is extremely important in governing fermentation performance. Furthermore, the dogma that alcohol productivities and tolerances vary widely between different yeasts no longer holds (see Ingledew, 1993 for further discussion).

Yeast process developments for bioethanol production

In addition to yeast genetic engineering and cell engineering as discussed above, aspects of *fer-*

Table 6.27. Some important physiological factors influencing ethanol production by *S. cerevisiae*.

Factor	Comments	References
Assimilable nitrogen	Some industrial fermentation media may not be optimized with respect to their availability of metabolizable nitrogenous compounds	Jones and Ingledew (1994)
Magnesium ions	Mg^{2+} supplementations increase ethanol yields and the ethanol tolerance of *S. cerevisiae*	Dombek and Ingram (1986); D'Amore *et al.* (1988); Walker *et al.* (1996) and Table 4.28
Oxygen	O_2 is required for fatty acid and sterol biosynthesis which helps to maintain plasma membrane structural integrity in the presence of ethanol	See Chapters 3 and 4 and Rose (1986)
Lipid compounds	Supplementation of fermentation media with membrane lipids enhances yeast cell tolerance to ethanol	See Table 4.27
Temperature	Spheroplast fusion has increased the tolerance of *S. cerevisiae* cells to ferment at 40°C	D'Amore *et al.* (1989)

menter engineering will also play important roles in maximizing ethanol production by yeasts.

Several fermentation strategies based on batch, fed-batch and continuous processes have been considered for bioethanol production on a large scale (Shama, 1988). Generally speaking, systems which augment yeast cell densities (e.g. immobilization and cell recycle) in fermenters will lead to higher ethanol productivities. The performance of high cell density systems in lignocellulosic hydrolysate fermentations has been reviewed by Olsson and Hahn-Hägerdal (1996) and Table 4.15 provides examples of immobilized cell systems for ethanol production. However, no single system can be regarded as universally applicable for all the different fermentable substrates and yeast strains. An important point relating to process engineering aspects in bioethanol production was made by Olsson and Hahn-Hägerdal (1996): 'Industrial processes generally do not permit optimal conditions for the microorganism since environmental parameters are determined by technical and economic considerations'.

6.3.2.2 Other Alcohols from Yeasts

Besides ethanol, other industrially useful alcohols can be produced by yeast fermentation including: higher alcohols, and polyhydric alcohols such as glycerol, xylitol, sorbitol, arabinitol, erythritol and mannitol.

Higher alcohols, or fusel oils, are produced during amino acid metabolism in yeasts (see Chapter 5) and several of these compounds constitute important flavour chemicals (congeners) in alcoholic beverages (e.g. isobutanol, isoamyl alcohol, phenylethanol and isopropanol). In the production of potable spirits like Scotch whisky, the distillation process can effectively concentrate and remove some of these fusel oils which can then be sold commercially for use as solvents (e.g. isoamyl alcohol in perfumes).

The production of glycerol from yeast glycolysis and the role of glycerol as a compatible solute in osmostress tolerance of yeast cells, were discussed in Chapters 5 and 4, respectively. Glycerol is an important flavour component in several alcoholic beverages, particularly in Japanese

Shochu (see Omori *et al.*, 1995). Glycerol has also been produced industrially by 'steered' fermentations in the manufacture of nitroglycerine-based explosives (see Chapter 5, section 5.2.1.1). Glycerol also finds uses in the production of synthetic resins, pharmaceuticals, cosmetics and toothpastes. However, bioglycerol from yeast fermentation is insignificant nowadays compared with chemically produced glycerol.

Other sugar alcohols produced by yeasts include xylitol. This compound is finding increasing uses in the food industry as a sweetener and as a sugar-substitute for insulin-dependent diabetics. Several yeast species can synthesize xylitol as a metabolic intermediate of D-xylose metabolism (see Singh and Mishra, 1995). Species of *Candida* are the best xylitol producers, including *C. tropicalis*, *C. guilliermondii* and *C. mogii* (see Sirisansaneeyakul *et al.*, 1996). Another *Candida* species, *C. entomaea*, can produce L-arabinitol from L-arabinose and this yeast is a primary candidate for the production of arabinitol from hemicellulosic materials such as corn fibre (Saha and Bothhast, 1996). Arabinitol, like other sugar alcohols, has uses as a low-calorie bulking agent, in biosurfactants, edible coatings and as a carrier for pharmaceutical agents. Other polyols produced by yeasts include erythritol (*Candida zeylanoides*) and mannitol (*Yarrowia lipolytica*).

6.3.3 Yeast Biomass-Derived Products

Yeast biomass, mainly in the form of baker's yeast, represents the largest bulk production of any single-celled microorganism throughout the world. Several million tons of fresh *S. cerevisiae* cells are produced annually for human food use. Since the middle of the 19th century, the industrial manufacture of baker's yeast has been an industry separate and distinctive from brewing, winemaking and distilling. Before that, bakers relied upon spent yeast from alcoholic beverage fermentations and bakeries and breweries were joint enterprises, often conducted from the same premises. In modern times, production of yeast for baking (and for alcoholic beverages) has evolved into a sophisticated biotechnology and some of the developments relating to this important industry are discussed below.

In addition to the use of live yeast biomass for the leavening of bread dough, many other applications of yeast cells and yeast cell extracts have emerged, and these are summarized in Table 6.28. Further information on some of these applications can be found in Reed and Nagodawithana (1991).

Most yeast biomass for industrial use is derived from *S. cerevisiae*, but other yeasts have specific uses and may be grown on a range of substrates unavailable to *S. cerevisiae* (Table 6.29).

6.3.3.1 Baker's Yeast

Scientific and technological aspects of the production of baker's yeast have been reviewed by Beudeker *et al.* (1990), Nagodawithana and Trivedi (1990), and Rose and Vijaylakshmi (1993). Note that the large-scale production of brewing, winemaking and distilling strains of *S. cerevisiae* follows the same principles that are employed in baker's yeast manufacture. In comparison with baker's yeast, however, the production of active pure starter cultures of wine and other beverage yeasts is a relatively recent technology (Rosen, 1989).

The production of baker's yeast involves a multi-stage propagation (progressive build-up of yeast cell density) of specially selected *S. cerevisiae* strains on sugar cane and sugar beet molasses. The sucrose-rich molasses is supplemented with additional sources of nitrogen, phosphorus and essential mineral ions such as magnesium. The later stages of production are highly aerobic processes and the molasses medium is delivered incrementally to the growing cells in a fed-batch manner to avoid the Crabtree effect and maximize respiratory growth. The regulation of carbohydrate metabolism in relation to the Crabtree effect was discussed previously in Chapter 5. The fed-batch method of baker's yeast propagation is sometimes referred to as the *Z-method* in which Z stands for the *Zulaufverfahren*

Table 6.28. Industrial uses of yeast biomass.

	Type of yeast product	Examples of uses
Whole-cell products	Compressed baker's yeast/ active dried yeast	Baking, brewing, winemaking and distilling
	Yeast cream	Baking and distilling
	Fodder yeast/ single-cell protein	Animal feed
	Biotherapeutic/growth factor yeasts	Human/animal probiotics
	'Reagent' yeasts	Biocatalysts in organic chemistry
	Biosorbent yeasts	Heavy metal sequestration
	'Mineral yeasts'	Nutritional trace elements (Cr, Se) source
	Cosmetic/pharmaceutical yeasts	Skin respiratory factor
	Pigmented yeasts	Feed colorants
	Biological control yeasts	Antifungal agents in agriculture
	Pollution control yeasts	BOD reductions
Extracted-cell products	Yeast extracts	Food use and microbiological growth media
	Yeast RNA derivatives	Flavour enhancers and pharmaceutical use
	Yeast cell walls	Food and pharmaceutical use
	Yeast B-complex vitamins	Capsules/tablets for dietary supplements
	Yeast enzymes	Invertase and lactase for food use
	Recombinant yeasts	Therapeutic proteins

Table 6.29. Non-*S. cerevisiae* yeast biomass with biotechnological uses.

Yeast	Uses/potential uses of biomass
Kluyveromyces marxianus and *K. lactis*	Animal feed yeast biomass from whey lactose. Sources of lactase
Candida utilis	Single-cell protein (SCP) from sulphite waste liquor and wood sugars
Phaffia rhodozyma	Carotene pigment (red food colorant)
Saccharomyces boulardii	Biotherapeutic agent
Pichia pastoris and *Hansenula polymorpha*	SCP and recombinant proteins from methanol
Yarrowia lipolytica and *Candida paraffinica*	SCP from *n*-alkanes
Schwanniomyces castellii	SCP from starch
Pichia stipitis and *Candida shehatae*	SCP from lignocellulosic biomass
Rhodotorula glutinis, *Lipomyces lipofer*, *Cryptococcus curvatus* and *Candida* spp. (e.g. *C. palmioleophila*)	Single cell oil (SCO), as substitutes for edible and non-edible oils from cheap carbon sources (e.g. whey, molasses, *n*-alkanes)

process (or zero alcohol, to indicate the undesirability of fermentative metabolism by the cells).

Several points concerning cell physiology should be noted with regard to baker's yeast propagation and subsequent utilization. For example, baker's yeasts are grown on an entirely different medium (i.e. sucrose-rich molasses) compared with the one in which they are destined to grow and metabolize (i.e. maltose-rich bread dough). Indeed, the goals of the baker's yeast producers are to maximize growth and minimize fermentation, whereas the goals of the baker are the converse. As well as these physiological discrepancies, baker's yeasts are also confronted with additional environmental stresses before, during and after their production which requires cells to possess fairly unique characteristics. The desirable properties of an 'ideal' baking strain of *S. cerevisiae* are manifold and are outlined in Table 6.30.

There are therefore several areas for the potential genetic improvement of baking strains of *S. cerevisiae*, and some of those are outlined in Table 6.31. Angelov *et al.* (1996) and Benitez *et al.* (1996) have discussed some of the biotechnological developments in the characterization, growth and dough fermentation properties of baker's yeast. The desirability of propagating strains of *S. cerevisiae* in molasses which display both high productivity (biomass produced per unit time) and high yields has been emphasized by Benitez *et al.* (1996). At present, yeast growth rate in fed-batch systems is quite low (generally 0.15–0.20 h^{-1}) but no baking strains exist which

Table 6.30. Desired characteristics of baking yeast strains.

Desired Characteristics	Comments
High glycolytic activity	Especially with regard to CO_2 evolution rate in dough fermentations (i.e. gassing power).
No maltose fermentation lag	Maltose is the predominant sugar in cereal flour/water doughs. Strains that have the ability to utilize maltose quickly, under glucose-repressing conditions, are desirable.
Osmotolerance	Yeasts with high trehalose levels and an ability to accumulate glycerol are desirable, together with resistance to high sugar and salt concentrations. Strains with low invertase activity would be preferred for leavening sweet doughs.
Cryotolerance	Better freeze-tolerance is required in yeasts incorporated into frozen doughs for 'instant' baking. This may be achieved by cells with high trehalose concentrations.
Chemical tolerance	Resistance to bread preservatives (e.g. propionate) and the presence of sodium chloride.
Melibiose utilization	Relevant in baker's yeast propagation in molasses containing raffinose (present at 8% w/v); invertase activity yields fructose and melibiose from raffinose, but baker's strains do not usually have melibiase activity.
Good storage ability	The viability and activity of stored baker's yeast may be enhanced by elevated levels of cellular trehalose. Therefore, during the latter stages of fed-batch propagation, cultures are generally aerated without nutrients (i.e. 'ripened') to stimulate trehalose biosynthesis and to synchronize cells into the stationary growth phase.
Non-agglomeration	'Grittiness' (described in Chapter 2) is undesirable in baker's yeast since it may deleteriously affect dough leavening capability.

Table 6.31. Biotechnological developments related to baker's yeast.

Development	Comments	References
Strain characterization	Genetic fingerprinting techniques based on chromosome polymorphisms detectable by electrophoretic karyotyping have been successfully used to differentiate baking strains of *S. cerevisiae*.	Codon and Benitez (1995)
Storage stability	Freeze-tolerant variants of *S. cerevisiae* have been isolated. Additionally, some freeze-tolerant non-*Saccharomyces* yeasts (e.g. *Torulaspora* spp) possess both freeze–thaw resistance and high leavening abilities.	Oda *et al.* (1986); Oda and Tonomura (1993)
Growth rates/yields	Recombinant melibiase-positive (MEL^+) strains have been constructed which are able to fully utilize raffinose in molasses and therefore fully utilize available sugars for improved growth.	Gasent-Ramirez *et al.* (1995)
Fermentation performance	Avoidance of a maltose-lag (see Table 6.30) achieved by rare mating or by inserting strong constitutive promoters (from another strain) upstream from the maltase and maltose permease genes. Derepression of invertase and maltase synthesis has also been achieved by transforming multiple copies of *SUC* and *MAL* regulatory regions into baking strains.	Oda and Ouchi (1990); Aldhous (1990); Gozalbo (1992)
Bread flavour	Isobutyl alcohol-producing mutants of baker's yeast have been isolated and shown to produce bread with distinctive flavours.	Watanabe *et al.* (1990)

can maximize cell yield while achieving very high growth rates. This is an area of potential future development in the genetics and physiology of baker's yeast.

6.3.3.2 Other Applications of Whole-Cell Yeast Biomass

Besides their direct food use such as in baking, yeasts may also be propagated as a source of protein for human and animal nutrition. Many studies have been conducted into the growth of yeasts on cheap, readily available carbon sources for the production of **Single Cell Protein** (SCP). However, some of the early promises of this technology in the 1970s never materialized, for a variety of political and economic, rather than scientific and technological reasons. A well-quoted example of this was the abandoned Toprina process originally established by British Petroleum which was based on the growth of yeasts on hydrocarbons (see Faust, 1987). Indeed Harrison (1993) concluded rather gloomily: 'The primary commercial production of yeast as an economically viable protein carrier is unlikely in the foreseeable future'.

Nevertheless, the provision of whole-cell yeast biomass for non-SCP applications has received increasing attention in recent years (Table 6.32).

6.3.3.3 Extracted Yeast Cell Products

Yeasts are rich sources of proteins, nucleic acids, vitamins and minerals but with negligible levels

Table 6.32. Novel applications of whole-cell yeast biomass.

Application	Comments	References
Livestock growth factor	In ruminants, live cultures of *S. cerevisiae* stabilize the rumen environment and improve nutrient availability to increase animal growth or milk yields. The yeasts may be acting to scavenge O_2 and prevent oxidative stress to rumen bacteria, or they may provide malic and other dicarboxylic acids to stimulate rumen bacterial growth.	Lyons *et al.* 1993); Newbold *et al.* (1996)
Biotherapeutic agent	*S. cerevisiae* may have uses as an anti-acne agent and in treatment of premenstrual distress. *S. boulardii* has been reported to act as a prophylactic and therapeutic agent against several intestinal diseases (e.g. *Clostridium difficile* colitis) and as an anti-*Candida albicans* agent. *S. boulardii* is patented and lyophilized in capsules by Laboratories Biocodex, France.	Weber *et al.* (1989); Facchinetti *et al.* (1997); McFarland and Elmer (1995)
Chemical 'reagent'	Organic chemists use *S. cerevisiae* (baker's yeast) as a reagent in enantioselective oxidation of lactones and stereospecific bioreductive catalyses of several aldehydes and ketones. Some processes have now been industrialized (e.g. production of trimegestrone by Roussel Uclaf in France for the treatment of postmenopausal disease).	Servi (1990); Stewart *et al.* (1996); Kometani *et al.* (1996); Crocq *et al.* (1997)
Food pigment	*Phaffia rhodozyma* produces a red carotene pigment (astaxanthin) which is the principal colorant in crustaceans, salmonids and flamingos. There is current interest in using *P. rhodozyma* in aquaculture to impart desired red pigmentation in farmed salmon and shrimps. Astaxanthin levels can be augmented by yeast mutagenesis or by recombinant DNA technology.	Johnston and Gil-Hwan (1991); Adrio and Veiga (1995); Fang and Chiou (1996); Johnson and Schroeder (1996)
Biocontrol agent	*S. cerevisiae* has potential as a phytoallexin elicitor in cereals. Several yeasts (e.g. *Debaryomyces hansenii, Candida* spp. and *Metschnikowia pulcherrima*) may be used in biocontrol of fungal fruit diseases.	Reglinski *et al.* (1994); Filonow *et al.* (1996)
Biosorbent/ bioremediation agent	Spent brewer's yeast can effectively remove heavy metals (e.g. Ag, U, Co, Cu, Cd etc.) from industrial effluents. *Kluyveromyces marxianus* can also effectively biosorb uranium. *Brettanomyces lambicus* can extract zinc from filter dusts. Several yeasts (e.g. *Candida utilis*) can effectively remove carbon and nitrogen from organic waste water and reduce BOD levels of effluents. *Candida, Lipomyces* and *Rhodotorula* spp. can detoxify certain pollutants (e.g. removal of herbicides, phenols and formaldehyde). *Candida* spp. (e.g. *C. hellenica*) can effectively treat oily wastewaters.	Simmons *et al.* (1995); Singleton and Simmons (1996); Omar *et al.* (1996); Wenzl *et al.* (1990); Bustard *et al.* (1996); Chanda and Chakabarti (1996); Ortiz *et al.* (1997); Wainwright 1992; Chigusa *et al.* (1996)
Biosensor	*S. cerevisiae* may be useful in immobilized cell biosensors for toxicity testing of environmental samples.	Campanella *et al.* (1995)
Bioelectrical fuel cells	Electron-generating yeast-powered fuel cells and yeast-based quantum semiconductors have potential in bioelectronics.	Bennetto *et al.* (1987): Coghlan (1992)

of triglycerides. When extracted by acids (to produce hydrolysates) or enzymes (autolysates) or salt (plasmolysates), yeasts develop characteristic savoury flavours and aromas which are utilized in numerous processed and convenience foods where a 'meaty flavour' is sought (e.g. dried soups, gravy granules, flavoured potato snacks, 'Marmite' and 'Vegemite'). Yeast extracts are normally produced from spent (and de-bittered) brewer's yeast, but increasingly on a larger scale from baker's yeast propagated by the fed-batch process referred to previously (see Reed and Nagodawithana, 1991). Other uses of yeast extracts are in the preparation of microbiological growth media (e.g. YEPD).

Yeast savoury flavours derive from the degradation of cell protein and RNA to yield taste enhancers such as glutamic acid, peptides and ribonucleotides. Yeast technologists are now able to modify the natural autolytic processes occurring in yeasts to develop more refined flavours (see Hay, 1993). This is achieved by heat-denaturation of the cell's natural hydrolytic proteases and nucleases followed by treatment with commercial ribonucleases (e.g. from malt rootlets as by products of the malting industry) to convert yeast RNA (present at around 10% per cell dry weight) into flavour-enhancing 5′-ribonucleotides such as 5′-inosine monophosphate (IMP) and 5′-guanosine monophosphate (GMP). Other derivatives from yeast RNA (e.g. 5′-cytidine monophosphate, CMP and 5′-uridine monophosphate, UMP) find applications in pharmaceuticals (e.g. antiviral agents) and in fine chemicals for biological research.

Yeast cell walls (or 'hulls') represent by-products from yeast extract manufacture. This material, referred to as yeast *glycan*, has potential uses in the food industry as a stabilizing and non-nutritive bulking agent. Cell wall mannoprotein of *S. cerevisiae* also has applications as a bioemulsifier in food products (Torabizadeh *et al.*, 1996). There may also be possible uses of yeast hulls as biosorbents for removal of heavy metals from industrial wastewaters. Bohn and BeMiller (1995) have also discussed the utility of yeast cell wall glucans in enhancing the human immune system. These immunomodulating properties are associated with the (1-3)-β-D-glucan moieties of the *S. cerevisiae* cell wall and Jamas *et al.* (1991) have been able to increase these properties through recombinant DNA technology. The modified glucan (termed PGG glucan) enhanced microbicidal activity by monocycte and neutrophil phagocytosis and increased the levels of cytokines, colony-stimulating and inflammatory factors. Importantly, PGG glucan was shown to produce significant protection against acute bacterial and fungal sepsis *in vivo*.

6.3.4 Industrial Enzymes and Chemicals

Compared with certain fungi (e.g. *Aspergillus* spp.) and bacteria (e.g. *Bacillus* spp.), yeasts are not particularly rich sources of industrially useful enzymes. Nevertheless, a few enzymes are prepared from yeast fermentations which have commercial applications, particularly in the food processing industries. Recombinant DNA technology can increase the range of enzymes secreted by yeasts and has also enabled *S. cerevisiae* to assimilate polymeric carbon sources in the production of bioethanol and biomass. For example, the construction of amylolytic and xylanolytic *S. cerevisiae* strains adds starchy and woody based wastes to the list of fermentation substrates available for bioethanol production by this yeast.

Table 6.33 summarizes the uses of some yeast-derived enzymes. Other enzymes not listed include peptidases, pectinases, glucanases, melibiase, cytochromes P450, alcohol dehydrogenase and phenylalanine ammonia-lyase (PAL). This latter enzyme, which catalyses the deamination of phenylalanine, can be prepared from *Rhodotorula glutinis* and has potential in the diagnosis and treatment of the human inborn error of metabolism, phenylketonuria. Further information on yeast enzymes can be found in Finkleman (1990) and Reed and Nagodawithana (1991).

Regarding the production of industrially useful chemicals by yeasts, several organic,

Table 6.33. Applications of yeast-derived enzymes.

Substrate	Enzyme	Applications
Starch	α-Amylase and glucoamylase	Bioethanol and biomass from conversion of starchy wastes. Production of low-carbohydrate beers. Glucoamylase genes from *Schwanniomyces* spp. and *S. cerevisiae* var. *diastaticus* have been cloned into brewer's yeasts.
Sucrose	Invertase	The hydrolysis of sucrose to glucose and fructose by yeast invertase (available from autolysis of baker's yeast) has applications in the confectionery industry (e.g. production of soft-centred chocolates) and in the production of invert sugar.
Inulin	Inulinase	*Kluyveromyces* spp. are particularly good sources of inulinase for the hydrolysis of polyfructans and levans (e.g. from Jerusalem artichoke tubers) in the production of ethanol and high-fructose syrups.
Lactose	Lactase	*Kluyveromyces* spp. are particularly good sources of lactase for the hydrolysis of lactose in milk and dairy wastes into ethanol and yeast biomass protein. (The commercialization of ethanol from cheese whey by *K. marxianus* has been achieved in Ireland, New Zealand and the USA.) Yeast lactase is also very useful in processed dairy foods, particularly lactose-reduced milk which is suitable for lactose-intolerant individuals. Lactose-utilizing genes from *Kluyveromyces* have been cloned into *S. cerevisiae*.
Fats	Lipase	The hydrolysis of triglycerides to free fatty acids and glycerol by lipases from oleaginous yeasts such as *Yarrowia lipolytica*, *Candida rugosa* and *Rhodotorula glutinis* has many industrial applications (e.g. digestive aids, flavour modifications, interesterification of oils etc.).
Cellulose and hemicellulose	Cellulase, β-glucosidase, xylanase	Some yeast-like fungi (e.g. *Aureobasidium pullulans*) have cellulolytic activity. *Dekkera intermedia* and *Kluyveromyces cellobiovorans* have β-glucosidase activity and can ferment cellobiose (major hydrolysis product from cellulose). Some yeasts (e.g. *Cryptococcus* and *Trichosporon* spp.) have extracellular xylanase activity and can release xylobiose and xylotriose from xylan. The cloning of cellulose and hemicellulose-utilizing genes from various organisms into *S. cerevisiae* is of great interest, particularly in bioethanol production from renewable biomass.
Casein	Chymosin	Recombinant *Kluyveromyces* spp. are employed to produce a calf rennet-substitute for milk casein coagulation in the manufacture of cheese.

amino and fatty acids can be synthesized by a variety of non-*Saccharomyces* species. Citric acid is quantitatively one of the most important fermentation acids produced and has many uses in the food, pharmaceutical and chemical industries. Other chemicals produced by yeasts include vitamins, polysaccharides, etc. and these are listed together with yeast-derived acids, in Table 6.34.

6.3.5 Therapeutic Proteins

Since the first commercialization of a human therapeutic protein from yeast (Hepatitis B 'Recombivax' vaccine by Merck in 1986), several other products have gained clearance from regulatory bodies (e.g. the US FDA) for clinical use (e.g. leukine GM-CSF by Immunex for autologous bone marrow transplant; human

Table 6.34. Industrially useful chemicals produced by yeasts.

Chemical	Examples	Yeasts
Organic acids	Citric	*Yarrowia lipolytica* and *Candida guilliermondii*
	Itaconic	*Candida* and *Rhodotorula* spp.
	Malic	*Candida utilis*
	D-gluconic	*Saccharomycopsis* spp., *Aureobasidium pullulans*
	L(+) Isocitric	*Candida brumptii*
	α-Ketoglutaric	*Candida hydrocarbofumarica*
	Brassylic	*Torulopsis, Candida*
	Sebacic	*Torulopsis, Candida*
	Fumaric	*Candida hydrocarbofumarica*
Fatty acids	Stearic	*Cryptococcus curvatus* (cocoa butter equivalents rich in stearic acid – see Hassan *et al.*, 1994)
	Long-chain dicarboxylic	*Candida tropicalis* (see Picataggio *et al.*, 1992)
Amino acids	Lysine	*Saccharomyces cerevisiae* and *Candida utilis* (using 5-formyl-2-ketovaleric acid as precursor)
	Tryptophan	*Candida* and *Hansenula* spp. (using anthranillic acid as precursor)
	Phenylalanine	*Rhodotorula rubra*
	Glutamic acid	*S. cerevisiae* (3-chloroaniline-resistant mutants)
	Methionine	*S. cerevisiae* (ethionine-resistant mutants)
Vitamins	Riboflavin	*Candida flaveri, C. guilliermondii, Eremothecium ashbyii*
	Pyridoxine	*Pichia* spp.
	D-Erythro-ascorbic acid	*Candida, Kluyveromyces, Torulopsis*
Sterols	Ergosterol	*Saccharomyces cerevisiae*
	Steroid precursors	*Hansenula, Kloeckera, Pichia, Rhodotorula* spp.
Polysaccharides	Pullulan	*Aureobasidium pullulans*
	Phosphomannan gums	*Hansenula, Pichia and Pachysolen* spp.
	Glycolipids	*Yarrowia lipolytica and Torulopsis bombicola*

Information from Burden and Eveleigh (1990); Mattey (1992); Wainwright (1992) and Stahmann (1997).

insulin by Novo-Nordisk for treatment of diabetes; tumour necrosis factor by Chiron) and many more are in clinical trials or at the research and development stage. The latter include therapeutic HIV vaccine development by British Biotechnology Ltd. (see Glaser, 1996). Table 6.35 lists some of the heterologous genes which can successfully be expressed in yeast cells and which have pharmacological applications and potential. Note that non-*S. cerevisiae* yeasts are playing increasingly important roles in the industrial production of human therapeutic proteins. The particular advantages which *Pichia pastoris* and *Hansenula polymorpha* possess over *S. cerevisiae* as production organisms for recombinant biopharmaceuticals have been discussed previously in this chapter and have been reviewed, respectively, by Romanos (1995) and Gellissen and Melber (1996). Other yeasts, notably *Kluyveromyces lactis*, *Schizosaccharomyces pombe*, *Schwanniomyces occidentalis* and *Yarrowia lipolytica* also have potential uses in the production of therapeutic products by recombinant DNA technology.

Table 6.35. Examples of cloned therapeutic proteins synthesized by yeasts.[a]

Donor DNA source	Examples of gene products
Prokaryotic	Tetanus toxin fragment C; streptokinase; whooping cough antigen
Viral	Various genes encoding surface antigens, enzymes and other proteins from the following viruses: Hepatitis B; Herpes simplex; HIV; Foot and Mouth; Influenza; Polio; Bovine Leukaemia; Polyoma; Epstein–Barr; Oncogenic retroviruses
Protozoal	Malaria antigen (see Bathurst, 1994)
Animal	Leech hirudin; viper echistatin; porcine interferon; rabbit α-globin; porcine urokinase; bovine pancreatic trypsin inhibitor (aprotinin); rat glia-derived nexin; bovine and mouse interleukin
Human	**Hormones**: Insulin; parathyroid hormone; human chorionic gonadotrophin; human growth hormone; somatostatin **Antibodies**: functional antibodies and Fab fragments; chorismate mutase 'abzyme' antibody; IgE receptor **Growth factors**: insulin-like growth factor (IGF1); nerve growth factor; epidermal growth factor; tissue factor; platelet-derived endothelial growth factors; interleukins; macrophage-colony stimulating factor; leukine GM-CSF; tumour necrosis factor **Interferons**: leukocyte interferon-alpha(D); interferon-alpha2, -beta1 and hybrid X-430 **Blood proteins**: haemoglobin; factors VIII and XIII; erythropoietin; serum albumin; antithrombin III; alpha1-antitrypsin; tissue plasminogen activator; fibrinogen; lactoferrin **Enzymes/inhibitors**: proteinase inhibitor 6; gastric lipase; thyroid peroxidase; pro-urokinase; salivary α-amylase; lysozyme; elastase inhibitor (Elafin); liver epoxide hydrolase; Cu, Zn, superoxide dismutase; cytochrome P450 **Miscellaneous**: Oestrogen receptor; cystic fibrosis transmembrane conductance regulator (CFTR); *CDC 28* and G1 cyclin homologues; cancer cell surface antigens; β-endorphin

Information mainly obtained from: Hadfield *et al.* (1993); Wiseman (1996) and Rallabhandi and Yu (1996) who quote original references for the particular yeasts and cloning technology employed.
[a] *S. cerevisiae* and non-*Saccharomyces* yeast species.

Yeasts have therefore become increasingly popular host organisms in the production of recombinant pharmaceuticals by the health-care biotechnology sector. This popularity is likely to be sustained as cell physiologists, molecular geneticists and fermentation technologists strive to maximize yields of nature-authentic, bioactive therapeutic proteins from recombinant yeast cells. Such collaborative efforts will hopefully lead to novel approaches in the future prevention and treatment of several human diseases.

6.4 YEASTS IN BIOMEDICAL RESEARCH

In addition to the aforementioned practical uses of yeasts in the provision of novel human therapeutic agents, yeast cells are also extremely valuable as experimental models in biomedical research. This is especially the case in aspects of oncology, pharmacology, toxicology, virology and human genetics (Table 6.36). Two yeasts in particular have greatly assisted fundamental studies in these areas: namely, *Saccharomyces cerevisiae* and *Schizosaccharomyces pombe*. These organisms are among the best understood eukaryotic unicells with the entire genome of the former having been completely sequenced (in 1996 – see Dujon, 1996; Johnston, 1996; Oliver, 1996a) and that of the latter currently, at the time of writing (April, 1997), ongoing with completion expected in 1998 (see the web site at http://www.nih.gov/sigs/yeast/fission.html). Functional analysis of the approx. 1000 'orphan' genes (for which no function has yet been

Table 6.36. Value of yeasts in biomedical research.

Research field	Examples of the value of yeast research	References
Oncology	• Molecular mechanisms of cell cycle control at G1–S and G2–M checkpoints are being determined using budding and fission yeasts. These controls are key to our understanding of cancer cell division.	See Chapter 4 and references therein.
	• The biochemical regulation of certain human oncogenes (e.g. *ras*) can be studied in *S. cerevisiae.*	See Nurse (1985); Pfeiffer (1997)
	• p53 gene mutations, which are implicated in tumorigenesis, can be studied using a yeast functional assay.	Ishioka *et al.* (1995); Tada *et al.* (1996)
	• The function of telomere-binding proteins in *Sch. pombe* are providing insight into how cancer cells maintain a constant average telomere length.	Cooper *et al.* (1997)
	• *S. cerevisiae* is useful in studies of apoptotic cell death.	Bemis *et al.* (1995)
Pharmacology and Toxicology	• Yeasts are used in studies of the genetics and biochemistry of multidrug resistance.	Balzi and Goffeau (1994); Shallom and Golin (1996); Goffeau *et al.* (1997)
	• Studies on the mode of drug action (e.g. anti-tumour agents) is feasible in yeast cells.	Delitheos *et al.* (1995); Popolo *et al.* (1996)
	• Drug metabolism, drug–drug interactions and pharmacokinetics can be studied in yeasts.	Friedberg and Wolf (1996)
	• Yeasts expressing human P450 are useful in drug assays.	Pompon *et al.* (1995)
	• Drugs can be screened and mutagens/genotoxic agents rapidly tested in *S. cerevisiae* or *Sch. pombe.*	Wiseman (1987); McAthey (1996); Staleva *et al.* (1996)
	• Mycotoxins, xenoestrogens, etc. can be assayed using yeasts.	Routledge and Sumpter (1996); Madhyastha *et al.* (1994)
	• Toxic chemicals in pollutant samples can be detected using yeast biosensors.	Camponella *et al.* (1995)
Virology	• *S. cerevisiae* contributes to human virology by providing: a virus model (e.g. killer plasmids and retrotransposons); anti-viral components (surface antigens for vaccines); system for studying viral protein action; two-hybrid system to study self-interaction of viral proteins.	Ward and Macreadie (1996)
Genetics	• Yeast artificial chromosomes (YACs) are invaluable in cloning human DNA and in mapping the human genome.	Resnick *et al.* (1995); Markie (1996)
Neurodegenerative disease	• Research into yeast prions is offering new insights into the molecular biology of protein-based inheritance and possible therapies for certain neurodegenerative human diseases (e.g. kuru, Creutzfeld-Jacob, fatal familial insomnia).	Lindquist (1996)

assigned) in *S. cerevisiae* is now underway in Europe (through EUROFAN, the European Functional Analysis Network – see Oliver, 1996b) and throughout the world (see Goffeau *et al.*, 1996). Elucidation by yeast cell physiologists of the biological function of all of these genes (and complete analyses of the yeast *proteome* – see Payne and Garrels, 1997) will not only lead to the understanding of how a simple eukaryotic cell works, but will also hopefully provide insight into aspects of human genetics. For example, there is some degree of conservatism between humans and yeasts in terms of genetic regulation and several human gene homologues have been found in *S. cerevisiae* (e.g. genes for adrenoleukodystrophy and cystic fibrosis on chromosome XI and for ataxia telangectasia on chromosome II – see Dujon, 1996). Human genes can be functionally expressed in *S. cerevisiae*; for example, the human cystic fibrosis transmembrane conductance regulator (CFTR) has been studied in yeast by Huang *et al.* (1996). Decottignes and Goffeau (1997) have reviewed ABC (ATP-binding cassette) proteins in *S. cerevisiae* and their human homologues, as revealed by the yeast genome project. Studies of yeast ABC proteins may provide insight into the fundamental molecular biology of several human diseases which are related to deficient ABC transporter functions including: cystic fibrosis, adrenoleukodystrophy, Zellweger syndrome, infantile hyperinsulinaemia, Dubin–Johnson syndrome, Behçet's syndrome, bare lymphocyte syndrome type I, insulin-dependent diabetes mellitus and possibly multiple sclerosis (see Decottignes and Goffeau, 1997). In practical terms, knowledge of yeast genome structure and homologies with the human genome may provide novel approaches (through specific drugs or gene therapy) in the treatment of human heritable diseases.

6.5 SUMMARY

This chapter has focused on several traditional, modern and emerging biotechnologies which exploit the activities of yeasts. As with previous chapters, where diversity of yeast nutrition, growth and metabolism was highlighted, the present discussion has emphasized the great diversity of yeast industrial applications. Thus, yeast biomass and yeast products find uses in many sectors including: alcoholic beverages, foods, biofuels, industrial chemicals, pharmaceuticals and in pollution control. Yeast cells are additionally beneficial as experimental model organisms in biomedical research. Nevertheless, the primary commercial impact of yeast has been, and will remain, in food and fermentation processes where the association between yeast and beer and bread is immediately recognizable to the general public. Human health, as well as human nutrition, is likely to become another recognizable role for yeast in the future, especially when one considers the increasing involvement of yeasts in the production of pharmaceuticals and in cancer research.

Another theme of this chapter has been the increasing utility of non-*Saccharomyces* yeasts in biotechnology. Particular attributes of yeasts such as *Pichia stipitis* and *Hansenula polymorpha* in recombinant DNA technology are now well recognized and these species, along with *Kluyveromyces*, *Candida*, *Schizosaccharomyces*, *Yarrowia* and *Schwanniomyces* species, should no longer be regarded as 'non-conventional'. These yeasts have uses and potential uses in the production of numerous biotechnological commodities from readily available and inexpensive growth substrates such as methanol, alkanes, whey and cellulosic materials. The methylotrophic yeasts, in particular, are better suited than *S. cerevisiae* for the secretion and modification of several mammalian proteins of pharmaceutical interest. Nevertheless, *S. cerevisiae* remains the most exploited microorganism known and is still the primary yeast species responsible for producing potable and industrial ethanol – quantitatively and economically the world's premier biotechnological commodity. Yeast strain improvement strategies, based on classical and modern genetic approaches, have been discussed in relation to yeast fermentation performance in the alcohol industries. Recombinant DNA technology in asexual yeasts, such as brewing strains of *S. cerevisiae*, is now able to amplify and control gene expression in a way that is impossible using classic genetic approaches. Molecular genetic improvements of industrial yeasts have widened the range of fermentable substrates available to such strains. With regard to fuel ethanol production, biotechnological developments are helping to enhance the competitiveness of bioethanol from renewable substrates. However, the range of improvements is often restricted by lack of knowledge of yeast cell physiology and there is still much mileage to be gained in exploiting physiological aspects of yeast growth and metabolism. For example, by closely addressing the nutritional and physical growth requirements of yeasts during fermentation it may be possible to significantly increase cellular ethanol productivities.

Questions relating to cell physiology are also very pertinent when considering the production of recombinant proteins by *S. cerevisiae* and other yeasts. It is important to remember that genetically engineered yeasts are coerced into producing physiologically alien proteins at abnormally high levels. In fact, certain heterologous genes when present at a high dosage in yeasts may be very toxic to cells (see Daniel, 1996). Therefore, the influence of the cell's growth environment on aspects such as plasmid stability and regulated gene expression need to be as well understood as the influence of genetic elements. Physiological factors such as nutrient- and temperature-regulation of heterologous gene expression and the control of growth rate in scaled-up fermentations are important considerations for industrialists exploiting recombinant yeasts. Collaboration between yeast cell physiologists, molecular biologists and bioprocess technologists is therefore seen as a major prerequisite for the commercial success of any yeast biotechnology.

Emerging yeast technologies are especially evident in the environmental and health-care sectors of biotechnology. In the former, yeasts are likely to play increasing roles in areas such as the bioremediation of industrial wastes and in biological control of fungal pests in agriculture. In several aspects of health-care biotechnology, yeasts are having a great impact. For example, in the prevention and treatment of several human diseases, recombinant yeast products are already playing important medical roles. Many new therapeutic proteins are under development and await clinical use. Finally, the contribution of yeasts in assisting our fundamental understanding of human diseases, such as cancer and inborn errors of metabolism should not be underestimated. In particular, it is hoped that research into yeast genomes and proteomes will find practical applications in the diagnosis and therapy of heritable human disorders in the not too distant future.

6.6 REFERENCES

Adrio, J.L. and Veiga, M. (1995) Transformation of the astaxanthin-producing yeast *Phaffia rhodozyma*. *Biotechnology Techniques*, **9**, 509–512.

Aldhous, P. (1990) Modified yeast fine for food. *Nature*, **344**, 186.

Angelov, A.I., Karadjov, G.I. and Roshkova, Z.G. (1996) Strains selection of bakers' yeast with improved technological properties. *Food Research International*, **29**, 235–239.

Asghari, A., Bothast, R.J., Doran, J.B. and Ingram, L.O. (1996) Ethanol production from hemicellulose hydrolysates of agricultural residues using genetically engineered *Escherichia coli* strain KO11. *Journal of Industrial Microbiology*, **16**, 42–47.

Balzi, E. and Goffeau, A. (1994) Genetics and biochemistry of yeast multidrug resistance. *Biochimica et Biophysica Acta*, **1187**, 152–162.

Barr, P.J., Brake, A.J. and Valenzuela, P. (1989) (eds) *Yeast Genetic Engineering*. Butterworths, Boston, USA.

Barre, P., Vezinhet, F., Dequin, S. and Blondin, B. (1993) Genetic improvement of wine yeast. In *Wine Microbiology and Biotechnology* (ed. G.H. Fleet), pp. 265–287. Harwood Academic Publishers, Chur, Switzerland.

Barron, N., Marchant, R., McHale, L. and McHale A.P. (1995) Studies on the use of a thermotolerant strain of *Kluyveromyces marxianus* in simultaneous saccharification and ethanol formation from cellulose. *Applied Microbiology and Biotechnology*, **43**, 518–520.

Barth, G. and Gaillardin, C. (1996) *Yarrowia lipolytica*. In *Nonconventional Yeasts in Biotechnology. A*

Handbook (ed. K. Wolf), pp. 313–388. Springer-Verlag, Berlin and Heidelberg.

Bashir, S. and Lee, S. (1994) Fuel ethanol production from agricultural lignocellulosic feedstocks – a review. *Fuel Science and Technology International*, **12**, 1427–1473.

Bathurst, I.C. (1994) Protein expression in yeast as an approach to production of recombinant malaria antigens. *American Journal of Tropical Medicine and Hygiene*, **50**, 20–26.

Beggs, J.D. (1978) Transformation of yeast by a replicating hybrid plasmid. *Nature*, **275**, 104–109.

Bemis, L.T., Geske, F.J. and Strange R. (1995) Use of the yeast two hybrid system for identifying the cascade of protein interactions resulting in apoptotic cell death. In *Methods in Cell Biology Vol. 46: Cell Death* (eds L.M. Schwartz and B.A. Osborne), pp. 139–151. Academic Press, San Diego, USA.

Benitez, T., Gasent-Ramirez, J.M., Gastrejor F. and Codon, A.C. (1996) Development of new strains for the food industry. *Biotechnology Progress*, **12**, 149–163.

Bennetto, H.P., Box, J., Delaney, G.M., Mason, J.R., Roller, S.D., Stirling, J.L. and Thurston, C.F. (1987) Redox-mediated electrochemistry of whole microorganisms: from fuel cells to biosensors. In *Biosensors: Fundamentals and Applications* (eds A.P.F. Turner, I. Karube and G.S. Wilson), pp. 291–314. Oxford University Press, Oxford.

Berry, D.R., Russell, I. and Stewart, G.G. (1987) (eds) *Yeast Biotechnology*. Allen and Unwin, London.

Beudeker, R.F., van Dam, H.W., van der Plaat, J.B. and Vellenga, K. (1990) Developments in Baker's yeast production. In *Yeast: Biotechnology and Biocatalysis* (eds H. Verachtert and R. DeMot), pp. 103–146. Marcel Dekker, New York and Basel.

Bilinski, C.A., Hatfield, D.E., Sobczak, J.A., Russell, I. and Stewart, G.G. (1987) Analysis of sporulation and segregation in a polyploid brewing strain of *Saccharomyces cerevisiae*. In *Biological Research on Industrial Yeasts. Vol. II* (eds G.G. Stewart, I. Russell, R.D. Klein and R.R. Hiebsch), pp. 37–47. CRC Press, Boca Raton, Florida, USA.

Bligh, H.F.J. and Kelly, S.L. (1990) Observations of an extreme form of late eccentric lytic fission in *Schizosaccharomyces pombe*. *FEMS Microbiology Letters*, **68**, 69–72.

Boekhout, T., van Belkum, A., Leenders, A.C.A.P., Verbrugh, H.A., Mukamurangwa, P., Swinne, D. and Scheffers, W.A. (1997) Molecular typing of *Cryptococcus neoformans*: taxonomic and epidemiological aspects. *Internati2nal Journal of Systematic Bacteriology*, **47**, 432–442.

Bohn, J.A. and BeMiller, J.N. (1995) (1-3)-β-D-Glucans as biological response modifiers: a review of structure–functional activity relationships. *Carbohydrate Polymers*, **28**, 3–14.

Brake, A.J. (1989) Secretion of heterologous proteins directed by the yeast α-factor leader. In *Yeast Genetic Engineering* (eds P.J. Barr, A.J. Brake and P. Valenzuela), pp. 269–280. Butterworths, Boston, USA.

Brauer, D. (1995) (ed.) *Biotechnology Vol. 12: Modern Biotechnology: Legal, Economic and Social dimensions*. VCH Publishers, Weinheim, Germany.

Burden, D.W. and Eveleigh, D.E. (1990) Yeast – diverse substrates and products. In *Yeast Technology* (eds J.F.T. Spencer and D.M. Spencer), pp. 198–227. Springer, Berlin.

Bustard, M., Bonnellan, N., Rollan, A., McHale, L. and McHale, A.P. (1996) The effect of pulse field strength on electric field stimulated biosorption of uranium by *Kluyveromyces marxianus* IMB 3. *Biotechnology Letters*, **18**, 479–482.

Butzke, C.E. and Bisson, L.F. (1996) Genetic engineering of yeast for wine production. *Agro Food Industry Hi-Tech*, **7**, 26–30.

Campanella, L., Favero, G. and Tomasseti, M. (1995) Immobilized yeast cells biosensor for total toxicity testing. *The Science of the Total Environment*, **171**, 227–234.

Cantwell, B.A. and McConnell, D.J. (1987) Transformation of brewing yeast strains. In *Biological Research in Industrial Yeast. Vol. II* (eds G.G. Stewart, I. Russell, R.D. Klein and R.R. Hiebsch), pp. 77–88. CRC Press, Boca Raton, Florida, USA.

Cantwell, B., Brazil G., Hurley, J. and McConnell, D. (1985) Expression of the gene for the endo-β-1,3-1,4-glucanase from *Bacillus subtilis* in *Saccharomyces cerevisiae*. European Brewery Convention. Proceedings of the 20th Congress, Helsinki, 1985. pp. 259–266. IRL Press at Oxford University Press, Oxford, UK.

Casey, G.P. (1990) Yeast selection in brewing. In *Yeast Strain Selection* (ed. C.J. Panchal), pp. 65–111. Marcel Dekker Inc., New York and Basel.

Cashin, M.-M. (1996) Comparative studies of five porous supports for yeast immobilization by adsorption/attachment. *Journal of the Institute of Brewing*, **102**, 5–10.

Chan, E.-C., Ueng, P.P., Eder, K.L. and Chen, L.F. (1989) Integration and expression of the *Escherichia coli* xylose isomerase gene in *Schizosaccharomyces pombe*. *Journal of Industrial Microbiology*, **4**, 409–418.

Chanda, S. and Chakabarti, S. (1996) Plant origin liquid waste: a resource for single cell protein production by yeast. *Bioresource Technology*, **57**, 51–54.

Chen, Y.L., Cino, J., Hart, G., Freedman, D., White, C. and Komives, E.A. (1997) High protein expression in fermentation of recombinant *Pichia pastoris* by a fed batch process. *Process Biochemistry*, **32**, 107–111.

Chigusa, K., Haswgawa, T., Yamamoto, N. and Watanabe, Y. (1996) Treatment of wastewater from oil manufacturing plant by yeast. *Water Science and Technology*, **34**, 51–58.

Christensen, B.E. (1987) Cross-breeding of distillers' yeast by hybridization of spore derived clones. *Carlsberg Research Communications*, **52**, 253–262.

Codon, A.C. and Benitez, T. (1995) Variability of the physiological features and of the nuclear genomes of bakers' yeasts. *Systematic and Applied Microbiology*, **18**, 343–352.

Coglan, A. (1992) Yeast gives semiconductors a lift. *New Scientist*, 25 January, 30.

Cooper, J.C., Nimmo, E.R., Allshire, R.C. and Cech, T.R. (1997) Regulation of telomere length and function by a Myb-domain protein in fission yeast. *Nature*, **385**, 744–747.

Crocq, V., Masson, C., Winter, J., Richard C., Lemaitre, G., Lenay, J., Vivat, M., Buendia, J. and Prat, D. (1997) Synthesis of trimegestrone: the first industrial application of baker's yeast mediated reduction of a ketone. *Organic Process Research and Development*, **1**, 2–13.

Curran, B.P.G. and Bugeja, V.C. (1993) Yeast cloning and biotechnology. In *Molecular Biology and Biotechnology*, 3rd edn (ed. J.M. Walker and E.B. Gingold), pp. 85–102. Royal Society of Chemistry, London.

D'Amore, T., Panchal, C.J., Russell, I. and Stewart, G.G. (1988) Osmotic pressure effects and intracellular accumulation of ethanol in yeast during fermentation. *Journal of Industrial Microbiology*, **2**, 365–372.

D'Amore, T., Ceoletto, G., Russell, I. and Stewart, G.G. (1989) Selection and optimization of yeast suitable for ethanol production at 40°C. *Enzyme and Microbial Technology*, **11**, 411–416.

Daniel, J. (1996) Measuring the toxic effects of high gene dosage in yeast cells. *Molecular and General Genetics*, **253**, 393–396.

Davies, S.E.C. and Brindle, K.H. (1992) Effects of overexpression of phosphofructokinase on glycolysis in the yeast *Saccharomyces cerevisiae*. *Biochemistry*, **31**, 4729–4735.

Decottignes, A. and Goffeau, A. (1997) Complete inventory of the yeast ABC proteins. *Nature Biotechnology*, **15**, 137–145.

Delitheos, A., Karavokyros, I. and Tiligada, E. (1995) Response of *Saccharomyces cerevisiae* strains to antineoplastic agents. *Journal of Applied Bacteriology*, **79**, 379–383.

Di Marco, S., Fendrich, G., Meyhack, B. and Grütter, M.G. (1996) Refolding, isolation and characterization of crystallizable human interferon-alpha 8 expressed in *Saccharomyces cerevisiae*. *Journal of Biotechnology*, **50**, 63–73.

Dohmen, R.J. and Hollenberg, C.P. (1996) *Schwanniomyces occidentalis*. In *Nonconventional Yeasts in Biotechnology*. A Handbook (ed. K. Wolf), pp. 117–137. Springer-Verlag, Berlin and Heidelberg.

Dombek, K.M. and Ingram, L.O. (1986) Magnesium limitation and its role in apparent toxicity of ethanol during yeast fermentation. *Applied and Environmental Microbiology*, **52**, 975–981.

Dujon, B. (1996) The yeast genome project – what did we learn? *Trends in Genetics*, **12**, 2263–270.

Echart, M.R. and Bussineau, C.M. (1996) Quality and authenticity of heterologous proteins synthesized in yeast. *Current Opinion in Biotechnology*, **7**, 525–530.

Erlichman, J. (1990) Taking a rise out of 'mutant' bread. *The Guardian* (UK daily newspaper). March 4th, 1990.

Evans, I. and McAthey, P. (1991) Comparative genetics of important yeasts. In *Genetically Engineered Proteins and Enzymes from Yeast: Production Control* (ed. A. Wiseman), pp. 11–74. Ellis Horwood Ltd., Chichester, UK.

Faber, K.N., Harder, W., Ab, G. and Veenhuis, M. (1995) Review – methylotrophic yeasts as factories for the production of foreign proteins *Yeast*, **11**, 1331–1344.

Facchinetti, F., Nappi, R.E., Sances, M.G., Neri, I., Grandinetti, G. and Genazzani, A. (1997) Effects of a yeast-based dietary supplementation on premenstrual syndrome. *Gynecologic and Obstetric Investigation*, **43**, 120–124.

Fang, T.J. and Chiou, T.Y. (1996) Batch cultivation and astaxanthin production by a mutant of the red yeast, *Phaffia rhodozyma* NCHU-FS501. *Journal of Industrial Microbiology*, **16**, 175–181.

Faust, U. (1987) Production of microbial biomass. In *Fundamentals of Biotechnology* (ed. P. Praeve), pp. 234–251. UCM Publishing Co., New York.

Fell, D. (1997) *Understanding the Control of Metabolism*. Portland Press. London and Miami.

Filonow, A.B., Vishniac, H.S., Anderson, J.A. and Janisiewicz, W.J. (1996) Biological control of *Botrytis cinerea* in apple by yeasts form various habitats and their putative mechanisms of antagonism. *Biological Control*, **7**, 212–220.

Finkleman, M.A.J. (1990) Yeast strain development for extracellular enzyme production. In *Yeast Strain Selection* (ed. C.J. Panchal), pp. 185–223. Marcel Dekker Inc., New York and Basel.

Fleet, G.H. (1993) (ed.) *Wine Microbiology and Biotechnology*. Harwood Academic Publishers, Chur, Switzerland.

Fleet, G.H. and Heard, G.M. (1993) Yeast – growth

during fermentation. In *Wine Microbiology and Biotechnology* (ed. G.H. Fleet), pp. 27–51. Harwood Academic Publishers, Chur, Switzerland.

Fogel, S. and Welch, J.W. (1987) Yeasts in biotechnology. In *Biological Research on Industrial Yeasts*. Vol. I (eds G.G. Stewart, I. Russell, R.D. Klein and R.R. Hiebsch), pp. 99–110. CRC Press, Boca Raton, Florida, USA.

Friedberg, T. and Wolf, C.R. (1996) Recombinant DNA technology as an investigative tool in drug metabolism research. *Advances in Drug Development*, **22**, 187–213.

Frommer, W.B. and Ninnemann, O. (1996) Heterologous expression of genes in bacterial, fungal, animal and plant cells. *Annual Reviews in Plant Physiology and Plant Molecular Biology*, **46**, 419–444.

Fujii, T., Nagasawa, N., Iwamatsu, A., Bogaki T., Tamai, Y and Hamachi, M. (1994) Molecular cloning, sequence analysis, and expression of the yeast alcohol acetyltransferase gene. *Applied and Environmental Microbiology*, **60**, 2786–2792.

Gasent-Ramirez, J.M., Codon, A.C. and Benitez, T. (1995) Characterization of genetically transformed *Saccharomyces cerevisiae* bakers' yeasts able to metabolize melibiose. *Applied and Environmental Microbiology*, **61**, 2113–2121.

Gellissen, G. and Melber, K. (1996) Methylotrophic yeast *Hansenula polymorpha* as production organism for recombinant pharmaceuticals. *Drug Research*, **46**, 943–948.

Glaser, V. (1996) Biotech firms focus toward therapeutic HIV vaccine development. *Genetic Engineering News*, **16**, 6.

Godia, F., Casas, C. and Sola, C. (1991) Application of immobilised yeast cells to sparkling wine fermentation. *Biotechnology Progress*, **7**, 468–470.

Goffeau, A., Barrell, B.G., Busey, A., Davis, R.W., Dujon, B., Feldmann, A., Galibert, F., Hoheisel, J.D., Jacq, C., Johnston, M., Louis, E.J., Mewes, H.W., Murakami, Y., Philippsen. P., Tettelin, H. and Oliver, S.G. (1996) Life with 6000 genes. *Science*, **274**, 546–567.

Goffeau, A., Park, J., Paulsen, I.T., Jonniaux, J.-L., Dinh, T., Mordant, P. and Saier, M.H. (1997) Multidrug-resistant transport proteins in yeast: complete inventory and phylogenetic characterization of yeast open reading frames within the major facilitator superfamily. *Yeast*, **13**, 43–54.

Gozalbo, D. (1992) Multiple copies of *SUC4* regulatory regions may cause partial de-repression of invertase synthesis in *Saccharomyces cerevisiae*. *Current Genetics*, **21**, 437–442.

Hadfield, C., Raina, K.K., Shashi-Menon, K. and Mount, R.C. (1993) The expression and performance of cloned genes in yeast. *Mycological Research*, **97**, 897–944.

Hall, M.N. and Linder, P. (1993) (eds) *The Early Days of Yeast Genetics*. Cold Spring Harbor Laboratory Press, Cold Spring Harbor, New York.

Hammond, J.R.M. (1993) Brewer's yeasts. In *The Yeasts. 2nd edn. Vol. 5: Yeast Technology* (eds A.H. Rose and J.S. Harrison), pp. 7–67. Academic Press, London.

Hammond, J.R.M. (1995) Genetically modified brewing yeast for the 21st Century – progress to date. *Yeast*, **11**, 1613–1627.

Hansen, H. and Hollenberg, C.P. (1996) *Hansenula polymorpha* (*Pichia angusta*). In *Nonconventional Yeasts in Biotechnology. A Handbook* (ed. K. Wolf), pp. 293–312. Springer-Verlag, Berlin and Heidelberg.

Hansen, J. and Kielland-Brandt, M.C. (1996a) Modification of biochemical pathways in industrial yeasts. *Journal of Biotechnology*, **49**, 1–12.

Hansen, J. and Kielland-Brandt, M.C. (1996b) Inactivation of *MET10* in brewers' yeast specifically increases SO_2 formation during beer production. *Nature Biotechnology*, **14**, 1587–1591.

Harrison, J.S. (1993) Food and fodder yeasts. In *The Yeasts. 2nd edn. Vol. 5: Yeast Technology* (eds A.H. Rose and J.S. Harrison), pp. 399–433. Academic Press, London.

Hassan, M., Blanc, P.J., Pareilleux, A. and Goma, G. (1994) Selection of fatty acid auxotrophs from the oleaginous yeast *Cryptococcus curvatus* and production of cocoa butter equivalents in batch culture. *Biotechnology Letters*, **16**, 819–824.

Hay, J.D. (1993) Novel yeast products from fermentation processes. *Journal of Chemical Technology and Biotechnology*, **58**, 203–205.

Hilt, W. and Wolf, D.H. (1992) Stress induced proteolysis in yeast. *Molecular Microbiology*, **6**, 2437–2442.

Hinchliffe, E. (1991) Strain improvement of brewing yeast. In *Applied Molecular Genetics of Fungi* (eds J.E. Perberdy, C.E. Caten, J.E. Ogden and J.W. Bennett), pp. 129–145. Cambridge University Press, Cambridge, UK.

Hinnen, A., Hicks, J.B. and Fink, G.R. (1978) Transformation of yeast. *Proceedings of the National Academy of Sciences* (*USA*), **75**, 1929–1933.

Hinnen, A., Buxton, F., Chandhuri, B., Hein, J., Hottiger, T., Meyhack, B. and Pohlig, G. (1995) Gene expression in recombinant yeast. In *Gene Expression in Recombinant Micro-organisms* (ed. A. Smith), pp. 121–193. Marcel Dekker Inc., New York.

Hirsch, H.H., Suarez Rendueles, P. and Wolf, D.H. (1989) Yeast (*Saccharomyces*) proteinases: structure, characteristics and function. In *Molecular and Cell Biology of Yeasts* (eds E.F. Walton and G.T. Yarranton), pp. 134–200. Blackie, London.

Hodgson, J. (1990) UK sweet on engineered yeast. *Bio/technology*, **8**, 281.

Hough, J.S., Briggs, D.E., Stevens, R. and Young, T.W. (1982) *Malting and Brewing Science*. 2nd edn. Chapman & Hall, London.

Huang, P., Stroffekova, K., Cuppoletti, J., Mahanty, S.K. and Scarborough, G.A. (1996) Functional expression of the cystic fibrosis transmembrane conductance regulator in yeast. *Biochimica et Biophysica Acta*, **1281**, 80–90.

Ingledew, W.M. (1987) *Schwanniomyces*: a potential super yeast? *Critical Reviews in Biotechnology*, **5**, 159–176.

Ingledew, W.M. (1993) Yeasts for production of fuel ethanol. In *The Yeasts. 2nd edn. Vol. 5: Yeast Technology* (eds A.H. Rose and J.S. Harrison), pp. 245–291. Academic Press, London.

Innis, M.A., McCabe, P.C., Cole, G.E., Whitman, V.P., Tal, R., Gelfand, B.H., Holland, M.J., Ben-Bassat, A., McRae, J., Inlow, D. and Meade, J.H. (1987) Expression of glucoamylase in yeast for fermentation of liquefied starch. In *Biological Research on Industrial Yeasts. Vol. I* (eds G.G. Stewart, I. Russell, R.D. Klein and R.R. Hiebsch), pp. 149–154. CRC Press. Boca Raton, Florida, USA.

Inose, T. and Murata, K. (1995) Enhanced accumulation of toxic compounds in yeast cells having high glycolytic activity. A case study on the safety of genetically engineered yeast. *International Journal of Food Science and Technology*, **30**, 141–146.

Ishioka, C., Englert, C., Winge, P., Yan, Y X., Englestein, M. and Friend, S.H. (1995) Mutational analysis of the carboxy terminal portion of p53 using both yeast and mammalian cell assays *in vivo*. *Oncogene*, **10**, 1485–1492.

Jackson, R.S. (1994) *Wine Science. Principles and Applications*. Academic Press, San Diego, USA.

Jamas, S., Easson, D.B., Ostroff, G.R. and Onderdonk, A.B. (1991) PGG-Glucans. A novel class of macrophage-activating immunomodulators. *American Chemical Society Symposia Series*, **469**, 44–51.

Javadekar, V.S., Siva Raman, H. and Gokhale, D.V. (1995) Industrial yeast strain improvement – construction of a highly flocculant yeast with a killer character by protoplast fusion. *Journal of Industrial Microbiology*, **15**, 94–102.

Johnson, E.A. and Schroeder, W.A. (1996) Biotechnology of astaxanthian production in *Phaffia rhodozyma*. *American Chemical Society Symposium Series*, **637**, 39–50.

Johnston, E.A. and Gil-Hwan, A. (1991) Astaxanthin from microbial sources. *Critical Reviews in Biotechnology*, **11**, 297–326.

Johnston, J.R. (1990) Brewing and distilling yeasts. In *Yeast Technology* (eds J.F.T. Spencer and D.M. Spencer), pp. 5–104. Springer-Verlag, New York.

Johnston, M. (1996) Towards a complete understanding of how a simple eukaryotic cell works. *Trends in Genetics*, **12**, 242–243.

Jones, A.M. and Ingledew, W.M. (1994) Fuel alcohol production: appraisal of nitrogenous yeast foods for very high gravity wheat mash fermentation. *Process Biochemistry*, **29**, 483–488.

Kavanagh, K. and Whittaker, P.A. (1996) Application of protoplast fusion to the nonconventional yeast. *Enzyme and Microbial Technology*, **18**, 45–51.

Kielland-Brandt, M.C., Nilsson-Tillgren, T., Gjermansen, C., Holmberg, S. and Pedersen, M. (1995) Genetics of brewing yeasts. In *The Yeasts. 2nd edn. Vol. 6: Yeast Genetics* (eds A.E. Wheals, A.H. Rose and J.S. Harrison), pp. 223–254. Academic Press, London.

Kirsop, B. (1982) Developments in beer fermentation. *Topics in Enzyme and Fermentation Biotechnology*, **6**, 79–131.

Klein, R.D. and Zaworski, P.G. (1990) Transformation and cloning systems in non-*Saccharomyces* yeasts. In *Yeast Strain Selection* (ed. C.J. Panchal), pp. 245–310. Marcel Dekker Inc., New York and Basel.

Knowles, J., Penttila, M., Teeri, T., Andre, L., Lehtovaara, P. and Salovuori, I. (1987) The development of cellulolytic yeasts and their possible applications. In *Biological Research in Industrial Yeasts. Vol. I* (eds G.G. Stewart, I. Russell, R.D. Klein and R.R. Hiebsch), pp. 189–199. CRC Press, Boca Raton, Florida, USA.

Kodama, Y., Fukui, N., Ashikari, T., Sibano, T., Morioka-Fujimoto, K., Hikari, Y and Nakatani, K. (1995) Improvement of maltose fermentation efficiency: constitutive expression of *MAL* genes in brewing yeasts. *Journal of the American Society of Brewing Chemists*, **53**, 24–29.

Kometani, I., Yoshii, H. and Matsuno, R. (1996) Large-scale production of chiral alcohols with bakers' yeast. *Journal of Molecular Catalysis B: Enzymatic*, **1**, 45–52.

Kukuruzinska, M.A., Bergh, M.L. and Jackson, B.J. (1987) Protein glycosylation in yeast. *Annual Reviews of Biochemistry*, **56**, 915–944.

Kunze, G. and Kunze, I. (1996) *Arxula adeninivorans*. In *Nonconventional Yeasts in Biotechnology. A Handbook* (ed. K. Wolf), pp. 389–409. Springer-Verlag, Berlin and Heidelberg.

Kwon, K.S., Song, M.Y and Yu, M.H. (1995) Purification and characterization of alpha (1)-antitrypsin secreted by recombinant yeast *Saccharomyces diastaticus*. *Journal of Biotechnology*, **42**, 191–195.

Lancashire, W.E. and Hadfield, C. (1986) *Tagging of micro-organisms*. European Patent Application No. 86309272.2.

Lea, A.H. and Piggott, J.R.P. (1995) *Fermented Bever-*

age Production. Blackie Academic and Professional, Glasgow, UK.

Lewis, M.J. and Young, T.W. (1995) *Brewing*. Chapman & Hall, London.

Lindquist, S. (1996) Mad cows meet mad yeast: the prion hypothesis. *Molecular Psychiatry*, **1**, 376–379.

Lyons, T.P., Jacques, K.A. and Dawson, K.A. (1993) Miscellaneous products from yeast. In *The Yeasts. 2nd edn. Vol. 5: Yeast Technology* (eds A.H. Rose and J.S. Harrison), pp. 293–324. Academic Press, London.

Macreadie, I.G., Castelli, L.A., Lucantoni, A. and Azad, A.A. (1995) Stress- and sequence-dependent release into the culture medium of HIV-1 Nef produced in *Saccharomyces cerevisiae*. *Gene*, **162**, 239–243.

Madhyastha, M.S., Marquardt, R.R., Frohlich, A.A. and Borga, J. (1994) Optimization of yeast bioassay for trichothecene mycotoxins. *Journal of Food Protection*, **57**, 490–495.

Markie, D. (1996) (ed.) *Methods in Molecular Biology. Vol. 54: YAC Protocols*. Humana Press, Totowa, New Jersey, USA.

Martini, A. (1993) Origin and domestication of the wine yeast *Saccharomyces cerevisiae*. *Journal of Wine Research*, **4**, 165–176.

Masschelein, C.A., Ryder, D.S. and Simon, J.-P. (1994) Immobilized cell technology in beer fermentation. *Critical Reviews in Biotechnology*, **14**, 155–177.

Mattey, M. (1992) The production of organic acids. *Critical Reviews in Biotechnology*, **12**, 87–132.

Mauersberger, S., Ohkuma, M., Schunck, W.-H. and Takagi, M. (1996) *Candida maltosa*. In *Nonconventional Yeasts in Biotechnology. A Handbook* (ed. K. Wolf), pp. 411–580. Springer-Verlag, Berlin and Heidelberg.

Mawson, A.J. (1994) Bioconversions for whey utilization and waste abatement. *Bioresource Technology*, **47**, 195–203.

McAthey, P (1996) Genotoxicity in *Schizosaccharomyces pombe*. In *Methods in Molecular Biology. Vol. 53: Yeast Protocols* (ed. I.H. Evans), pp. 343–353. Human Press, Totowa, New Jersey, USA.

McFarland, L.V. and Elmer, G.W. (1995) Biotherapeutic agents: past, present and future. *Microecology and Therapy*, **23**, 46–73.

Meaden, P.G. (1990) DNA fingerprinting of brewers' yeast: current perspectives. *Journal of the Institute of Brewing*, **96**, 195–200.

Meaden, P.G. (1996) DNA fingerprinting of brewers' yeast. *Ferment*, **9**, 267–272.

Meinander, N., Hallborn, J., Keränen, S., Ojamo, S., Penttilä, M., Walfridsson, M. and Hahn-Hägerdal, B. (1994) Utilization of xylose with recombinant *Saccharomyces cerevisiae* harbouring genes for xylose metabolism from *Pichia stipitis*. *Proceedings of the 6th European Congress in Biotechnology*. pp. 1143–1146. Elsevier Science B.V., Amsterdam.

Mendoza-Vega, O., Sabatié, J. and Brown, S.W., (1994) Industrial production of proteins by fed-batch culture of the yeast *Saccharomyces cerevisiae*. *FEMS Microbiology Reviews*, **15**, 369–410.

Moses, V. and Moses, S. (1995) *Exploiting Biotechnology*. Harwood Academic Publishers, Chur, Switzerland.

Nagodawithana, T.W. and Trivedi, N.B. (1990) Yeast selection for baking. In *Yeast Strain Selection* (ed. C.J. Panchal), pp. 139–184. Marcel Dekker Inc., New York and Basel.

Naumov, G.I., Naumova, E.S., Lantto, R.A., Louis, E.J. and Korhola, M. (1992) Genetic homology between *Saccharomyces cerevisiae* and its sibling species *S. paradoxus* and *S. bayanus*: electrophoretic karyotypes. *Yeast*, **8**, 599–612.

Ness, F., Lavallée, F., Dubourdieu, D., Aigle, M. and Dulau, L. (1993) Identification of yeast strains using the polymerase chain reaction. *Journal of the Science of Food and Agriculture*, **62**, 89–94.

Neujahr, H.Y. (1990) Yeasts in biodegradation and biodeterioration processes. In *Yeast: Biotechnology and Biocatalysis* (eds H. Verachtert and R. DeMot), pp. 321–348. Marcel Decker, New York and Basel.

Newbold, C.J., Wallace, R.J. and McIntosh, F.M. (1996) Mode of action of the yeast *Saccharomyces cerevisiae* as a feed additive for ruminants. *British Journal of Nutrition*, **76**, 249–261.

Newswatch (1990) A yeast release. A genetic engineering first for the food industry. *The Genetic Engineer and Biotechnologist*, **May/June**, 24–25.

Norton, S. and D'Amore, T. (1994) Physiological effects of yeast cell immobilization: applications for brewing. *Enzyme and Microbial Technology*, **16**, 365–375.

Nurse, P. (1985) Yeast aids cancer research. *Nature*, **313**, 631–632.

Oakley Gutowski, K.M., Hawthorne, D.B. and Kavanagh, T.E. (1992) Application of chromosome fingerprinting to the differentiation of brewing yeasts. *Journal of the American Society of Brewing Chemists*, **50**, 48–52.

Oda, Y. and Ouchi, K. (1990) Hybridization of baker's yeast by the rare mating method to improve leavening ability in dough. *Enzyme and Microbial Technology*, **12**, 989–993.

Oda, Y. and Tonomura, K. (1993) Selection of a novel baking strain from the *Torulaspora* yeasts. *Bioscience Biotechnology and Biochemistry*, **57**, 1320–1322.

Oda, Y., Uno, K. and Ohta, S. (1986) Selection of yeast for bread-making by the frozen-dough method. *Applied and Environmental Microbiology*, **52**, 941–943.

O'Kennedy, R., Houghton, C.J. and Patching, J.W. (1995) Effects of growth environment on recombinant plasmid stability in *Saccharomyces cerevisiae* grown in continuous culture. *Applied and Environmental Biotechnology*, **44**, 126–132.

Oliver, S.G. (1996a) From DNA sequence to biological function. *Nature*, **379**, 597–600.

Oliver, S.G. (1996b) A network approach to the systematic analysis of yeast gene-function. *Trends in Genetics*, **12**, 241–242.

Olsson, L and Hahn-Hägerdal, B. (1996) Fermentation of lignocellulosic hydrolysates for ethanol production. *Enzyme and Microbial Technology*, **18**, 312–331.

Omar, N.B., Merroun, M.L.I., Gonzalez-Munoz, M.T. and Arias, J.M. (1996) Brewing yeasts as a biosorbent for uranium. *Journal of Applied Bacteriology*, **81**, 283–287.

Omori, T., Ogawa, K. and Shimoda, M. (1995) Breeding of high glycerol producing *shochu* yeast (*Saccharomyces cerevisiae*) with acquired salt-tolerance. *Journal of Fermentation and Bioengineering*, **79**, 560–565.

Oppenoorth, W.F.F. (1959) Modification of the hereditary character of yeast by ingestion of cell-free extracts. *European Brewery Convention. Proceedings of the 7th Congress, Rome 1959*. pp. 180–207. Elsevier Science Publishing Co., Amsterdam, The Netherlands.

Ortiz, C.P., Steyer, J.P. and Bories, A. (1997) Carbon and nitrogen removal from waste water by *Candida utilis* – kinetics aspects and mathematical modelling. *Process Biochemistry*, **32**, 179–189.

Panchal, C.J. and Tavares, F.C.A. (1990) Yeast strain selection for fuel ethanol production. In *Yeast Strain Selection* (ed. C.J. Panchal), pp. 225–243. Marcel Decker Inc., New York and Basel.

Parent, S.A. and Bostian, K.A. (1995) Recombinant DNA technology: yeast vectors. In *The Yeasts. 2nd edn. Vol. 6: Yeast Genetics* (eds A.E. Wheals., A.H. Rose and J.S. Harrison), pp. 121–178. Academic Press, London.

Payne, W.E. and Garrels, J.I. (1997) Yeast protein database (YPD): a database for the complete proteome of *Saccharomyces cerevisiae*. *Nucleic Acids Research*, **25**, 57–62.

Pedersen, M.B. (1986) DNA sequence polymorphisms in the genus *Saccharomyces*. III. Restriction endonuclease fragment patterns of chromosomal regions in brewing and other yeast strains. *Carlsberg Research Communications*, **51**, 163–183.

Pedersen, M.B. (1987) Practical use of electro-karyotypes for brewing yeast identification. European Brewery Convention. Proceedings of the 21st Congress, Madrid 1987. pp. 489–496. IRL Press at Oxford University Press, Oxford, UK.

Pedersen, M.B. (1994) Molecular analyses of yeast DNA – tools for pure yeast maintenance in the brewery. *Journal of the American Society of Brewing Chemists*, **52**, 23–27.

Perry, C. and Meaden, P. (1988) Properties of a genetically engineered dextrin-fermenting strain of brewer's yeast. *Journal of the Institute of Brewing*, **94**, 64–67.

Pfeiffer, N. (1997) New yeast discovery hints at improved cancer treatment. *Genetic Engineering News*, **17**, 1, 15.

Picataggio, S., Rohrer, T. Deanda, K., Lancing, D., Reynolds, R., Mielenz, J. and Eirich, L.D. (1993) Metabolic engineering of *Candida tropicalis* for the production of long-chain dicarboxylic acids. *Bio/technology*, **10**, 894–898.

Piper, P.W and Kirk, N. (1991) Inducing heterologous gene expression in yeast as fermentations approach maximal biomass. In *Genetically Engineered Proteins and Enzymes from Yeast: Production Control* (ed. A. Wiseman), pp. 147–189. Ellis Horwood, Chichester, UK.

Pompon, D., Perret, A., Bellamine, A., Laine, R., Gautier, J.C. and Urban, P. (1995) Genetically engineered yeasts and their applications. *Toxicology Letters*, **82/83**, 815–822.

Popolo, L., Vigano, F., Erba, E., Mongelli, N. and D'Incalci, M. (1996) Effects of a novel DNA-damaging agent on the budding yeast *Saccharomyces cerevisiae* cell cycle. *Yeast*, **12**, 349–359.

Porro, D., Martegani, E., Ranza, B.M. and Alberghina, L. (1992) Development of high cell density cultures of engineered *Saccharomyces cerevisiae* able to grow on lactose. *Biotechnology Letters*, **14**, 1085–1088.

Priest, F.G. and Campbell, I. (1996) (eds) *Brewing Microbiology*. 2nd edn. Chapman & Hall, London.

Querol, A. and Ramon, D. (1996) The application of molecular techniques in wine microbiology. *Trends in Food Science and Technology*, **7**, 73–78.

Rallabhandi, P. and Yu, P.-L. (1996) Production of therapeutic proteins in yeasts: a review. *Australasian Biotechnology*, **6**, 230–237.

Reed, G. and Nagodawithana, T.W. (1991) Yeast-derived products. In *Yeast Technology*. Chapter 8. pp. 369–412. AVI/Van Nostrand Reinhold, New York.

Reglinski, T., Lyon, G.D. and Newton, A.C. (1994) Induction of resistance mechanisms in barley by yeast-derived elicitors. *Annals of Applied Biology*, **124**, 509–517.

Reiser, J., Ochsner, U.A., Kälin, M., Glumoff, V. and Fiechter, A. (1996) *Trichosporon*. In *Nonconventional Yeasts in Biotechnology. A Handbook* (ed. K.

Wolf), pp. 581–606. Springer-Verlag, Berlin and Heidelberg.

Resnick, M.A. Bennett, C., Perkins, E., Porter, G. and Priebe, S.D. (1995) Double-strand breaks and recombinant repair: the role of processing, signalling and DNA homology. In *The Yeasts. 2nd edn. Vol. 6: Yeast Genetics* (eds A.E. Wheals, A.H. Rose and J.S. Harrison), pp. 341–410. Academic press, London.

Romanos, M. (1995) Advances in the use of *Pichia pastoris* for high level gene expression. *Current Opinion in Biotechnology*, **6**, 527–533.

Romanos, M.A., Scorer, C.A. and Clare, J.J. (1992) Foreign gene expression in yeast: a review. *Yeast*, **8**, 423–488.

Rose, A.H. (1986) Alcohol production and alcohol tolerance. In *Proceedings of the 2nd Aviemore Conference on Malting, Brewing and Distilling* (eds I. Campbell and F.G. Priest), pp. 92–106. Institute of Brewing, London.

Rose, A.H. and Harrison, J.S. (1993) (eds) *The Yeasts. 2nd edn. Vol. 5: Yeast Technology*. Academic Press, London.

Rose, A.H. and Vijaylakshimi, G. (1993) Baker's yeasts. In *The Yeasts. 2nd edn. Vol. 5. Yeast Technology* (eds A.H. Rose and J.S. Harrison), pp. 357–397. Academic Press, London.

Rose, M.D. (1995) Modern and post-modern genetics in *Saccharomyces cerevisiae*. In *The Yeasts. 2nd edn. Vol. 6: Yeast Genetics* (eds A.E. Wheals, A.H. Rose and J.S. Harrison), pp. 69–120. Academic Press, London.

Rosen, K. (1989) Preparation of yeast for industrial use in the production of beverages. In *Biotechnology Applications in Beverage Production* (eds C. Cantarelli and G. Lanzarini), pp. 169–187. Elsevier Applied Science, London.

Routledge, E.J. and Sumpter, J.P. (1996) Oestrogen activity of surfactants and some of their degradation products assessed using a recombinant yeast screen. *Environmental Toxicology and Chemistry*, **15**, 241–248.

Russell, I. and Stewart, G.G. (1979) Spheroplast fusion of brewer's yeast strains. *Journal of the Institute of Brewing*, **85**, 95–98.

Russell, I. and Stewart, G.G. (1995) Brewing. In *Biotechnology. Vol. 9: Enzymes, Biomass, Food and Feed* (eds G. Reed and T.W. Nagodawithana), pp. 419–462. VCH, Weinheim, Germany.

Russell, I., Jones, R.M and Stewart, G.G. (1987) Yeast – the primary industrial micro-organism. In *Biological Research on Industrial Yeasts. Vol. I* (eds G.G. Stewart, I. Russell, R.D. Klein and R.R. Hiebsch), pp. 1–20. CRC Press, Boca Raton, Florida, USA.

Saha, B.C. and Bothast, R.J. (1996) Production of L-arabinitol from L-arabinose by *Candida entomaea* and *Pichia guilliermondii*. *Applied Microbiology and Biotechnology*, **45**, 299–306.

Salek, A., Schnettler, R. and Zimmerman, U. (1992) Stably inherited killer activity in industrial yeast strains obtained by electrotransformation. *FEMS Microbiology Letters*, **96**, 103–110.

Samuel, D. (1996) Archaeology of ancient Egyptian beer. *Journal of the American Society of Brewing Chemists*, **54**, 3–12.

Schaaf, I., Heinisch, J. and Zimmermann, F.K. (1989) Overproduction of glycolytic enzymes in yeast. *Yeast*, **5**, 285–290.

Schofield, M.A., Rowe, S.M., Hammond, J.R.M., Molzahn, S.W. and Quain, D.E. (1995) Differentiation of brewery yeast strains by DNA fingerprinting. *Journal of the Institute of Brewing*, **101**, 75–78.

Schreuder, M.P., Deen, C., Boersma, W.J.A., Pouwells, P.H. and Klis, F.M. (1996) Yeast expressing hepatitis B virus surface antigen determinants on its surface: implications for a possible oral vaccine. *Vaccine*, **14**, 383–388.

Servi, S. (1990) Baker's yeast as a reagent in organic synthesis. *Synthesis*, **January**, 1–25.

Shallom, J.M and Golin, J. (1996) The unusual inheritance of multidrug resistance factors in *Saccharomyces*. *Current Genetics*, **30**, 212–217.

Shama, G. (1988) Developments in bioreactors for fuel ethanol production. *Process Biochemistry*, **23**, 138–145.

Shinohara, T., Mamiya, S. and Yanagida F. (1997) Introduction of flocculation property into wine yeast (*Saccharomyces cerevisiae*) by hybridisation. *Journal of Fermentation and Bioengineering*, **83**, 96–101.

Sibirny, A.A. (1996a) *Pichia guillliermondii*. In *Nonconventional Yeasts in Biotechnology. A Handbook* (ed. K. Wolf), pp. 255–275. Springer-Verlag, Berlin and Heidelberg.

Sibirny, A.A. (1996b) *Pichia methanolica* (*Pichia pinus* MH4). In *Nonconventional Yeasts in Biotechnology. A Handbook* (ed. K. Wolf), pp. 277–291. Springer-Verlag, Berlin and Heidelberg.

Simmons, P., Tobin, J.M. and Singleton, I. (1995) Considerations on the use of commercially available yeast biomass for the treatment of metal-containing effluents. *Journal of Industrial Microbiology*, **14**, 240–246.

Singh, A. and Mishra, P. (1995) *Microbial Pentose Utilization. Current Applications in Biotechnology*. Elsevier, Amsterdam, The Netherlands.

Singleton, I. and Simmons, P. (1996) Factors affecting silver biosorption by an industrial strain of *Saccharomyces cerevisiae*. *Journal of Chemical Technology and Biotechnology*, **65**, 21–28.

Sirisananeeyakul, S., Staniszewski, M. and Rizzi, M. (1996) Screening of yeasts for production of xylitol from D-xylose. *Journal of Fermentation and Bioengineering*, **80**, 565–570.

Siso, M.I.G. (1996) The biotechnological utilization of cheese whey: a review. *Bioresource Technology*, **57**, 1–11.

Sreekrishna, K. and Dickson, R.C. (1985) Construction of strains of *Saccharomyces cerevisiae* that grow on lactose. *Proceedings of the National Academy of Sciences (USA)*, **82**, 7909–7913.

Sreekrishna, K. and Kropp, K.E. (1996) *Pichia pastoris*. In *Nonconventional Yeasts in Biotechnology. A Handbook* (ed. K. Wolf), pp. 203–253. Springer-Verlag, Berlin and Heidelberg.

Stahmann, K.-P. (1997) Vitamins, amino acids. In *Fungal Biotechnology* (ed. T. Anke), pp. 81–90. Chapman & Hall, Weinheim, Germany.

Stewart, G.G. and Russell, I. (1986) One hundred years of yeast research and development in the brewing industry. *Journal of the Institute of Brewing*, **92**, 537–538.

Stewart, G.G. and Russell, I. (1995) Brewers' yeast. In *Food Biotechnology* (eds Y.H. Hui and G.G. Khachatourians), pp. 847–871. VCH, New York.

Stewart, J.D., Reed, K.W., Zhu, J., Chen, G. and Kayser M.M. (1996) A 'designer yeast' that catalyzes the kinetic resolutions of 2-alkyl substituted cyclohexanones by enantioselective Baeyer–Villiger oxidations. *Journal of Organic Chemistry*, **61**, 7652–7653.

Subden, R.E. (1990) Wine yeast: selection and modification. In *Yeast Strain Selection* (ed. C.J. Panchal), pp. 113–137. Marcel Dekker Inc., New York and Basel.

Sudbery, P.E. (1994) The non-*Saccharomyces* yeasts. *Yeast*, **10**, 1707–1726.

Sudbery, P.E. (1995) Genetics of industrial yeasts other than *Saccharomyces cerevisiae* and *Schizosaccharomyces pombe*. In *The Yeasts. 2nd edn. Vol. 6: Yeast Genetics* (eds A.E. Wheals, A.H. Rose and J.S. Harrison), pp. 255–283. Academic Press, London.

Szczodrak, J. and Fiedurek, J. (1996) Technology for conversion of lignocellulosic biomass to ethanol. *Biomass and Bioenergy*, **10**, 367–375.

Tada, M., Iggo, R.D., Ishii, N., Shinde, Y., Sakuma, S., Estreicher, A., Sawamura, Y. and Abe, H. (1996) Clonality and stability of the p53 gene in human astrocytic tumour cells – quantitative analysis of p53 gene mutations by yeast functional assay. *International Journal of Cancer*, **67**, 447–450.

Tada, S., Takeuchi, T., Sone, H., Yamano, S., Schofield, M.A., Hammond, J.R.M. and Inoue, T. (1995) Pilot scale brewing with industrial yeasts which produce the α-acetolactate decarboxylase of *Acetobacter aceti* ssp. *xylinium*. European Brewery Convention. Proceedings of the 25th congress, Brussels, 1995. pp. 369–376. IRL Press at Oxford University Press, Oxford, UK.

Tavares, F.C.A. and Echeverrigaray, S. (1987) Yeast breeding for fuel ethanol production. In *Biological Research on Industrial Yeasts. Vol. I* (eds G.G. Stewart, I. Russell, R.D. Klein and R.R. Hiebsch), pp. 59–80. CRC Press. Boca Raton, Florida, USA.

Tezuka, H., Mori, T., Okumura, Y., Kitabatake, K. and Tsumura, Y. (1992) Cloning of a gene suppressing hydrogen sulphide production by *Saccharomyces cerevisiae* and its expression in a brewing yeast. *Journal of the American Society of Brewing Chemists*, **50**, 130–133.

Thomas, K.C. and Ingledew, W.M. (1992) Production of 21% ethanol by fermentation of very high gravity wheat mashes. *Journal of Industrial Microbiology*, **10**, 61–68.

Thomas, K.C., Hynes, S.H. and Ingledew, W.M. (1996) Practical and theoretical considerations in the production of high concentrations of alcohol by fermentation. *Process Biochemistry*, **31**, 321–331.

Torabizadeh, H., Shojaosadati, S.A. and Tehrani, H.A. (1996) Preparation and characterization of bioemulsifier from *Saccharomyces cerevisiae* and its application in food products. *Lebensmittel Wissenschaft und Technologie*, **29**, 734–737.

Tubb, R.S. (1986) Amylolytic yeasts for commercial applications. *Trends in Biotechnology*, **4**, 98–104.

Van der Vaart, J.M., TeBiesebeke, R., Chapman, J.W., Toschka, H.Y., Klis, F.M. and Verrips, C.T. (1997) Comparison of cell wall proteins of *Saccharomyces cerevisiae* as anchors for cell surface expression of heterologous proteins. *Applied and Environmental Microbiology*, **63**, 615–620.

Vasavada, A. (1995) Improving productivity of heterologous proteins in recombinant *Saccharomyces cerevisiae* fermentations. *Advances in Applied Microbiology*, **41**, 25–54.

Vaughan-Martini, A., Martini, A. and Cardinali, G. (1993) Electrophoretic karyotyping as a taxonomic tool in the genus *Saccharomyces*. *Antonie van Leeuwenhoek*, **63**, 157–163.

Wainwright, M. (1992) *An Introduction to Fungal Biotechnology*. John Wiley and Sons, Chichester, UK.

Walker, J.M. and Gingold, E.B. (1993) (eds) *Molecular Biology and Biotechnology*. 3rd edn. Royal Society of Chemistry, London.

Walker, G.M., Birch, R.M., Chandrasena, G. and Maynard, A.I. (1996) Magnesium, calcium and fermentative metabolism in industrial yeasts. *Journal of the American Society of Brewing Chemists*, **54**, 13–18.

Walmsley, R.M. (1994) DNA fingerprinting of yeast. *Ferment*, **7**, 231–234.

Walmsley, R.M., Wilkinson, B.M. and Kong, T.H.

(1989) Genetic fingerprinting for yeasts. *Bio/technology*, **7**, 1168–1170.

Ward, A.C. and Macreadie, I.G. (1996) Yeasts and human virology. *Today's Life Science*, **August**, 46–53.

Watanabe, M., Fukuda, K., Asano, K. and Ohta, S. (1990) Mutants of baker's yeast producing a large amount of isobutyl alcohol or isoamyl alcohol, flavour components of bread. *Applied Microbiology and Biotechnology*, **34**, 154–159.

Watari, J., Nomura, M., Sahara, H., Koshino, S. and Keranen, S. (1994) Construction of flocculent brewer's yeast by chromosomal integration of the yeast flocculation gene *FLO 1*. *Journal of the Institute of Brewing*, **100**, 73–77.

Watson, D.C. (1993) Yeasts in distilled alcoholic-beverage production. In *The Yeasts. 2nd edn. Vol. 5: Yeast Technology* (eds A.H. Rose and J.S. Harrison), pp. 215–244. Academic Press, London.

Watts, J.W. and Stacey N.J. (1991) Novel methods of DNA transfer. In *Applied Molecular Genetics of Fungi* (eds J.F. Peberdy, C.E. Caten, J.E. Ogden and J.W. Bennett), pp. 44–65. Cambridge University Press, Cambridge, UK.

Weber, G., Adamczyk, A. and Freytag, S. (1989) Treatment of acne with a yeast preparation. *Fortschr. Med.*, **107**, 563–566.

Wenzl, R., Burgstaller, W. and Sckinner, F. (1990) Extraction of zinc, copper and lead form a filter dust by yeasts. *Biorecovery*, **2**, 1–13.

Wésolowski-Louvel, M., Breunig, K.D. and Fukuhara, H. (1996) *Kluyveromyces lactis*. In *Nonconventional Yeasts in Biotechnology. A Handbook* (ed. K. Wolf), pp. 139–201. Springer-Verlag, Berlin and Heidelberg.

Wightman, P., Quain, D.E. and Meaden, P.G. (1996) Analysis of production brewing strains of yeast by DNA fingerprinting. *Letters in Applied Microbiology*, **22**, 90–94.

Wiseman, A. (1987) (ed.) *Enzyme Induction, Mutagen Activation and Carcinogen Testing in Yeast*. Ellis Horwood Ltd., Chichester, UK.

Wiseman, A. (1996) Therapeutic proteins and enzymes from genetically engineered yeasts. *Endeavour*, **20**, 130–132.

Wolf, K. (1996) (ed.) *Nonconventional Yeasts in Biotechnology. A Handbook*. Springer-Verlag, Berlin and Heidelberg.

Young, T.W. (1981) The genetic manipulation of killer character into brewing yeast. *Journal of the Institute of Brewing*, **87**, 292–295.

Young, T.W. and Hosford, E.A. (1987) Genetic manipulation of *Saccharomyces cerevisiae* to produce extracellular protease. European Brewery Convention. Proceedings of the 21st Congress, Madrid 1987. pp. 521–528. IRL Press at Oxford University press, Oxford, UK.

Zhang, Z., Moo-Young, M. and Chisti, Y. (1996) Plasmid stability in recombinant *Saccharomyces cerevisiae*. *Biotechnology Advances*, **14**, 401–435.

Zimmerman, M. and Fournier, P. (1996) Electrophoretic karyotyping of yeast. In *Nonconventional Yeasts in Biotechnology. A Handbook* (ed. K. Wolf), pp. 101–116. Springer-Verlag, Berlin and Heidelberg.

Zimmerman, M. and Sipiczki, M. (1996) Protoplast fusion of yeast. In *Nonconventional Yeasts in Biotechnology*. A Handbook (ed. K. Wolf), pp. 83–99. Springer-Verlag, Berlin and Heidelberg.

Index

ABC proteins 310
ABT 14
acceleration phase 133
acetaldehyde 165-166
Acetobacter spp. 172
acetylation 275
N-acetylglucosamine phosphate mutase 224
Aciculoconidium
  taxonomy 2
acid phosphatase 37
acidification power 142
acidification power test 84
Acinetobacter 254
F-actin 31
G-actin 31
actin cytoskeleton 105
actin-directed secretory vesicles 102
actin filament 114
actin immunofluorescence 14
actin patches 114
active dried yeast 301
active transport 64-65
acyclic polyols 74
adaptive oxidant stress response 167
adaptive stress response 167
ADE2 276
adenylate cyclase 249
adenylate energy charge 142
ADH 272
ADH 1 275
ADH 2 275, 279
adjuncts 73
adrenoleukodystrophy 310
Advisory Committee on Novel Foods and Processes (ACNFP) 282
Aequorea victoria 13
aequorin 88
AFT 1 gene 289
Agave tequilana 294
agglomeration 27-28
AIDS 8
air lift culture 145
alanine transporter 79
alcian blue 14
alcohol dehydrogenase 88, 205, 223, 225
alcohol electrode 288
alcohol metabolism 231-234
alcohol oxidase 29, 233
alcoholic beverages 283-294
alcoholic fermentation 206, 220
aldehydes 207
ale yeast strains 286
alginate 146
alginate beads 111
aliphatic hydrocarbons 165
alkaline and acid phosphatases 249, 250
n-alkane metabolism 75, 234-235
alkylating agents 168
allosamidin 178
allylamines 178
alpha 1-antitrypsin 308
alternative respiration 211
Ambrosiozyma
  taxonomy 2
American whiskey 290, 292, 294
amino acid families 244
amino acid metabolism 244-246
amino acid transporters 79
9-aminoacridine 14
aminopeptidase 245
ammonium catabolite repression 78
ammonium ion repression 78-79
ammonium ions assimilation 241-243
amphibolic pathway 209, 222

amphotericin B 172
amyloglucosidase 276
amylolytic yeasts 54, 62
analytical profile index 3
anamorphic 2
anaplerotic reactions 209
anhydrobiosis 154
aniline blue 14
antagonism 169
anti-cancer drug 123
anti-*Candida albicans* agent 304
anti-*Candida* drugs 80
antidesiccation 182
antifreeze peptides 153
anti-mitotic drugs 123
antimycotics 177-178
antioxidant defence 168
antioxidant enzymes 152, 167
antioxidant genes 169
antioxidant in beer 253
antioxidation 182
antithrombin III 308
α1-antitrypsin 281
anti-yeast agents 85, 177-178
antizymosis 172
antizymotic agents 172, 178, 238
AOX 275
AOX 1 279
apoptosis 175
apoptotic cell death 180
aprotinin 308
aquaporin 66
aquavit 292, 294
arabinitol 74
arabinitol in osmoregulation 159
arabinose fermentation 231
D-arabinose transport 70
ARE 275
arginine transporters 79
Armagnac 292, 294
Arthroascus
  taxonomy 2
Arthroascus javanensis
  in yeast predation 174
Arxiozyma
  taxonomy 2
Arxula
  taxonomy 2
Arxula adeninivorans
  carbon sources 54
  industrial uses 267
Arxula terrestre
  carbon sources 54
Ascoidea
  taxonomy 2
Ascomycotina *2*
Ashbya
  taxonomy 2
Ashbya gossypii
  plant pathogen 170
asparagine transporter 79
Aspergillus awamori 298
Aspergillus oryzae 294
Aspergillus spp. 289, 305
astaxanthin 13, 169
astral microtubules 108-110
asynchronous batch culture 139
asynchronous continuous culture 139
ataxia telangectasia 310
atomic force microscopy 16-17
ATP bioluminescence 141-142
ATP citrate lyase 236
ATP hydrolases 249
ATP synthase 210
ATP synthetases 249
H^+-ATPase 83-84
ATPase
  proton pumping 20, 178
  temperature effects 150
atypical budding 150, 168
Aureobasidium pullulans
  cellulolytic activity 306
  colour 13
  D-gluconic acid production 307
  hemicellulose utilization 229
  pullulan production 307
autolysis 175, 177-178
autolytic mutants 25
autonomously replicating sequences (ARS) 33, 272
autophagic death 179
autophagy 179
auxanography 3
axial budding pattern 105
azoles 178

Bacillus spp 305
Baeyer-Villiger oxidations 254
baker's yeast 219, 280
  drying 154
  elemental analysis 52
  grittiness 27-28
  in organic synthesis 254
  invertase 306
  maltose uptake 72-73
  production 300-303
  propagation 134
  respiratory growth 55
  trehalose levels 159
ballistoconidiogenesis 103
bare lymphocyte syndrome type I 310
bark beetles 171

barotolerant yeasts 160
Basidiomycetous yeasts 55
Basidiomycotina 2
batch culture 145
batch growth curve 131
batch synchronous cultures 138-139
BCM/Allev 2.00 System 3
bead beater 161
beer 146, 207, 268, 283-284
beet molasses 61
Behçet's syndrome 310
Belgian beer 172
BEM 1 105
Bensingtonia 2
benzene metabolism 252
betaine 157
binary fission 103
biocide metabolism 252
biodiversity 3-4
bioelectrical fuel cells 304
bioemulsifying yeasts 62
bioethanol 5, 294-299
biofilms 130-132
biological control 5
biological control yeasts 301
biopolymer metabolism 228-229
biorecovery of copper 89
bioremediation 89, 91
biosensors 288
biosorbent yeasts 301
biosorbents 305
biosorption 91
biosurfactants 238
biotechnologically important yeasts 267
biotheraputic yeasts 301
biotransformations 162
bipolar budding 103, 105
birth scar 104
birth scars 23
black knot disease 170
Blastomyces dermatidis 8, 171
Blastomycosis 8
blue stain fungi 172
Botryoascus
  taxonomy 2
Botryoascus synaedendrus 174
Botryotina fuckeliana 173
Botrytis cinerea 171
bottom fermenters 287
bottom yeasts 25
bovine leukaemia 308
bovine prolactin 277
Bovine Spongiform Encephalopathy (BSE) 34
brandy 292
Braun homogenizer 161
Brettanomyces
  cell shape 13
  nitrate assimilation 77
  preservation 140
  taxonomy 2
  cell shape 13
  taxonomy 2
Brettanomyces lambicus
  as biosorbent 304
Brettanomyces naardenensis
  starch utilization 229
  xylose fermentation 230
Brettanomyces spp
  Custers effect 216, 219
  organic acid utilization 239
brewers' wort 60-61, 72, 226, 288
brewer's yeast
  hybridization 269
  mutation 269
  rare mating 269
  selection 269
  single-chromosome transfer 269
  spheroplast fusion 269
  transformation 269
brewing 145, 283-290
brewing fermentations 79, 143
brewing strains
  auxotrophic mutants 273
  transformant selection 273
brewing yeast
  acid washing 162
  amino acid transport 79, 80
  as chromium source 91
  biofilms 130
  cell size 12
  CO_2 pressure effects 160
  cold shock 153-154
  derepressed mutants 73
  ethanol stress 166-167
  ethanol tolerance 21
  flocculation 25-27
  flocculation characteristics 170-181
  glycogen as vitality indicator 226
  growth media 59
  hydrophobicity 26-27
  inoculum development 143
  isolation 140
  maltose transport 72-73
  mitochondria 40-41
  mitochondrial genome 41
  mycotoxin inhibition 173
  oxygen requirements 148
  petite mutations 41
  polyploid nature 124
  selection 89
  senescence 180-181
  starvation 163

brewing yeast (*cont.*)
 sterol biosynthesis 226
 storage conditions 154
 strain differentiation 287
 transport of higher saccharides 74
 trehalose levels 159
 wild yeasts in 59
brewing yeasts 283-290
 genetic fingerprinting 268
 genetic research and development 269
 genetical attributes 270
 tetrad analysis 268
brightfield stains 141
British Biotechnology Ltd 307
bud fission 103
BUD genes 105
bud scar 104
bud scar count 181
bud scars 23, 180-181
budding 102-107
budding from stalks 103
budding index 109
bud-site selection genes 105
Bullera
 colour 13
 taxonomy 2
 vegetative reproduction 103
Bullera sinensis 170
Bulleromyces
 taxonomy 2

C230 276
Ca^{2+} - dependent mutants 88
Ca^{2+} detoxification 89
Ca^{2+} - H^+ antiporter 88
Ca^{2+} ions
 cell cycle regulation 88, 123
 in transformation 278
cal mutants 88
calcium
 requirements of yeast 53
calcium alginate 131
calcium ions
 homeostasis 65
 in flocculation 27
 in meiosis 28
 in sporulation 28
 requirements 57
calcofluor 104
calcofluor white 14, 181
calmodulin 88
cAMP
 as a second messenger 218
 dependent protein kinase 121
 dependent protein kinases 249
 in cell cycle control 121
 in glucose transport 69
 in glycogen degradation 226
 in pseudohyphal growth 112
 phosphodiesterase 249
 regulation of carbon metabolism 218
 synthesis and degradation 249
campesterol 167
CaMV 275
canavanine^R 276
cancer 113
cancer cell surface antigen 308
Candida
 as plant pathogens 170
 capsules 29
 emerging pathogens 8
 flocculation 26
 food spoilage 177
 fungal parasitism 173
 growth on short-chain alkanes 75
 habitats 4
 killer yeast 170
 lactose transport 73
 nitrate assimilation 77
 predatory characteristics 174, 176
 taxonomy 2
 vitamin requirements 58
Candida albicans
 acid protease 229
 acquired thermotolerance 152
 adherence 27
 amino acid transport 78
 antizymotic agents 178-179
 as opportunistic pathogen 171
 biofilms 131-132
 candidosis 7
 cell shape 13
 chromosomes 33
 dimorphic transitions 250
 dimorphism 113
 effects of ethanol 164
 ergosterol biosynthesis 178
 filamentous growth 110-113
 galvanotropic behaviour 162
 germ tubes 110
 GOGAT 243
 growth on solid media 128-132
 industrial uses 267
 ion assimilation 89
 laser light susceptibility 161
 nosocomial infections 5
 peptide transporters 80
 sterol biosynthesis 238
 thigmotropism 130
 vegetative reproduction 103
 yeast-to-hyphal transitions 111

Candida antarctica
cellular fatty acids 76
starch utilization 229
Candida blankii
xylose fermentation 230
Candida boidinii
industrial uses 267
methanol utilization 231
methylotrophic yeast 232
Candida brumptii
L(+) isocitric acid production 307
Candida curvata
oleaginous yeast 239
Candida cylindraceae
lipid utilization 229
Candida diddensiae
oleaginous yeast 239
Candida diddensii
inhibitory plant chemicals 170
Candida entomaea
xylitol production 300
Candida ernobii
wood habitats 170
Candida fabanii
nitrile metabolism 252
Candida famata
industrial uses 267
purine metabolism 252
Candida glabrata
pathogen 8
Candida guilliermondii
citric acid production 307
lipid utilization 229
methylotrophic yeast 232
xylitol production 300
in biocontol 173
Candida hellenica
in pollution control 304
Candida hydrocarbofumarica
organic acid production 307
Candida intermedia
alkane utilization 235
Candida kefyr
pectin utilization 229
Candida krusei
as a pathogen 8, 171
intestinal parasite 171
Candida lipolytica
alkane utilization 235
Candida lusitaniae
as a pathogen 8, 171
cellibiose transport 73
Candida maltosa
cloning technology 271
growth media 58
industrial uses 267
methylotrophic yeast 232
phenol metabolism 252
Candida mogii
osmotolerance 148
xylitol production 300
Candida nitratophila
nitrate uptake 78
nitrite transport 243
Candida oleophila
in biocontol 173
Candida palmioleophila
biomass uses 301
Candida paraffinica
biomass uses 301
Candida parapsilosis
as a pathogen 171
intestinal parasite 171
pathogen 8
Candida pelliculosa
industrial uses 267
Candida pintolopesii
fermentative capacity 149
habitat 171
Candida pseudotropicalis
pectin utilization 229
Candida rugosa
fat breakdown 235
lipase 306
Candida shehatae
biomass uses 301
fuel ethanol production 294-296
glucose transport 68
hemicellulose utilization 229
industrial uses 267
respirofermentation 219
xylitol pathway 208-209
xylose fermentation 228, 230-231
xylose transport 71
Candida sloofii
habitats 4, 171
temperature limits 147
Candida solani
pectin utilization 229
Candida spp.
amino acid production 307
as biocontrol agent 304
in chemical detoxification 304
methylamine utilization 244
modes of sugar catabolism 214
nitrate assimilation 243
organic acid production 307
vitamin production 307
Candida stellata
in wine fermentations 290
Candida tenuis
xylose fermentation 230

Candida tropicalis
alkane utilization 235
dicarboxylic acid production 238
fatty acid production 307
high pressure effects 160
industrial uses 267
intestinal parasite 171
methylotrophic yeast 232
pathogen 8
phosphate transport 81
xylitol production 300
Candida tsukubaensis
starch utilization 229
Candida utilis
ammonium-limited chemostat cultures 77
biomass propagations 231
biomass uses 301
citric acid transport 76
colonial growth 127
Crabtree effect 217
divalent metal ion transport 86
ethanol influx 74
fermentative capacity 149
glucose transport 68
glycerol catabolism 233
glycerol transport 74
growth contaminant 62
hexose monophosphate pathway 208
industrial uses 267
K^+ - limited chemostats 56
Kluyver effect 216, 220
K_s values for glucose 60
lactate accumulation 75-76
lactic acid transport 76
malic acid production 307
methylotrophic yeast 232
nitrate assimilation 77
pentose metabolism 228
peroxisomes 29-30
sulphur metabolism 252
wild yeasts in baking 1
xylose isomerase 231
xylose transport 71
Zn^{2+} accumulation 88
Candida wickerhamii
cellibiose transport 73
fuel ethanol production 295
Candida zeylanoides
spoilage yeast 177
xylitol production 300
candidosis 8, 171
capacitance probe 141
capsules 28-29
carbohydrate biosynthesis 221-227
carbon
requirements of yeast 53
carbon dioxide fixation 53
carbon metabolism 205-241
carbon sources for yeast 52-54
carboxyfluorescein 141
carboxypeptidase Y 245
cardinal growth temperatures 146
cardinal water potentials of growth 148
carrageenan 131, 146
cassava 295
catabolite inactivation 218
in glucose transport 69
of maltose transporter 72
catabolite inactivator 72
catabolite repression 217
in glucose transport 69
catabolite repressor 72
catalase 29, 151-152
catalase A 168
catalase T 168
cationic dyes 85
Cd toxicity 163
CDC 24 105
CDC 28 120
CDC 42 105
cdc mutants 137
cdc13 122
cdc2 122-123
cdc25 122
CDC28 homologues 308
CDCFDC 14
CDK1 122
cell cycle
blockage 104
checkpoints 117, 119
control 112-124
ethanol production 138
fluctuations
landmarks 113
Mg^{2+} fluctuations 87
nutrient stress 111
regulation by Mn^{2+} ions 87
cell cycle-related oscillations 220-221
cell death 153, 175-181
cell division cycle (cdc) mutants 113
cell engineering 298
cell membrane
role in nutrient transport 63-65
cell polarity 105
cell size
in vitality assessments 142
cell size control 117-119
cell wall mutants 24-25
biotechnological significance 26
flocculins 26
hydrophobicity 27
medical significance 26

porosity 63
role in nutrient transport 63
cell walls
biosorption 26
in idealized cell 19
isolation 18
marker enzymes 18
structure and function 22-28
cellular ageing 175, 179-181
cellulase 298
centromeres 31, 33
centromeric sequence (CEN) 272
Cephaoloascus
taxonomy 2
ceramic microspheres 131
Ceratocystis montia 172
CFDA 14
chaperone function 151
chaperoned protein folding 166
chaperonin 151
checkpoint control genes 179
checkpoint gene *chk 1* 161
cheese whey 61, 73, 220, 227, 238, 294
chemical mutagens 163
chemical stress 162-169
chemical tolerance 302
chemiosmotic coupling 78
chemiosmotic principles 210
chemostat 177
chemostats 136, 137, 145
in yeast nutrition 59
chemotropism 129
chicken lysozyme 277
Chinosphaera
taxonomy 2
Chiron 307
chitin
in bud scars 23
in cell walls 21-23
in yeast budding 102-108
chitin ring 102
chitin synthase 178
chitin synthesis 224-225
chitin synthetase 102
chitinase 178-179
chloramphenicol 172
chloramphenicol^R 276
chloride uptake system 82-83
cholesterol uptake 75
chorismate mutase abzyme antibody 308
chromatin 32
chromium utilization 91
chromosome fingerprinting 287
chromosome length polymorphisms (CLPs) 287
chromosomes 32-33
cider 283, 290
cilofungin 179
Citeromyces
taxonomy 2
citric acid cycle 207, 209-210
citrulline 78
CKIs 120
classification 3
Clavispora
taxonomy 2
clevage plane 103
cloning 272
Clostridium difficile 304
CMAC 14
CMP 305
CO_2 evolution 142
CO_2 production 144
Coccidiascus
taxonomy 2
Coccidioides immitis 171
coccidioidomycosis 171
cocoa 172
cocoa butter 238
codon usage 270
Cognac 292
cold-shock protein 153
cold-shock sensitivity 153
cold stress 153-154
colonial dimorphism 127
commensalism 169
compatible solutes 85, 157
competent 278
competition 169
compressed baker's yeast 301
ConA-FITC 16
Concanavalin A 15
confocal microscopy 14-15, 40
congeners 299
conjugation 124-125
continuous culture 135-137, 139, 145
continuous selection 166-167
continuous stirred tank reactor 145
continuous synchronous cultures 138-139
contre-effect Pasteur 216
copper
requirements of yeast 53
copper biomineralization 89
copper-dependent ferroxidase 90
copper ion homeostasis 89, 169
copper-metallothionein 89
copper resistance 89
copper toxicity 89, 163
corn steep liquor 227, 295
Cornelius vessel 143
cortical membrane patches 114
cosmetic/pharmaceutical yeasts 301
cottonseed oil 167

Coulter counter 144
cows milk 295
CoZn superoxide dismutase 168
Crabtree effect 40, 56, 148, 216-219, 300
Crabtree - negative yeasts 69, 149
Crabtree - positive yeasts 69, 149
cream liqueurs 294
Creutzfeld-Jacob 309
Creutzfeld-Jacob disease 34
critical cell size 103, 117-119
Crohn's disease 7, 171-172
crop protectants 171
cross-linked gelatin 131, 146
crude oil 238
cryo-damage 153
cryopreservation 140, 153
cryoprotectant 140, 152-153
cryoprotection 182
β-cryoptogein 171
cryotolerance 302
cryptic size control 118
Cryptoccoccus cereanus 13
Cryptoccoccus curvatus
  biomass uses 301
  stearic acid production 307
Cryptoccoccus neoformans
  capsules 29
  chromosomes 33
  pathogen 8
cryptococcosis 8, 171
Cryptococcus
  capsules 28-29
  fungal parasitism 173
  habitats 4
  taxonomy 2
Cryptococcus albidus
  hemicellulose utilization 229
  K_s value for thiamine 60
  oleaginous yeast 239
Cryptococcus humicola
  killer yeast 173
Cryptococcus laurentii
  capsules 29
  killer yeast 173
  oleaginous yeast 239
  spoilage yeast 177
Cryptococcus skinneri
  wood habitat 170
Cryptococcus spp.
  modes of sugar catabolism 214
  xylanase activity 306
crystal violet 141
CTA1 genes 169
CTR1 gene 89
CTT genes 169
cultivation strategies 143-145
cultural yeast growth 139-145
culture fluid viscosity 144
CUP1 gene 89, 169, 273, 275, 276, 279
Custers effect 148, 216, 219-220
CWH 51 224
CWH 52 224
CYC/GRE hybrid 279
cyclic AMP-dependent protein kinase 218
cyclin-dependent kinase-cyclin complex 114
cyclin-dependent kinases 105, 120-123
cyclins 105, 120-124, 126, 247
cyclohexanone monooxygenase 254
cycloheximide 163, 172
cylindroconical fermenter 145, 160, 287
Cyniclomyces
  taxonomy 2
  temperature limits 147
Cyniclomyces guttulatus
  habitat 4, 171
cysteine transporter 79
cystic fibrosis 310
cystic fibrosis transmembrane conductance regulator
    (CFTR) 308
Cystofilobasidium
  taxonomy 2
cytochemical dyes 14
cytochrome C oxidase 212
cytochrome C peroxidase 168
cytochrome C pool 212
cytochrome C reductase 212
cytochrome P450 164, 166, 308
  in alkane metabolism 75
cytochrome P450 monoxygenase 235
cytochrome P450 reductase 235
cytoduction 34
cytofluorescent dyes 14
cytokinesis 103, 107
cytoplasm
  in idealized cell 19
  marker enzymes 18
  structure and function 29-31
cytoplasmic pH 164
cytoskeletal cables 114
cytoskeleton
  structure and function 29, 31
cytosolic Ca^{2+} 88
cytosolic pH 29

DAB 14
DAPI 14
DASPMI 14, 39-40
deacidified wine 146
DEAE-cellulose 131, 146
death rate 134
Debaryomyces

cell shape 13
habitats 4
killer yeast 170
nitrate assimilation 77
random budding 104-405
taxonomy 2
vegetative reproduction 103
Debaryomyces castellii
maltose utilization 220
Debaryomyces hansenii
as biocontrol agent 173, 304
glycerol accumulation 74
halotolerance 154, 157
halotolerant yeast 56
in high salt environments 75
osmotolerance 148
Debaryomyces nepalensis
xylose fermentation 230
Debaryomyces polymorpha
xylose fermentation 230
Debaryomyces yamadae
Kluyver effect 220
deceleration phase 134
dehydration 73
Dekkera
cell shape 13
Custers effect 216, 219
taxonomy 2
Dekkera intermedia
β-glucosidase activity 306
deoxynivalenol 173
desiccation 73, 140
detergents 85
detoxification 182
Deuteromycotina 2
dextrins 72
diauxic growth 134
diauxie 134, 217
dielectric permittivity 136, 141
diffusion channels 64-65
dihydroxyacetone synthase 233
dilution rate 135
dimethyl sulphide
in lager beers 252-253
dimethyl sulphoxide
in malt wort 253
dimorphism 110-113
$DiOC_6$ 14
diploidization 124, 278
direct microbial conversion (DMC) 295
distilled spirits 283
$diuron^R$ 276
2μm DNA 270, 274
DNA damage 163
DNA-damage checkpoints 161
DNA-division cycle 115
DNA fingerprinting 287
DNA plasmids 173
DNA repair 163
dolichols 223, 238
Drosophila spp. 4, 170
dsRNA killer plasmid 34
dsRNA plasmids 173-174
Dubin-Johnson syndrome 310
DUK1 85
duplicate-pore K channel 85
Dutch elm disease 170
D-xylose
transport 70-71
dynein 108, 114

Eaton Press 161
Echinocandia 178
Edinburgh Minimal Medium 59
EFG10 289
Ehrlich pathway 244
elafin 308
electrical stress 162
electrochemical proton gradient 78
electrofusion 162
electron microscopy 16
electron transfer 205, 209
electron transport chain 210-212
electropermeabilization 162
electrophoretic karyotype analysis 268, 287
electrophoretic karyotyping 32-33
electroporation 162, 278
electrosensor 142
ellipsoidal unicellular yeast forms 111
Emil Christian Hansen 140
ENA1 gene 157
ENA1/PMR2 gene 85
endocytic mutants 63
endocytic pathways 21
endocytosis 20-21, 35
endophytic yeasts 170
endoplasmic reticulum
in idealized cell 19
marker enzymes 18
β-endorphin 308
endosomes 20-21, 63
energy generation 205
energy metabolism 205-213
ENO 272, 275
enolase 223
entrained growth oscillations 138
epidermal growth factor 308
Epstein-Barr 308
equatorial tubulin rings 108, 110
ER cisternal proteases 37
ER-derived vesicles 18

Eremothecium
taxonomy 2
Eremothecium ashbyii
riboflavin production 307
ergosterol 21, 55, 76, 167
ergosterol biosynthesis 178, 238-240
Erlenmeyer vessels 55
Erysiphe graminis 171
erythritol
transport 74
erythritol in water stress 158
Erythrobasidium
taxonomy 2
erythropoietin 308
Escherichia coli 163, 272, 295, 298
esters 207
ETG1 224
ethanol
efflux 74
from immobilized yeast 146
influx 74
stress 152, 163-167
stress proteins 166
toxicity 163-168
transport 74
ethanol tolerance 298
ethanol toxicity 21
role of trehalose 74
ethylcarbamate 243
EUROFAN 310
Europhium clavigenum 172
exocytosis 20, 35
exoglucanase 37
exponential phase 133-134
expression 272
extracellular acidification 84
extracellular pH 54, 81, 142
extrachromosomal elements 34-35

Fab fragments 308
facilitated diffusion 64
α factor 277
α-factor leader 275
factors VIII and XIII 308
facultative anaerobe 148-149
facultatively fermentative yeasts 149
fatal familial insomnia 34, 309
fatty acid carrier systems 76
fatty acid desaturase 236-237
fatty acid esters 170
fatty acid metabolism 235-238
fatty acid synthase 236
fatty acid synthesis 237
F-C ConA 14
fecosterol 238
fed-batch cultivation 145
fed-batch culture 134, 145, 219
fed-batch fermentation 15
fed-batch technology 280
Fellomyces
taxonomy 2
fermentable growth medium induced pathway
134-135
fermentation
capacity 142
interactions with bacteria 172
Mg^{2+} uptake 87
phosphate uptake 81
tests 3
use in classifcation 149
fermentation yeasts 69
fermenter engineering 299
ferric reductase 90-91
FET3 90
fibrillar glucan network 25
fibrinogen 308
fig wasps 171
filamentation 110-113
filamentous growth 110-113
fill-and-draw system 280
Filobasidium
taxonomy 2
fimbriae 28
fission 106-110
fission scars 108
FITC 15
FKS1 224
FLO genes 26-27
FLO1 genes 289
flocculation 25-27
flotation 27
flow cytometry 15-16, 142
fluconazole 238
flucytosine 178
fluidized bed culture 145
fluorescence-activated cell sorting (FACS) 15
fluorescent Ca^{2+} indicators 88
fluorescent probes 142
fluorochrome stains 141
fluorochromes 142
fluorochromic dyes 13-14
fodder yeast 301
Food Advisory Committee (FAC) 282
Food and Drug Administration (FDA) 282
food poisoning 7
food spoilage 7
food spoilage yeasts 75, 148
foot and mouth 308
forced synchronization 138
FPS1 74
FPS1 gene 157
fragile mutants 24-25

FRE1 90-91
free diffusion 63-64
free radicals 167-169
freeze-thaw stress 153
French Press 161
fructose 1,6-bisphosphatase 218, 223
fruit flies 171
fuel alcohol 292
fuel ethanol-producing yeasts 294-299
functional antibodies 308
fungivorous insects 171
FUN^{TM} 14
Fusarium spp. 173
Fusarium toxin 173
fusel alcohols 207
fusel oils 244, 299
fuzzy interference 144

G 418^{R} 276
G0 arrest 135
G1 cyclin homologues 308
G1-S control 120-122
G2-M control 122-124
GAL 273, 279
GAL genes 218
GAL1, 10 275
GAL2 69
galactitol catabolism 233-234
galactitol transport 74
galactokinase 70
Galactomyces
  taxonomy 2
β-galactosidase 298
galvanitropic behaviour 162
gamma irradiation 161
GAP 79-80
GAP 491 275
gastric lipase 308
gcs1 phenotype 136
gene isolation 272
gene therapy 310
general amino acid permease 78
General Glucose Sensor 69
generalistic yeasts 62
genetic tagging 287
genetically modified baking yeast 282-283
genome
  size 32-33
genotoxic agents 309
Geotrichum spp. 13
germ tube inducers 111
germ tubes 110
German beer 172
GFP 14
gin 292, 294
glass wool 146
GLR1 168
glucagon 226
glucan synthesis 223
glucan synthetase 102
β-glucanase 289
glucanases 179
glucans
  in cell walls 22-26
glucitol catabolism 233-234
glucoamylase 298
glucokinase 67
gluconeogenesis 221-223
glucose 6-phosphate dehydrogenase 207
glucose poisoning 72
glucose sensing 69
glucose-sensitive yeasts 219
glucose sensors 69
glucose transport 67-69
glutamate dehydrogenase 241
glutamate synthase 242-243
glutamate transporter 79
glutamine in nitrogen and carbon metabolism 242
glutamine pathway 243
glutamine synthetase 161-162, 241-243
glutathione
  in antioxidant defences 168
  in sulphur assimilation 82
glutathione reductase 168
glycan 26
glyceraldehyde 3-phosphate dehydrogenase 223
glycero-phosphate shuttle 211
glycerol
  as compatible solute 157-158
  efflux 74
  facilitator 157
  homeostasis 157
  in freeze thaw stress 153
  in heat stress 152
  transport 74-75
glycerol 3-phosphate dehydrogenase 233
glycerol catabolism 233
glycerol kinase 233
glycerol production 205-207, 299
glycine transporter 79
glycogen
  in stationary phase cells 135
  in viability assessment 141
  in vitality assessments 142
glycogen phosphorylase 226
glycogen synthase 226
glycogen synthesis 225-226
glycolysis 205-207, 222
glycolytic flux rate 141
glycosylation 275
glyoxylate cycle 209-211, 240
glyoxy-peroxisome 31

glyoxysomes 29-31
GMP 305
GOGAT 242-243
Golgi apparatus
  in idealized cell 19
  in secretion 34-38
  master enzymes 18
  Mn^{2+} ion sequestration 87
Golgi membranes isolation 18
GP400 151
gradostat 145
Grappa 292, 294
GRAS status 267
gravitational stress 161-162
GRE 275
Green Fluorescent Protein 14, 41
growth cycle 115
growth factor requirements 57-58
growth factor yeasts 301
growth-limiting nutrient 59-60
growth-limiting substrate 136-137
growth yields 143
GTPase activity 105, 125
Guilliermondella
  taxonomy 2
Guilliermondella selenospora
  chromosomes 32-33
  predatory characteristics 174

H_2S excretion 252
H_2S production 82
habitats for yeasts 4-5
haemocytometer 144
haemoglobin 90, 308
haemoproteins 90
HAL3 genes 85
halotolerance genes 85
halotolerant yeasts 56, 157
Hanseniaspora
  cell shape 13
  taxonomy 2
  vegetative reproduction 103
Hanseniaspora urarum
  in wine fermentations 290
Hansenula
  chromosomes 33
  emerging pathogens 8, 171
  inhibitory plant chemicals 170
  nitrate assimilation 77
  nitrogen assimilation 55
  phosphomannans 29
  sphingolipids 29
  taxonomy 2
  vitamin requirements 58
Hansenula anomala
  malic acid transport 75-76
  pathogen 8
Hansenula capsulata
  capsule 29
  fungal interaction 172
Hansenula holstii
  fungal interaction 172
Hansenula polymorpha
  biomass uses 301
  capsules 29-30
  cloning technology 271
  fermentations 280, 281
  growth media 58
  human polypeptide secretion 37
  industrial uses 267
  methanol metabolism 232-233
  methanol transport 74
  MOX 275, 279, 281
  PMA1 275
  pyruvic acid production 241
  taxonomy 2
  therapeutic proteins 307
  transcriptional activation 273
Hansenula spp.
  amino acid production 307
  nitrate assimilation 243
  polysaccharide production 307
  sterol production 307
Harden-Young effect 216
haustoria 174
haustoria-mediated predation 174, 179
Hayflick limit 180
HCl-Giemsa 14
heat evolution 144
heat shock 149-154
  proteins 151-153
  response 165
heat shock element (HSE) 152
heat shock genes 135
heat shock transcription factor (HSF) 152
heavy metal ion stress 152
heavy metal toxicity 163
heavy metals 85
Helicase 24, 171
Helix pomatia 171
hemicellulosic hydrolysates 70
hepatitis B 280, 308
hepatitis B vaccine 306
Herpes simplex 308
heterologous gene expression 268-283
heterologous proteins
  secretion 273
  signal sequences 274
hexitol dehydrogenase 233
hexokinase 67, 222
hexose monophosphate pathway 208
Hibiscus tree 174

high cell densities 144
high cell density fermentations 280
high gravity fermentations 163
high gravity wort 159
high pressure stress 160-161
high voltage pulses 135
higher alcohols 244-245, 299
hirudin 276
Hirudo medicinalis 276
HIS3 276
histidine transporters 79
histones 31-32
Histoplasma capsulatum
  chromosome 33
  dimorphism 113
  pathogenicity 8, 171
histoplasmosis 8, 171
HIV 7, 179, 308
HIV vaccine 307
HKR1 224
HMG CoA reductase 238
HOG pathway 158
HOG1 gene 158
Holzer's enzyme competition theory 214
Hormoascus
  taxonomy 2
HSA 277
HSE 275, 279
Hsp 104 151-152
Hsp 12 151
Hsp 26 151
Hsp 30 151
Hsp 35 151
Hsp 48 151
Hsp 60 151
Hsp 70 family 151
Hsp 83 151
human chorionic gonadotrophin 308
human gastrin 277
human growth hormone 308
human mycoses 178
human P450 309
human therapeutic proteins 275
HXT family 72
HXT genes 69
hybrid plasmids 279
hybridizations 297
hydrocarbon metabolism 234
hydrogen
  requirements of yeast 53
hydrogen peroxide 167
hydrolases 37
hydrophobicity 165
hydrostatic pressure 160
hydroxyalkyl methacrylate gels 131
hydroxyl radical 167
hygromycin^{R} 276
hyperosmotic shock 154
hyphae 111
hyphal growth 111
hypoosmotic shock 154
hypotonic shock 157-159
hypotonicity 157
HYR1 112

identification 3
IgE receptor 308
IGF1 308
ilicit transport 80
image analysis 15, 144
imidazoles 238
immobilization of industrial enzymes 280
immobilization of yeasts 91, 112, 129-131
immobilized culture 145
Immunex 306
immunity derterminant 174
immunocompromised patients 171
immunoelectron microsopy 16
IMP 305
immunofluorescence microsopy 14
India ink 14
induced thermotolerance 150
industrial alcohols 294-300
industrial fermentations 112, 130, 152, 163
industrial yeast propagations 143-144
industrial yeasts
  sulphur metabolism 253
infantile hyperinsulinaemia 310
influenza 308
inorganic pyrophosphatase 250
inositol 1,4,5-triphosphate
  in signal transduction 250
inspissosis 170
insulin 226, 308
insulin-dependent diabetes 310
interferon -α 8 281
interferon-alpha 2 308
interferon-beta 1 308
interleukins 308
intestinal yeast parasites 171
intracellular acidification 84
intracellular compartmentalization 86
intracellular pH 53-54, 75, 83-84, 141-142, 148, 150, 151, 177
intracellular phosphate 56
intrinsic thermotolerance 150
inulin hydrolysis 146
inulinolytic yeasts 62
invagination 19, 21
invertase 21, 37, 71
iodine 14

Irish whiskey 292
Isaatchenkia
  taxonomy 2
isocitrate lyase 209-210
isoprenylation 275
Itersonilia
  taxonomy 2

Janus green 14
Jerusalem artichokes 295

K^+
  in osmoregulation 157-159
K^+ efflux
  in sulphate transport 82
K^+ - efflux channel 64-65
K^+ - H^+ - symporter 84
K^+ ion channels 84
K^+ ion expulsion
  in amino acid uptake 78
  in nitrate uptake 76
K^+ ion extrusion 72
K^+ ion uptake 75
K^+ ions
  in phosphate transport 81
  in proton exchange 84
K^+ leakage 85
K^+ - selective electrodes 85
K^+ transporter 84
K1 killer toxin 292
K1 toxin 174
K2 toxin 174
K28 toxin 174
karyogamy 31, 124
karyokinesis 115
ketoconazole 178
KEX2 275
KIL k1 276
killer plasmids 34, 36, 173-175, 274
killer toxin
  receptor 23-24, 174
  secretion 62
  types 174
killer toxins 37, 175, 179, 277
killer viruses 174
killer yeasts 173-175
kinetochores 31
KLac 55
Kloeckera
  habitats 62
  taxonomy 2
  vegetative reproduction 103
Kloeckera apiculata
  in biocontrol 173
  in wine fermentations 290-291
Kloeckera javanica 75
Kloeckera spp.
  alkane utilization 235
  sterol production 307
Kluyver effect 216, 220
Kluyveromyces
  as plant pathogens 170
  chromosomes 33
  flocculation 26
  food spoilage 177
  immobilization 146
  inulinase 54
  killer plasmids 12, 34, 36
  nitrate assimilation 77
  taxonomy 2
  vegetative reproduction 103
  vitamin requirements 58
Kluyveromyces bulgaricus
  immobilization 146
Kluyveromyces cellobiovorans
  β-glucosidase activity 306
Kluyveromyces cellobiovorus
  xylose fermentation 230
Kluyveromyces fragilis
  xylose transport 71
Kluyveromyces lactis
  GRAS status 270, 271
  growth media 58
  human polypeptide secretion 37
  immobilization 146
  industrial uses 267
  killer plasmids 274
  killer toxin 277
  killer yeast 170
  lactose transport 73
  malate transport 75
  mannosylation 276
  respiratory chain 212
  succinic acid transport 76
  therapeutic proteins 307
  trehalose in glucose metabolism 227
Kluyveromyces marxianus
  biomass uses 301, 304
  cheese whey fermentations 306
  cloning technology 271
  Crabtree effect 217
  dimorphism 111, 113
  electropermeabilization 162
  fuel ethanol production 295-296
  galactose transport 68
  glucose transport 68
  industrial uses 267
  inulin utilization 229
  lactic acid transport 76
  lactose fermentation 290
  lactose transport 71
  morphology 15

pectin utilization 229
photosensitive dye treatment 161
polyphosphates 249
pseudomycelia 127
sucrose transport 71
thermoduric strains 147
xylose fermentation 230
xylose transport 71
Kluyveromyces spp.
calf-rennet substitute 306
fuel alcohol production 298
in wine fermentations 290
inulinase 306
lactase 306
modes of sugar catabolism 214
protein utilization 229
vitamin production 307
Kluyveromyces wickerhamii
Kluyver effect 220
KNR4 224
kombucha 172
KRE series 224
Krebs cycle 209
K_s 59-60
K_T 62, 67, 68, 70-76
kuru 34, 309

L-drying 140
L-rhamnose transport 70
LAC12 73
β-lactamase 276
H^+-lactate symport 75
Lactococcus lactis 241
lactoferrin 308
lactose permease 73, 298
lactose proton symport 73
lag phase 132-133
lager yeast strains 286
lanosterol 55, 238
laser-flow cytometry 144
lectin hypothesis 26-27
leech hirudin 308
LEU2 276
leucine transporter 79
Leucosporidium
taxonomy 2
Leucosporidium frigidum 147
leukine GM-CSF 306, 308
leukocyte interferon-alpha (D) 308
lifespan 180
lignocellulose-related sugars 70
lignocellulosic hydrolysates 294-296
limited respiratory capacity 216, 219
linear killer plasmid 34
Lineweaver-Burke relationship 82
linoleic acid 167
γ-linolenic acid 237
linoleyl fatty acid residues 21
Lin's medium 59
linseed oil 167
lipid biosynthesis 239
lipid metabolism 235-238
lipid particles 29
lipolysis 164
lipolytic yeasts 62, 235
Lipomyces
capsules 29
taxonomy 2
vegetative reproduction 103
Lipomyces kononenkoae
starch utilization 229
Lipomyces lipofer
biomass uses 301
oleaginous yeast 239
Lipomyces spp.
in chemical detoxification 304
Lipomyces starkeyi
herbicide metabolism 251
oleaginous yeast 239
starch utilization 229
lipopeptides 75
liposan 75
liqueurs 283-284
liquid nitrogen 140, 153
liver eposide hydrolase 308
Loderomyces
taxonomy 2
lomofungin 14, 172
low-alcohol beer 146
low temperature stress 153-154
Lucifer yellow 14
lyophilization 140, 153
LYS2 276
lysine transporters 79
lysozyme 308
Lyticase 24

maceration 291
macrophage colony stimulating factor 308
magnesium
as cell size transducer 123
in dimorphism 111
in ethanol stress 165, 167
in mitochondrial morphology 39
in molasses 60
in phosphate transport 81
in thermal protection 152
requirements of yeast 53
magnesium ion
in phosphorus metabolism 249
magnesium ions
in metabolic flux control 214

Maillard condensation 61
MAL 303
MAL genes 72, 218
MAL61 genes 289
malaria antigen 308
Malassezia
  as a pathogen 171
  taxonomy 2
  vegetative reproduction 103
Malassezia furfur 8, 113
malate shuttle 211
malate synthase 209
malic acid permease 75
malic enzyme 240
malolactic fermentation 75, 240
malt wort 61, 72, 227
maltase 72
maltose-fermenting cells 72
maltose permease 72
maltose transporter 72
maltotetraose transport 72
maltotriose transport 72
maltotriose transporter 74
manganese
  requirements of yeast 53
mannanases 179
mannans
  in cell walls 22
mannitol
  in water stress 158
  transport 74
mannitol catabolism 233-234
mannitol cycle 234
mannitol dehydrogenase 234
mannoprotein synthetase 178
mannoproteins 174
  in cell walls 22
mannosylation 276
Marmite 305
mating pheromone signal transduction pathway 112
mating pheromones 158
  in G1 progresion 125
  in sexual reproduction response 124-127
  response 124-127
mating signal transduction 88
mating sterility 180
mating type
  budding pattern 105-106
mating type switching 124
maximum specific growth rate 59
mechanical stress 161-162
MEL$^+$ strains 303
MEL1 275, 277, 279
membrane biogenesis 75
membrane intrinsic protein 66, 74
membrane trafficking 63
Menkes disease 89
mesophiles 147
mesophilic yeasts 147
MET10 253
MET10 genes 289
MET2 253
MET25 273, 275, 279
metabolic engineering 297
metabolic heat 141
metabolic oscillations 220
metallothioneins 163, 168
methanol
  transport 74
methanol utilization 231-233
methanotrophic yeasts 231
methionine
  in sulphur storage 82
  transporter 79
methionyl aminopeptidase 275
methotrexate^R 276
methylamnnonium ions 77
methylation 275
methylene blue 14, 141
methyl-glyoxal (MG) 283
methylmethanesulphonate 163
methylotrophic yeasts 54, 231-233, 279
  Crabtree-negative 281
  fermentations 280
  GRAS status 282
  methanol transport 74
Metschnikowia
  in predation 176
  killer yeast 173
  osmophilic yeast 154
  taxonomy 2
Metschnikowia reukaufii
  xylose transport 71
Metschnikowia bicupsidata 148
Metschnikowia pulcherrima
  as biocontrol agent 173, 304
mevalonic acid 238, 240
MFα 1 279
MFα 1/MFa 1 275
Mg^{2+} ion release 141
Mg^{2+} ions
  and ethanol production 299
  in molasses 298
Mg^{2+} transport mutants 87
Mg-ANS 141
miconazole 178, 238
microbial mining 91
microbodies 29
  in alkane assimilation 75
microcalorimetry 141
microfilaments 31
microgravity bioreactors 162

microsomes
  in alkane utilization 75
  isolation 18
microtubule organizing centres (MTOC) 109, 114-115
microtubules 31
milk sugar 73
mineral elements
  requirements for yeast 56-57
mineral yeasts 301
Ministry of Agriculture Fisheries and Food (MAFF) 282
mitochondria
  in idealized cell 19
  morphology and 3D structure 38-41
  phosphate transport 82
mitochondrial marker enzymes 18
mitochondrial DNA 34
mitochondrial electron transport 212
mitochondrial ferrochelatase 90
mitochondrial NADH pool 212
mitochondrial proteases 245
mitochondrial reticula 39
mitochondrial superoxide dismutase 152, 166
mitogen-activated protein kinase (MAPK) 112, 126, 158
Mn SOD 150
Mn superoxide dismutase 168
Mn^{2+} homeostais 169
modification 272
molasses 71, 219-220, 243, 278, 280, 294-295, 298, 300
  carmelization 61
  composition 61
molybdenum
  requirements of yeast 53
Monilinia laxa 173
Monod equation 59
monopolar budding 103
monoterpenes 170
morphogenetic hierarchy 105-107
morpholines 178
motor protein 114
MOX 275
MOX1 279
MPF 114
mRNA synthesis 163
Mucor pusillus rennin 277
multilateral budding 103
multiple sclerosis 310
mutagenesis 297
mutagens 309
mutualism 169
mycoparasitizing fungi 179
mycoses 171
mycotoxins 173, 309
myristrylation 275
Myxozyma 71

Na^+-H^+ antiport 85
Na^+
  in osmoregulation 157
Na^+ toxicity 85
NADH-CoQ reductase 212
NADH fluorescence 141, 144
Nadsonia
  taxonomy 2
  vegetative reproduction 103
naftifine 178
natural crosses 297
near infrared high power laser 161
necrotrophic mycoparasitism 173
Nematospora
  habitat 4
  taxonomy 2
Nematospora coryli 170
nerve growth factor 308
neutral red 14
nexin 308
NHA1 85
nickel requirements of yeast 53
nikkomycin 178
nitrate ion assimilation 243
nitrate proton symporter 78
nitrate reductase 243
nitrate utilization 77-78
nitrification 243
nitrile metabolism 252
nitrite reductase 78, 243
nitrogen catabolite control 77
nitrogen catabolite repression 78
nitrogen-limited chemostast 77
nitrogen metabolism 241-247
nitrogen mustard 163
nitrogen requirements of yeast 53
non-agglomeration 302
non-conventional yeasts 269
non-fermentable carbon source 124
non-osmotolerant yeasts 148
non-*Saccharomyces* yeasts
  biotechnological uses 301
  cloning technology 271
  fermentations 280
  glucose transport 68
  in wine fermentations 290-291
  karyotype analysis 268
  therapeutic proteins 307
non-thermotolerant yeasts 150
nosocomial pathogen 171
Novo-Nordisk 307
Novozyme 24, 171
nuclear envelopes 18

nuclear lamins 122
nuclei
  isolation 18
nucleolar organizer 32
nucleolus 32
nucleoplasm 32
nucleoprotein 32
nucleus
  in idealized cell 19
  marker enzymes 18
  structure and function 31-35
nutrient foraging 112
nutrient-modulated size control 123
nystatin 85, 172, 178

obligately fermentative yeasts 149
obligately psychrophilic yeasts 147
oestrogen receptor 308
oil pollution 234
oil-recovery 234
oleaginous yeasts 54, 75-76, 233, 236, 239
oleic acid 54-55, 166
oligoterpenes 170
oligotrophic environments 62
oligotrophic growth 62
oligotrophic yeasts 62
omeprazole 178-179
oncogenic retroviruses 308
Oosporidium
  taxonomy 2
Oosporidium spp.
  colour 13
Opaque-2 275
Ophiostoma ulmii 4, 113, 170
oral vaccines 279
organic acid metabolism 239-241
organic acids 207
orphan genes 308
orthophosphate 56
orthophosphate transport 81-82
osmoduric yeasts 147-148, 154
osmolytes 156
osmophilic yeasts 147, 154-159
osmoregulation 66, 83, 85
osmosensors 158
osmostress response 156-160
osmotic dehydration 154-160
osmotic potential 66
osmotic stress 85, 154-160
  glycerol efflux 74
osmotolerance 302
osmotolerant yeasts 75, 147-148
Ouzo 292, 294
overflow reaction 219
oxidants 167-169
β-oxidation 235-237
oxidative phosphorylation 209-212
oxidative stress 167-169
oxido-reductive metabolism 219
oxygen consumption 144
oxygen free radicals 164
oxygen requirements of yeast 53
oxygen uptake 142

P-type ATPases 84-87, 157
P-type Ca^{2+}-ATPase 88
P150 151
$p34^{cdc2}$ 122-123
p53 gene 309
p56-B-type cyclin 122
Pachysolen
  nitrate assimilation 77
  phosphomannans 29
  taxonomy 2
  vegetative reproduction 103
Pachysolen spp.
  polysaccharide production 307
Pachysolen tannophilus
  fuel ethanol production 296
  industrial uses 267
  modes of sugar catabolism 214
  xylose fermentation 228, 230-231
Pachytichospora
  taxonomy 2
packed bed culture 145
PAGE 287
palm oil 294
palmitic acid 166-167
panthothenate deficiency 252
papain 179
paper 295
papilliae 127
papulacaudrin 178
parasitic yeasts 170
parasitism 169, 179
parathyroid hormone 308
Pasteur effect 69, 148-149, 213-216
patch-clamp technique 65
pathogenic yeast biofilms 131-132
pathogenic yeasts 6-9, 171-182
  karyotype analysis 268
Pediococcus damnosus 172
pentose-fermenting yeasts 166, 294
pentose phosphate cycle 208
pentose sugar metabolism 228-231
periplasm
  in idealized cell 19
  maltose binding 72
  structure and function 21
periplasmic enzymes 21, 36
periplasmic invertase 71
periplasmic space 21

permeability mutants 25
permittistat 137
peroxisomes 233, 279
  biogenesis 29-31
  in idealized cell 19
  isolation 18
  marker enzymes 18
  structure and function 29-31
petites 217
PFK1 genes 214
PGG glucan 305
PGK 272, 279
PGK/ARE hybrid 279
PGK1 275
Phaecoccomyces spp.
  colour 13
Phaffia
  taxonomy 2
Phaffia rhodozyma
  astaxanthin 13, 169
  biomass uses 301, 304
  industrial uses 267
Phaffia spp.
  colour 13
phased culture 138-139
phenol metabolism 252
phenylalanine ammonia-lyase (PAL) 305
phenylketonuria 305
pheromone-receptor 125-127
pheromone response pathway 126
pheromone signals 124-127
pheromones 124-127
PHO genes 82
PHO15 273
PHO5 275, 277, 279
phosphatases 249-250
phosphatidate phosphatases 250
phosphatidylcholine 167, 250
phosphatidylethanolamine 250
phosphatidylinositol 166, 250
phosphatidylserine 167, 250
phosphoenolpyruvate carboxykinase 209
phosphofructokinase 205, 214, 222, 297
phosphoketolase pathway 208
phospholipases 250
phosphorus
  requirements of yeast 53
phosphorus metabolism 247-250
phosphorylation 275
photoactivated psoralens 163
photosensitive chromogenic molecules 161
phytoalexin response 170-171
phytoalexins 170
phytopathogenic elicitor 171
Pichia
  flocculation 26
  food spoilage 177
  fungal parasitism 173
  habitats 63
  inhibitory plant chemicals 170
  killer plasmids 36
  nitrate assimilation 55, 77
  phosphomannans 29
  taxonomy 2
  vegetative reproduction 103
  vitamin requirements 58
Pichia acaciae
  killer yeast 173
Pichia angophorae
  xylitol fermentation 234
Pichia angusta
  taxonomy 2
Pichia anomala
  starch utilization 229
Pichia guilliermondii
  industrial uses 267
Pichia holestii
  starch utilization 229
Pichia kluyveri
  killer yeast 173
Pichia membranefaciens
  vitamin requirements 58
Pichia methanolica
  cloning technology 271
  industrial uses 267
Pichia minuta
  wood habitat 170
Pichia miso
  halotolerant yeast 56
  lipid utilization 229
Pichia pastoris
  AOX 275, 279, 281
  biomass uses 301, 304
  cloning technology 271
  efficient secretion 275
  fermentations 280, 281
  glycosylation 276
  growth media 58
  HSA prepo 277
  human polypeptide secretion 37
  industrial uses 267
  methanol metabolism 231-233
  methanol transport 74
  peroxisomes 29
  phytopathogenic elicitor expression 171
  pyruvic acid production 241
  secretion 36
  therapeutic proteins 307
Pichia pastoris (*cont.*)
  transcriptional activation 273
Pichia pini
  fungal interactions 172

Pichia pinus
methylotrophic yeast 232
Pichia segobiensis
xylose fermentation 230
Pichia sorbitophila
in high salt environments 85
Pichia spp.
alkane utilization 235
in wine fermentations 290
methylamine utilization 244
polysaccharide production 307
pyridoxine production 307
sterol production 307
Pichia stipitis
biomass uses 301
fuel ethanol production 294-296
glucose transport 68
hemicellulose utilization 229
modes of sugar catabolism 214
xylitol pathway 208-209
xylose fermentation 228, 230-231
xylose transport 71
fuel alcohol production 298
Pichia vini
lipid utilization 229
pigmented yeast 301
pilot-scale bioreactor 142
pitching rate 143
Pityrosporum
cell shape 13
vegetative reproduction 103
Pityrosporum spp.
habitat 171
plant hemicellulose 228
plant pathogenic fungi 173
plasma membrane
proton-pumping ATPase 20, 65
plasma membrane ATPase 83-84, 151, 179, 249
plasma membrane channels 85
plasma membranes
isolation 18
marker enzymes 18
phospholipids 19-20
proteins 20
structure and function 19-21
vesicles 18
plasmid instability 278
plasmid stability 278-280
plasmid stabilization 279
plasmid vectors 272, 274
plasmids
2μm 32, 34
plasmolysis 155
plasmolyzing agents 179
plate count 141
platelet-derived growth factors 308
PMA 275
PMA1 84
Pneumocystis carinii 8
pneumocystosis 8
POF1 286
polarity-establishment genes 106
polio 308
pollution control yeasts 301
polyamines 57, 152, 168, 244
polybreeding 297
polycyclic hydrocarbon metabolism 252
polyene macrolides 172
polyenes 178
polyethylene glycol (PEG) 278
polygalacturonidase-secreting yeasts 62
polymerase chain reaction (PCR) 268, 287
polymeric Mg-orthophosphate 86
polymyxin B 85
polyoma 308
polyoxin 178
polyphosphates 249
polyubiquitination 151
porcine interferon 308
porcine urokinase 308
porous ceramic 146
porous scintered glass 131
post-translational processing 275
potassium
requirements of yeast 53
potassium ion transport 84-85
potatoes 295
powdery mildew infection 171
predation 174, 176, 179
preformed cellulose 146
preprotoxin 174
Primuline yellow 141
prions 34, 36
probiotic 171
proline transporter 79
promoters 272
protamines 32
proteasomes 31, 38, 245
G protein 121, 125-126
protein kinase superfamilies 249
protein kinases 249-251
protein metabolism 244-247
protein phosphatases 250
protein-phospholipid complex 167
proteinase inhibitor 6 308
proteinases A and B 245
proteolytic yeasts 62
proteome 310-311
proton exchange 84
proton motive force 65, 211
proton-pumping ATPase 54, 83-84
proton pumping mechanisms 83-84

protoplast fusion 16
pro-urokinase 308
provacuoles 37
pseudohyphae 110-112
pseudohyphal growth 111, 180
pseudomycelia 103, 134
pseudostationary phase 129
[Psi], [Eta], [URE 3] 34
psychrophiles 147
psychrophillic yeasts 146-147, 149
pulse-field techniques 287
purine metabolism 252
putrescine 244
PYK 275
pyrolysis-gas chromatography 287
pyrophosphorylase 223
pyruvate carboxylase 209, 220
pyruvate decarboxylase 205
pyruvate dehydrogenase 209
pyruvate kinase 205, 222

quorum sensing 143

rabbit α-globin 308
rad 6 mutants 127
RAD genes 179
RAD9 gene 161
radial growth rate 127
radial growth rate constant 127
radiation stress 161
Raki 294
RAPD-PCR 287, 292
rare matings 297
RAS-adenylate cyclase 150
RAS/cAMP pathway 112, 121-122, 134-135
rate change points (RCPs) 117
R-C P 14
reagent yeasts 301
recombinant brewing yeasts 289
recombinant DNA technology 144, 152
recombinant human proteins 269
recombinant pharmaceuticals 144
recombinant strains
  preservation 140
recombinant yeasts 136, 301
  batch culture 281
  continuous culture 281
  fed-batch culture 281
  growth systems 281
  immobilized culture 281
recombivax vaccine 306
redox balance 220
redox pigments 89
redox potential 144
redox state 168
replicative deactivation 164
respiration
  phosphate transport 81
respiratory chain 89
respiratory quotient 142
respiratory yeasts 69
respirofermentative metabolism 215, 219
retrotransposons 33-35
reverse transcriptase 34
RFLP 292
RFLPs 268, 287
Rhizoctonia solani 171
Rhizopus stolonifer 173
Rhodamine 123 14-15, 141
Rhodosporidium
  colour 13
  taxonomy 2
Rhodosporidium toruloides
  chromosomes 33
  industrial uses 267
  mating pheromone response 88
  siderophores 90
Rhodotorula
  capsules 28-29
  colour 13
  concentric collars 104
  emerging pathogens 8, 172
  food spoilage 177
  fungal parasitism 173
  habitats 4, 62
  nitrate assimilation 77
  taxonomy 2
  vitamin requirements 58
Rhodotorula glutinis
  biomass uses 301
  glucose transport 68
  killer yeast 170
  lipase 306
  oleaginous yeast 239
  PAL 305
  xylose transport 71
Rhodotorula gracilis
  xylose isomerase 231
Rhodotorula graminis
  oleaginous yeast 239
Rhodotorula mucilaginosa
  chromosomes 33
Rhodotorula rubra
  as a pathogen 8
  fermentation capacity 149
  killer yeast 170
  K_s value for phosphate 60
  phenylalanine production 307
  pyruvate kinase 56
Rhodotorula spp.
  alkane utilization 235
  benzene metabolism 252

Rhodotorula spp. *(cont.)*
in chemical detoxification 304
itaconic acid production 307
modes of sugar catabolism 214
organic acid utilization 239
protein utilization 229
sterol production 307
Rhodotorulic acid 90
ribitol transport 74
D-ribose transport 70
ribosomes 29
20s and 23s RNA 34
rotting fruit 170
rough ER membranes 18
RQ 142
rum 292, 294
rum1 120
Russian teakvass 172
rye bread 172

S-adenosyl-L-methionine transporter 79
S-phase promoting factor (SPF) 120
Sabouraud's medium 59
Saccharomyces
dimorphism 112-113
genetics of industrial yeasts 268
nitrate assimilation 77
vegetative reproduction 103
Saccharomyces bayanus
fructose transport 70
immmobilization 146
industrial uses 267
taxonomy 2
T_{-max} values 146
Saccharomyces boulardii
biomass uses 301, 304
biotherapeutic agent 171, 304
industrial uses 267
Saccharomyces carlsbergensis
carbon sources 54
cross-breeding 269
fructose transport 70
immobilisation 146
protein utilization 229
taxonomy 284, 286
vitamin requirements 58
Saccharomyces cerevisiae
α-factor leader 277
α-factor pheromone 63
academic strains 268
aged population 104
allantonin metabolism 249
amino acid biosynthesis 244
amino acid production 307
amino acid utilization 55
amylolytic strains 305
as chemical reagent 304
as livestock growth factor 304
asymmetrical and symmetrical cell division 103-104
auxotrophy for oleic acid, ergosterol 148
barotolerant mutant 160
biomass production 62
biosensors 304
calcium transport 88
cell cycle regulation 112-124
cell envelope 19-29
cell shape 13
cell types 13
cell wall porosity 63
cellular ageing 177-181
chromosomes 32-33
colour 13
continuous growth 135-137
Crabtree effect 216-219
cytosolic Mg^{2+} levels 87
dehydration 154-156
diauxic growth 134
dietary antigens 172
dimorphism 113
distiller's strains 166, 269
divalent metal ion transport 86
ergosterol production 307
exploitation history 7
fermentative capacity 149
fimbriae 28
flocculant strains 131
flotation behaviour 131
freeze-tolerant variants 303
fuel ethanol production 295-299
galactose transport 69-70
genetic fingerprinting 287
genome 270
glucose repression pathway 217
glucose transport 62, 67-69
glutathione-deficient mutants 168
glycerol catabolism 233
glyoxylate cycle 209-210
GOGAT 243
GRAS status 270
habitats 4
halotolerant mutant 85
hexose monophosphate pathway 208
hexose transporters 67
human polypeptide secretion 37
hydrophobic strains 130
hyperosmotic shock 154-159
immobilization 130, 146
in biomedical research 308-310
in wine fermentations 290-291
industrial strains 87
industrial uses 267
iron transport 88-91

K^+ efflux 65
kamikaze mutants 179
killer plasmids 12, 36
killer yeast 170
K_s values for glucose 60
K_s values for glycerol phosphate 60
K_s values for Mg^{2+} 60
laboratory reference strains 268
lactic acid transport 76
lactose-utilization 73
life cycle 124
logical cell cycle map 118
low water potential tolerance 148
malic acid reduction 75
malic acid transport 76
mating 268
mating pheromone response 88
mating pheromones 37
mean lifespan 180
meiosis 28
membrane protein genes 66
minor fermentation products 207
mitochondrial mutants 41
mitosis spindle dynamics 131
nitrogen limitation 214
nitrogen starvation 63
N-linked glycosylation 276
oscillatory metabolism 220
osmotic stress 154-160
Pasteur effect 213-216
peptide transport 80
petite mutants 41
pheromone response pathway 126
phosphatase biosynthesis 249
phosphate storage 65
phosphate transport 81-82
phosphlipid biosynthesis 250, 251
plasmids 36
polymorphism 62
polyphosphates 249
protein degradation 245
proteolytic activity 229
pseudohyphal growth 103, 111-113, 127
P-type ATPase 84-85
radiation stress 161
recombinant DNA technology 268-283
regulation of sugar catabolism 211-221
respiratory bottleneck 219
respiratory chain 213
respiratory-deficient mutants 217
secretion 35-37
size 12
sodium transport 85
spontaneous synchrony 220
sugar transport 66-74
sugar utilization 52-54
sulphate assimilation 252-253
sulphate transport 82
sulphite transport 82
taxonomy 2
telomeres 33
tetrad analysis 268
thermal stress 150-153
thermoprotection 152
thermotolerance 151
transformation 270
translation 247
trehalose futile cycle 226
trehalose in stationary phase cells 158
tryptophan overproduction 244
turgor potential 66
types of glucose limitation 215
urea transport 78
vegetative reproduction 102-107, 110-124
virgin cells 181
vitamin requirements 58
wine strains 268
xylanolytic strains 305
xylose transport 71
Zn^{2+} deprivation 88
Saccharomyces cerevisiae var carlsbergensis 284
Saccharomyces cerevisiae var diastaticus 229, 289-290, 297-298
glucoamylase genes 306
industrial uses 267
Saccharomyces cerevisiae var turbidans 177
Saccharomyces diastaticus
food spoilage 177
Saccharomyces kluyveri 54
Saccharomyces monacensis 286
Saccharomyces paradoxus
as plant pathogen 170
fructose transport 70
taxonomy 2
T_{max} values 146
Saccharomyces pastorianus
fructose transport 70
taxonomy 2
Saccharomyces pyriformis
industrial uses 267
Saccharomyces sensu stricto
sugar transport 70
T_{max} values 147
taxonomy 2
Saccharomyces telluris
fermentative capacity 149
temperature limits 147
Saccharomyces uvarum
ethanol tolerant variants 166-167
spheroplast fusion 297
Saccharomycodes
ringed ridges 104

Saccharomycodes (*cont.*)
shape 13
taxonomy 2
vegetative reproduction 103
Saccharomycodes ludwigii
sulphate and sulphite transport 82
Saccharomycopsis
budding 103
carbon sources 54
vegetative reproduction 103
Saccharomycopsis fibuligera
dimorphism 113
industrial uses 267
predation 174
starch, pectin, protein utilization 229
vegetative reproduction 103
Saccharomycopsis spp.
alkane utilization 235
D-gluconic acid production 307
Saccharomycopsis vini
pectin utilization 229
Saitoella
taxonomy 2
saké 283, 290
saké yeasts 166
salivary α-amylase 308
Salmonella tester strain 283
Salmonella typhimurium 87
saprophytic yeasts 170
saturated fatty acids 77
Saturnispora
taxonomy 2
sawdust 295
scar plugs 108
Schizoblastosporion
taxonomy 2
Schizosaccharomyces
chromosomes 33
DAPI-stained motochondria 15
fission scars 23
flocculation 26
nitrate assimilation 77
scar plugs 23
shape 13
taxonomy 2
vegetative reproduction 103
Schizosaccharomyces octosporus
osmotolerance 148
predation 174
Schizosaccharomyces pombe
bipolar growth 107-110
calcium transport 88
cell cycle regulation 112-124
cell septum 106-110
cell walls 23-25
cloning technology 271
continuous centrifugation 161
Crabtree effect 216
dimorphism 113
fission 106-110
fuel alcohol production 298
G protein 127
glucose transport 68
glyoxylate cycle 210
GOGAT 243
growth medium 59
high pressure effects 160
human polypeptide secretion 37
immobilization 146
in biomedical research 308-310
in situ temperatures 150
industrial uses 267
iron transport 90
K_s value for Mg^{2+} 60
life cycle 125
lipid utilization 229
malic acid transport 76
malo-ethanolic fermentations 240-241
meiosis 28
microtubular dynamics 109
modes of sugar metabolism 214
morphogenetic transitions 110
pheromone response pathway 127
P-type ATPase 84
radiation stress 161
ras1 protein 127
respiratory chain 213
sodium transport 85
SV40 275
synchronized cultures 117
therapeutic proteins 307
trehalose in glucose metabolism 227
ultracentrifuged cells 162
vitamin requirements 58
wee mutant 118-119
xylose fermentation 230
xylose transport 71
Schwanniomyces
vitamin requirements 58
Schwanniomyces alluvius
starch utilization 229
Schwanniomyces castellii
biomass uses 301
fuel ethanol production 295-296
industrial uses 267
starch utilization 229
Schwanniomyces occidentalis
cloning technology 271
growth media 58
industrial uses 267
starch utilization 229
therapeutic proteins 307

Schwanniomyces spp.
  glucoamylase genes 306
Scotch whisky 172, 290, 292, 294, 299
scrapie 34
second messengers 20, 88
secretion 272
secretory mutants 36
secretory system 35-38
secretory vesicle
  in idealized cell 19
selectable markers 273, 276
selection 272
self-synchronized culture 138
senescence factor 180
separate hydrolysis and fermentation (SHF) 295
septin-ring genes 106
septins 103
septum 103
septum formation 108-110
serial transfer 140
serine transporter 79
serum albumin 308
sexual aggregation 23
sexual conjugation 28
sexual yeasts 63
SFA 276
Shochu 294, 300
shuttle plasmids 272
sialic acid 275
SIC1 120
siderophores 62, 90
signal sequences 277
signal transduction 20, 125-127
  in stationary phase cells 135
  second messenger functions 83
signalling 272
simultaneous saccharification and fermentation (SSF) 295
single-cell oil 228, 238
single-cell protein 228, 324, 245, 301, 303
sinusoidal periodic fluctuations 221
Sirobasidium
  taxonomy 2
size analyzer 144
SLF1 89
slide count 141
SLN 158
SMF1 87
SMR1 276
SNAREs 36
SNF3 69, 71
SNZ1 135
SOD1, SOD2 genes 169
sodium ions 85
  transport 85
somatostatin 308
sophorose lipids 75
sorbitol
  in water stress 158
sorbitol catabolism 233-234
sorbitol transport 74
sour dough 172
soy sauce 146, 172
soyabean oil 167
sparkling wine 146
specialized metabolism 254
specialized yeast communities 62
specific growth rate, μ 59
spectrofluorimetry 141
spermidine 244
spermine 244
sphaerosomes 29
spheroplast fusion 297
spheroplast transformation 278
spindle microtubules 108-109
spindle pole bodies
  isolation 18
  review 31
spindle pole bodies (SPBs) 106, 115-116
spoilage yeasts 84, 177
spontaneous fermentation 172
spontaneous synchrony 138
Sporidiobolus
  taxonomy 2
Sporidiobolus pararoseus
  killer yeast 173
Sporobolomyces
  capsules 28
  habitats 4
  taxonomy 2
  vegetative reproduction 103
Sporopachydermia
  taxonomy 2
Sporothrix schenckii 113
sporotrichosis 8
squalene 55, 166, 238
squalene epoxide 178
SSF 73
STA genes 27, 289
STA2 gene 283
stainless steel wire 131
start 114-115, 117-121, 125
starter cultures 140
stationary phase 134-136
  heat-resistance 177
STE13 275
Stephanoascus
  taxonomy 2
Sterigmatomyces
  cell shape 13
  taxonomy 2
  vegetative reproduction 103

Sterigmatosporidium
taxonomy 2
sterol biosynthesis 238
sterol content
in vitality assessment 142
sterol transporter 76
Stickland reaction 244
stover 295
straw 295
streptokinase 308
Streptomyces 172
Streptomyces spp. 178
stress-damaged proteins 151
stress defence proteins 151-154
stress protectant 73
stress-response elements (STREs) 158
stress signalling 158
stretch-activated channels 157
stringent response 163
stuck fermentations 88, 163
subgravitational forces 162
suboptimal growth temperatures 149
SUC 303
SUC genes 218
SUC2 71, 275, 277
succinate dehydrogenase 212
succinic acid 207-208
sucrose-binding protein 71
sucrose proton symporter 71
Sudan black 14
sugar alcohol metabolism 233
sugar catabolism 205-221
sugar crops 295
sugarcane juice 71, 161
sulphate-proton symport 82
sulphate transporter proteins 82
sulphate transporters 82
sulphite waste liquor 227, 278, 295
sulphiting agents 178
sulphur
requirements of yeast 53
sulphur amino acids 55-56
sulphur dioxide 178
sulphur metabolism 250-253
superoxide anion 167
superoxide dismutase 135, 308
superoxide radicals 152
supersecreters 276
supraoptimal growth temperatures 149
sustained oscillatory growth 138
SUT 76
SV40 275
sweet sorghum syrup 295
Sympodiomyces
taxonomy 2
Sympodiomycopsis
taxonomy 2
synchronization indices 138
synchronous batch culture 139
synchronous continuous culture 139
synchronous culture 137-139
synthetic sponge 131

tannins 170
taxonomy 1
telemorphic 2
telomere-binding proteins 309
telomeres 33
temperature stress 149-154
Tequila 294
tetanus toxin 308
tetrazolium dyes 141
therapeutic proteins 306-308
thermal death 164
thermal death kinetics 176-177
thermal death rate 176
thermophilic yeasts 147, 149
thermoprotectant 152
thermoprotection 182
thermotolerance 150-152, 165-166
theronine transporter 79
thigmotropism 129-130
thiocarbamates 178
thionin peptides 170
thioredoxin 168
thrombomodulin 281
thyroid peroxidase 308
Tilletiaria
taxonomy 2
Tilletiopsis
taxonomy 2
tissue factor 308
tissue plasminogen activator 308
tolnaftate 178
toluidine blue 14
tonoplast 37
tonoplast Ca^{2+} transporters 88
top fermenters 286
top-fermenting yeasts 130
top yeasts 25
Toprina process 303
Torulaspora
taxonomy 2
vegetative reproduction 103
Torulaspora delbrueckii
in wine fermentations 290
Torulaspora spp.
freeze-thaw resistance 303
Torulopsis bombicola
glycolipid production 307
Torulopsis bovina
thermophile 147

Torulopsis (Candida) pintolopesii
modes of sugar catabolism 214
Torulopsis ernobii
lipid utilization 229
Torulopsis glabrata
lipid utilization 229
Torulopsis psychrophila 147
Torulopsis sonorensis
methylotrophic yeast 232
Torulopsis spp. 147, 173, 177
alkane utilization 235
organic acid production 307
vitamin production 307
TP1/α2 operator 279
TPI 275
transcriptional promoters 275, 279
transformant cells 273
transformation 272, 297
transformation strategies 278
transmembrane proton gradient 83
transport of
alcohols 74-75
anions 81-83
calcium ions 88
cations 83-91
cellobiose 73
chloride 82-83
copper ions 89-91
disaccharides 71-74
divalent metal cations 86-91
fatty acids 76-77
fructose 69-70
galactose 69-70
glucose 66-69
heavy metals 90-91
hydrocarbons 75
iron ions 89-91
lactose 73
magnesium ions 86-87
maltose 71-73
manganese ions 87
mannose 69
nitrogenous compounds 77-81
oligosaccharides 74
organic acids 75
pentose sugars 70-71
phosphate 81-82
polysaccharides 74
potassium ions 84-85
protons 83-84
purines 80-81
pyrimidines 80-81
sodium ions 85
sterols 75-76
sucrose 71
sugars 66-74
sulphate 82
sulphite 82
trehalose 73-74
vitamins 80-81
water 66
zinc ions 88
tree exudates 170
trehalose
as stress metabolite 158-160
as stress protectant 73-74
carriers 73
effects of heat 152-154
H^+-symporter 73
in ethanol stress 166
in osmoregulataion 157-160
in stationary phase cells 135
in viability assessments 141
in yeast physiology 227
metabolism 73-74
transport 73-74
trehalose 6-phosphate sythetase 226
trehalose futile cycle 226
trehalose synthesis 225-227
Tremella
taxonomy 2
triacylglycerol metabolism 238
triazoles 238
Trichoderma reesei 298
Trichosporon
capsules 29
emerging pathogens 8, 171
taxonomy 2
Trichosporon beiglii
as pathogen 8
Trichosporon cutaneum
biopolymer utilization 229
Crabtree effect 217
growth media 58
industrial uses 267
intestinal parasite 171
oleaginous yeast 239
phenol metabolism 252
purine metabolism 252
Trichosporon pullulans
immobilization 146
starch utilization 229
Trichosporon scotti
psychrophile 147
Trichosporon spp.
xylanase activity 306
Trigonopsis
cell shape 13
taxonomy 103
vegetative reproduction 103
TRITC 15
TRK genes 84

tRNA synthesis 163
TRP1 276
Tsuchiyaea
  taxonomy 2
tuber plants 295
tubulin 114
  α and β 31
tubulin immunofluorescence 14
tumour necrosis factor 308
tunicamycin 172, 178
TUN^R 276
turbidostat 137, 145
turgor potential 66
turgor pressure 102, 135
Ty elements 34-35, 274
Ty1 retrotransposon 287
Ty-VLPs 34-35

ubiquinone 238
ubiquinone pool 212
ubiquitin 151, 247
ubiquitin pathway 31
ubiquitin system 276
ubiquitinated proteins 38
ubiquitination system 245, 247
ultrafiltration membranes 131
ultrafreezing 140
ultrasonic stress 162
ultrasound 162
ultraviolet irraidation 161
uncouplers 85
unipolar budding 103
unsaturated fatty acids 77
URA3 276
urea aminohydrolase 243
urea utilization 243
US FDA 306
UTH genes 179-180

vaccine production 35
vacuolar compartment 37
vacuolar enzymes 38
vacuolar ions 38
vacuolar metabolites 38
vacuolar proteases 38
vacuolar proteolysis 72, 218
vacuole
  Ca^{2+} uptake 88
  cation storage 86
  iron storage 90
  magnesium transport 87
  Mn^{2+} accumulation 87
vacuoles
  component functions 38
  image analysis 15
  in idealized cell 19
  in nutrient storage 65
  in secretion 34-38
  isolation 18
  marker enzymes 18
  phosphate storage 81
  phosphate transport 82
vectorization 272
Vegemite 305
very high cell density fermentations 280
viability 84-85, 141-142
viper echistatin 308
virus-like particles 174
vital stains 141
vitality 84, 141-142
vodka 292, 294
voltage-gated diffusion pores 85

Wallerstein Laboratories Nutrient 59
Warburg-Dickens pathway 208
water potential 66, 147
water stress 154-160
weak acid preservatives 75, 84, 148, 163, 177, 239
wee 1 118-119
wheat mash fermentations 244
wheat mashes 298
whey 295
whey beverages 290
whey lactose hydrolysis 146, 207, 293
whisky 284, 290, 293
whooping cough antigen 308
Wickerhamia
  taxonomy 2
  vegetative reproduction 103
Wickerhamiella
  taxonomy 2
wild *Saccharomyces* yeasts 59
wild yeasts
  detection of 59
  in brewing 1
  non-*Saccharomyces* 59
Williopsis
  taxonomy 103
  vegetative reproduction 103
Williopsis californica
  nitrification 243
Williopsis mrakii
  killer yeast 173
Williopsis saturnus
  killer yeast 173
  respiratory chain 213
Wilson's disease 89
wine 146, 207, 283-284, 290
wine fermentations
  organic acid metabolism 75
wine must 61
wine production 291